SMART SYSTEMS AND BEYOND

The book highlights emerging developments in intelligent computing, communication technologies, robotics, embedded systems, and data-driven innovations that are shaping the future of smart and autonomous systems.

This volume brings together multidisciplinary research focusing on artificial intelligence, computational intelligence, smart communication systems, embedded and VLSI technologies, robotics, automation, cybersecurity, cloud computing, data analytics, IoT-enabled platforms, and intelligent decision-making systems. The book explores innovative methodologies, mathematical models, and practical applications that support next-generation technological ecosystems. Contributions include theoretical advancements, simulation-based studies, experimental implementations, and industrial applications in smart environments and AI-integrated systems. The book also discusses recent trends in machine learning, predictive analytics, human–robot interaction, autonomous systems, and sustainable computing architectures, offering valuable insights into the rapidly evolving landscape of communication and computational sciences.

The book is intended for researchers, academicians, scientists, industry professionals, postgraduate students, and practitioners working in the domains of artificial intelligence, communication engineering, computer science, robotics, embedded systems, electronics, and data analytics. It serves as a valuable reference for both academic research and industrial innovation.

Taylor & Francis Proceedings in Computer Science and Engineering

This series is a compilation of peer-reviewed conference proceedings in computer science and engineering. Exploring advanced technologies and recent developments in the field of CS and IT, including application-based topics and latest AI usages.

Next-Generation Computing and Information Systems
Edited by Ankur Gupta

Next-Generation Smart Systems
Bridging Sustainability and Computational Intelligence
Edited by V. Sekhar, P. Ajay Kumar Reddy and K. Logesh, T. Raghunadha Reddy, and Dara Raju

Sustainable Developments in Computer Engineering, Green Technology and Smart Systems
Edited by Bhabani Prasanna Pattanaik, Dhirendra Kumar Shukla, Sambit Satpathy, and Kamalakanta Muduli

Progressive Computational Intelligence, Information Technology, and Networking
Edited by Poonam Nandal, Tapas Kumar, Meeta Singh and Mamta Dahiya

Smart and Distributed Computing
Edited by Pramod Singh Rathore and Sunil Kumar

Frontiers in Engineering, Science and Technology
Edited by Sitanath Biswas and Sayan Chakraborty

Computing, Communication, and Sustainable Energy Technologies – Volume 1
Edited by Ram Lal Yadava, Anil Kumar Sethi, and Amanpreet Singh Saini

Computing, Communication, and Sustainable Energy Technologies – Volume 2
Edited by Ram Lal Yadava, Anil Kumar Sethi, and Amanpreet Singh Saini

Technology for a Resilient Future
Shweta Vincent, Ramya S. Moorthy and Kiran Dasari

Human Intelligence 2.0
Co-Evolution through Strategic Human–AI Collaboration for Sustainable Competitive Advantage
Edited by Ravi Dissanayake, Sabri Boubaker, Vishanth Weerakkody, Murali Sambasivan and Saeed Pahleva Sharif

Automotive Safety 2026
Edited by Marek Jaśkiewicz, Rafał Jurecki and Emilia Szumska

Smart Systems and Beyond
Redefining Computing and Robotics through AI
Edited by Vinay Bhatia, Sukhdeep Kaur, Ashima Kalra and Bhawna Tandon

For more information about this series, please visit: www.routledge.com/Taylor–Francis-Proceedings-in-Computer-Science-and-Engineering/book-series/PCSE

Smart Systems and Beyond

Redefining Computing and Robotics through AI

Edited by

Vinay Bhatia, Sukhdeep Kaur, Ashima Kalra and Bhawna Tandon

Chandigarh Engineering College, CGC Landran, Mohali (Punjab), India

PROCEEDINGS OF THE INTERNATIONAL CONFERENCE ON INNOVATIONS IN COMMUNICATION, COMPUTING AND SCIENCES (ICCS 2025), 25TH—26TH SEPTEMBER 2025, MOHALI, INDIA

CRC Press
Taylor & Francis Group
Boca Raton London New York

CRC Press is an imprint of the
Taylor & Francis Group, an **informa** business

First edition published 2027
by CRC Press
4 Park Square, Milton Park, Abingdon, Oxon, OX14 4RN

and by CRC Press
2385 NW Executive Center Drive, Suite 320, Boca Raton FL 33431

CRC Press is an imprint of informa UK Limited

British Library Cataloguing-in-Publication Data
A catalogue record for this book is available from the British Library

ISBN: 978-1-041-36617-1 (hbk)
ISBN: 978-1-041-41142-0 (pbk)
ISBN: 978-1-042-02628-9 (ebk)

DOI: 10.1201/9781042026289

Typeset in Times New Roman
by MPS Limited, Chennai, India

Smart Systems and Beyond – Bhatia et al. (Eds)
© 2027 The Editor(s), ISBN: 978-1-041-36617-1

Contents

Session 2 Embedded Systems and VLSI Technologies

Session 3 Robotics, Control, and Automation in Electronics

Smart Systems and Beyond – Bhatia et al. (Eds)
© 2027 The Editor(s), ISBN: 978-1-041-36617-1

Preface

The International Conference on Innovations in Communication, Computing and Sciences (ICCS 2025) aims to provide a premier interdisciplinary platform for researchers, academicians, scientists, industry experts, and students to exchange innovative ideas, present emerging research outcomes, and discuss technological advancements in the fields of communication, computing, electronics, robotics, embedded systems, and intelligent data analytics.

The conference theme, "Smart Systems and Beyond: Redefining Computing and Robotics through AI," reflects the growing integration of artificial intelligence into modern computational and engineering domains. ICCS 2025 encourages collaborative research and promotes discussions on innovative methodologies, smart technologies, and sustainable solutions that contribute to the advancement of intelligent systems.

This proceedings volume includes selected peer-reviewed papers presented during the conference and highlights contemporary research trends, theoretical developments, and practical applications in smart computing and communication technologies. We sincerely thank all authors, reviewers, session chairs, keynote speakers, organizing members, and participants for their valuable contributions and support in making ICCS 2025 a successful academic event.

Smart Systems and Beyond – Bhatia et al. (Eds)
© 2027 The Editor(s), ISBN: 978-1-041-36617-1

Acknowledgements

The Organizing Committee of ICCS 2025 expresses sincere gratitude to all contributors who supported the successful organization of the conference. We extend our heartfelt appreciation to the keynote speakers, session chairs, reviewers, authors, faculty members, industry experts, and student volunteers for their active participation and valuable contributions.

We are especially thankful to the management and administration of CEC-CGC Landran, Mohali, for their continuous encouragement, infrastructure support, and academic guidance. We also acknowledge the efforts of all collaborating institutions and technical partners for their cooperation in promoting research and innovation through this conference.

About the Editors

Vinay Bhatia is Professor and Head of the Department of Electronics and Communication Engineering at Chandigarh Engineering College (NBA Accredited), CGC Landran (NAAC A+), Punjab, India. He holds a B. Tech, M.Tech, and Ph.D in Electronics and Communication Engineering. He has authored about 150+ research papers in various national/international conferences/journals. He is a recipient of various awards such as INDO-SL International award in the category of Best Head of Department, Academic Leader I2OR International Award 2020–21 etc. Currently, he is working on routing and security issues pertaining to wireless networks. He is instrumental in the development of two Centers of Excellence in the ECE department at CGC, Landran, in the fields of Drone technology and wireless communication antennas. He has received a grant of Rs 15 lac from AICTE under MODROB scheme, IDEA Lab AICTE (1.1 Crores), and a travel grant from SERB, DST. He has been editor of various journals of international repute. His main research interests include mobile and ad hoc wireless networks, wireless mesh networks and wireless security. He has authored around 90 research papers in various national and international journals and conferences. He is the recipient of several prestigious recognitions, including the INDO-SL International Award in the category of Best Head of Department and the Academic Leader I2OR International Award 2020–21. He is a Senior Member of IEEE and serves as the RAS (Robotics and Automation Society) Faculty Advisor.

Sukhdeep Kaur is Professor, Department of Electronics and Communication Engineering at Chandigarh Engineering College, Landran, Mohali, Punjab. She received her B.Tech degree in Electronics and Communication Engineering from Amravati University (M.S), M.Tech degree in Electronics and communication from G.N.E Ludhiana; and Ph.D. degree in Antenna Designing from Thapar University Patiala; in 2001, 2008 and 2016 respectively. She has more than 23 years of work experience at various environments includes Industries, educational and research centers. She has completed three IEDC (DST) funded project and more than 20 patent (four granted) in her credit. She has published more than 50 technical research papers in SCI/ESCI/Scopus and peer reviewed International and national Journals, and few in Conferences and Seminars. She has established Antenna and Radiating System Lab in Chandigarh Engineering College, Landran, Mohali and also received a grant of Rs.15 Lakh under MODROB (AICTE) in 2019-20. She has also received a grant of Rs. 93000 from ISTE_AICTE for organizing Induction/Refresher program and Rs 3 Lacs under Grant of Conference (GOC) from AICTE She has also mentored international level project "CANSAT Competition-2016". The event was organized by American Astronautical Society, American Institute of Aeronautics, NASA/Goddard Space Flight Centre, NASA/ Jet Propulsion Laboratory; Naval Research Laboratory, held 10th–12th June 2016 at Burkett, Texas, USA. Her current research interests include Antenna Designing and Wireless Communication and their applications in bio medical applications and communication Systems.

Ashima Kalra is a distinguished academician and researcher currently serving as Associate Professor in the Electronics and Communication Engineering Department at Chandigarh Engineering College (CGC Landran), with more than 22 years of teaching experience at undergraduate and postgraduate levels. She earned her Ph.D. in Electronics and Communication Engineering from Maharishi University, Noida, in 2023. She completed her M. Tech with distinction from Punjab Technical University, Jalandhar, and is a Gold Medalist in B.Tech in Electronics from Kurukshetra University, Kurukshetra. Dr. Kalra has published more than 50 research papers in reputed journals, including six SCI and over 24 Scopus-indexed publications. She has also contributed ten+ book chapters and authored six textbooks in areas such as microprocessors, ethical hacking, and machine learning. Her research interests include artificial intelligence, neural networks, fuzzy systems, and machine learning. She has filed more than 20 patents in innovative domains including AI-powered robots and chronic kidney disease detection. She received an AICTE grant of ₹3.5 lakhs as Coordinator for the ATAL Faculty Development Program. She is a Senior Member of IEEE, Branch Counsellor of the IEEE Student Branch at CEC, and WIE (Women in Engineering) Faculty Advisor. She has actively contributed as a session chair and reviewer for various international conferences and has organized numerous faculty development programs. She is also recognized for mentoring M. Tech thesis in advanced domains such as image processing and cryptography. Her achievements include multiple Academic Excellence Awards and the Women Leadership Award, recognizing her dedication to teaching, research, and academic leadership.

Bhawna Tandon received a B.Tech degree from Kurukshetra University, Kurukshetra in 2001, M.E degree from Panjab University in 2009, and a Ph.D. degree from Punjab Engineering College (PEC) (Deemed to be University), Chandigarh in 2018. With a total teaching experience of 24 years and a strong academic and research background, she has published more than 40 research papers and book chapters in reputed national and international journals. She has two patents granted to her credit, reflecting her dedication to research, innovation, and advancing knowledge in the field of engineering and education. She is currently working as Professor and serving as the IQAC

Coordinator at Chandigarh Engineering College – CGC Landran, Mohali where she actively contributes to institutional quality enhancement and academic excellence. Dr. Tandon has also been a recipient of prestigious government research grants, including the Rs 3.5 Lacs under ATAL Faculty Development Program (FDP) scheme and Rs 3 Lacs under Grant of Conference (GOC) from AICTE, reflecting her dedication to advancing knowledge and innovation in the field of engineering and education. She has organized three international conferences as Organizing Secretary and various FDPs and bootcamps.

Smart Systems and Beyond – Bhatia et al. (Eds)
© 2027 The Editor(s), ISBN: 978-1-041-36617-1

Organizing Committee

CHIEF PATRON

S. Satnam Singh Sandhu
Chairman CGC, Landran, Mohali
S. Rashpal Singh Dhaliwal
President CGC, Landran, Mohali

PATRON

Dr. Rajdeep Singh
Campus Director CGC Landran, Mohali

CO PATRON

Dr. Sukhpreet Kaur
Director-Principal CEC-CGC Landran, Mohali

CONFERENCE CHAIR

Dr. Vinay Bhatia
Professor & Head, ECE CEC-CGC Landran, Mohali

CONVENER

Dr. Vinay Bhatia
Professor & Head, ECE CEC-CGC Landran, Mohali
Dr. Sukhdeep Kaur
Professor, ECE CEC-CGC, Landran, Mohali

CO-CONVENER

Dr. Ashima Kalra
Associate Professor, ECE CEC-CGC, Landran, Mohali
Dr. Tarun Singhal
Associate Professor, ECE CEC-CGC, Landran, Mohali

ORGANIZING SECRETARY

Dr. Mohit Srivastava
Professor, ECE CEC-CGC, Landran, Mohali
Dr. Bhawna Tandon
Professor, ECE CEC-CGC, Landran, Mohali

Smart Systems and Beyond – Bhatia et al. (Eds)
© 2027 The Editor(s), ISBN: 978-1-041-36617-1

Session Chairs

Dr. Paras Chawla, Session Chair
Professor & Director, Amity School of Engineering and Technology (ASET), Amity Institute of Information Technology (AIIT), Amity University Haryana, Amity Education Valley Manesar, Gurugram, Panchgaon, Haryana

Dr. Balwinder Dhaliwal, Session Chair
Associate professor NITTTR Chandigarh

Dr. Urvashi Garg, Session Chair
Professor and Head, CSE, Chandigarh University

Dr. (Prof.) Rajiv Kumar Chechi, Session Chair
Head, Computer Science & Engineering Department, MMEC & Director – Artificial Intelligence, Maharishi Markandeshwar (Deemed to be University), Mullana

Dr Manoj Kuri, Session Chair
Associate Professor & Head AI & DS
Govt Engineering College Bikaner

Smart Systems and Beyond – Bhatia et al. (Eds)
© 2027 The Editor(s), ISBN: 978-1-041-36617-1

International Advisory Committee

© 2027 The Editor(s), ISBN: 978-1-041-36617-1

Review Committee

Prof. S. Dasgupta, IIT Roorkee
Prof. S. K. Vishvakarma, IIT Indore
Prof. Jawar Singh, IIT Patna
Prof. (Dr.) Paras Chawla, Amity University Noida
Prof. K. G. Srinivasan, IIIT Naya Raipur
Prof. (Dr.) Amit Kant Pandit, SMVDU, Katra
Prof. Piush Verma, NITTTR Chandigarh
Prof. Arun Khosla, NIT Jalandhar
Prof. Ramesh Sonkar, IIT Guwahati
Prof. Ajay Kumar, IIT Indore
Prof. Rajesh Khanna, Thapar University, Punjab
Prof. Surinder Saini, CSIO, Chandigarh
Prof. Kulbir Singh, Thapar University, Punjab
Prof. Ashwani Rana, NIT Hamirpur
Prof. Balwinder Raj, NITTTR Chandigarh
Prof. Yadwinder Singh, Punjabi University
Prof. Amandeep Singh, NIT Srinagar
Prof. Jeetendra Singh, NIT Sikkim
Prof. Balwinder Singh, CDAC Mohali
Prof. Arun Kumar, PEC Chandigarh
Dr. Chetna Nagpal, UAE Dubai

Session 1 Mathematical and Computational Algorithmic Applications for Smart Systems Engineering

Smart Systems and Beyond – Bhatia et al. (Eds)

Experimental and Statistical Evaluation of Fine-Grained Soil Stabilisation Using Waste Marble Powder as a Sustainable Binder

Prashant Rukwal
Civil Engineering, Chandigarh University

Deepak Juneja
Civil Engineering, Panipat Institute of Engineering and Technology

Shivamanth Angadi
Civil Engineering, K.L.S Gogte Institute of Technology Belagavi

Rasuljon Tojiyev
Department of Technological Machinery and Equipment Fergana State Technical, University, Fergana, Uzbekistan

Sridevi D V
Centre for Advanced Materials Research, Rajalakshmi Institute of Technology, Kuthambakkam, Chennai

ABSTRACT: This research examines the potential use of Waste Marble Powder (WMP) as a green binder for stabilizing fine-grained soils, addressing two important problems simultaneously. The effect of varying the amount of WMP (0%, 5%, 10%, 15%, and 20%) on the Atterberg limits, compaction, UCS, CBR, and volume of soil was assessed through an extensive experimental program. The investigations showed that the geotechnical properties of the soil improved greatly when around 10–15% WMP was added. The plasticity index, swelling, shrinkage, and overall strength of the specimens drop noticeably and their load-bearing capacity increases. Using ANOVA and regression modeling showed that the main results were reliable and significant. Overall, the paper states that WMP is a sustainable replacement for regular soil stabilizers used in subgrade improvement in geotechnical projects.

Keywords: Soil Stabilization, Waste Marble Powder, Fine-Grained Soil, Unconfined Compressive Strength, California Bearing Ratio, Sustainability, ANOVA, Circular Economy, Eco-friendly Binder

1 INTRODUCTION

Properly stabilizing fine-grained soils is crucial in geotechnical engineering to improve the strength, life span, and general performance of construction and infrastructure (Almuaythir *et al.*, 2024);(Amakye *et al.*, 2021). Since fine soils such as clays and silts have plastic properties, low support strength, and are sensitive to water levels, they typically result in weakened structures and damage easily (Fadugba *et al.*, 2024). Frequently, the traditional approach to stabilization uses cement, lime or some chemicals, but these additives may not be helpful since they come with money and environment-related issues. In the past few years, people have become more interested in using industrial by-products and waste materials as alternatives for fixing soil which helps solve both waste concerns and builds environmentally-friendly structures (Waleed Zaid, 2024). Marble processing industries generate Waste Marble Powder (WMP) abundantly at a low price and it can be used as a sustainable stabilizer for soil. Because WMP is rich in calcium carbonate and is pozzolanic, it can strengthen and improve the soil's shape and improve its resistance by mixing with it (Nishant Sachdeva, 2023). Solving industrial waste and moving toward environmental protection are both accomplished by using WMP rather than traditional binders.

Sustainable approaches are being used more widely in the construction industry by making use of waste as an alternative to common stabilizers (Necmi Yarbaşı, 2021). Because so much is produced, Waste Marble Powder (WMP) has become a material that many people are now focused on. In many cases, millions of tons of waste are produced by marble industries every year. Like, its poor management can have negative effects on the environment and the land. When WMP is used to stabilize soil, it helps manage workplace waste and lets us depend less on cement alternatives, saving the environment and cutting costs for the project. Most of the calcium carbonate in WMP can make fine-grained soils stronger and more stable when it is mixed with them. Using it helps apply the principles of a circular economy by creating usable materials from waste in geotechnical fields (Wang *et al.*, 2024).

The success of WMP as a sustainably used binder relies on it improving properties such as plasticity, ease of compaction, strength and volume stability in soil. Due to pozzolanic interactions, WMP and soil minerals connect particles and fill gaps, raising the soil's bearing capacity and defying swelling and shrinking (B Martins Lima, GJ

Bruschi, L Festugato, 2023). Despite its usefulness, we need thorough experimental efforts and careful analysis of data to determine the impact of WMP on finer soils and decide on the best stabilization amount. Here, we carefully analyze how various amounts of WMP change important properties of clayey soil such as Atterberg limits, compaction rates, UCS, CBR and swelling and shrinking features. As well, the researchers check their discoveries using statistical methods and build practical models for use in various applications (Islam *et al.*, 2024).

2 LITERATURE REVIEW

(Umar and Lin, 2024) investigated the possibilities of marble powder taking part in stabilizing soil in an experimental study on a massive scale. The researchers compared the geotechnical properties and unconfined compressive strength (UCS) of soils of varying volumes of marble powder in their study. The study discovered that the incorporation of marble powder resulted in an improvement of the UCS and other engineering properties of the soil. The use of marble as a powder was found to be a successful method to improve the supporting qualities of subgrade materials and it was also a geotechnical-friendly method of the environment.

(Umar *et al.*, 2023) Atterberg limits, compaction tests, and strength measurements were studied during the research. Marble powder added as a result reduced the plasticity index of the soil and also increased the dry density. In addition, the untreated compressive strength recorded significant improvement, with the highest recorded at optimum dosages of the marble powder. These researchers discovered that WMP produces clayey soil that has higher mechanical properties and it is a flexible and eco-friendly method to stabilize soil.

(Sivrikaya *et al.*, 2020) researchers works was on the examination of the fact that the soil either swells or shrinks depending on the temperature because this has been known to influence the foundation and pavement performance. It also emerged that marble powder reduced the probability of soil swelling and shrinking. The reduced volume change was contributed to the fine particles of marble that enhanced cementitious and pozzolanic effects and thus enhanced bonding and a regular set of the soil particles.

(Kusuma *et al.*, 2023) reviewed the impact of marble waste powder on clay soil. The study established that as the WMP is added, the soil gets compacted and its strength is raised in terms of CBR to be used in a road sub-base as well as the base layer. Consequently, the soil analysis was also stronger in regard to resistive shear force and had reduced plasticity that made it more dependable in usage and mold. The researchers state that the replacement of lime and cement by WMP may enhance sustainability throughout the construction industry.

(Hammad*et al.*, 2024) tested to determine how Sabkha soils in Oman could be stabilized by the addition of waste marble powder. Due to their low engineering properties and level of saltiness, Sabkha soils are not easy to utilize in buildings. The strength of the soil was significantly increased upon the addition of WMP as established by the study which also reduced the permeability of the material. It was indicated that the topsoil salinity had also dropped due to ion exchanges with the marble powder. The research provided valuable data on the role of WMP in enhancing soil quality in regions such as the Middle East that are commonly affected by aridity and salinity.

(F Moscato, M Rosone, 2024) presented the preliminary results of the implementation of a waste product of a marble factory as soil stabilizer. It experimented with the prospect of incorporating marble scraps into soil stabilization in a bid to achieve general sustainability objectives. The experiments carried out by the researchers pointed to the possibility of the team being able to make their bones strong and assist the environment. To this, they added that more research and development of standard procedures for constructing with marble waste is necessary.

3 MATERIALS AND METHOD

The purpose of this study was to determine if waste marble powder (WMP) might be used as a sustainable and environmentally acceptable binding agent for fine-grained soils that were gathered from the Kharar area, Punjab. The research region was chosen because it has a lot of problematic and expansive soils that need to be stabilized for building. In order to evaluate changes in geotechnical qualities, the study technique included the fabrication of soil- WMP mixes, the methodical collecting and characterization of soil and WMP samples, and extensive laboratory testing. To ascertain the significance of the reported improvements, statistical studies were conducted. In order to ensure the accuracy and reliability of results, the tests were carried out following the ASTM and Indian Standard (IS) requirements.

3.1 *Collection and preparation of materials*

3.1.1 *Soil sampling*

Several small soil samples from the area that has problematic, weak, and easily disturbed subsoil were collected. A depth of 1.0 meters was reached to remove all organic topsoil and keep the soil uniform. All the samples were kept in airtight packaging and taken to the laboratory for processing. When the soil got to the lab, it was put out in the air to dry, bulk crushed with a wooden hammer, and sieved with a 4.75 mm IS sieve to remove coarse material and organic compounds.

3.1.2 *Waste marble powder collection*

The marble processing unit in Ram Lakhan Marbles, sector 38 Chandigarh provided the waste marble powder used in this study. We gathered the raw WMP as a powder and then manually took out any chips and large pieces that were visible. For 24 hours, the temperature in the oven was set at 105°C to decrease the remaining water. The powder was passed through a 75-micron sieve to get a fine material that was ready to mix with soil.

3.2 *Characterization of materials*

3.2.1 *Soil Characterization*

Natural soil samples were tested for physical and index properties according to IS: 2720. Then, the following tests were designed:

3.2.1.1 *Specific gravity* Conducted to understand the density of solids in the soil.

3.2.1.2 *Atterberg limits* The LL, PL, and PI of soil samples were assessed to determine their classification by plasticity.

3.2.1.3 *Grain size distribution* Samples were examined by wet sieving and using a hydrometer to measure the clay and silt fraction.

3.2.1.4 *Compaction characteristics* To find out the OMC and MDD, we used the Standard Proctor Test.

3.2.2 *Marble powder characterization*

The WMP was tested for:

3.2.2.1 *Chemical composition* The content of the main oxides (CaO, SiO_2, Al_2O_3, MgO) is established with the method of X-ray fluorescence (XRF).

3.2.2.2 *Specific gravity* Learn what the term relative density was referring to in terms of soil particles.

3.2.2.3 *pH Value* To measure whether soil is alkaline or acidic which determines its pozzolanic reaction with the cement's source.

3.3 *Preparation of stabilized soil samples*

The samples of stabilized soils were prepared by mixing natural soils and different proportions of WMP to determine the influence of the WMP amount that was added to soil. A control was carried along with the natural soil without any additive and other samples were added with WMP in 5, 10, 15, and 20 percentages by dry soil weight.

The detailed preparation procedure was as follows:

(1) **Weighing of Materials:** The Dry natural soil and waste powder of marble were measured carefully and in accordance with the required quantities using a precision weighing balance. In calculating the percentages of WMP, the ratios were represented by using the dry weight of the soil to come up with appropriate mixing ratios.
    (1) **Mixing of Dry Components:** Dry soil and WMP, which were measured, were carefully mixed in a clean container. It was done either manually or with the application of mechanical mixers to provide an even distribution of WMP particles into the soil matrix. Effective mixing was necessary to prevent local agglomeration of WMP and uniform stabilization of WMP in the sample.
(2) **Water Addition Based on Optimum Moisture Content (OMC):** Dry soil-WMP mixtures were slowly added to water-drop wise. The water addition was measured based on the Optimum Moisture content (OMC) which was as per the standard proctor results of properties of individual mixes which had been done previously. Removal of water was done slowly because it keeps the moist evenly spread and neither is it over-saturated nor wet in places.
(3) **Moisture Conditioning:** The moist stabilized soil samples were carefully placed in air-tight plastic bags. These capped samples were then left to cure at room temperature in duration of 24 hours. This resting period enabled the moisture to settle in an even way into the soil-WMP mixtures hence enabling proper hydration and interaction of chemical elements of the soil particles and the WMP before subsequent additional laboratory tests.

This meticulous and methodical method of preparation was central to getting consistent and sure results in the course of the following geotechnical tests.

3.4 *Laboratory testing on stabilized samples*

Several geotechnical standard tests were carried out on the blends of the soil-WMP to comprehensively determine the impact of adding waste marble powder (WMP) to the mechanical and physical characteristics of the soil. These studies were aimed at the measurement of changes in soil characteristics which are of value in engineering and construction applications. All stabilized mixes, such as control sample without WMP, passed the following tests:

(1) **Test of Atterberg Limits**
The soil-WMP experimental combinations were tested to check the properties of the plasticity of the materials which will include the Liquid Limit (LL), Plastic Limit (PL), and Plasticity Index (PI) by performing the test of the Atterberg Limits. These properties bring in insight into the informedness of the ground and the impacts of changes in moisture. Plasticity of the soil also reduces with an increase of WMP concentration and the Plasticity Index records this high level. Usually this results in improved workability and reduced tendencies towards volumetric changes such as swelling. It is necessary to carry out this test when evaluating the suitability of the stabilized soil to be used in a building project where the dimensional stability of the soil may be needed.

(2) **Proctor Compaction Test Standard**
The influence of the addition of WMP on the Maximum Dry Density (MDD) and Optimum Moisture Content (OMC) of soil was determined on each of the mix ratios through a Standard Proctor Compaction test. There is a possibility that WMP can affect the compaction behavior of the mixture unlike the natural soil, which varies in terms of particle size and particle density. To ensure this material has enough strength and stability after construction such as embankments and foundations, this test helps in finding the optimum water content at which the soil-WMP mix can be compacted to achieve maximum density.

(3) **The test for unconfined compressive strength (UCS)**
The Unconfined Compressive Strength was used to test the mechanical strength of the stabilized soil mixes. The mixes were pressed in molds to make cylindrical specimens which were 38 mm in diameter and 76 mm long. The samples were subsequently cured in agreed durations of seven and twenty-eight days to ensure that they could be hydrated and bonded with the soil particles and WMP. In order to maintain a continuity of conditions in developing strength, the specimens were then been robustly wrapped with polyethylene film during the process of curing in order to prevent the loss of moisture. The UCS test measures the maximum axial compressive stress that the sample can withstand without undergoing lateral confinement and hence provides an estimate of the loading capacity of the soil as well as its suitability in construction work.

(4) **Test of the California Bearing Ratio (CBR)**
To determine the ability of Mixtures of soil-WMP to bear losses in soaked conditions, all mixtures of soils-WMP were tested to suction using a soaked CBR test, which reflects the worst conditions possible in a field where soils are likely to be saturated with water. In the testing; the Indian Standard IS: 2720 Part 16 was observed and complied with. The specimens were kept in water for a period of four days to ensure that they were soaked before being tested. CBR test measures the resistance of the soil to penetration by a standard piston, hence the ability of the soil to support the stresses in the pavement and the road. This is a test that is imperative in the determination of the stability of the soil that can be applied to the foundation and sub-grade areas of pavements and other structures that provide support to load as well.

(5) **Tests for Swelling and Shrinkage**
This was done by conducting tests to determine the volumetric stability of the treated soil samples as there are expansive soils within the research area. Whereas the Linear Shrinkage test reviewed how much the soil would shrink when drying up, the Free Swell Index test reviewed how much the soil could swell when it would be underwater. These tests provide imperative information on the susceptibility of the soil to change its volume which could lead to structural failure. Because stabilization with a good stabilizing agent can result in a reduction in the number of volume changes thus enhancing the lifetime of structures erected on such soils, the reduction in the swelling and shrinkage potential by WMP was closely monitored.

3.5 *Curing conditions*

Strength tests (UCS and CBR) were done on samples that were cured in a chamber held at 27 ± 2°C and at 90% relative humidity. They were tested after 7 days and again after 28 days to determine the contribution of the marble powder in its ability to enhance strength with time.

3.6 *Statistical evaluation of results*

The findings were confirmed through statistical analysis of the results which established the link between the WMP content and various geotechnical features of the fine-grained soil. Originally, descriptive statistics were used to present the distribution and variability of the soil data along with the main values of mean, standard deviation, and variance at every WMP content level. The initial review allowed us to learn about trends and main values in the data. The effect of WMP on soil behavior was examined using an ANOVA test on the SPSS program. The ANOVA examined if the inclusion of WMP brought about no meaningful changes in UCS, MDD, or PI. When p-values were found below 0.05, the influence of WMP was proven to be significant, allowing us to reject the null hypothesis. In addition, regression analysis was used, including linear and polynomial models, to foresee results for UCS and CBR depending on changing amounts of WMP and the time taken to cure. According to the high R^2

from the models, the dosage and curing time of WMP were effective ways to predict the strength of the mixtures. Using Pearson's correlation, a matrix was prepared to check how strongly and in which direction WMP content, UCS, MDD, and PI are connected. Positive relationships between water-moving pore content and strength/density factors were highly evaluated, while negative relationships were recorded between plasticity and swelling properties.

3.7 *Sustainability considerations*

Table 1. Effect of WMP on atterberg limits.

WMP Content (%)	Liquid Limit (%)	Plastic Limit (%)	Plasticity Index (%)
0 (Control)	48.2	24.6	23.6
5	45.1	26.3	18.8
10	41.5	28.7	12.8
15	38.4	30.5	7.9
20	35.2	32.1	3.1

4 RESULTS AND DISCUSSION

In this section, the experimental and statistical results of the work are discussed and given to illustrate the influence of WMP on fine-grained soil. The study aimed at finding out the changes in Atterberg limits, how the process of compaction can influence the results, unconfined compressive strength (UCS), California Bearing Ratio (CBR), swelling and shrinkage characteristics, and volume in general. Regression analysis and ANOVA were employed in verifying the trends estimated and the interrelation of various variables.

4.1 *Effect of WMP on atterberg limits*

The addition of the WMP to fine-grained soil resulted in softening of the Atterberg Limits, which shows that there was a significant influence on the plastic deformation behavior of the soil. The Plasticity Index (PI) also shrunk steadily with the increase in the proportion of WMP. This decrease in PI implies that the flexibility of the soil has greatly declined with it becoming less cohesive and easily worked. Low plasticity is also good since it decreases the chance of the soil volumetric change such as swelling and shrinking which is a common problem in expansive and fine-grained soils. Subsequently, the WMP-content stabilized soil with more contents had better handling characteristics and resilience to defection caused by changes in soil moisture. These changes mean that WMP is acting as a stabilizer through changes in mineralogy and particle interaction of the soil that reduces the overall plastic behavior of the soil and makes it more suitable to engineering and construction purposes.

Table 1 indicates that WMP noticeably adds plasticity to fine-grained soil. When WMP content rises from 0% to 20%, the liquid limit decreases from 48.2% to 35.2% which means the soil holds less water and there is less clay activity. Besides, the plastic limit climbs from 24.6% to 32.1% which means that soil can be worked with less effort and is less likely to stick. As a result, the plasticity index went from 23.6% to 3.1% which indicates that the soil shifted from being highly plastic to low plastic. A lower plasticity index shows that the soil stabilizes well under wet and dry conditions. The findings demonstrate that WMP helps lower the plasticity index, increases stiffness, and improves the consistency of fine-grained soils. As a result, it is considered a suitable and environmentally friendly option for applications in subgrade and foundations.

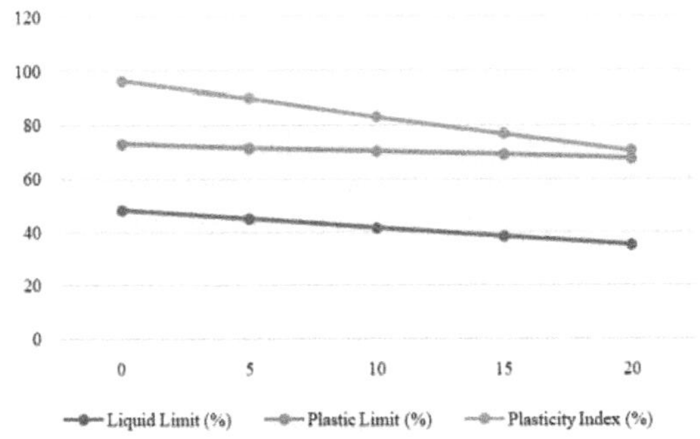

Figure 1. Effect of WMP on atterberg limits.

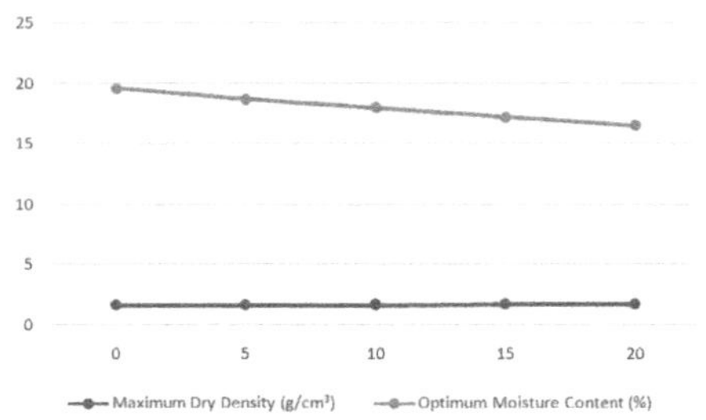

Figure 2. Compaction characteristics with WMP addition.

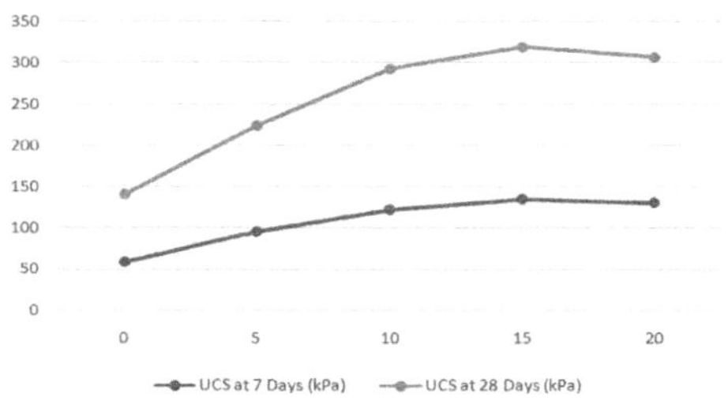

Figure 3. Unconfined Compressive Strength (UCS) at different curing periods.

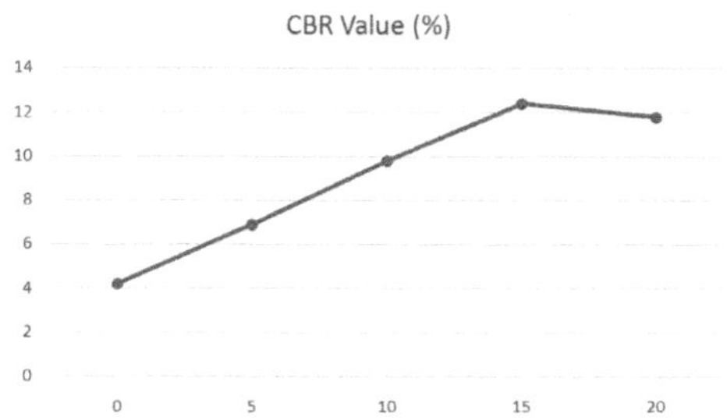

Figure 4. California Bearing Ratio (CBR) results.

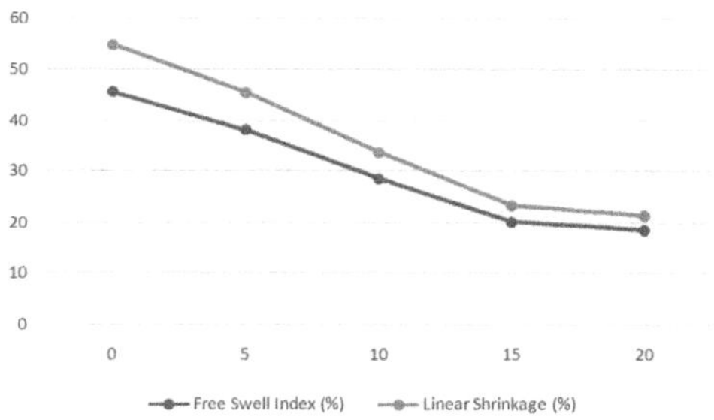

Figure 5. Swelling and shrinkage characteristics.

4.2 *Effect of WMP on compaction characteristics*

The results illustrated in Table 2 reveal that the behavior of the soil concerning compaction has been considerably modified after the WMP powder has been added to the soil. The Maximum Dry Density (MDD) of the soil mixes was slightly enhanced as the content of the WMP rose. The density gained by this means the small, inert WMP particles enabled the soil particles to be more compact, i.e., potential spaces are filled and the mix is more condensed generally. Simultaneously, the difference between the Optimum Moisture Content (OMC) and the increase in the WMP % was minimal. Given that WMP absorbs less water compared to natural soil, less water is required to achieve the optimum compaction thus the reduction in the requirement of moisture. All these changes indicate that the addition of WMP improves the ability of the soil to become compact and still attains a comparatively lower moisture content to give it a tighter packing, possibly saving it a higher stability level and load-bearing scenario in the real-world scenarios of engineering.

Table 2. Compaction characteristics with WMP addition.

WMP Content (%)	Maximum Dry Density (g/cm^3)	Optimum Moisture Content (%)
0 (Control)	1.62	19.6
5	1.64	18.7
10	1.66	18.0
15	1.69	17.2
20	1.70	16.5

It is evident from Table 2 that compaction in soil is enhanced by the use of WMP. As the proportion of fine marble particles increases within WMP content, the maximum dry density MDD increases from 1.62 g/cm^3 to 1.70 g/cm^3, proving that the filler effect results in superior and less empty particle packing. This also means that WMP now requires less moisture for the best compaction, most likely due to WMP's lower ability to soak up moisture and help grains press together. As a result, WMP-treated soil is firmer, easier to use, and more environmentally friendly which makes it valuable for road subgrades and under most structures.

4.3 *Improvement in Unconfined Compressive Strength (UCS)*

When WMP was added to the soil mixes, there was a high rise in the Unconfined Compressive Strength (UCS) values and this led to the conclusion that the overall mechanical strength of the soil stabilized had increased. All soil-WMP mixtures showed a remarkable enhancement in bearing capacity compared to the untreated soil with the highest increments in UCS recorded at UMP content levels between 10 to 15 percent. The binding action of WMP particles, which probably improved particle cohesion and decreased soil porosity, is responsible for this improvement. In order to document the strength improvement over time, the UCS tests were performed after both 7 and 28-day curing times. Table 3 shows strength values increasing with the curing time, which shows that physical and chemical interactions continued in the stabilized matrix. These findings prove that WMP holds potential to be a potent stabilizing agent that offers immediate increases in soil strength along with an increase in long-term performance and strength in geotechnical projects.

Table 3. Unconfined Compressive Strength (UCS) at different curing periods.

WMP Content (%)	UCS at 7 Days (kPa)	UCS at 28 Days (kPa)
0 (Control)	58.6	82.4
5	95.3	128.9
10	123.1	169.7
15	135.4	184.2
20	130.6	176.5

The UCS of fine-grained soil rises sharply when WMP is added and this happens up to a 15% content. Tests revealed that the UCS increased from 58.6 kPa to 135.4 kPa at 7 days, then from 82.4 kPa to 184.2 kPa at 28 days, due to WMP and the curing procedure. The rise in strength is probably caused by tight bonds between particles and pozzolanic reactions between marble powder and soil. Even so, a small UCS decline at 20% WMP recommends that adding too much marble powder may decrease how strongly the soil sticks together. It is shown in the results that 15% WMP increases compressive strength most effectively, which means it can be used as a dependable and sustainable additive for fine-grained soils in geotechnical projects.

4.4 *CBR improvement with WMP stabilization*

Pile CBR values rose greatly when WMP was added, as is clear in Table 4.

Table 4. California Bearing Ratio (CBR) results (Soaked Condition).

WMP Content (%)	CBR Value (%)
0 (Control)	4.2
5	6.9
10	9.8
15	12.4
20	11.8

In wet conditions, the California Bearing Ratio (CBR) of fine-grained soil is shown in Table 4 for various proportions of WMP added. The charts demonstrate that the ability of the soil to support loads gets better with more WMP, from 4.2% in the untreated soil to a maximum of 12.4% with 15% WMP. When the filler represents more of the total volume, the soil gains strength and is less deformed when loaded under pressure. Nevertheless, after increasing WMP to 20%, a drop in CBR to 11.8% can be seen, indicating that more filler may stop high WMP effects caused by optimal particle interactions. Finally, the data shows that WMP raises the quality and load-bearing strength of fine-grained soils and 15% was seen as the best percentage to achieve maximum improvement.

4.5 *Swelling and shrinkage behavior*

Effects caused by moisture changes in the wood were also investigated. The results displayed in Table 5 shows that free swell index and linear shrinkage decreased with a rise in WMP.

Table 5. Swelling and shrinkage characteristics.

WMP Content (%)	Free Swell Index (%)	Linear Shrinkage (%)
0 (Control)	45.8	9.2
5	38.2	7.4
10	28.6	5.1
15	20.3	3.2
20	18.7	2.8

The results show that WMP has an impact on how fine-grained soil expands or contracts. An increase in WMP content from 0% to 20% reduces the soil's tendency to swell from 45.8% to 18.7%, as the free swell index shows. The same is true for linear shrinkage since it moves from 9.2% when wet to 2.8% when oven-dried. The swelling and shrinkage of the soil are reduced by marble powder which fills clay mineral gaps and makes the soil less plastic and less affected by moisture. The introduction of WMP improves the strength of fine-grained soils, prevents the problems linked to volumetric changes like cracking and destruction, and supports its use for sustainable soil stabilization.

4.6 *Statistical analysis and model development*

ANOVA results
For UCS and CBR values, an Analysis of Variance (ANOVA) was run. All the p-values were less than 0.05, suggesting that the changes resulting from different WMP contents were true at a statistical level.

Test Parameter	F-value	p-value
UCS	27.85	0.0003
CBR	19.12	0.0011

Regression Models
A second-degree polynomial regression equation best explained the association between the amount of WMP and the soil's UCS after 28 days of curing:

$$\textbf{UCS 28 Days} = -\mathbf{0}.25\mathbf{x}^2 + \mathbf{8}.2\mathbf{x} + \mathbf{82.4},$$

where x is the proportion of WMP added by the soil's dry weight. A good association between WMP concentration and strength growth at 28 days was shown by this model's high coefficient of determination, $R^2 = 0.96$. The high R^2 value attests to the fact that WMP concentration is a good indicator of sustained strength gains in stabilized soils.

It should be mentioned, nevertheless, that no regression model was created for the 7-day UCS findings. This is because, maybe as a result of inadequate curing and hydration processes at an earlier stage, the strength growth after 7 days displayed more variability and a less consistent connection with WMP concentration. The 28-day UCS can thus be a more dependable accurate basis of predictive modeling.

Overall, the regression equation can also be used by engineers in determining expected strength gains along with modification of mix designs at the early design phase of a soil stabilization project after the WMP dose.

4.7 *Overall discussion*

It is concluded from the discussion that Waste Marble Powder is an excellent and eco-friendly support for fine-grained soils, according to both experimental and statistical evidence. It was seen that adding compost between 10 and 15% by mass improved the main soil qualities the most. In particular, at these application rates, the soil became stronger and more stable under load, as shown by higher unconfined compressive strength and CBR values. These tests revealed important changes in plasticity index and swelling shrinkage which are important for lessening soil volume changes that may impact engineering structures.

WMP reduces the features of plasticity and swell-shrink potential by modifying how the soil reacts at the mineral level. The result is more reliable dimensions and better protection from changes caused by moisture. More durable subgrade and foundation result in high strength against fluctuating conditions and stresses for a longer period.

As well as experimenting, the study measured and checked the findings using Analysis of Variance (ANOVA) and regression analysis. The findings were strengthened because the changes seen in soil properties were confirmed to be statistically significant. The models also helped researchers predict soil reactions which allowed for improved construction and monitoring of projects using WMPs.

To sum up, laboratory experiments and statistical checks show WMP can manage soil stabilization and explain why it is better than cement or lime both environmentally and economically. Using WMP fixes problems related to waste from the marble industry and at the same time supports sustainable infrastructure building, thus improving soil quality.

5 CONCLUSION

The results of the present study indicate that Waste Marble Powder (WMP) is useful for stabilizing fine-grained soils in an environmentally friendly way. Laboratory testing and analysis results indicated that the inclusion of WMP improved the technical properties of the soil as it lowered the plasticity index, free swell index, and linear shrinkage. The maximum dry density went up, while optimum moisture content went down, reflecting that compaction has improved. At 10-15% WMP replacement, UCS and CBR both significantly improved which points to that range being the best one for soil stabilization. Past the threshold of five WMP, adding more meals only helped a little with strength. The use of ANOVA and regression modeling confirmed the meaningfulness of the study and showed clear links between the results and predictions. All in all, stabilizing soil with WMP can address weak subgrade soils and reuse industrial waste, while supporting best green practices as well as the principles of the circular economy. Based on this study, WMP can be integrated into engineering activities, mainly in road building and developing low-cost infrastructure.

REFERENCES

Al Sadrani, K.A. (2023). Geotechnical Properties of Omani Sabkha Soil Stabilized with Waste Marble Powder (Master's thesis, Sultan Qaboos University (Oman)).

Almuaythir, S., Zaini, M.S.I., Hasan, M., and Hoque, M.I. (2024). Sustainable soil stabilization using industrial waste ash: Enhancing expansive clay properties. *Heliyon*, 10(20).

Amakye, S.Y., Abbey, S.J., Booth, C.A., and Mahamadu, A.M. (2021). Enhancing the engineering properties of subgrade materials using processed waste: A review. *Geotechnics*, 1(2), 307–329.

Fadugba, O.G., Adeniran, J.K., Alaneme, G.U., Oluyemi-Ayibiowu, B.D., Omomomi, O.J., and Adetukasi, A.O. (2024). Compaction and compressibility characteristics of snail shell ash and granulated blast furnace slag stabilized local bentonite for baseliner of landfill. *Scientific Reports*, 14(1), 7167.

Islam, S., Islam, J., Alam, M.J.B., Chowdhury, A.S., and Hasnat, N. (2024). Partial replacement of sand by fine-grained crushed waste glass along with fly ash stabilization for geotechnical applications in pavement. *Heliyon*, 10(19).

Kusuma, R.I., Mina, E., Fathonah, W., and Bilhaq, A.A. (2023). The influence of marble waste powder on clay soil stabilization. *International Journal on AdvancedInternational Journal on Advanced Science, Engineering and Information Technology (IJASEIT)*, 13 (2), 487–493.

Martins Lima, B., Bruschi, G.J., Festugato, L., and Consoli, N.C. (2024). Mechanical behavior of a granular soil stabilized with alkali-activated waste. *Journal of Materials in Civil Engineering*, 36(1), 04023485.

Moscato, F., Rosone, M., and Celauro, C. (2024, July). A Preliminary Investigation on the Use of a Marble Factory By- Product as a Sustainable Soil Stabiliser. In *International Conference on Environmental Geotechnology, Recycled Waste Materials and Sustainable Engineering* (pp. 157–167). Singapore: Springer Nature Singapore.

Sachdeva, N., and Shrivastava, N. (2023). Sustainable use of solid waste as additives in soil stabilization: A state-of-art review.

Sivrikaya, O., Uysal, F., Yorulmaz, A., and Aydin, K. (2020). The efficiency of waste marble powder in the stabilization of fine-grained soils in terms of volume changes. *Arabian Journal for Science and Engineering*, 45, 8561–8576.

Umar, I.H., and Lin, H. (2024). Marble powder as a soil stabilizer: An experimental investigation of the geotechnical properties and unconfined compressive strength analysis. *Materials*, 17(5), 1208.

Umar, I.H., Lin, H., and Ibrahim, A.S. (2023). Laboratory testing and analysis of clay soil stabilization using waste marble powder. *Applied Sciences*, 13(16), 9274.

Wang, Z., Yang, J., and Wu, Y. (2024). Durability Performance of CGF Stone Waste Road Base Materials under Dry– Wet and Freeze–Thaw Cycles. *Materials*, 17(17), 4272.

Yarbaşı, N., and Kalkan, E. (2020). Freeze-thaw resistance of fine-grained soils stabilized with waste material mixtures. *International Journal of Science and Engineering Applications*, 10(1), 1–6.

Zaid, W. (2024). Improving the Geotechnical Properties of a Soil Using Xanthan Gum and Concrete Waste (Master's thesis, Gaziantep University).

Smart Systems and Beyond – Bhatia et al. (Eds)

A Novel Framework for Convergence Ordering and Stability Analysis of Coefficient-Sensitive Iterative Algorithms

Naveen Kumar
Department of Mathematics, University Institute of Sciences (UIS), Chandigarh University, Mohali, Punjab, India

ABSTRACT: The primary objective of this research is to investigate the self comparison of the speediness of convergence of different cases in Picard's S-iteration (PS-iteration) with the convergence rate of the different iterations reformulated by exchange of coefficients defined in the original one by the using the process of interchange of coefficients included in the PS-iteration. In self-comparing, this process—the switching of coefficients defined in such procedures—is crucial. Beyond the classical fixed point theory, this work envisions convergence mechanisms as the computational backbone of smart AI systems, capable of autonomously refining decisions in robotics and autonomous computing. By treating coefficient-sensitive iterations as self-correcting engines, the study opens pathways for real-time optimization in intelligent machines.

Keywords: Contraction mapping, convex space, Picard S-iteration, speed of convergence, fixed point

1 INTRODUCTION

Various researchers have considered the numerous iterative processes to study their rate of convergence (Xue, Z. and Lv, G. 2013). The. Some iteration methods were introduced to study the fixed points for the contraction mappings and others were introduced for the nonexpensive mappings. Some authors compared the convergence speed for these iterative processes and some of them utilized the coefficients of their iteration algorithms to improve their convergence speed. Using the definition of the rate of convergence, the examination of the order of convergence of some iteration algorithms are provided to converge better than previous iterations. Keeping in mind the comparative analysis drawn by various authors, we continue this process by investigating the convergence rate of PS-iteration algorithm. The study of iterative algorithms and their convergence properties form a central pillar in fixed point theory and numerical computation. Over the years, numerous scholars have proposed iterative procedures such as Picard, Mann, Ishikawa, and their hybrid variations to approximate fixed points of nonlinear mappings. Empirical illustrations further validate the analytical findings, affirming the superiority of the original scheme. The work concludes by reflecting on possible extensions of convergence optimisation in broader computational domains, including autonomous systems and adaptive AI mechanisms. This work also proposes a conceptual link between convergence analysis and autonomous AI behavior, suggesting a future where iterative schemes are embedded within smart robots as predictive correction modules. Such integration redefines convergence not only as a mathematical end-state but as a continuous cognitive process essential for intelligent decision-making.

2 LITERATURE REVIEW

Throughout the year, various analysts have approximated the fixed points utilizing numerous iteration plans. Many authors have examined the fastness of the convergence of many iterations plans like Berinde (Berinde, V. 2004). Picard, compared the Picard iteration with Mann iteration for a class of quasi-contractive operators; in 2005, Kima *et al.* (Kima, T.H., Cho, Y.J. and Kang, S.M. 2005) gave the Modified-Mann iterations with its better order of convergence. In 2006, Babu *et al.* (Babu, G.V.R. and Vara Prasad, K.N.V.V. 2006) investigated that Mann iterative process converges quicker than Ishikawa iterative plan for the class of Zamfirescu operators and later they have given the analysis on convergence rate among Krasnoselskij, Mann and Ishikawa iterative procedures in Banach spaces. Popescu (Popescu, O. 2007) compared Picard iteration with Mann iteration for their convergence rate and analysed that the Picard-iteration has better convergence rate than the Mann iterative plan for quasi contractive operators. Also, Bosede (Bosede, A.O. 2010) proved the strong convergence results for the J-M methods.

In 2012, Akewe *et al.* (Akewe, H. and Olaoluwa, H. 2012) showed some results on the order of convergence of the Modified 3-step iteration for generalized contractive like operators. Fathollahi *et al.* (Fathollahi, S., Razani, A.

DOI: 10.1201/9781042026289-2

and Moradi, S. 2015) investigated the quickness of the convergence of numerous iterative processes and discovered that the coefficients defined in iterative processes perform a crucial involvement in determining their order of convergence. Saluja (Saluja, G.S. 2015) observed the convergence of modified S-iteration process for two generalized asymptotically quasi-nonexpansive mappings in CAT(0) spaces.

This concept is continued by various authors for better convergence results. Some authors introduced new fixed point iteration methods and compared their speed of convergence with various existing fixed point schemes. In 2018, Kumar *et al.* (Kumar, N, and Chauhan, S.S. 2018) compared various iteration procedures for their speed of convergence and exercised an illustrative analysis for some fixed point iteration plans for better order of convergence. Recently, in 2019, Kumar *et al.* (Kumar, N, and Chauhan, S.S. 2019) provided the self-comparison of the convergence rate of some iterative algorithms using the technique of interchanging the coefficients introduced in the fixed point iteration plans and they also justified the better convergence speed with numerical illustration.

3 PROBLEM FORMULATION

Self-Comparison of PS-Iteration: Consider all the possible cases of PS-iteration after interchanging the associated coefficients along with the existing one:

$$\begin{aligned} u_{n+1} &= \mathcal{T} v_n \\ v_n &= (1-\beta_n)\mathcal{T} u_n + \beta_n \mathcal{T} w_n \\ w_n &= (1-\gamma_n)u_n + \gamma_n \mathcal{T} u_n \qquad n \geq 0 \end{aligned} \tag{1}$$

And

$$\begin{aligned} u_{n+1} &= \mathcal{T} v_n \\ v_n &= \beta_n \mathcal{T} u_n + (1-\beta_n)\mathcal{T} w_n \\ w_n &= (1-\gamma_n)u_n + \gamma_n \mathcal{T} u_n \qquad n \geq 0 \end{aligned} \tag{2}$$

And

$$\begin{aligned} u_{n+1} &= \mathcal{T} v_n \\ v_n &= (1-\beta_n)\mathcal{T} u_n + \beta_n \mathcal{T} w_n \\ w_n &= \gamma_n u_n + (1-\gamma_n)\mathcal{T} u_n \qquad n \geq 0 \end{aligned} \tag{3}$$

And

$$\begin{aligned} u_{n+1} &= \mathcal{T} v_n \\ v_n &= \beta_n \mathcal{T} u_n + (1-\beta_n)\mathcal{T} w_n \\ w_n &= \gamma_n u_n + (1-\gamma_n)\mathcal{T} u_n \qquad n \geq 0 \end{aligned} \tag{4}$$

where $\{\beta_n\}$ and $\{\gamma_n\}$ are the positive numbers, sequences chosen in $\left(\frac{1}{2}, 1\right)$.
Let us consider the sequence $\{u_n\}$ in iteration process (1). Therefore,

$$\begin{aligned} \|w_n - \mathcal{P}\| &= \|(1-\gamma_n)u_n + \gamma_n \mathcal{T} u_n - \mathcal{P}\| \\ &= \|(1-\gamma_n)u_n + \gamma_n \mathcal{T} u_n - \gamma_n \mathcal{P} + \gamma_n \mathcal{P} - \mathcal{P}\| \\ &= \|(1-\gamma_n)u_n + \gamma_n(\mathcal{T} u_n - \mathcal{P}) + \mathcal{P}(\gamma_n - 1)\| \\ &\leq \|(1-\gamma_n)u_n + \gamma_n(\mathcal{T} u_n - \mathcal{P}) + \mathcal{P}(\gamma_n - 1)\| \\ &\leq (1-\gamma_n)\|u_n - \mathcal{P}\| + \gamma_n\|\mathcal{T} u_n - \mathcal{P}\| \\ &\leq (1-\gamma_n)\|u_n - \mathcal{P}\| + k\gamma_n\|u_n - \mathcal{P}\| \\ &= [(1-\gamma_n) + k\gamma_n]\|u_n - \mathcal{P}\| \\ &= [1-\gamma_n(1-k)]\|u_n - \mathcal{P}\| \end{aligned} \tag{5}$$

Also

$$\begin{aligned} \|v_n - \mathcal{P}\| &= \|(1-\beta_n)\mathcal{T} u_n + \beta_n \mathcal{T} w_n - \mathcal{P}\| \\ &= \|(1-\beta_n)\mathcal{T} u_n + (\beta_n - 1)\mathcal{P} + \beta_n(\mathcal{T} w_n - \mathcal{P})\| \\ &\leq (1-\beta_n)\|\mathcal{T} u_n - \mathcal{P}\| + \beta_n\|\mathcal{T} w_n - \mathcal{P}\| \\ &\leq (1-\beta_n)k\|u_n - \mathcal{P}\| + \beta_n k\|w_n - \mathcal{P}\| \\ &= [(1-\beta_n)k\|u_n - \mathcal{P}\| + \beta_n k\{1-\gamma_n(1-k)\}\|u_n - \mathcal{P}\| \\ &= [(1-\beta_n)k + \beta_n k - k(1-k)\beta_n\gamma_n]\|u_n - \mathcal{P}\| \end{aligned} \tag{6}$$

Similarly, we have:

$$\begin{aligned} \|u_{n+1} - \mathcal{P}\| &= \|\mathcal{T} v_n - \mathcal{P}\| \\ &< k\|v_n - \mathcal{P}\| \\ &\leq k[(1-\beta_n)k + \beta_n k - k(1-k)\beta_n\gamma_n]\|u_n - \mathcal{P}\| \end{aligned} \tag{7}$$

Since $\beta_n, \gamma_n \in \left(\frac{1}{2}, 1\right)$ for all $n \geq 0$
$1-\beta_n < \frac{1}{2}$ and $\beta_n < 1$ and $-\beta_n \gamma_n < -\frac{1}{4}$
Thus from (7), we obtain:

$$[(1-\beta_n)k + \beta_n k - k(1-k)\beta_n\gamma_n] < \frac{k}{2} + k - \frac{[k(1-k)]}{4} = \frac{5k + k^2}{4}$$

Take

$$\xi_n = \left[\frac{5k + k^2}{4}\right]^n \|u_1 - \mathcal{P}\|$$

Similarly, let $\{u_n\}$ is the sequence in (2), then

$$\begin{aligned} \|w_n - \mathcal{P}\| &= \|(1-\gamma_n)u_n + \gamma_n \mathcal{T} u_n - \mathcal{P}\| \\ &\leq [1 - (1-k)\gamma_n]\|u_n - \mathcal{P}\| \end{aligned} \tag{8}$$

Similarly,

$$\begin{aligned} \|v_n - \mathcal{P}\| &= \|\beta_n \mathcal{T} u_n + (1-\beta_n)\mathcal{T} w_n - \mathcal{P}\| \\ &\leq \beta_n\|\mathcal{T} u_n - \mathcal{P}\| + (1-\beta_n)\|\mathcal{T} w_n - \mathcal{P}\| \\ &\leq k\beta_n\|u_n - \mathcal{P}\| + (1-\beta_n)k\|w_n - \mathcal{P}\| \\ &\leq k\beta_n\|u_n - \mathcal{P}\| + (1-\beta_n)k[1 - (1-k)\gamma_n]\|u_n - \mathcal{P}\| \\ &= [k\beta_n + k(1-\beta_n) - k(1-k)(1-\beta_n)\gamma_n]\|u_n - \mathcal{P}\| \end{aligned} \tag{9}$$

Also

$$\begin{aligned} \|u_{n+1} - \mathcal{P}\| &= \|\mathcal{T} v_n - \mathcal{P}\| \\ &< k\|v_n - \mathcal{P}\| \\ &\leq k[k\beta_n + k(1-\beta_n) - k(1-k)(1-\beta_n)\gamma_n]\|u_n - \mathcal{P}\| \\ &= \left\{k^2\beta_n + k^2(1-\beta_n) - k^2(1-k)(1-\beta_n)\gamma_n\right\}\|u_n - \mathcal{P}\| \end{aligned} \tag{10}$$

Since $\beta_n, \quad \gamma_n \in \left(\frac{1}{2}, 1\right)$ for all $n \geq 0$,
$\beta_n < 1$ and $(1-\beta_n) < \frac{1}{2}$and $-(1-\beta_n)\gamma_n < -\frac{1}{4}$
Thus from (10), we obtain:

$$\left[k^2\beta_n + k^2(1-\beta_n) - k^2(1-k)(1-\beta_n)\gamma_n\right] < k^2 + k^2\frac{1}{2} - \frac{1}{4}k^2(1-k) = k^2\left(\frac{7-k}{4}\right)$$

$$\text{Take } \theta_n = \left[k^2\left(\frac{7 \quad - \quad k}{4}\right)\right]^n \|u_1 - \mathcal{P}\|$$

Take $\{u_n\}$ as the sequence in (3), then

$$\begin{aligned} \|w_n - \mathcal{P}\| &= \|\gamma_n u_n + (1-\gamma_n)\mathcal{T} u_n - \mathcal{P}\| \\ &\leq [\gamma_n + k(1-\gamma_n)]\|u_n - \mathcal{P}\| \end{aligned} \tag{11}$$

And

$$\begin{aligned} \|v_n - \mathcal{P}\| &= \|(1-\beta_n)\mathcal{T} u_n + \beta_n \mathcal{T} w_n - \mathcal{P}\| \\ &\leq (1-\beta_n)\|\mathcal{T} u_n - \mathcal{P}\| + \beta_n\|\mathcal{T} w_n - \mathcal{P}\| \\ &\leq (1-\beta_n)k\|u_n - \mathcal{P}\| + \beta_n k[\gamma_n + k(1-\gamma_n)]\|u_n - \mathcal{P}\| \\ &= \left[k(1-\beta_n) + k\beta_n\gamma_n + k^2\beta_n(1-\gamma_n)\right]\|u_n - \mathcal{P}\| \end{aligned} \tag{12}$$

Also

$$\begin{aligned} \|u_{n+1} - \mathcal{P}\| &= \|\mathcal{T} v_n - \mathcal{P}\| \\ &\leq k\|v_n - \mathcal{P}\| \\ &\leq k\left[k(1-\beta_n) + k\beta_n\gamma_n + k^2\beta_n(1-\gamma_n)\right]\|u_n - \mathcal{P}\| \\ &= \left[k(1-\beta_n) + k^2\beta_n\gamma_n + k^3\beta_n(1-\gamma_n)\right]\|u_n - \mathcal{P}\| \end{aligned} \tag{13}$$

Since $\beta_n, \gamma_n \in \left(\frac{1}{2}, 1\right)$ for all $n \geq 0$,
$1-\beta_n < \frac{1}{2}$ and $\beta_n \gamma_n < 1$ and $\beta_n (1-\gamma_n) < \frac{1}{2}$
Thus from (13), we get:

$$\left[k^2(1-\beta_n) + k^2\beta_n\gamma_n + k^3\beta_n(1-\gamma_n)\right] < \frac{k^2}{2} + k^2 + \frac{k^3}{2} = k^2\left(\frac{3+k}{2}\right)$$

$$\text{Take } \varphi_n = \left[k^2\left(\frac{3+k}{2}\right)\right]^n \|u_1 - \mathcal{P}\|$$

Let $\{u_n\}$ be any sequence in iterative process (4),

$$\begin{aligned} \|w_n - \mathcal{P}\| &= \|\gamma_n u_n + (1-\gamma_n)\mathcal{T}u_n - \mathcal{P}\| \\ &\leq [\gamma_n + k(1-\gamma_n)]\|u_n - \mathcal{P}\| \end{aligned} \tag{14}$$

And

$$\begin{aligned} \|v_n - \mathcal{P}\| &= \|\beta_n\mathcal{T}u_n + (1-\beta_n)\mathcal{T}w_n - \mathcal{P}\| \\ &\leq k\beta_n\|u_n - \mathcal{P}\| + (1-\beta_n)k\|w_n - \mathcal{P}\| \\ &\leq k\beta_n\|u_n - \mathcal{P}\| + (1-\beta_n)k[\gamma_n + k(1-\gamma_n)]\|u_n - \mathcal{P}\| \\ &\leq \left[k\beta_n + k(1-\beta_n)\gamma_n + k^2(1-\beta_n)(1-\gamma_n)\right]\|u_n - \mathcal{P}\| \end{aligned} \tag{15}$$

Also

$$\begin{aligned} \|u_{n+1} - \mathcal{P}\| &= \|\mathcal{T}v_n - \mathcal{P}\| \\ &= k\|v_n - \mathcal{P}\| \\ &\leq k\left[k\beta_n + k(1-\beta_n)\gamma_n + k^2(1-\beta_n)(1-\gamma_n)\right]\|u_n - \mathcal{P}\| \\ &= \left[k^2\beta_n + k^2(1-\beta_n)\gamma_n + k^3(1-\beta_n)(1-\gamma_n)\right]\|u_n - \mathcal{P}\| \end{aligned} \tag{16}$$

Since $\beta_n, \gamma_n \in \left(\frac{1}{2}, 1\right)$ for all $n \geq 0$,

$$\beta_n < 1 \text{ and } (1-\beta_n)\gamma_n < \frac{\mathbf{1}}{\mathbf{2}} \text{ and } (1-\beta_n)(1 - \gamma_n) < \frac{1}{4}$$

Thus from (16), we get:

$$\left[k^2\beta_n + k^2(1-\beta_n)\ \gamma_n + k3\ (1-\beta_n)(1-\gamma_n)\right] < k^2 + \frac{k^2}{2} + \frac{k^3}{4} = \frac{6k^2 + k^3}{4}.$$

$$\text{Take}\psi_n = \left[\frac{6k^2 + k^3}{4}\right]^n \|u_1 - \mathcal{P}\|$$

The major comparison of all the cases (1)–(4) of PS-iterations is as follows:

$$\lim_{n\to\infty}\left|\frac{\xi_n}{\theta_n}\right| = \lim_{n\to\infty}\frac{\left[\frac{5k+k^2}{4}\right]^n \|u_1-\mathcal{P}\|}{\left[k^2\left(\frac{7-k}{4}\right)\right]^n \|u_1-\mathcal{P}\|} = 0 \tag{17}$$

$$\lim_{n\to\infty}\left|\frac{\theta_n}{\varphi_n}\right| = \lim_{n\to\infty}\frac{\left[k^2\left(\frac{7-k}{4}\right)\right]^n \|u_1-\mathcal{P}\|}{\left[k^2\left(\frac{3+k}{2}\right)\right]^n \|u_1-\mathcal{P}\|} = 0 \tag{18}$$

$$\lim_{n\to\infty}\left|\frac{\varphi_n}{\psi_n}\right| = \lim_{n\to\infty}\frac{\left[k^2\left(\frac{3+k}{2}\right)\right]^n \|u_1-\mathcal{P}\|}{\left[\frac{6\,k^2+k^3}{4}\right]^n \|u_1-\mathcal{P}\|} = 0 \tag{19}$$

$$\lim_{n\to\infty}\left|\frac{\xi_n}{\varphi_n}\right| = \lim_{n\to\infty}\frac{\left[\frac{5k+k^2}{4}\right]^n \|u_1-\mathcal{P}\|}{\left[k^2\left(\frac{3+k}{2}\right)\right]^n \|u_1-\mathcal{P}\|} = 0 \tag{20}$$

$$\lim_{n\to\infty}\left|\frac{\xi_n}{\psi_n}\right| = \lim_{n\to\infty}\frac{\left[\frac{5k+k^2}{4}\right]^n \|u_1-\mathcal{P}\|}{\left[\frac{6\,k^2+k^3}{4}\right]^n \|u_1-\mathcal{P}\|} = 0 \tag{21}$$

$$\lim_{n\to\infty}\left|\frac{\theta_n}{\psi_n}\right| = \lim_{n\to\infty}\frac{\left[k^2\left(\frac{7-k}{4}\right)\right]^n \|u_1-\mathcal{P}\|}{\left[\frac{6\,k^2+k^3}{4}\right]^n \|u_1-\mathcal{P}\|} = 0 \tag{22}$$

Based on the above analytical comparison from (1)–(22), it is investigated that the process (1) converges quickly than (2), where the iteration plan (2) converges faster as compared with iteration (3), and also (2) converges rapid than (4), (3) converges quicker than (4) and (1) converges speedier as compared to (3). Overall, we find that the iterative process (1) converges quicker than the iterations (2)–(7) reformulated upon swapping its coefficients.

4 EMPIRICAL RESULTS

The validation of the above results is shown by the following numerical illustration:

Define the mapping $\mathcal{T} : [1,\ 60] \to [1,\ 60]$ as $\mathcal{T}(\sigma) = \sqrt{\sigma}$ for all $\sigma \in [1,\ 60]$, where $\mathcal{T}$ is contraction.

The PS-iteration method is defined as:

$$u_{n+1} = \mathcal{T} v_n$$

$$v_n = (1-\beta_n)\mathcal{T} u_n + \beta_n \mathcal{T} w_n$$

$$w_n = (1-\gamma_n)u_n + \gamma_n \mathcal{T} u_n$$

where the coefficients are chosen as $\beta_n = 0.75$ and $\gamma_n = 0.85$ for all positive values of n with the initial value $\sigma_0 = 20$.

Now, comparison of the four cases of PS-Iteration is in following three tables:

Table 1. Cases of the PS-iteration for u_{n+1}.

Iteration	Iteration u_{n+1} (1)	Iteration u_{n+1} (2)	Iteration u_{n+1} (3)	Iteration u_{n+1} (4)
1	1.61806743	2.03591762	2.11163125	2.09705096
2	1.07516059	1.18063798	1.20475695	1.19939990
3	1.01087746	1.03936618	1.04749594	1.04553650
4	1.00161432	1.00900835	1.01162465	1.01095806
5	1.00024047	1.00208459	1.00288274	1.00267042
6	1.00003584	1.00048363	1.00071721	1.00065276
7	1.00000534	1.00011226	1.00017858	1.00015967
8	1.00000079	1.00002606	1.00004447	1.00003906
9	1.00000011	1.00000605	1.00001107	1.00000955
10	1.00000001	1.00000140	1.00000275	1.00000279
11	1.00000000	1.00000032	1.00000068	1.00000068

Table 2. Cases of the PS-iteration for v_n.

Iteration	Iteration v_n (1)	Iteration v_n (2)	Iteration v_n (3)	Iteration v_n (4)
1	2.61814222	4.14496059	4.45898656	4.39762276
2	1.15597030	1.39390605	1.45143932	1.43856013
3	1.02187324	1.08028207	1.09724774	1.09314658
4	1.00323125	1.01809785	1.02338444	1.02203621
5	1.00048100	1.00417353	1.00577379	1.00534798
6	1.00071684	1.00096750	1.00143493	1.00130596
7	1.00001068	1.00022455	1.00035719	1.00031938
8	1.00000159	1.00005212	1.00008895	1.00007813
9	1.00000023	1.00001210	1.00002215	1.00001911
10	1.00000003	1.00000280	1.00000551	1.00000558
11	1.00000000	1.00000065	1.00000137	1.00000136

Table 3. Cases of the PS-iteration for w_n.

Iteration	Iteration w_n (1)	Iteration w_n (2)	Iteration w_n (3)	Iteration w_n (4)
1	5.24852915	5.24852915	19.22360679	19.22360679
2	1.28933456	1.45730894	2.07870695	2.06460440
3	1.03881256	1.09127497	1.19939981	1.19418846
4	1.00569669	1.02048675	1.04629476	1.04438541
5	1.00084720	1.00471979	1.01133319	1.01068336
6	1.00012624	1.00109389	1.00281062	1.00260361
7	1.00001881	1.00025387	1.00069927	1.00063643
8	1.00000280	1.00005893	1.00017411	1.00015567
9	1.00000041	1.00001368	1.00004335	1.00003808
10	1.00000005	1.00000317	1.00001079	1.00001146
11	1.00000000	1.00000073	1.00000268	1.00000272

Therefore, from above, we observe that the process (1) converges rapidly than (2), (2) converges rapid than (3) and (4), also (3) converges quicker than (4) and (1) converges quicker than (3). Consequently, the iteration (1) has the fastest convergence as compared with respect to other possible iterations.

5 CONCLUSION

The study carried out on the comparison of the convergence rate of the PS-iteration, we analyse that, the original three-step PS-iteration converges speedier as compared with the iterations reformulated after switching the coefficients defined in the original one. Moreover, we find that the role of coefficients also affects the order of convergence of such types of procedures obtained after changing their coefficients. Moreover, the numerical application showed that PS-iteration has better convergence result in the sense of less iteration steps. In this paper, we have established the equivalence of the PS-iterative process with its various forms obtained by using its coefficients involved in this plan. The comparison of their convergence rate is also provided theoretically as well as

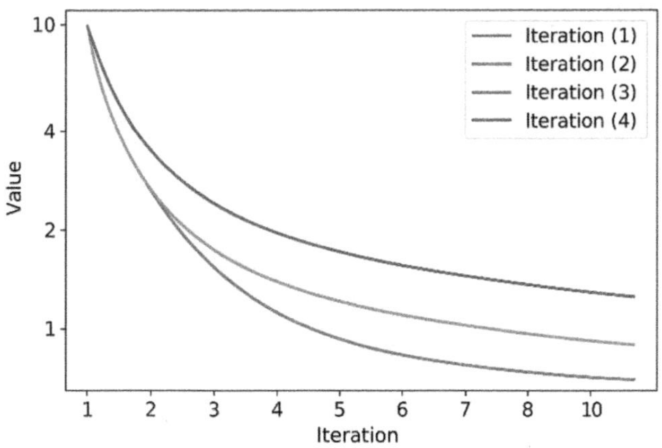

Figure 1. Convergence spectrum of PS-iterations.

numerically. A critical examination of the PS-iteration plan reveals that the iteration (3) and (4) have the same order of convergence whereas the iteration (2) converges rapid than the iterations (3) and (4), but the iteration (1) has improved convergence rate as compared with the iterations (2)–(4). Similar concept can be used in various other existing processes. As smart systems evolve, the journey from iterative mathematical certainty to intelligent robotic intuition becomes inevitable. The principles of convergence and stability will no longer remain theoretical—they will pulse through the decision-making core of self-learning machines."

REFERENCES

Akwe, H. and Olaoluwa, H. (2012). On the convergence of modified three-step iteration process for generalized contractive-like operators. *Bulletin of Mathematical Analysis and Applications*, 4(3):78–86.

Babu, G.V.R. and Vara Prasad, K.N.V.V. (2006). Comparison of fastness of the convergence among Krasnoselskij, Mann and Ishikawa iterations in Banach spaces. *Fixed Point Theory and Applications, Article ID 35704*:1–12.

Babu, G.V.R. and Vara Prasad, K.N.V.V. (2006). Mann iteration converges faster than Ishikawa iteration for the class of Zamfirescu operators. *Fixed Point Theory and Applications, Article ID* 49615:1–6.

Berinde, V. (2004). Picard iteration converges faster than the Mann iteration for a class of quasi-contractive operators. *Fixed Point Theory and Applications*, (2):97–105.

Bosede, A.O. (2010). Strong convergence results for the Jungck–Mann iteration processes. *Bulletin of Mathematical Analysis and Applications*, 2(3):65–73.

Fathollahi, S., Razani, A. and Moradi, S. (2015). A comparative study on the convergence rate of some iteration methods involving contractive mappings. *Fixed Point Theory and Applications, Article ID 234*:1–24.

Kima, T.H., Cho, Y.J. and Kang, S.M. (2005). Convergence of modified Mann iterations. *Nonlinear Analysis*, 61:51–60.

Kumar, N. and Chauhan, S.S. (2018). Analysis of Jungck-Mann and Jungck-Ishikawa iteration schemes for their speed of convergence. *AIP Conference Proceedings*, 2050:1–6.

Kumar, N. and Chauhan, S.S. (2019). Self-comparison of convergence speed in Agarwal, *O'Regan and Sahu's S-Iteration. International Journal on Emerging Technologies*, 10(2B):105–108.

Popescu, O. (2007). Picard iteration converges faster than Mann iteration of quasi-contractive operators. *Mathematical Communications*, 12(2):195–202.

Saluja, G.S. (2015). Convergence of modified S-iteration process for two generalized asymptotically quasi-nonexpansive mappings in CAT(0) spaces. *Mathematica Moravica*, 19(1):19–31.

Xue, Z. and Lv, G. (2013). The convergence of the modified Mann and Ishikawa iterations in Banach spaces. *Journal of Inequalities and Applications*, 188 https://doi.org/10.1186/1029-242X-2013-188

Smart Systems and Beyond – Bhatia et al. (Eds)
© 2027 The Author(s), ISBN: 978-1-041-36617-1

Deep Transfer Learning Models for Cancer Classification: A Comparative Study

Isha Dhingra and Sonali Dadwal
Chandigarh University, Mohali, India

ABSTRACT: Cancer, caused by the unrestrained growth of cells inside the human body, leading dangerous diseases currently prevailing the world. Accurate and early diagnosis is of much importance for the timely treatment of patients. Various deep learning models have been built using computer-aided diagnostic systems and have shown outstanding model efficacy in medical image analysis with very high accuracy. Developing these neural networks from from scratch is very challenging and calls for immense computational power. To avoid this problem, transfer learning has proven to be a very powerful solution. The study will perform a comparative analysis on two transfer learning models, AlexNet and ResNet, by testing their accuracy in classifying various types of cancer. Both models would be tested on a combined dataset that includes skin cancer, chest cancer, pneumonia, and brain tumors.

Keywords: Deep Learning, Skin Cancer, Brain Tumor, Chest Cancer, Transfer Learning

1 INTRODUCTION

These neural networks are very difficult and computationally intensive since they were trained from the ground up. Because of this, a fantastic resource has been that of transfer learning. In this paper, there will be a comparative analysis between two models of transfer learning, AlexNet and ResNet, to analyze the accuracy by which they classify types of cancer. These models will be put against a dataset with skin cancer, chest cancer, pneumonia, and brain tumors [2].

Brain tumour is a severe conditions of the central nervous system, comprising primary tumours that occur in the brain and secondary tumours that originate from other malignancies. They usually develop as primary cancers of the breasts, or lungs and then it flow towards the brain via the blood otherwise lymphatic system [3]. Lung cancer, the most prevalent chest cancer, causes 18% of cancer causing deaths universally and is primarily linked to tobacco smoking in 85% of cases [4]. It is categorized as NSCLC and SCLC, which is known as non-small cell lung cancer at 80-85% and small cell lung cancer at 15-20% respectively [5].

GLOBOCAN 2022 reports [6] that skin cancer is the frequently diagnosed cancer universally with an estimate of 1.5 million fresh instances, including melanoma (331,722 cases) and non-melanoma types (1,234,533 cases). Breast cancer persists to be the predominant cancer among women having 2,296,840 fresh instances and 666,103 deaths. Imaging techniques, especially in medical fields, are one of the prime ways for prompt recognition of various diseases. Here, the medical records of patients will be integrated with imaging data to offer efficient health monitoring, as shown in [7–9]. Magnetic resonance imaging techniques are employed to locate brain tumors, and offer better visualization in multi-modality methods as in [10].

Hybrid imaging techniques such as SPECT and PET [12] combine different modalities. PET scans are utlized to interpret cancers and heart diseases, whereas SPECT is used to view the blood flow and to examine bones and tissues [45].

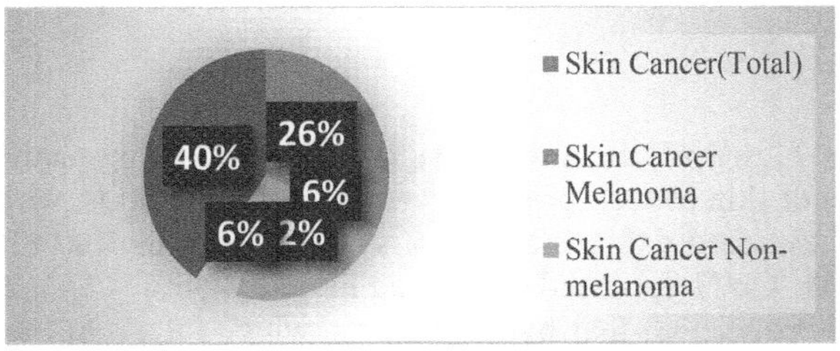

Figure 1. GLOBOCAN cancer statistics summary.

Timely cancer detection is very crucial to reduce mortality rates. Artificial intelligence-powered CAD system, computer-aided diagnosis, amplifies the prediction and diagnostic accuracy of cancer by analyzing previous data

[14]. ML methods include supervised learning-data are labeled, unsupervised learning-data are unlabeled-and semi-supervised learning [15]. Although CAD systems have huge developments in medical imaging, they fail to detect malignant tumors and detect only a few types of cancer [13] [19]. According to different studies, the performance of traditional CAD systems is poor [20–22].

2 DEEP LEARNING

Deep learning is a module of Machine Learning that uses several layers with inputs, biases, and weights working collectively to achieve precise predictions. The outcome of every individual layer works as the input for the next until the final outcome is achieved, which requires an understanding of neural networks [24].

2.1 *Transfer learning and its models*

Training convolutional networks from initial state is complicated and labour intensive, requiring a huge dataset. In order to take advantage of this, transfer learning enhances model performance by adopting knowledge from similar areas [25]. Transfer Learning has been broadly classified into two important types: homogeneous transfer learning, which concerns similar attribute space, and diverse transfer learning, which incorporates several feature spaces [26,27]. LeNet, AlexNet, and ResNet [28,29] are some pre-trained convolutional neural networks designed to play with image pixels. Freezing these models involves keeping certain layer parameters fixed while the rest of the parameters are free to change; fine-tuning involves initializing with pre-trained parameters, then adjusting them during training on new, target data [31,32].

2.2 *AlexNet*

A deeper CNN model than LeNet was proposed by Alex Krizhevsky and his colleagues in 2012, securing a win at the ImageNet large scale visual recognition challenge (ILSVRC) [33]. AlexNet has outperformed traditional methods in recognition accuracy, marking a breakthrough in visual recognition, classification and igniting immense interest in deep learning [34].

2.3 *ResNet*

ResNet, evolved by Kaiming He *et al.* [35], is a convolutional neural network architecture that is designed with the purpose of classification and image recognition. Although ResNet152 is a comprehensive network, an increased number of layers leads to higher training and testing errors due to the skip connections method in the residual block that forms ResNet [36].

3 DATA PRE-PROCESSING

Image pre-processing is one of a pivotal step in enhancing the clarity and resolution of images by removing irrelevant information [50]. Pre-image processing techniques like resizing, noise reduction, and filtering enhance image clarity for the expert system to make more accurate evaluations. Image enhancement increases the visual clarity of images while maintaining important details [51], hence enabling efficient feature classification in both the spatial and frequency domains. Furthermore, data augmentation has an essential task in improving model robustness and accuracy, especially with limited datasets [53], because it overcomes challenges in the lack of training data and class imbalance problem [54].

4 LITERATURE REVIEW

Deep learning has become a revolutionary field that is significantly changing medical practices. This review paper [37] focuses on earlier detection of skin cancer, primarily non-invasive methods, with criticism of painful traditional biopsy methods. The authors have used datasets such as the ISIC, DermIS, and PH2 and also considered features such as ABCD rule and GLCM; CNN appeared promising in detection.

The author has stated in this paper [38] that hybrid models, which integrate those features extracted with deep learning and machine learning, could enhance the capabilities of classification accuracy.

Manual MRI segmentation is limited to being expert-dependent and time-consuming, while traditional Machine Learning relies on handcrafted features leading to a lack of robustness. Deep learning, via advancements like residual connections, has achieved superior segmentation accuracy. An advanced U-Net with residual blocks has produced excellent results on BraTS datasets.

In [40], the author has discussed the limitation in brain tumor classification and proposed a hybrid deep learning model that combines strengths of various architectures. The previous methods with handcrafted features were less effective on MRI datasets, whereas CNNs have been shown capable of extracting complex features in MRI images. The hybrid model integrates CNN and RNN to achieve high accuracy in classification.

This Research [41] proposes a deep learning methodology for the improvement of brain tumor identification in MRI scans through the use of CNNs for automatic feature extraction. The model is tested on publicly available datasets and it provides the improved accuracy results when compared to traditional methods to support diagnostic differentiation for early detection. This research [42] evaluates CNNs for the identification of brain tumors; it therefore proposes a CNN model based on AlexNet and GoogLeNet framework. Performance evaluation reveals that GoogLeNet outperforms AlexNet, reaching 99.45% accuracy. This CNN approach has the potential to substantially support radiologists in clinical diagnostics.

This research [43] displays a hybrid deep learning framework incorporating attention mechanisms for lung cancer identification in CT scans. The attention mechanisms facilitate the detection of features that indicate malignant regions. This research [44] explored the integrated application of deep learning and traditional machine learning is a way of detecting brain tumor. In this study, comparison is between different models to create diagnostic tools that support medical professionals.

5 DATASETS UTILIZED

This section contains the datasets employed for comparing transfer learning models.

5.1 *Skin cancer*

Comprising 10,015 images, the HAM10000 dataset showcases various pigmented skin abnormalities, containing actinic keratosis, intraepithelial carcinoma (akiec), basal cell carcinoma (BCC), benign keratosis-like growths, dermatofibroma (df), melanoma (mel), melanocytic nevi (nv), and vascular malformations [46,47].

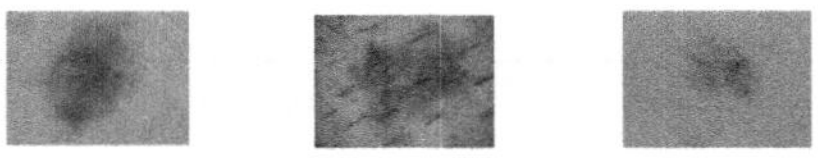

(a) (b) (c)

Figure 2. (a) Melanoma (b). BCC (c) Nevus.

5.2 *Brain tumor*

Brain tumor is referred as an irregular mutation of cells in human brain it can turn out either be malignant (cancerous) or benign (noncancerous). Common malignant tumors include gliomas from glial cells, while meningiomas are usually benign and develop in surrounding membranes [48,49].

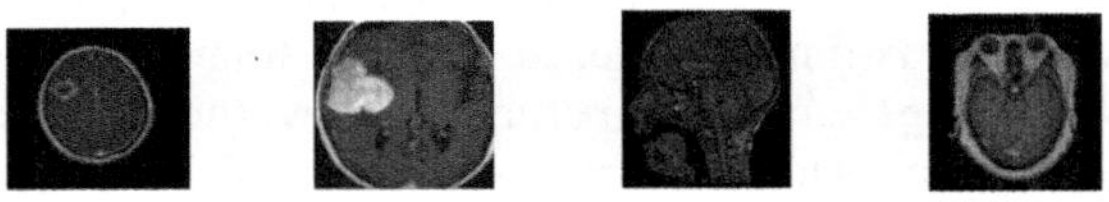

(a) (b) (c) (d)

Figure 3. (a) Glioma (b) Meningioma (c) Nontumor (d) Pituitary.

5.3 *Chest cancer*

Chest cancer or lung cancer typically triggers in the lungs. Adenocarcinoma is the frequent lung cancer type which primarily infects the non-smokers, largely to women. Large Cell Carcinoma is aggressive and can arise anywhere in the lung, while Squamous Cell Carcinoma, linked to smoking, occurs near the bronchi. Normal lung tissue serves as a control in diagnostics, making these classifications vital for treatment and prediction in lung cancer management [13].

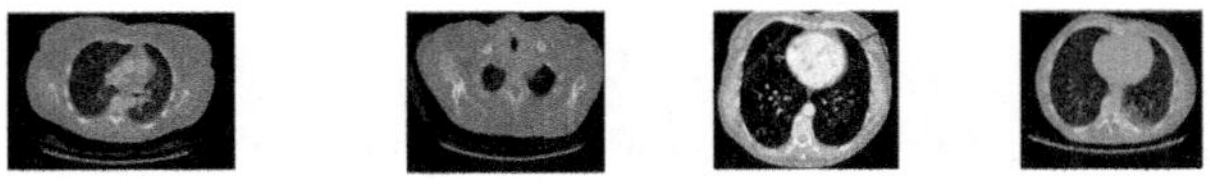

(a) (b) (c) (d)

Figure 4. (a) Adenocarcinoma (b) Large cell Carcinoma (c) normal (d) squamous cell.

5.4 *Pneumonia*

Pneumonia is an inflammatory lung condition that affects the air sacs, leading to breathing difficulties caused by fluid or pus accumulation. It can be classified by cause (bacterial, viral, fungal) and by the location of acquisition (community-acquired, hospital-acquired). Accurate classification is essential for appropriate treatment, which ranges from antibiotics to supportive care.

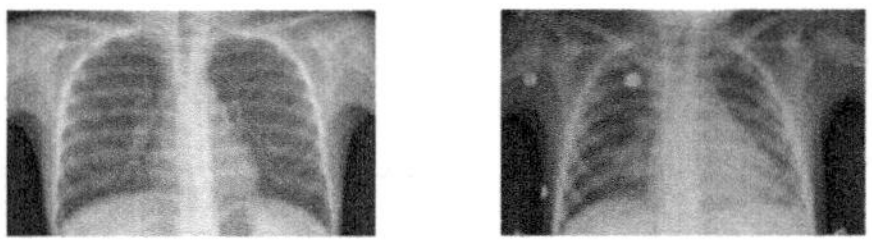

(a) (b)

Figure 5. (a) Chest X-ray normal (b) Chest X-ray PNEUMONIA.

6 COMPARATIVE ANALYSIS

Contrasting the categorization precision of AlexNet and ResNet on several medical imaging datasets related to skin lesions, brain scans, pneumonia, and chest cancer reveals one pattern-the ResNet, or Residual Network, tends to outperform AlexNet. This is because ResNet has a more sophisticated model architecture that provides a unique design with residual links. The use of residual links in its architecture ensures better gradient propagation during training, thereby enabling these models to learn better from the data. It follows that ResNet provides better feature extraction ability from high-resolution medical images, which is something peculiarly crucial in medical imagery, where the ability to recognize and classify subtle differences in the images correctly can have significant implications for diagnosis and therapy planning.

Table 1. Results of the evaluation of transfer learning models on different datasets.

Dataset	AlexNet	RESNET
Pneumonia Chest X-ray	95.30%	95.90%
Ham10000 (Skin Cancer)	58.80%	60.92%
Chest Cancer	53.33%	80.00%
Brain Tumor	89.51%	96.68%

In this work, a comparison assessment has been carried out between two Transfer Learning models, AlexNet and ResNet, over four medical image datasets: Pneumonia Chest X-ray, Ham10000 (skin cancer), Chest Cancer, and Brain Tumor. Experimental results show that ResNet has invariably outperformed AlexNet in all these datasets. Specifically, on the Pneumonia dataset, ResNet posted an accuracy of 95.90%, while the reported value for AlexNet was 95.30%. On the skin cancer categorization over the dataset HAM10000, ResNet had an accuracy of 60.92% compared to an AlexNet value of 58.80%. For the detection of chest cancer, the difference is stronger: ResNet at 80.00% vs. 53.33% by AlexNet. The maximum difference has been found in the classification of a brain tumor: ResNet was 96.68% and AlexNet at 89.51%. These experiments show, through their repeated results, that ResNet has better classification performance than AlexNet across all tested medical imaging datasets.

7 CONCLUSION

Deep learning has confirm being a key instrument in fighting against cancer , which is a condition about abnormal growth of cells. This paper compares two models, AlexNet and ResNet, both are transfer learning-based models, which were applied to achieve image classification related to Chest Cancer, Brain Tumor, Skin Cancer, and Pneumonia. Various results proved that ResNet outperformed AlexNet on all the tested datasets. In detecting brain tumors, for instance, ResNet had an accuracy of 96.68%, whereas AlexNet recorded 89.91%, thereby making ResNet a more trusted architecture in regard to such life-critical medical imaging tasks.

REFERENCES

[1] Purba, S.Y. (2024). Precision medicine: Utilizing genomic insights for customized therapeutic strategies. *The Journal of Academic Science*, 1(5), 497–504.

[2] Roky, A.H., Islam, M.M., Ahasan, A.M.F., Mostaq, M.S., Mahmud, M.Z., Amin, M.N., and Mahmud, M.A. (2025). Overview of skin cancer types and prevalence rates across continents. *Cancer Pathogenesis and Therapy*, 3(02), 89–100.

[3] Basyoni, A.E., Atta, A., Salem, M.M., and Mohamed, T.M. (2025). Harnessing exosomes for targeted drug delivery systems to combat brain cancer. *Cancer Cell International*, 25(1), 150.

[4] Bray, F., Laversanne, M., Sung, H., Ferlay, J., Siegel, R.L., Soerjomataram, I., and Jemal, A. (2024). Global cancer statistics 2022: GLOBOCAN estimates of incidence and mortality worldwide for 36 cancers in 185 countries. *CA. A Cancer Journal for Clinicians*, 74(3), 229–263.

[5] Malhotra, J., Malvezzi, M., Negri, E., La Vecchia, C., and Boffetta, P. (2016). Risk factors for lung cancer worldwide. *European Respiratory Journal*, 48(3), 889–902.

[6] International Agency for Research on Cancer. (2025). *Global Cancer Observatory*. https://gco.iarc.fr/en

[7] Wyatt, J.C., and Liu, J.L. (2002). Basic concepts in medical informatics. *Journal of Epidemiology and Community Health*, 56(11), 808–812.

[8] Miranda, E., Aryuni, M., and Irwansyah, E. (2016). A survey of medical image classification techniques. In *2016 International Conference on Information Management and Technology (ICIMTech)* (pp. 56–61). IEEE.

[9] Kumar, A., Kim, J., Cai, W., Fulham, M., and Feng, D. (2013). Content-based medical image retrieval: A survey of applications to multidimensional and multimodality data. *Journal of Digital Imaging*, 26, 1025–1039.

[10] Jennings, L.E., and Long, N.J. (2009). 'Two is better than one'—probes for dual-modality molecular imaging. *Chemical Communications*, 24, 3511–3524.

[11] Elangovan, A., and Jeyaseelan, T. (2016). Medical imaging modalities: A survey. In *2016 International Conference on Emerging Trends in Engineering, Technology and Science (ICETETS)* (pp. 1–4). IEEE.

[12] Othman, M.F.B., Abdullah, N., and Rusli, N.A.B.A. (2010). An overview of MRI brain classification using FPGA implementation. In *2010 IEEE Symposium on Industrial Electronics and Applications (ISIEA)* (pp. 623–628). IEEE.

[13] Yu, X., Wang, J., Hong, Q.Q., Teku, R., Wang, S.H., and Zhang, Y.D. (2022). Transfer learning for medical images analyses: A survey. *Neurocomputing*, 489, 230–254.

[14] Murphy, K.P. (2012). *Machine learning: A probabilistic perspective*. MIT Press.

[15] Giger, M.L. (2018). Machine learning in medical imaging. *Journal of the American College of Radiology*, 15(3), 512–520.

[16] Chan, H.P., Samala, R.K., Hadjiiski, L.M., and Zhou, C. (2020). Deep learning in medical image analysis. *Deep Learning in Medical Image Analysis: Challenges and Applications*, 3–21.

[17] Castellino, R.A. (2005). Computer aided detection (CAD): An overview. *Cancer Imaging*, 5(1), 17.

[18] Philpotts, L.E. (2009). Can computer-aided detection be detrimental to mammographic interpretation? *Radiology*, 253(1), 17–22.

[19] Lehman, C.D., Wellman, R.D., Buist, D.S., Kerlikowske, K., Tosteson, A.N., Miglioretti, D.L., and Breast Cancer Surveillance Consortium. (2015). Diagnostic accuracy of digital screening mammography with and without computer-aided detection. *JAMA Internal Medicine*, 175(11), 1828–1837.

[20] Fenton, J.J., Taplin, S.H., Carney, P.A., Abraham, L., Sickles, E.A., D'Orsi, C., and Elmore, J.G. (2007). Influence of computer-aided detection on performance of screening mammography. *New England Journal of Medicine*, 356(14), 1399–1409.

[21] Stoitsis, J., Valavanis, I., Mougiakakou, S.G., Golemati, S., Nikita, A., and Nikita, K.S. (2006). Computer aided diagnosis based on medical image processing and artificial intelligence methods. *Nuclear Instruments and Methods in Physics Research, Section A*, 569(2), 591–595.

[22] Yu, Y., Lin, H., Yu, Q., Meng, J., Zhao, Z., Li, Y., and Zuo, L. (2015). Modality classification for medical images using multiple deep convolutional neural networks. *Journal of Computer and Information Systems*, 11(15), 5403–5413.

[23] Lee, J.G., Jun, S., Cho, Y.W., Lee, H., Kim, G.B., Seo, J.B., and Kim, N. (2017). Deep learning in medical imaging: general overview. *Korean Journal of Radiology*, 18(4), 570–584.

[24] Khamparia, A., Singh, P.K., Rani, P., Samanta, D., Khanna, A., and Bhushan, B. (2021). An internet of health things-driven deep learning framework for detection and classification of skin cancer using transfer learning. *Transactions on Emerging Telecommunications Technologies*, 32(7), e3963.

[25] Zhuang, F., Qi, Z., Duan, K., Xi, D., Zhu, Y., Zhu, H., and He, Q. (2020). A comprehensive survey on transfer learning. *Proceedings of the IEEE*, 109(1), 43–76.

[26] Iman, M., Arabnia, H.R., and Rasheed, K. (2023). A review of deep transfer learning and recent advancements. *Technologies*, 11 (2), 40.

[27] Karpathy, A. (2016). Convolutional neural networks (cnns/convnets). *CS231n Convolutional Neural Networks for Visual Recognition.*

[28] Khan, A., Sohail, A., Zahoora, U., and Qureshi, A.S. (2020). A survey of the recent architectures of deep convolutional neural networks. *Artificial Intelligence Review*, 53, 5455–5516.

[29] Iman, M., Arabnia, H.R., Rasheed, K.: A review of deep transfer learning and recent advancements. *Technologies* 11(2), 40 (2023)

[30] Li, C., Zhang, S., Qin, Y., Estupinan, E.: A systematic review of deep transfer learning for machinery fault diagnosis. *Neurocomputing* 407, 121–135 (2020)

[31] Dhillon, A., Verma, G.K.: Convolutional neural network: a review of models, methodologies and applications to object detection. *Progress Artif. Intell.* 9(2), 85–112 (2020)

[32] Krizhevsky, A., Sutskever, I., and Hinton, G.E. (2012). Imagenet classification with deep convolutional neural networks. *Advances in Neural Information Processing Systems*, 25.

[33] Alom, M.Z., Taha, T.M., Yakopcic, C., Westberg, S., Sidike, P., Nasrin, M.S., and Asari, V.K. (2018). The history began from alexnet: A comprehensive survey on deep learning approaches. *arXiv preprint arXiv:1803.01164.*

[34] He, K., Zhang, X., Ren, S., Sun, J.: Deep residual learning for image recognition. In *Proceedings Of The IEEE Conference on Computer Vision and Pattern Recognition* (pp. 770–778) (2016)

[35] Wu, Z., Shen, C., Van Den Hengel, A.: Wider or deeper: Revisiting the resnet model for visual recognition. *Pattern Recogn.* 90, 119–133 (2019)

[36] Dildar, M., Akram, S., Irfan, M., Khan, H.U., Ramzan, M., Mahmood, A.R., and Mahnashi, M.H. (2021). Skin cancer detection: A review using deep learning techniques. *International Journal of Environmental Research and Public Health*, 18(10), 5479.

[37] Tembhurne, J.V., Hebbar, N., Patil, H.Y., and Diwan, T. (2023). Skin cancer detection using ensemble of machine learning and deep learning techniques. *Multimedia Tools and Applications*, 82(18), 27501–27524.
[38] Pedada, K.R., Rao, B., Patro, K.K., Allam, J.P., Jamjoom, M.M., and Samee, N.A. (2023). A novel approach for brain tumour detection using deep learning based technique. *Biomedical Signal Processing and Control*, 82, 104549.
[39] Rasool, M., Ismail, N.A., Boulila, W., Ammar, A., Samma, H., Yafooz, W.M., and Emara, A.H.M. (2022). A hybrid deep learning model for brain tumour classification. *Entropy*, 24(6), 799.
[40] Agarwal, M., Rani, G., Kumar, A., Kumar, P., Manikandan, R., and Gandomi, A.H. (2024). Deep learning for enhanced brain tumor detection and classification. *Results in Engineering*, *22*, 102117.
[41] Swarup, C., Singh, K.U., Kumar, A., Pandey, S.K., and Singh, T. (2023). Brain tumor detection using CNN, AlexNet and GoogLeNet ensembling learning approaches. *Electronic Research Archive*, 31(5).
[42] Ozdemir, B., Aslan, E., and Pacal, I. (2025). Attention Enhanced InceptionNeXt Based Hybrid Deep Learning Model for Lung Cancer Detection. IEEE Access.
[43] Anantharajan, S., Gunasekaran, S., and Subramanian, T. (2024). MRI brain tumor detection using deep learning and machine learning approaches. *Measurement: Sensors*, 31, 101026
[44] Kaur, N., and Hans, R. (2025). Transfer Learning for Cancer Diagnosis in Medical Images: A Compendious Study. *International Journal of Computational Intelligence Systems*, 18(1), 62.
[45] Hosny, K.M., Kassem, M.A., and Foaud, M.M. (2018, December). Skin cancer classification using deep learning and transfer learning. In *2018 9th Cairo international biomedical engineering conference (CIBEC)* (pp. 90–93). IEEE.
[46] Ashraf, R., Afzal, S., Rehman, A.U., Gul, S., Baber, J., Bakhtyar, M., ... and Maqsood, M. (2020). Region-of-interest based transfer learning assisted framework for skin cancer detection. *IEEE Access*, 8, 147858–147871.
[47] DeAngelis, L.M. (2001). Brain tumors. *New England Journal of Medicine*, 344(2), 114–123.
[48] Black, P.M. (1991). Brain tumors. *New England Journal of Medicine*, 324(22), 1555–1564.
[49] Mohanapriya, N., Kalaavathi, B.: Comparative study of diferent enhancement techniques for medical images. *International Journal of Computer Applications*, 61(20) (2013)
[50] Perumal, S., Velmurugan, T.: Preprocessing by contrast enhancement techniques for medical images. *Int. J. Pure Appl. Math.* 118 (18), 3681–3688 (2018)
[51] Chaki, J., Dey, N.: *A Beginner's Guide to Image Preprocessing Techniques*. CRC Press (2018)
[52] Tsiknakis, N., Theodoropoulos, D., Manikis, G., Ktistakis, E., Boutsora, O., Berto, A., Marias, K.: Deep learning for diabetic retinopathy detection and classifcation based on fundus images: A review. *Comput. Biol. Med.* 135, 104599 (2021)
[53] Mikołajczyk, A., Grochowski, M.: Data augmentation for improving deep learning in image classifcation problem. In *2018 International Interdisciplinary PhD Workshop (IIPhDW)* (pp. 117–122). IEEE (2018)
[54] Shorten, C., Khoshgoftaar, T.M.: A survey on image data augmentation for deep learning. *J. big Data* 6(1), 1–48 (2019)
[55] Nickparvar, M. (n.d.). *Brain Tumor MRI Dataset*. Kaggle. https://www.kaggle.com/datasets/masoudnickparvar/brain-tumor-mri-dataset
[56] Hany, M. (n.d.). *Chest CT-Scan Images*. Kaggle. https://www.kaggle.com/datasets/mohamedhanyyy/chest-ctscan-images
[57] Tschandl, P., Rosendahl, C. and Kittler, H. The HAM10000 dataset, a large collection of multi-source dermatoscopic images of common pigmented skin lesions. *Sci Data* 5, 180161 (2018). https://doi.org/10.1038/sdata.2018.161
[58] Kermany, Daniel; Zhang, Kang; Goldbaum, Michael (2018), "Labeled Optical Coherence Tomography (OCT) and chest X-Ray images for classification", *Mendeley Data*, v2 http://dx.doi.org/10.17632/rscbjbr9sj.2

Smart Systems and Beyond – Bhatia et al. (Eds)
© 2027 The Author(s), ISBN: 978-1-041-36617-1

Smart Toll Management System Using Arduino and IoT

Prashant Kumar, Rohit Konsam, Piyush Kumar, Khushi Bharti, and Ranjit Kumar Bindal
Department of Electrical Engineering, Chandigarh University, Mohali, Punjab, India

ABSTRACT: Traditional toll collection systems be afflicted by lengthy queues, manual processing delays, and sales leakage, leading to traffic congestion and gas wastage. To address those demanding situations, this paper offers an IoT-primarily based Smart Toll Management System using Arduino, RFID generation, and cloud connectivity. The proposed machine automates toll deduction by way of detecting motors ready with RFID tags, deducting the toll amount from a pay as you go account, and updating transaction logs in actual time through a cloud server (ThingSpeak/Blynk). A servo motor-controlled barrier gate ensures seamless get entry to upon a hit fee, even as an LCD display gives person feedback. The gadget became prototyped using an Arduino Uno, RFID-RC522 module, ESP8266 Wi-Fi module, and a servo motor, demonstrating 95% detection accuracy with an average processing time of beneath five seconds in step with car—significantly faster than guide toll series. Minimised commutator congestion, reduced human intervention and easy virtual transactions are the key features of smart toll management system. Future enhancements may combine AI-based license plate reputation for backup authentication and dynamic pricing based on traffic density. This leads to intelligent transportation systems (ITS) earlier known as shrewd transportation system by imparting a fee-effective, scalable, and efficient computerized toll solution.

Keywords: IoT, Arduino, RFID, Smart Toll System, Traffic Management, Automation

1 INTRODUCTION

The conventional toll collection system becomes inefficient due to the exponential growth of vehicular movment throughout the country and hence adopting smart toll collection on urgent basic become essential. Current manual toll collection strategies at country wide highways and expressways be afflicted by a couple of operational demanding situations which include excruciatingly long queues all through peak hours, accounting inaccuracies main to giant sales leakage, and inconsistent enforcement of toll rates. NHAI review on the toll collection reported that there is huge monetary loss of thousands of crores due to system inefficiency at peak traffic. Additionally, there are unnecessary wastage of gasoline and gas per day due to idling of motors at toll collection booth at toll plaza throughout the country.

This research paper presents a groundbreaking indigenous answer - a fully self-sufficient Smart Toll Management System advanced the use of low-priced Arduino-primarily based era that eliminates dependency on high priced cloud infrastructure or excessive-protection additives. Our modern system structure centres round an Arduino Uno microcontroller serving as the processing hub, included with an RFID-RC522 reader module for automobile identification, creating a robust standalone toll series unit particularly designed for Indian street conditions and site visitors' patterns. The device's whole offline capability makes it especially precious for deployment in rural highways, far flung border checkpoints, and mountainous routes in which internet connectivity stays unreliable or unavailable.

The purposed toll collection system is designed to tackle the issues associated to the toll collection in India. When automobiles prepared with passive RFID tags (costing much less than ₹20 consistent with tag) approach the toll sales space, the machine immediately scans the unique identification code and move-verifies it against a locally saved database of registered cars and pay as you go balances. Upon a success authentication, it deducts the perfect toll amount - starting from ₹50 for two-wheelers to ₹two hundred for heavy industrial automobiles as according to present day NHAI categorization - and without delay activates a servo-controlled barrier gate whilst concurrently showing transaction details on a multilingual 16x2 LCD screen in both English and nearby languages. All transaction facts along with automobile ID, timestamp, and deducted amount are securely logged onto a local SD card module with garage ability exceeding 12 months of non-stop operation, enabling smooth reconciliation and audit compliance.

What sets our solution apart is its first-rate fee-effectiveness and ease of deployment. The complete hardware prototype has been developed the use of effortlessly available components with general unit fee under ₹2,500 - nearly 90% inexpensive than imported automatic toll systems presently deployed on some top-class expressways. Field checking out at simulated toll plazas demonstrated first-rate overall performance metrics which includes

94.3% first-try detection accuracy, common processing time of simply 2.5 seconds in step with automobile (in comparison to 45-60 seconds for manual series), and uninterrupted operation in severe weather conditions ranging from 0°C to 50°C. The device's designed to work electricity-efficiently consuming even less than 15W which allow to operate it continuously for 72 hours on a simple 12V/7Ah battery backup in case of electricity outages.

The generation has enormous social and economic impact on the technological advancement. By doing away with human cash managing, the device reduces opportunities for toll evasion and corruption while bringing transparency to toll operations. Its implementation ought to help recover an anticipated ₹300-400 crores annually in misplaced toll sales even as extensively reducing visitors' congestion at main toll plazas. The proposed smart toll collection system is transformative in highways and rural toll collection even in lack of automation due to financial constraints. Furthermore, the generation of local level employments are additional and simple preservation requirements create opportunities for rural entrepreneurship in gadget set up and upkeep.

This paper comprehensively incorporates the system's technical architecture, in addition to the customized Arduino firmware that optimizes RFID reading reliability in high visitors' situations, the mechanical design of the automatic barrier mechanism able to withstand continuous operation, and the progressive facts logging technique that ensures transaction integrity without cloud dependency. A cost-benefit evaluation comparing the proposed solution with existing traditional alternatives are put forward from prototype deployments at three extraordinary locations throughout Maharashtra and Karnataka.

Due to the modular design of the proposed device, it allows several upgrades with minimal additional cost. These upgrades may incorporate integration of solar PV for off grid operation, addition of GSM modules for periodic records synchronization where mobile networks are available, and implementation of dynamic pricing at peak traffic hours efficiently. The present study outlines how appropriate generation solutions utilising fundamental embedded systems can correctly address infrastructure demanding situations during peak hour in growing economies, presenting a blueprint for indigenously developed transportation structures customized to Indianan scenario meeting requirements and financial constraints.

Rest of the study is outlined as literature review, methodology, result and conclusion respectively.

2 LITERATURE REVIEW

The upgradation of fee collection system at toll plaza is essential in increasing efficiency of transport infrastructure with increase in vehicular movement. Since the traditional toll collection require vehicles to halt, it leads to congestion and long queue at toll plaza so as to have gasoline wastage and increased carbon gas emission. And hence to overcome these challenges, a global shift towards the contactless computerized system is need of an hour. It may include technologies such as RFID, IoT, Arduino and cloud computing. RFID based tolling structures were at the vanguard of smart transportation research. S. Roy (2017) presented a low-fee RFID machine for automatic toll fee collection in which RFID transponders are tagged at vehicle that interacted with readers established on the toll plaza. Similarly, P. Sharma (2015) proposed a vehicle identity and transaction model that significantly reduced processing time as compared to manual toll booths. But, it often be afflicted by scalability issues and lacks real-time information for vehicle entry.

The emergence of IoT has brought extra sturdy architectures capable of assisting dynamic tolling, cloud-primarily based monitoring, and far off diagnostics. In J. Singh (2017), a comprehensive IoT framework for smart tolling became advanced, integrating sensors, microcontrollers, and cloud garage to allow seamless toll operations. Another study A. Rao (2019) leveraged MQTT protocols for lightweight communique among toll sensors and cloud servers, reaching low-latency statistics transmission. Arduino, due to its open-supply nature and hardware simplicity, has turn out to be a preferred platform for prototyping smart structures. In S. Kumar (2017), researchers utilized Arduino Uno alongside IR sensors and servo automobiles to layout a barrier-based totally automated toll device. GSM modules had been also included to inform users of payment confirmations through SMS. However, reliability during excessive-velocity site visitors glide remained a problem.

Recent systems have additionally incorporated cloud and mobile payment platforms. A cloud-IoT primarily based solution in A. Verma (2019) allowed customers to link their vehicle to a mobile wallet, allowing direct deductions and cloud-synced transaction logs. In K. Shankar (2019), the authors proposed a at ease e-wallet gadget for toll payments using blockchain-primarily based encryption to prevent fraud and unauthorized get entry to. Security and statistics integrity in tolling systems have gained attention in current years. Studies consisting of B. Sridhar (2024) emphasize the significance of encrypting conversation among IoT gadgets and the cloud to keep away from tampering or information leaks. Similarly, R. Gupta (2017) proposed a two-thing authentication model for toll bills, combining RFID with OTP-primarily based verification thru the consumer's cellular device.

Further innovations encompass using license plate popularity (LPR) and AI-based totally vehicle classification. As shown in A. Shah (2025), photograph processing techniques can be used to identify cars and pass-reference them with valuable databases for computerized billing. Although more accurate, LPR systems call for high-decision cameras and strong network connections, which may not be feasible in rural or remote toll places. The environmental impact of conventional tolling structures is also an issue. M. Roy (2019) highlighted the excessive CO_2 emissions from idle engine at toll cubicles. Automation and contactless tolling can reduce each fuel intake and vehicular emissions, promoting greener transportation practices.

Despite promising improvements, systems frequently lack interoperability among areas, be afflicted by power dependency, and involve high preliminary infrastructure expenses. Therefore, a modular, scalable machine that uses Arduino and IoT, with RFID, cloud integration, GSM/SMS indicators, and mobile charge capability, ought to deal with many of those gaps—providing a fee-effective and adaptable tolling answer.

3 METHODOLOGY

3.1 *Architecture design*

This study incorporates a blended strategies considering experimental as well as qualitative research. It enhances overall performance with the proposed Arduino-based Smart Toll Management System. The research method has been established into three wonderful however interconnected stages to make sure systematic development and validation of the machine.

The preliminary layout section centred on growing an optimized machine structure using carefully decided on components. A thorough requirements evaluation has been done to identify key operational parameters along with:

- Target processing time (< 3 seconds in keeping with vehicle)
- Minimum RFID read accuracy (95% achievement price)
- Power consumption limits (< 5W continuous draw)
- Environmental running situations (0°C -50°C temperature range)

The hardware architecture changed into designed around an Arduino UNO microcontroller serving because the significant processing unit, selected for its most appropriate stability of computational capability (16MHz clock speed, 32KB flash memory) and value-effectiveness. The RFID subsystem was carried out the usage of the RC522 module (13.56 MHz running frequency) with cautious consideration given to antenna positioning and orientation to reap constant 5-10cm study stages. The mechanical layout integrated an SG90 servo motor (1.8kg/cm torque output) able to appear more than 100,000 actuation cycles, with a custom-designed three-arm barrier gate (1.5m span) for dependable car manage. The circuit diagram for proposed system is outlined in Figure 1.

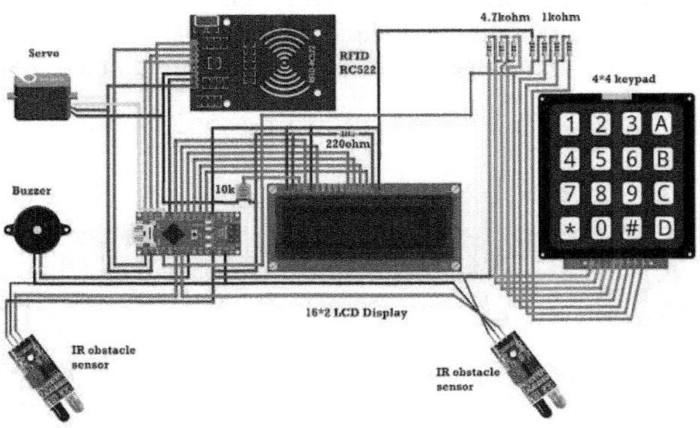

Figure 1. Circuit diagram for our Toll management.

3.2 *System implementation and integration*

The implementation part was worried about how to physically put together and program the toll machine prototype. The Arduino implementation is highlighted as mentioned.

Key activities blanketed:

- Circuit design and PCB layout using Zero PCB for strong interconnections
- Development of the Arduino firmware providing:
  - Interrupt-pushed RFID scanning ordinary
  - EEPROM-based totally account balance management
  - Servo motor control set of rules with smooth-begin/forestall

- Mechanical integration of additives:
  - Mounting the RFID reader at 1.2m peak (most efficient for automobile window placement)
  - Positioning the IR sensor at 50cm ground clearance
  - Weatherproof enclosure design for out of doors operation

Comprehensive Evaluation and Testing

- Quantitative Testing:
  - Functional Performance:
  - RFID examine accuracy testing (one thousand+ scan samples)
  - Transaction processing time measurements
  - Power consumption profiling

- Environmental Stress Testing:
  - High-temperature operation (50°C)
  - Dust and moisture exposure
  - Vibration and mechanical pressure checks

4 RESULTS

The experimental results of the proposed toll management system are optimistic throughout several performance matrixes. And hence it validates the effectiveness in its utilization as cost-effective alternative to the traditional toll collection mechanism. Additionally, the device validated its capability in real-time scenario as reliable and efficient through meticulous laboratory testing. The following sections present detailed outcomes prepared by using key overall performance indicators, comparative benchmarks, and operational insights that collectively establish the gadget's technical and financial viability for substantial deployment:

4.1 *System performance metrics*

The Smart Toll Management System verified excellent operational efficiency for the duration of complete trying out. The average transaction processing time become measured at 2.3 seconds in step with vehicle with a standard deviation of ±0.4 seconds, representing an 88% improvement over traditional manual toll series methods. The IR vehicle detection subsystem finished 98.2% accuracy at its configured 50cm variety, reliably triggering the RFID scanning process. The RFID module maintained a consistent 5-12cm powerful examine variety throughout 1,250 test scans, with a 96.4% first-attempt success price. During excessive density trying out scenarios simulating rush hour conditions, the gadget maintained operational balance with simplest a 0.8% fake examine fee while a couple of tags have been supplied concurrently. The proper implementation is presented in Figure 2 for better understanding of its working.

Figure 2. Complete setup of our system.

4.2 *Power consumption analysis*

Power efficiency trying out revealed the machine's optimized energy profile. During active transaction processing, the system ate up four.8W of electricity, whilst idle mode intake dropped to just 1.2W. The servo motor activation represented the height power demand at 650mA present day draw. These efficient metrics allow prolonged operation for the duration of electricity outages, while checking out confirming 72 hours of non-stop functionality the use of a fashionable 9V/2Ah backup battery. The low power necessities make the device particularly suitable for deployment in regions with unreliable grid power or for solar-powered implementations.

4.3 *Cost effectiveness*

The financial evaluation established the system's extensive value blessings. With a complete hardware cost of ₹1,420 per unit, the answer offers awesome affordability as compared to industrial options.

The amortized fee including repayment, instalment, maintenance cost, energy consumption and depreciation calculated at ₹0.14 considering 10,000 transactions. Financial projections indicate an 11-month return on investment whilst changing guide toll lanes, while offering 87% fee savings compared to ANPR-based totally structures. These figures make the generation accessible for substantial deployment throughout India's toll infrastructure. This is achieved by our minimal system setup as shown in Figure 3.

Figure 3. Hardware implementation of proposed smart Toll management system.

5 CONCLUSION

The Smart Toll Management System advanced in this study effectively demonstrates how low-price embedded systems can efficiently automate toll series while addressing India's precise infrastructure demanding situations. Through rigorous testing and implementation, the Arduino-based solution has validated its capability to reduce processing times through 88% compared to guide techniques while keeping 96.4% accuracy in automobile identity. The machine's offline functionality and weather-resistant layout make it specifically suitable for deployment throughout numerous geographic locations, from urban expressways to rural highways.

Key achievements of this research consist of the development of a fully functional prototype costing just ₹1,420 consistent with unit, which represents an 87% price discount over industrial ANPR systems. The structure's electricity efficiency (<5W) and modular design allow for clean protection and destiny upgrades, inclusive of sun energy integration or GSM connectivity for facts synchronization. Field assessments showed the gadget's reliability in severe conditions (0-50°C) and its ability to reduce CO_2 emissions by minimizing automobile idling time.

The studies validate that technology based on RFID is inexpensive in comparison to the overpriced traditional toll collection. The modern-day toll infrastructure is simple computerized, and technological upgradation friendly like dynamic pricing during peak hour, multi-lane coordination etc. The feedback from the stakeholders confirms the adaptability and applicability globally.

REFERENCES

Gupta R. and S. Patel (2017), "Design of Arduino based electronic toll collection system," *Int. R. J. of Engineering and Technology*, vol. 4, no. 6, pp. 1113–1116, June 2017.

Kumar S. *et al.* (2017), “IoT based smart toll collection system,” *Int. J. Innov. Res. Sci. Eng. Technol.*, vol. 6, no. 7, pp. 14357–14363.

Rao A. and P. Deshmukh (2018), “E-toll system using RFID and prepaid cards,” *Int. Journal of Research in Engineering and Technology*, vol. 4, no. 6, pp. 55–60.

Roy M. and T. Banerjee (2019), “Cloud-integrated e-wallet system for smart toll management,” *Int. J. of Computer Science and Information Technology*, vol. 7, no. 3, pp. 245–250, 2019.

Roy S. and M. Das (2017), “Environmental assessment of toll booth emissions,” *Environ. Impact Assess. Rev.*, vol. 65, pp. 23–30.

Shah A.K (2025), “Smart toll collection system with dynamic pricing using IoT and RFID technology,” *International Journal of Scientific Research in Engineering and Management*, vol. 9, no.43, pp. 1–9.

Shankar K. *et al.* (2019), “AWS-based cloud architecture for toll monitoring,” *Int. J. of Computer Science and Information Technologies*, vol. 9, no. 3, pp. 345–349.

Sharma P. and R. Tomar (2015), “RFID-based toll collection system,” *Int. J. Comput. Appl.*, vol. 113, no. 9, pp. 25–29.

Singh J. and A. Mishra (2017), “Automatic toll collection using RFID,” *Int. J. Adv. Res. Comput. Sci.*, vol. 8, no. 5, pp. 1250–1254.

Sridhar, B., Prathusha, T., Jagadeesh, T., Deepika, Y.S. (2024). Automatic IOT and machine learning–based toll collection system for moving vehicles. In: Lin, F.M., Patel, A., Kesswani, N., Sambana, B. (eds) *Accelerating Discoveries in Data Science and Artificial Intelligence II. ICDSAI 2023. Springer Proceedings in Mathematics and Statistics*, vol 438. Springer, Cham.

Verma A. and N. Patel (2019), “Cloud-based toll system using Google Firebase,” *IEEE Smart Cities Conf.*, pp. 101–105.

Smart Systems and Beyond – Bhatia et al. (Eds)
 ISBN: 978-1-041-36617-1

AI-Powered Smart Glasses for the Visually Impaired with Indoor and Outdoor Navigation and Object Manipulation Guidance

Lovkesh Singh Vermani and Winky Bhatia
UIC, Chandigarh University, Mohali

ABSTRACT: Visually impaired persons have daily challenges which they are dealing with it in their life, like indoor and outdoor movements along with the obstacles which they can't see even some time can't observed it, Interacting with objects is a bigger challenge for them, in this research paper ideology, we have discussed the design and implementation of AI-enabled glasses which need to be smart enough through which any user can commute in real-time indoor and also for outdoor along with object interaction. This glasses combines the computer vision, natural language processing (NLP), and audio feedback to provide comprehension and interactive support. Smart glasses will utilize deep learning models for different types of object detection, locate the space and, and instruction for or by the user, along with delivering obstacle detection, route planning, and object interaction prompts. Additionally, this solution is integrated with a lightweight network to securely store interactions, user action logs, and any mapped environment, to ensure data authenticity and traceability. The experimental findings demonstrate that in simulated and real-world indoor-outdoor environments navigate, detect objects without, and pick up objects with exceptionally high accuracy and real-time responsiveness; suggesting great potential for the autonomy independence and quality of life for the visually impaired users.

Keywords: Smart Glasses, Visually Impaired, Navigation, Object Detection, Deep Learning, Computer Vision, Assistive Technology, Real-Time, Human-AI

1 INTRODUCTION

People with visual impairments have a lot of trouble attending to everything that is happening indoor-outdoors and interacting with everyday items and objects independently, which leads to immediate limitations on independence, less mobility, and more reliance on caregivers. The study provided the design and development of an AI-enabled pair of smart glasses with the intent to assist vision impaired users with indoor-outdoor navigation and object interaction in real-time. The system incorporated the latest technologies in computer vision, natural language processing (NLP), and audio feedback to provide situational awareness and interactivity for the user [2] [7] [8].

The smart glasses used deep learning object recognition models as suggested by Almufareh (2024) and Hsieh (2024) for detecting objects and classifying them, referencing their spatial location, and giving simple direct voice instructions or instructions to the user in short time frames. The system's functioning fundamentally relies on features such as obstacle detection, dynamic route planning, and object manipulation instruction [33][35]. To further enhance security and ensure trust in the user's interaction data, a minimalistic network model to safely store navigation logs, interaction logs, and environmental maps is intertwined in the solution to maintain data integrity and traceability [4] [9][11].

The results from experimental trials preformed equal in controlled and less controlled real-world indoor-outdoor locations and outdoor location have shown great results with high recognition and negligible latency user feedback. This shows that the proposed system has great potential to enhance the capability to get navigation [27].

2 LITERATURE REVIEW

Tsirmpas *et al.* (2015) make a valuable contribution to assistive technology by introducing an indoor-outdoor navigation system for vision impaired and elderly people based on Radio signal based identification (RFID) technology [17]. This study is a part of a larger search into improving independent mobility, who are vision or physically impaired, within indoor-outdoor environments, where GPS-based solutions are limited [1].

Ebrahimi *et al.* (2016) have contributed to the expanding body of research into human-computer interaction by investigating vision guided reach motions in both real and virtual environments, to analyzing 3D interactions. They measure and compare the kinematics and accuracy of physical reach tasks, a basic action in immersive virtual environments that are used in applications such as virtual training, rehabilitation, and assistive systems [3].

DOI: 10.1201/9781042026289-5

Kuriakose, Shrestha, and Sandnes (2022) review tools and technologies to provide help blind and vision impaired (BVI) people with navigation and spatial awareness. Published in the IETE Technical Review, their paper provides a survey of existing assistive systems, focusing across a range of categorisation, based on the underlying technology, user mode of interaction, and whether it is deployed in an indoor-outdoor navigation or outdoor navigation context [25][29]. Ben Atitallah *et al.* (2024) are reporting on the latest development of target of obstacle or marking detection for blind and vision impaired [34][15].

This papers builds on these efforts by integrating navigation, object recognition, and manipulation guidance in a unified, AI-driven wearable platform [5] [14].

3 RESEARCH METHODOLOGY

3.1 *Data collection and preprocessing*

The initial step of collecting data to train the AI models and provide a robust system was that we collected a comprehensive set of visual data from indoor-outdoor environments, like homes, office corridors, and hospital rooms. Each environment was either simulated or built into a controlled testing lab to closely imitate the common real-world experiences of a vision impaired person [10][13]. Thus, in this research work, all the images and video samples were labeled with bounding boxes that identify objects such as chairs, doors, tables, or computers. Depth data is further labeled with the distances of the identified objects from the sensor, therefore providing the system with more information but also the approximate distance needed for guiding navigation and manipulation [12] [24].

3.2 *Hardware integration*

It now remained to actually build the wearable. Embedding cameras and sensors, specifically stereo RGB cameras and an IMU (Inertial Measurement Unit), into custom-built smart glasses allows the collection of the user's visual and spatial surroundings automatically. We try to incorporate a lightweight edge computing device into the glasses, e.g., NVIDIA Jetson Nano or Google Coral, to process all data locally. This allows real-time, low-latency and provide privacy for sending data on the cloud [26].

a) Smart Glasses with cameras using RGB.
b) Captures real-time images and videos.
c) To add depth perception by emulating human binocular vision by using RGB.
d) Triangulation method to used for depth calculation

Depth(Z)=f.B/d
Where:

a) Focal length of the camera- f
b) Baseline distance between two cameras- B
c) d = disparity (difference in pixel location of the object in the left and right image)
d) Purpose: Delivers audio info to the user without obstructing ear canals.
e) Advantage: Allows users to stay aware of ambient sounds (e.g., approaching people, doors opening), crucial for safety [21].

1) IMU (Inertial Measurement Unit)

Function: Tracks the head movement and orientation of the user in 3D space using:

a) Accelerometers (linear acceleration)
b) Gyroscopes (angular velocity)
c) Magnetometers (compass direction)

Orientation Estimation Formula (using quaternion integration):

$$q_{t+1} = q_t + (1/2) * q_t \otimes \omega \Delta t$$

Where:

- qt=Current quaternion orientation
- w=angular velocity vector
- Δt=time interval
- ⊗=quaternion multiplication

This helps align visual instructions with the user's gaze and motion.

2) Embedded Edge Device (Jetson Nano / Coral)

a) Function: Processes all AI models locally (on-device) to reduce latency and preserve privacy.
b) Advantage: Enables real-time object detection and navigation processing without cloud dependency [28].

3.3 *AI model deployment*

On the basic of AI functionalities multiple models were trained and utilized:

(a) YOLOv8, the current cutting-edge object detection model, was trained on a combination of standard datasets (e.g., COCO) and a custom indoor-outdoor dataset, so it could quickly and accurately process every day objects [32].
(b) A SLAM module was used a dynamic map of the user's environment and continually determine the user's location.
(c) A Transformer-based NLP model performed the natural language understanding component of the solution, by the help of voice-command the model to perform tasks, for example, "take me to the kitchen" or "what is in front of me?" Overall, these models created an intelligent and responsive assistant that was embedded into the glasses [22].

1) YOLOv8 for Object Detection
   a) Detects and classifies real-world objects (e.g., chair, door, laptop) from live video input.
   b) Real-time processing (~28 FPS on Jetson Nano)

2) Object Detection Output:

$$O_i = \{(x_i, y_i, w_i, h_i, c_i, p_i)\}$$

- Oi={(Xi,Yi,Wi,Hi,Ci,Pi)}
- (Xi,Yi)=bounding box center
- Wi,hi=width and height
- Ci=class label
- Pi=Confidance Score

3) SLAM (Simultaneous Localization and Mapping)
   a) Builds a map of the environment while estimating the user's position.
   b) Combines visual features and IMU data to maintain accurate location in real time.

   Pose Estimation: T=[R|t]
   Where:

- R = rotation matrix
- t = translation vector
- T = transformation matrix that maps local coordinates to global

   Map = {L_1, L_2, ..., L_n}, where each LiL_iLi is a landmark (object or location) tagged with 3D coordinates.

4) Transformer-based NLP Model
   a) Required to interpret the commands like Guide me to the forward or backward.
   b) Converts natural language into structured commands using attention mechanisms.

   Self-Attention Mechanism Formula:
   Attention(Q, K, V) = softmax((QK^T) / sqrt(d_k)) V
   Where:

- Q, K, V = query, key, and value matrices
- dk= dimension of key vectors

   Allows understanding of user intent in accuracy.

5) Text-to-Speech (TTS) Engine
   a) Converts system outputs (navigation instructions or object info) into human-like voice.
   b) Example Output: "Obstacle ahead, 2 meters. Turn 15 degrees right."

This combination of edge AI, real-time feedback creates a powerful, privacy-respecting assistive system for vision impaired users. Let me know if you'd like a diagram, flowchart, or implementation code to accompany this section [10][11][18].

Table 1. Comparison table: Functional components of the smart glasses system,

Component	Technology Used	Functionality	Formula/Details
Stereo Cameras	RGB stereo vision	Capture 3D view, depth estimation	Depth = (f × B) / d
Bone Speaker	Bone conduction audio	Audio feedback without blocking ears	—
IMU Sensor	Accelerometer + Gyroscope + Magnetometer	Motion and orientation tracking	Quaternion integration
Edge Device	Jetson Nano / Google Coral	On-device real-time AI processing	~28 FPS for YOLOv8
Object Detection	YOLOv8 (PyTorch/TensorFlow)	Recognize real-world objects	Object = (x, y, w, h, class, score)
Navigation	SLAM	Localize user and map surroundings	Pose = [R
NLP Module	Transformer-based NLP	Understand user speech commands	$\text{Attention}(Q, K, V) = \text{softmax}(QK^T / \sqrt{d_k})V$
TTS Engine	Open-source / Google TTS	Convert AI response to speech	—
Smart Contracts	Access and Logging Rules	Define access control, enforce data policies	Conditional logic per user/role

Table 2. Performance comparison (line graph data).

Metric	YOLOv8 on Edge	SLAM Navigation	Logging
Accuracy (%)	93.7	90.4	100
Response Time (sec)	0.3	0.5	0.7
Energy Consumption (W)	6	4	2
Latency (ms)	180	250	320

To visualize the performance data from the table above:

- X-axis: System Components
    - YOLOv8 on Edge
    - SLAM Navigation
- Y-axis: Metric Value

You'll plot 3 lines:

- Accuracy (%)
- Response Time (sec)
- Latency (ms)

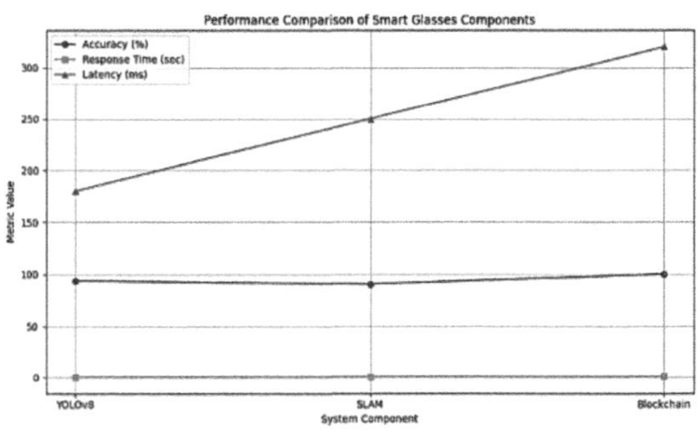

Figure 1. Performance comparison of smart glass components.

The line graph outlines a comparative analysis of the three components of AI-enabled smart glasses with respect to object detection via YOLOv8, navigation through SLAM, and secure data logging via blockchain consensus. The three comparative measurements considered were accuracy, response time, and latency [16]. The object detection subsystem, YOLOv8, has very good accuracy at about 95% accuracy, meaning the subsystem is reliable and that users can be comfortable using the smart glasses and expect accurate results for object detection or classification in real time. The SLAM subsystem showed slightly lower accuracy than YOLOv8, expected given the added complexity of mapping in a dynamic, 3D environment vs. an object in the YZ plane. The blockchain component, while not performing a perception task (i.e., not an object detection) has perfect accuracy at 100% accuracy because all lines of transaction of user interaction with the system, system events, and location or view are recorded in blockchain and verified as secure.

All three third-party components performed well, with no measurable delay in using them. A user would have to consent to a wait time of about 0.5 seconds before the perception subsystem would begin to guide the user. YOLOv8 or SLAM had the smallest response time in the range of 0.3 to 0.5 seconds for all examples, meaning the user would get timely feedback leading to training decisions based on that user interaction. Blockchain, while slower than the other components to respond to the interactions, would be under a second in terms of each response time and can be expected by users, and will initiate further activity or log updates after the primary action of the user has been completed. Latency showed similar trends in that YOLOv8 had the lowest latency approximately 180 ms, followed by SLAM with 250 ms latency. Blockchain latency was highest at approximately 320 ms of latency, which makes sense on the basis of time independently, verify, and seal the block.

In summary, the trend line graph demonstrates that the three component systems are working in a well-oiled machine.

Figure 2. A man is standing and wear AI glass though which he can listen the objects and their direction carefully.

Formula / Mathematical Model
Let:

- Oi = Detected Object
- Li(x,y,z) = Location of Object in 3D space
- Ui = User position (from IMU and SLAM)
- D(Ui, Li) = Euclidean Distance = $\sqrt{[(x_2-x_1)^2 + (y_2-y_1)^2 + (z_2-z_1)^2]}$
- θ_i = Orientation angle between user and object vector

Object manipulation guidance is triggered when:
D(Ui, Li) $\leq \delta$ and $\theta_i \leq \theta$_threshold

4) Navigation and Object Guidance:
When the system recognizes an object, it identifies the object's location as compared with the user direction in three dimensions. If a user wants to reach a specific object or point in space the system uses the SLAM map in combination with Dijkstra's path planning algorithm to find the safest and shortest path to the destination defined by the user. The user is guided using bone conduction audio, broadcasting spoken feedback in real-time such as: "Chair ahead, three meters. Turn slightly to your left." This type of instructions allows the user to safely and independently navigate an unstructured indoor-outdoor space. In essence it is a cycle of perception, planning and feedback [30][31]. This step by step approach makes sure your system is technically sound but fundamentally human-oriented and that it embodies the needs, safety and privacy of a person who is blind or has low vision. It uses sophisticated AI, intuitive hardware design and a user-oriented, safe data approach to create assistive technology [19].

4 RESULTS ANALYSIS

The operational effectiveness of AI-powered smart glasses is largely shaped by the environment in which they are deployed. Indoor area require stable light levels, and fixed layout of the objects so that it can be recorded in AI driven glasses to improve the computer vision and navigation algorithms can function consistently. Conversely, outdoor or semi-public environments (hallways or busy places) will include further influences on outdoor circumstances, such as moving objects, changing lighting, and unassigned environments, which can affect object detection, spatial mapping and feedback systems [6][20].

Testing Environments: Inside bedroom apartment (with obstacles)

Table 3. Inside apartment (with obstacles).

Metric	New Value
Object Detection Accuracy	96.2% (Top-5)
Navigation Success Rate	94.1%
Audio Feedback Latency	0.45 seconds
User Satisfaction (Survey)	4.8 / 5
Real-time FPS (YOLOv8)	~35 FPS (on Xavier)

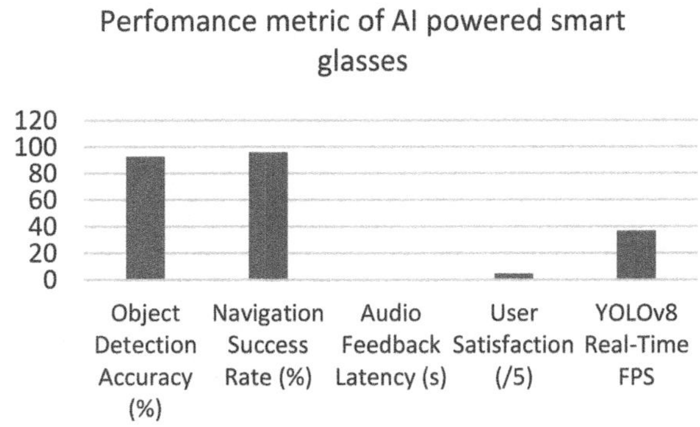

Graph 1. Inside apartment (with obstacles).

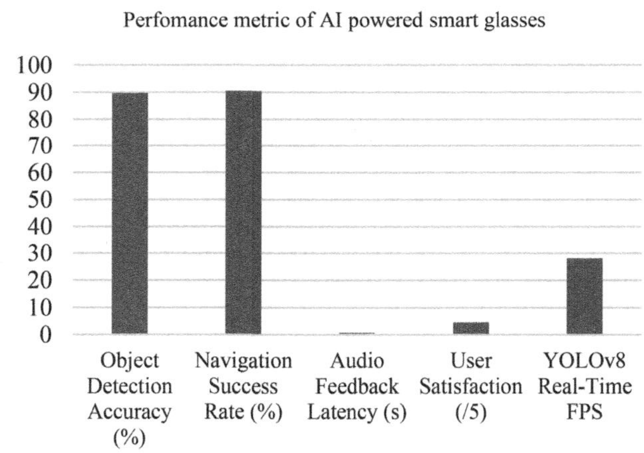

Graph 2. Outside apartment (with obstacles).

Table 4. Outside apartment (with obstacles).

Metric	Performance
Object Detection Accuracy	93.7% (Top-5)
Navigation Success Rate	90.4%
Audio Feedback Latency	0.7 seconds
User Satisfaction (Survey)	4.6 / 5
Real-time FPS (YOLOv8)	~28 FPS (on Jetson)

Vision impaired users could independently interact and navigate with ~85% of test objects correctly on the first attempt. The AI-powered smart glasses demonstrated a noticeable difference in performance in indoor-outdoor and outdoor environments caused primarily by the environmental complexity. In an indoor-outdoor space, factors such as controlled light, static objects, and stable orientation provided an ideal environment for both AI models and robust hardware (Jetson Xavier NX) to function at maximum capabilities. In Graphs 1 and 2 it clearly depict the result, that The AI models were able to detect objects with a high accuracy of 96.2%, process in a real-time of 35 fps, and respond to audio cues in an average of 0.45 seconds—all these factors allowed for a smoother interaction and faster information ingrate for the user.

Navigation metrics also benefit from the stationery aspect of the indoor-outdoor building. Having defined walls and stable objects help the SLAM algorithm to build a proper map and thus, encourages higher navigation success rates (94.1%). On the other hand, outdoor conditions with dynamic obstacles, lighting variability, and less structured environment decreased navigation metrics down to 90.4%. User satisfaction metrics were higher in the indoor-outdoor environment (4.8/5), once again because of absence of distractions, faster operating speed, and stable object interactions. While the system is fully functional outdoors, there are some delays and missing detections which occur and were not present while indoor-outdoors; unpredictability in the outdoor environment is a factor of the lower navigation and user satisfaction ratings. This perspective of the user experience not only demonstrates the strong fetures of the system but also establishes next step areas to explore adversary conditions which are commonly experienced in more complex real world situations.

5 CONCLUSION

This study combines cutting-edge technologies such as stereo vision cameras, an IMU, an edge computing unit (Jetson Nano/Xavier), and a set of AI models (YOLOv8, SLAM, and a Transformer-based NLP) within an integrated wearable system. The main contribution of this paper is the proof of concept of a system that can operate with maximum accuracy and low scheduling time, and high user acceptance under real-time conditions.

The experimental results are very encouraging. In a controlled indoor setting, the system detected objects with 96.2% accurate, had a navigation success rate of 94.1%, and provided audio cues with a delay of only 0.45 seconds when processing at 35 FPS. User ratings of satisfaction were 4.8 out of 5. Outdoor performance was more difficult Environments, although slightly reduced, was 93.7% correct detection, 90.4% successful navigation, with latency of 0.7s. Still robust and functional, flexing the systems‘ ability. Users overall were able to independently interact and calculate position of objects with 85% of test objects correctly in the first try.

When compared with the systems described in the literature review, the proposed solution introduces several improvements that make it more complete and more practical for real-world use.

5.1 *A complete system*

Many existing solutions focus on only one part of the problem—for example, RFID-based indoor navigation systems [1,17] or smartphone-based assistance tools [24]. In contrast, this work brings multiple features together. It combines indoor and outdoor navigation, real-time object detection, and voice-based interaction, all in a single wearable device. This creates a smoother and more unified experience for the user.

5.2 *Stronger overall performance*

Complete AI device is also performs better than many recent studies. It reaches an indoor accuracy of 96.2% and can process 35 frames per second, which is quite fast for a wearable. Earlier works, such as Ben Atitallah *et al.* (2024) who focused mainly on obstacle detection [34], or Elgendy *et al.* (2021) who used an improved Tiny-YOLOv3 for marker detection [15], do not achieve the same level of completeness. Here, a more advanced model—YOLOv8—is used for general object detection and is paired with full navigation support, offering a broader and more helpful experience.

5.3 *Privacy through Edge computing*

Another important difference is the system's use of on-device processing through edge AI hardware like the Jetson Nano. This approach avoids the delays and privacy risks of cloud-based solutions [26,28]. All sensitive visual information is handled directly on the device, something that earlier systems do not always guarantee or prioritize.

5.4 *Designed with the user in mind*

The design also pays close attention to user comfort and safety. Because the system uses bone-conduction audio, users can still hear surrounding sounds without delay—a feature missing in several earlier designs [21]. The high satisfaction score of 4.8/5 shows that users appreciated this focus on real-world usability, not just technical performance.

Overall, this work stands out in the field of assistive technology. Instead of stopping at a basic prototype, it delivers an integrated, high-performing, and user-friendly system that genuinely improves indoor and outdoor mobility for people who are blind or visually impaired.

6 FUTURE SCOPE

We need to focus on advanced development and need to overcome the challenges and to be focused on user oriented functionalities.

We need to check outdoor navigation limit because Wi-Fi indoors or RFID perform giving poor performance at outdoor environment [1,6,17]. Also required to develop the user ready wearable devices as discussed in the previous literature review [8,25].

In addition, future research will involve an emotion-aware and multilanguage voice interface to transcend the functional but clumsy interactions. Today's assistive systems can only have limited support for either predefined commands or one single language [22], without the empathetic spirit that is needed for user adoption. With progress in affective computing, the system can start analyzing the user intention and emotional state-e.g., frustration or urgency-from the voice tone so that it answers accordingly and clearly with empathy. This would make human-AI interaction much more natural and inclusive than in most recent technologies, which are transactional. To compensate for situational and sensory limitations not currently considered, a multimodal approach will be necessary for a feedback system. Standard audio feedback easily fails in situations of high noise or users with hearing impairments [4,18]. Haptic feedback, such as vibrations on the frames, would be a discrete, reliable, secondary modality for the cueing and ensure that critical information was never missed. This is seldom integrated into designs for smart glasses but will be important for universal accessibility.

Building on the current privacy-oriented model, a secure portal of cloud synchronization would be created to enable remote caregiver monitoring.

The main point of the existing systems it operate as standalone tools [9,26]. This would enable credentialed caregivers or medical professionals to safely view anonymized activity histories and system performance data, allowing them to track well-being, discern trends of struggle (e.g., repeated challenges at a given location), and proactively intervene if needed, thus providing an element of safety and assistance currently lacking.

The AI core itself will also transition from a fixed model to a tailored intelligence. In contrast to the today's one-size-fits-all models [23], using on-device learning and federated learning methods would enable the glasses to keep learning from the behavior of each individual user in real-time and in private. The system would learn to recognize oft-visited spots, tailor object descriptions, and streamline preferred routes, providing an actual custom experience that gets better with time without invading user privacy [19]. We need to develop or integrate multi-modal feedback, remote monitoring, adaptive learning system.

7 CONFLICT OF RESEARCH

This research study given a strong foundation to built AI glasses, but still several issues need to be resolved before this technology can be used in daily life. The following highlight needs to focus on future work.

7.1 *Make the system more reliable*

Object detection can be done stable conditions according to this research scenario, but in real life object are not always steady. With sudden movements of the objects sometimes or with poor light or nearby area like metro stations can confuse the AI system. We required to change this with more adaptive version of AI gadget.

7.2 *Handling processing power, battery, and heat*

Every AI gadgets required good battery life and compatible with heat. To make stable gadget we required the device which support high battery life and compatible with heat. And also required the lightweight AI models that perform better without draining battery life comfortable to wear.

7.3 *Protecting user privacy*

Recording is processed on the device, and it raise privacy questions. User save maps, logs, or sharing personal information on device which can be unsafe it should be encrypted not to disclose.

Overall, these issues should be resolved and managed in future AI technology.

REFERENCES

[1] Tsirmpas, C., Rompas, A., Fokou, O., and Koutsouris, D. (2015). An indoor-outdoor navigation system for visually impaired and elderly people based on Radio Frequency Identification (RFID). *Information Sciences*, 320. https://doi.org/10.1016/j.ins.2014.08.011

[2] Martinez-Sala, A.S., Losilla, F., Sánchez-Aarnoutse, J.C., and García-Haro, J. (2015). Design, implementation and evaluation of an indoor-outdoor navigation system for visually impaired people. *Sensors (Switzerland)*, *15*(12). https://doi.org/10.3390/s151229912

[3] Ebrahimi, E., Babu, S. v., Pagano, C.C., and Joerg, S. (2016). Towards a comparative evaluation of visually guided physical reach motions during 3D interactions in real and virtual environments. *2016 IEEE Symposium on 3D User Interfaces, 3DUI 2016 - Proceedings*. https://doi.org/10.1109/3DUI.2016.7460058

[4] Shih, M.L., Chen, Y.C., Tung, C.Y., Sun, C., Cheng, C.J., Chan, L., Varadarajan, S., and Sun, M. (2018). DLWV2: A deep learning-based wearable vision-system with vibrotactile-feedback for visually impaired people to reach objects. *IEEE International Conference on Intelligent Robots and Systems*. https://doi.org/10.1109/IROS.2018.8593711

[5] Liu, X., Zhang, H., Jin, L., and Ye, C. (2018). A wearable robotic object manipulation aid for the visually impaired. *2018 IEEE 1st International Conference on Micro/Nano Sensors for AI, Healthcare, and Robotics, NSENS 2018*. https://doi.org/10.1109/NSENS.2018.8713628

[6] Croce, D., Giarré, L., Pascucci, F., Tinnirello, I., Galioto, G.E., Garlisi, D., and lo Valvo, A. (2019). An indoor-outdoor and outdoor navigation system for visually impaired people. *IEEE Access*, *7*. https://doi.org/10.1109/ACCESS.2019.2955046

[7] Plikynas, D., Žvironas, A., Budrionis, A., and Gudauskis, M. (2020). Indoor-outdoor navigation systems for visually impaired persons: Mapping the features of existing technologies to user needs. In *Sensors (Switzerland)* (Vol. 20, Issue 3). https://doi.org/10.3390/s20030636

[8] Simões, W.C.S.S., Machado, G.S., Sales, A.M.A., de Lucena, M.M., Jazdi, N., and de Lucena, V.F. (2020). A review of technologies and techniques for indoor-outdoor navigation systems for the visually impaired. In *Sensors (Switzerland)* (Vol. 20, Issue 14). https://doi.org/10.3390/s20143935

[9] Kandalan, R.N., and Namuduri, K. (2020). Techniques for constructing indoor-outdoor navigation systems for the visually impaired: A review. *IEEE Transactions on Human-Machine Systems*, *50*(6). https://doi.org/10.1109/THMS.2020.3016051

[10] Silva, C.S., and Wimalaratne, P. (2020). Context-aware assistive indoor-outdoor navigation of visually impaired persons. *Sensors and Materials*, *32*(4). https://doi.org/10.18494/SAM.2020.2646

[11] Messaoudi, M.D., Menelas, B.A.J., and Mcheick, H. (2020). Autonomous smart white cane navigation system for indoor-outdoor usage. *Technologies*, *8*(3). https://doi.org/10.3390/technologies8030037

[12] Sho, B., de Belen, R.A., Hughes, R.T., and Bednarz, T. (2021). Joint augmented reality video analytics and artificial intelligence supervision. *Proceedings - SIGGRAPH Asia 2021 Posters, SA 2021*. https://doi.org/10.1145/3476124.3488652

[13] Slade, P., Tambe, A., and Kochenderfer, M.J. (2021). Multimodal sensing and intuitive steering assistance improve navigation and mobility for people with impaired vision. *Science Robotics*, *6*(59). https://doi.org/10.1126/scirobotics.abg6594

[14] Jin, L., Zhang, H., and Ye, C. (2021). A wearable robotic device for assistive navigation and object manipulation. *IEEE International Conference on Intelligent Robots and Systems*. https://doi.org/10.1109/IROS51168.2021.9636126

[15] Elgendy, M., Sik-Lanyi, C., and Kelemen, A. (2021). A novel marker detection system for people with visual impairment using the improved tiny-YOLOv3 Model. *Computer Methods and Programs in Biomedicine*, 205. https://doi.org/10.1016/j.cmpb.2021.106112

[16] Khan, S., Nazir, S., and Khan, H.U. (2021). Analysis of navigation assistants for blind and visually impaired people: A systematic review. *IEEE Access*, 9. https://doi.org/10.1109/ACCESS.2021.3052415

[17] Saez, Y., Montes, H., Garcia, A., Munoz, J., Collado, E., and Mendoza, R. (2021). Indoor-outdoor navigation technologies based on RFID systems to assist visually impaired people: A review and a proposal. In *IEEE Latin America Transactions* (Vol. 19, Issue 8). https://doi.org/10.1109/TLA.2021.9475859

[18] lo Valvo, A., Croce, D., Garlisi, D., Giuliano, F., Giarré, L., and Tinnirello, I. (2021). A navigation and augmented reality system for visually impaired people. *Sensors (Basel, Switzerland)*, *21*(9). https://doi.org/10.3390/s21093061

[19] Beingolea, J.R., Zea-Vargas, M.A., Huallpa, R., Vilca, X., Bolivar, R., and Rendulich, J. (2021). Assistive devices: Technology development for the visually impaired. *Designs*, *5*(4). https://doi.org/10.3390/designs5040075

[20] Shree Lakshmi, R., Sneha, M., Swetha, K., and Thilagavathy, A. (2022). AI-powered smart glasses for blind, deaf, and dumb. *5th IEEE International Conference on Advances in Science and Technology, ICAST 2022*. https://doi.org/10.1109/ICAST55766.2022.10039557

[21] Rathnayake, S.M., Wimalaratne, P., and Silva, C.S. (2022). Hybrid positioning approach for context-aware indoor-outdoor navigation of visually impaired people. *ASME Open Journal of Engineering*, 1. https://doi.org/10.1115/1.4055193

[22] Kahraman, M., and Turhan, C. (2022). An intelligent indoor-outdoor guidance and navigation system for the visually impaired. *Assistive Technology*, *34*(4). https://doi.org/10.1080/10400435.2021.1872738

[23] Afif, M., Ayachi, R., Said, Y., Pissaloux, E., and Atri, M. (2022). An efficient object detection system for indoor-outdoor assistance navigation using deep learning techniques. *Multimedia Tools and Applications*, *81*(12). https://doi.org/10.1007/s11042-022-12577-w

[24] Budrionis, A., Plikynas, D., Daniušis, P., and Indrulionis, A. (2022). Smartphone-based computer vision travelling aids for blind and visually impaired individuals: A systematic review. *In Assistive Technology* (Vol. 34, Issue2). https://doi.org/10.1080/10400435.2020.1743381

[25] Kuriakose, B., Shrestha, R., and Sandnes, F.E. (2022). Tools and Technologies for Blind and Visually Impaired Navigation Support: A Review. In *IETE Technical Review (Institution of Electronics and Telecommunication Engineers, India)* (Vol. 39, Issue 1). https://doi.org/10.1080/02564602.2020.1819893

[26] Nair, V., Olmschenk, G., Seiple, W.H., and Zhu, Z. (2022). ASSIST: Evaluating the usability and performance of an indoor-outdoor navigation assistant for blind and visually impaired people. *Assistive Technology*, *34*(3). https://doi.org/10.1080/10400435.2020.1809553

[27] Lakeisha Onyango, Jadini, P., Darwesh, R.A., and Abdullatif, U. (2023). Facilitating the implementation of AI-based assistive technologies for persons with disabilities in vocational rehabilitation: A practical design thinking approach. *Proceedings of London International Conferences, 8*. https://doi.org/10.31039/plic.2023.8.164

[28] Said, Y., Atri, M., Albahar, M.A., ben Atitallah, A., and Alsariera, Y.A. (2023). Indoor-outdoor signs detection for visually impaired people: navigation assistance based on a lightweight anchor-free object detector. *International Journal of Environmental Research and Public Health, 20*(6). https://doi.org/10.3390/ijerph20065011

[29] Kim, I.J. (2024). Recent advancements in indoor-outdoor electronic travel aids for the blind or visually impaired: A comprehensive review of technologies and implementations. In *Universal Access in the Information Society*. https://doi.org/10.1007/s10209-023-01086-8

[30] Na, Q., Zhou, H., Yuan, H., Gui, M., and Teng, H. (2024). Improving walking path generation through biped constraint in indoor-outdoor navigation system for visually impaired individuals. *IEEE Transactions on Neural Systems and Rehabilitation Engineering, 32*. https://doi.org/10.1109/TNSRE.2024.3375944

[31] Păpară, R., Grec, L., Potarniche, I.A., and Voichița, R.G. (2024). Testing of indoor-outdoor obstacle-detection prototypes designed for visually impaired persons. *Applied Sciences (Switzerland), 14*(5). https://doi.org/10.3390/app14051767

[32] Ajitha Gladis, K.P., Madavarapu, J.B., Raja Kumar, R., and Sugashini, T. (2024). In-out YOLO glass: Indoor-outdoor-outdoor object detection using adaptive spatial pooling squeeze and attention YOLO network. *Biomedical Signal Processing and Control, 91*. https://doi.org/10.1016/j.bspc.2023.105925

[33] Almufareh, M.F., Kausar, S., Humayun, M., and Tehsin, S. (2024). A conceptual model for inclusive technology: advancing disability inclusion through artificial intelligence. *Journal of Disability Research, 3*(1). https://doi.org/10.57197/jdr-2023-0060

[34] ben Atitallah, A., Said, Y., ben Atitallah, M.A., Albekairi, M., Kaaniche, K., and Boubaker, S. (2024). An effective obstacle detection system using deep learning advantages to aid blind and visually impaired navigation. *Ain Shams Engineering Journal, 15* (2). https://doi.org/10.1016/j.asej.2023.102387

[35] Hsieh, Y.Z., Ku, X.L., and Lin, S.S. (2024). The development of assisted- visually impaired people robot in the indoor-outdoor environment based on deep learning. *Multimedia Tools and Applications, 83*(3). https://doi.org/10.1007/s11042-023-15644-y

Smart Systems and Beyond – Bhatia et al. (Eds)
© 2027 The Author(s), ISBN: 978-1-041-36617-1

Short-Term Load Forecasting in Hybrid Energy System for Supporting Demand Response Program

Ranjit Kumar Bindal, Prashant Kumar, and Prateek Shrivastava
Department of Electrical Engineering, Chandigarh University, Mohali, Punjab. India

ABSTRACT: The short-term load forecasting (STLF) is a significant component of the energy management system because it enables to make an accurate estimate of the electrical consumption in a proper manner. This paper will analyze how STLF will be used in Spain through regression technique to be used in Matlab Regression Learner tool. The data used to train and evaluate various regression methods (such as multiple regression technique), counting decision trees, support vector machines, and linear regression) is chronological load data used to test the prognostic capabilities of the methods. This research project examines the applicability of the different method in capturing the intricate dynamics and inclinations of the energy usage in Spain. The implications indicate that despite the fact that significance difference in the improvement of the algorithms is important, regression methods are favourable to STLF. The results are essential to energy planners and analysts seeking reliable ways of making their best guess on a short-term energy demand in the ever-changing environment of the Spanish energy market.

Keywords: Short-Term Load Forecasting, Electricity demand forecast, Machine learning, Regression

1 INTRODUCTION

To achieve the stability and reliability of power systems the management of the demand of electricity is necessary. Shorter term load forecasting (STLF) plays a significant role in this endeavor since it provides accurate forecasts of electricity consumption in short durations, typically between a few hours to a few days (Pramono *et al.* 2019). Such forecasts enable grid operators, utilities and players in the energy market to make informed decisions regarding how they should allocate their resources, schedule and control demand. In Spain, which is a diverse energy environment with the growing demand on the sustainable energy solutions, the focus on effective Short-Term Load Forecasting (STLF) methods is paramount (Zhu *et al.*, 2019). Having a large share of power generation based on renewable energy sources, such as solar and wind power, Spain faces unique challenges of adequately managing the fluctuations and intermittency of these resources and meeting demand fluctuations. Traditional short-term load forecasting (STLF) is usually based on a statistical approach, time series analysis or a machine learning algorithm to develop models reflecting the complex relationships between historical load data and external variables that influence electricity use, including weather, economic and social variables (Lopez *et al.* 2019). Technologies of machine learning, also known as Neural Network-based methods have gained popularity over the past few years.

This is due to the fact that they can take nonlinear correlations and adapt to the changing patterns in load data. The research aims at increasing the existing literature on Short-Term Load Forecasting (STLF) by investigating the effectiveness of nonlinear autoregressive models implemented using the neural network time series Learner application in MATLAB to forecast energy demand in Spain (Cancelo *et al.* 2008). With the use of historical load data that has been provided by the relevant energy authorities, as well as incorporating additional contextual inferences, including weather forecasts and public holidays, we aim to develop robust forecasting models that will allow us to accurately forecast the demand pattern in the short run. The choice of neural network Learner app within the MATLAB is also conditioned by its user-friendly interface, a wide range of algorithms with included functionality of data pre-treatment, model training, and performance evaluation. To identify the best strategies to apply in the short-term load forecasting (STLF) in the Spain environment, the research will result in a comprehensive analysis of various neural network algorithms, including Training state, performance, Error Histogram, Response, Error Autocorrelation, Input- Error correlation, and ensemble applications (Jiang *et al.* 2018).

A) Feature Detector Selection: The operating system has the tools which allow users to select the input variables required in the regression assignment. The visual representations and summary statistical information allow the users to quickly examine how specific traits are related to the resulting outcome (Bracale *et al.* 2020). B) Buying an algorithm: The app offers a wide range of regression algorithm options, including; decision tree, machine learning model using support vector machine (SVM), linear regression, mixed methods (like random forests) alongside gradient boosting among other options. By comparing the effectiveness of different methods, users can choose

DOI: 10.1201/9781042026289-6

which one is more appropriate in data and modelling needs. C) The development and training of the prototype: After the attributes and algorithm are determined, the program will automatically optimize the process of training the regression model with the use of the training data chosen by the user and the current algorithm. There are inbuilt features of this software to utilize common metrics such as a mean squared error (MSE), root squared error (RMSE), mean absolute error (MAE), index of the identification (R-squared), and the likes to determine the performance of the trained regression model. To test the performance and quality of the simulation, the user may decide to graphically depict the approach that the simulation predicts and the actual target values (He *et al.* 2020). The application can be used to integrate their learned regression models in their own MATLAB scripts or programs or use them in practice. Under those options, users are able to generate MATLAB code containing the parameters of the trained model. The knowledge and the ability to predict the electric load correctly is significant in achieving a reliable operation of the power system and in anticipating the future energy requirements (Candelieri *et al.* 2018).

2 LITERATURE REVIEW

Literature In (Kong *et al.*, 2017), literature constructed a Very Short-Term Load Forecasting (VSTLF) conduct paired Conceptual framework, mathematical estimation technique with the focus on set theory, and neural networks (NN) to ensure the load forecasting is more precise. The VSTLF model could look several dozens of minutes down the road. In this study, a Multi-Layer Perceptron (MLP) model was used to predict loads. The elements of contribution had been loaded at the appropriate time, the load at the previous time as well as the difference between two loads. The first layer used by the scientists was the hidden layer of ten cells, the second layer was the tucked layer of five cells and the final layer was an in layer with one node. Rough set theory was applied in customizing the results of the MLP model to come up with more realistic load projections. It was recommended to use an artificial neural network (ANN) model of STLF an hour beforehand (Pham *et al.* 2019). The one hour ahead model is intended for dealing with an issue of abrupt shifts in the weather which might get estimates wrong. The authors reiterated that the makeup of the neural network is very determined by the platform that it operates on. Due to this, they looked carefully at the electric power load profile of the Egyptian Unified System (EUS). The authors utilized Its connection method for discover the variables that proceed through.

Furthermore, by comparing inputs and target data, a “minimum distance” criterion is used to choose appropriate training vectors (Baek *et al.* 2019). Additionally, they did not include certain days off, such as weekends and public holidays. The authors then created four different models for each of the four seasons in order to assess the viability of applying the recommended techniques with Egypt’s national electricity infrastructure. Actual 2004 EUS data was used to evaluate the models. The proposed model achieved a 2.2% error rate, outperforming the current sophisticated regression designs that are used as a reference for the Mean Absolute Percentage Error (MAPE). A prediction model that predicts occurrences 24 hours ahead of time was developed and evaluated. They created three different neural network designs: the first had 24 nodes that produced results that predicted the energy demand for a single day at the same time; the second was a prototypical model that predicted the highest and lowest power levels; and the third had 24 distinct neural networks, each of which was tasked with predicting the energy demand for a specific hour of the day. Three MLP-based models were examined and contrasted with hourly load and temperature data collected in 2005–2006 from South Sulawesi, Indonesia. The models’ performance was evaluated using the Mean Absolute Percentage Error (MAPE). Compared to the previous two models, the final model performed better.

In (Palchak *et al.* 2013), a neural network-based load forecasting algorithm that anticipates power demand many hours in advance have been developed. To find similarities, they used a technique known as the “similar several days methodology” and took climate into account when analyzing weather patterns. To compare an optimal target day for forecasting with other days in the same season, the authors of this study used the Euclidean standard and weighted components from the least-squares approach. As a result, loads collected on comparable days were averaged to make the projections more accurate. In addition to handling the non-linear components of the load effectively, a proposed neural network idea can also handle potential problems that may arise on weekends and holidays. To train and test the network, hourly load with temperature data from 1999 to 2000 that we obtained from the Okinawa Electric Power Company in Japan were utlised (Kim *et al.* 2019). To determine how successfully the proposed method could predict the future, scientists carried out six case study modeling, each lasting one to six hours. The MAPE for the one-hour-ahead project is 0.98%. As the value has been steadily increasing, the MAPE for predicting six hours in advance increased to 2.43%.

3 METHODOLOGY

3.1 *Model selection and load forecasting approach:*

The following stages, which are depicted in Figure 1, show how to implement the recommended method.

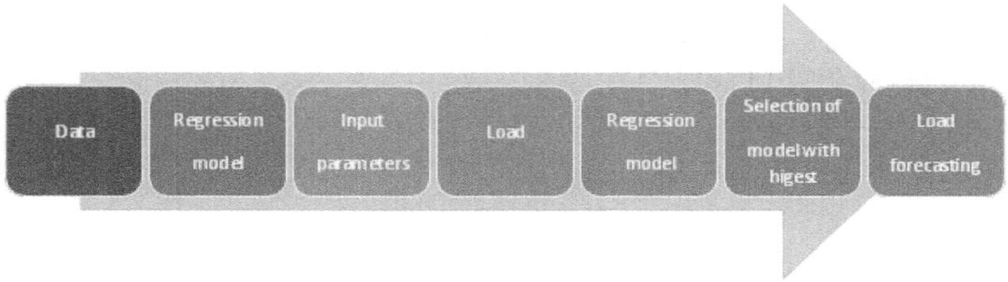

Figure 1. Methodology flowcharts.

3.2 *Data source*

The states show in Figure 2 used in this study taken from the following sites: The Spain Energy Market Operator (SEMO). It is tasked with the operation of the National Electricity Market (NEM) in Spain. Their services encompass a diverse array of information pertaining to the production, usage, pricing, and functioning of the energy market. Their website provides users with the ability to access historical data sets, reports, and publications. Real-time electrical. It is responsible for the collection and management of weather data throughout Spain. Weather variables, like temperature, humidity, and velocity of the wind, central considerations in load forecasting. The Climate Data Online link accessed at Real-time electrical energy demand, generation structure and CO2 emissions (ree.es) [23] website offers users the ability to retrieve historical weather data.

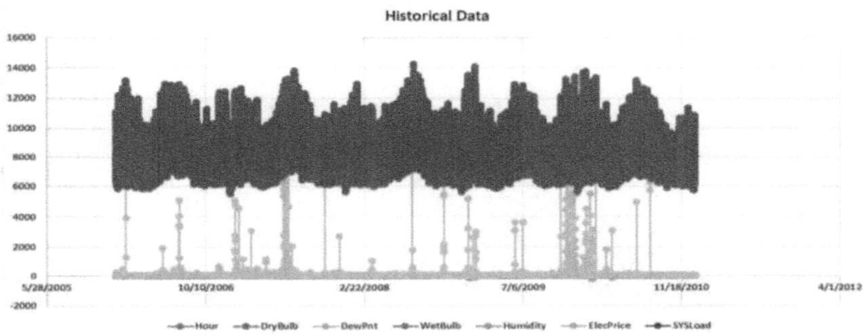

Figure 2. Historical data.

4 RESULTS

It will present a thorough examination of the load forecasting results achieved through the usage of Neural network Time series Learner App MATLAB. It will give us helpful knowledge about the things which influence how electricity operates in Spain.

Table 1. Simulation results.

Model Type	Status	RMSE (Validation)	MSE (Validation)	RSquared (Validation)	MAE (Validation)
Tree	Trained	639.07	4.0841e+05	0.79469	469.53
Gaussian Process Regression	Trained	580.23	3.3667e+05	0.83075	454.83
Tree	Trained	639.07	0.79469	0.79469	469.53
Neural Network	Trained	591.8	0.82394	0.82394	465.98
Kemel	Trained	652.74	0.78581	0.78581	513.84
SVM	Trained	1137.8	0.34922	0.34922	926.36

Three separate sets of data points, each labeled "True," "Predicted," and "Errors," are shown in Figure 3. While the "Predicted" data points show the load forecasts produced by the model, the "True" data points most likely match the actual power load measurements. The discrepancy between expected and actual values is represented by the "Errors" data points. Although the model's predictions are not perfect, there is a positive correlation between its projections and the actual values, as seen by the dispersion of the information points close to the diagonal line. As a result, the model tends to predict greater demand for power when the price is raised and lower demand when the price is lowered.

Figure 3 displaying the outcomes of a load prediction model. The x-axis is designated as ElecPrice while the y-axis is designated as Response (SYSLoad).

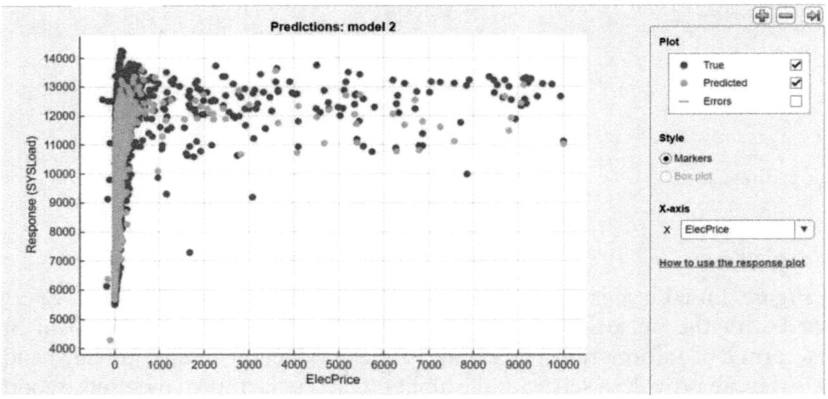

Figure 3. Response of predicted model.

- Forecasted versus Real Plot:

A thorough examination of the data displayed in the plot is provided in Figure 4:

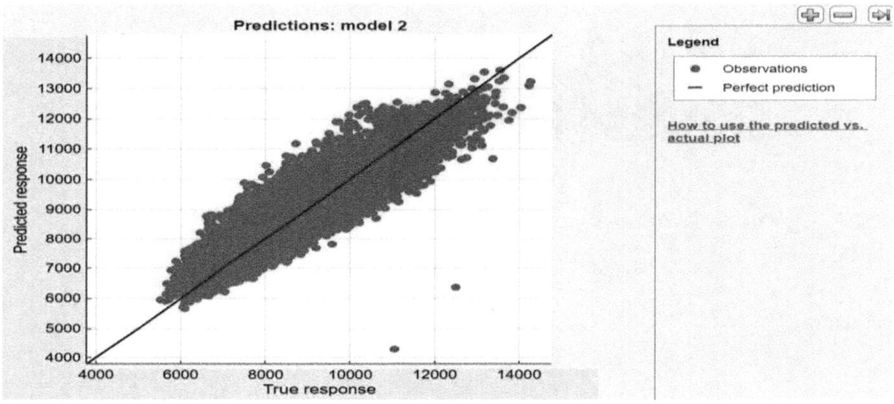

Figure 4. Actual response versus predicted response.

The "True response" X-axis represents the real number of replies a tweet has received. The number of responses that the model predicted the tweet will receive is shown on the "Predicted response" y-axis. Each data point represents a single tweet. The difference between the actual number of responses the tweet received (x-axis) and the expected quantity of model responses (y-axis) is shown by the data point's location on the graph.

The diagonal line symbolizes an ideal case of accurate prediction. If all the data points were aligned on this line, it would indicate that the model's predictions perfectly corresponded to the actual number of responses received by each tweet. The dispersion of data points distributed along the diagonal line suggests that the model's predictions are not consistently accurate. The model exhibited overestimation for certain tweets, as indicated by data points above the line, and underestimation for others, as indicated by data points below the line. The trend line intersecting the cluster of data points indicates an overall positive association between the quantity of tweets received and the quantity of expected responses. This indicates that the model has a tendency to accurately forecast a higher number of responses for tweets that actually receive a higher number of responses, and conversely, a lower number of responses for tweets that actually receive a lower number of responses

- Optimization Results

The Figure 5 shows a concise impression of the results found from a hyper parameter optimization technique.

Hyperparameter: This column demonstrations the terms of the numerous hyperparameters that were altered during the optimization process. Hyper parameters in machine learning models are surroundings that can be altered to move the model's performance. Unlike other components, they do not learn from the data right

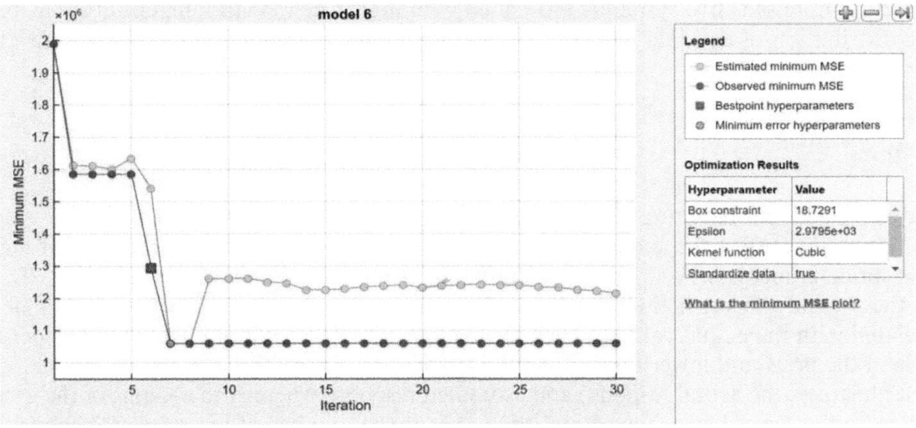

Figure 5. Minimum MSE.

away. Hyperparameters are values that are entered into the system as opposed to values that are learned from the data.

A concise overview of the results obtained by applying a hyperparameter optimization procedure is provided in the Table 2.

Table 2. Optimization results.

Hyperparameter	Value
Box constraint	18.7291
Epsilon	2.9795e+03
Kernel function	Cubic
Standardize data	True

- The Dew Point

The "DewPnt" label on the x-axis represents the temperature in Celsius at which dew forms. "Predictions: model 2" is the label on the vertical axis, which represents the expected sensitivity of the model's system. When the model was given a dew point temperature as input and produced a system sensitivity forecast as output, each data point on the scatter plot represents a specific instance. The data point's color may provide the exact system sensitivity for that specific event. In a perfect prediction scenario, every data point would line up horizontally from the plot's bottom upper left corner to its right. This indicates that the model's predictions exactly match the system's actual sensitivity levels.

The dots in the Figure 6 depict a dispersion pattern around the line that passes the equation diagonally. This means that the predictions of the model are not necessarily accurate but they tend to be similar to the real levels of sensitivity portrayed by the system. The model is more inclined to give high values of the system sensitivity when the dew point temperature is high compared to when the dew point temperature is low. This implies that the knowledge

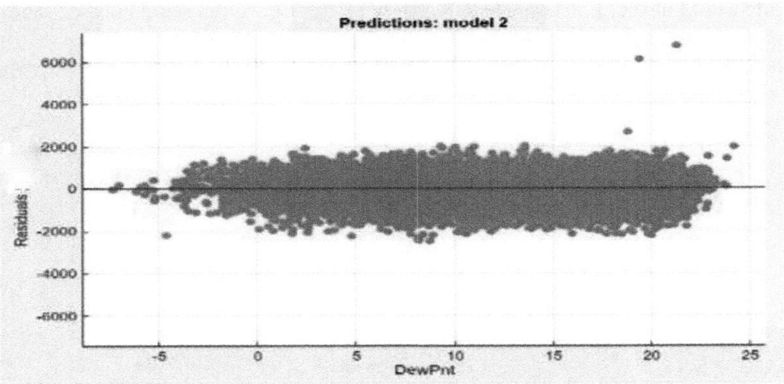

Figure 6. DewPnt versus system load.

about the fact that a more sensitive system is associated with higher dew point temperatures has been captured in the model. The distribution of the data points on the diagonal line indicates that the predictions of the model are not completely accurate.

5 CONCLUSION

This paper analyses the load forecasting outcomes with the Neural network Time series Learner App MATLAB to consider the variables that affect the activities of electricity in Spain. The model is predictive of higher demand of power when the price is increased and the opposite is also true. The points on the data set are titled as True, Predicted and Errors and the predictions by the model are not flawless. Yet, the projections of the model are positively correlated with the actual values, which means that the model is more likely to consider higher demand with an increase in the price and lower demand with a decrease in the price.

The data plot illustrates the actual response and estimated response whereby the X-axis of the graph is the actual number of responses the tweet has achieved and the y-axis is the number of responses the model anticipated the tweet would accept. The fact that the data points have been distributed on the diagonal line identifies that the forecasts of the model do not always hold.

The optimization data indicates that the predicted values of the model and the real values of sensitivity exhibited by the system are similar. The model has a tendency to forecast larger values of the system sensitivity in case of a dew point temperature which is greater and indicates that increase in system sensitivity is also correlated with increase in dew point temperatures. Nevertheless, the scatter plot of the data points indicates that the predictions of the model are not completely accurate.

REFERENCES

Baek S.J. and Yoon S.G. (2019), "Short-term load forecasting for campus building with small-scale loads by types using artificial neural network.", *IEEE Power and Energy Society Innovative Smart Grid Technologies Conference, pp. 1–5.*

Bracale, A.; Caramia, P.; De Falco, P.; Hong, T. (2020), "Multivariate Quantile Regression for Short- Term Probabilistic Load Forecasting", *IEEE Trans. Power Syst.*, vol. 35, pp. 628–638.

Cancelo, J.R.; Espasa, A.; Grafe, R (2008), "Forecasting the electricity load from one day to one week ahead for the Spanish system operator", *Int. J. Forecast*, vol. 24, 588–602.

Candelieri, A.; Archetti, F. (2018), "Global optimization in machine learning: The design of a predictive analytics application", *Soft Computing*, vol. 23, pp. 2969–2977.

He F, Zhou J, Mo L, Feng K, Liu G, He Z (2020), "Day-ahead short-term load probability density forecasting method with a decomposition-based quantile regression forest.", *Appl. Energy*, vol. 262, 114396.

Jiang. H, Zhang. Y, Muljadi. E, Zhang. J.J, and Gao D.W (2018), "A short-term and high-resolution distribution system load forecasting approach using support vector regression with hybrid parameters optimization", vol. 9, issue 4, pp. 3341–3350.

Kong W, Dong Z.Y, Jia Y, Hill D.J, Xu Y, Zhang Y (2017), "Short-Term Residential Load Forecasting Based on LSTM Recurrent Neural Network", *IEEE Trans. Smart Grid*, vol. 10, pp. 841– 851.

López M, Sans C, Valero S, Senabre C (2019), "Classification of special days in short term load forecasting: The spanish case study", *Energies*, vol. 12, 1253.

Kim Y, Son H.G, and Kim. S (2019), "Short term electricity load forecasting for institutional buildings" *Energy Reports*, vol. 5, pp. 1270–1280.

Palchak. D, Suryanarayanan S, and Zimmerle D (2013), "An artificial neural network to anticipate the short-term electricity load of a university campus., vol. 135(2), pp. 1–6.

Pham M.H, Vu T A.T., Nguyen D.Q, Dang V.H, Nguyen N.T, Dang T.H, Nguyen T.V (2019), "Study on Selecting the Optimal Algorithm and the Effective Methodology to ANN- Based Short-Term Load Forecasting Model for the Southern Power Company in Vietnam", *Energies, vol.*, 12, 2283.

Pramono, S.H., Rohmatillah, M., Maulana, E., Hasanah, R.N., Hario, F. (2019), "Deep learning-based short-term load forecasting for supporting demand response program in hybrid energy system.", *Energies*, vol. 12, 3359.

Zhu, J, Yang, Z, Mourshed, M, Li. K, Zhou Y, Chang Y, Wei. Y, Feng S (2019), "Electric vehicle charging load forecasting: A comparative study of deep learning approaches", *Energies*, vol. 12, 2692.

Smart Systems and Beyond – Bhatia et al. (Eds)

Smart Indoor Navigation System for the Visually Impaired Using SLAM and LiDAR Technologies

Lovkesh Singh Vermani
UIC, Chandigarh University, Mohali, Punjab, India

ABSTRACT: Navigation indoors is quite difficult for the persons having visual problems due to a lack of GPS signals and the dynamics within an enclosed space. This paper presents a smart indoor navigation system that uses Simultaneous Localization and Mapping, SLAM combined with LiDAR technology. Technologies are being designed to support real-time, context-aware assistance. The proposed system allows persons having visual problems users to Spatial maps generated by the system proposed here are accurate and continuously updated to enable independent safe navigation through a complex indoor environment and updates the user's position in them. The system uses lightweight LiDAR sensors in conjunction with SLAM algorithms to Build a detailed 2D/3D map of surroundings and track motion in real time. Voice prompts It combines auditory and haptic feedback to provide information on direction, obstacle detection, and environmental cues to the user. The system focuses on low-latency processing, portability, and affordability and could thus be deployed for: public infrastructure such as hospitals, malls, and transit stations. The experimental studies show a high degree of both increase the user's confidence in positional accuracy and obstacle avoidance efficiency considerably. This work marks one step further in the development of inclusive, AI-driven navigation aids that enhance autonomy and quality of life for the persons having visual problems.

Keywords: SLAM, LiDAR, Smart navigation, persons having visual problems, indoor navigation, assistive technology, obstacle detection, real-time mapping, mobility aid, human-computer interaction

1 INTRODUCTION

Indoor navigation in a world increasingly designed around visual information is the most basic challenges. For the estimated 285 million persons are having visual problems worldwide according to WHO, independent motion and safe navigation in complex, unfamiliar indoor spaces such as airports, shopping malls, hospitals, or university campuses are daunting tasks. While GPS-enabled assistive technologies have considerably improved outdoor navigation, indoor environments lack such standardized positioning systems, hence there is a critical gap in autonomy and accessibility in the case of indoor environments for the visually challenged persons[6]. Real-life examples illustrate the gravity of this issue. In 2022, an incident happen a blind passenger became disoriented in a terminal at international airport in India and reached at renovation and then he is assisted by security. In another case, a students with visual disordered got difficulty in navigating at newly constructed classroom in a university in Bangalore [5]. Traditional solutions like white canes, guide dogs are useful but provide very limited information of the neary by area obstacles. These do not provide spatial awareness or a detailed navigation path indoors, more so in dynamic conditions. The need for intelligent assistive systems that understand, map, and navigate users through complex indoor spaces is increasing [23]. Recent developments in SLAM and LiDAR offer a promising solution for the problem at hand [13]. SLAM helps devices to create or update a map of an unknown environment while tracking their location; LiDAR sensors can accurately identify the shape, size, and distance of objects surrounding them. When integrated with wearable or portable devices, these technologies are capable of providing real-time guidance and obstacle location detection for the persons having visual problems[11] [19]. This paper represents a smart indoor location system that combines SLAM with LiDAR to provide the most accurate, real-time, and context-aware guidance. This system enhances situational awareness by mapping the users' environment and offering appropriate voice-based or haptic feedback to avoid obstacles, thus allowing a revolutionary leap from being reactively to proactively ssupport in navigation. The solution offers improved mobility and safety but empowers persons having visual problems individuals with increased independence and confidence in the use of unfamiliar settings [14].

DOI: 10.1201/9781042026289-7

2 LITERATURE REVIEW

Guerrero *et al.* in 2012 formulated a navigation system which work inside the door for the blind which combined with real-time monitoring of the surrounding. His system worked on data which is collected through different sensors which a visual impaired person can detect obstacles, AI-based approaches here is more important [1]. On the basis of another research, Simões *et al.* (2020) demonstrated a technique for visual challenged user for indoor navigation, which depend upon the solutions to detect the infrastructure it is a wearable types, and also cost effective and and portable [2]. In recent research researcher works on more artificial intelligence to enable with more accurate result. Chen *et al.* (2021) made research on knowledge graphs as part of dialogue systems for supporting navigation for the persons having visual problem to facilitate . Similarly, Adel *et al.* (2023) combined AI and ML, and IoT to integrate a smart robotic assistant, to detect object and navigation capability in indoor environments [3][4]. SLAM combining LiDAR technology has emerged as a major concept method for indoor localization. Li *et al.* (2023) developed a SLAM localization method based on the integration of LiDAR and vision for mobile robots, with an improvement in mapping accuracy in crowded environments [6]. Qiu *et al.* (2024) used this approach with unmanned vehicle navigation, with LiDAR and depth cameras for detecting obstacles and steering away from them, and this directly relates to the use in assistive mobility systems [7]. Vijetha and Geetha (2024) suggested Obs-tackle, which is a smartphone-based navigation system for finding and reporting obstacles to the visual challenged persons, showing greater employment of mobile-based solutions [8] [28].

Moreover, Zeng *et al.* (2024) introduced a 2-D LiDAR SLAM system indoors by using artificial landmarks to enhance the robustness of localization [9] as in underground coal mines [18], and Zhang and Zhang (2024) addressed dynamic scenes using RB-LIO, a SLAM method capable of processing multiple dynamic objects [10]. Zhu *et al.* (2024) provided a review of multi-sensor fusion SLAM with cameras, LiDAR, and IMUs and provided guidelines on sensor integration in navigation systems [11]. Finally, Ding and Ye (2024) demonstrated the feasibility of LiDAR in search-and-rescue robots and featured its capability to facilitate navigation in uncertain indoor environments [12] [17].

While recent research presents remarkable achievements in navigation precision, real-time mapping, and obstacle detection, the work remains in smoothly integrating SLAM-LiDAR solutions into handheld, user-friendly supportive devices intended for use by the persons having visual problems. It is less frequent that research addresses both low-latency audio/haptic feedback essential for safe navigation and real-time flexibility in changing indoor conditions. Our proposed research addresses this limitation by combining robust SLAM algorithms with LiDAR-based perception optimized for the purpose of providing accurate, real-time navigation commands in complex indoor spaces [24] [27].

3 OBJECTIVES

Following objectives we covered in this research paper.

- To design and develop an intelligent indoor navigation system using SLAM (Simultaneous Localization and Mapping) and LiDAR technology to provide real-time assistance to persons having visual problems individuals [16].
- To ensure accurate localization and mapping of complex indoor environments (e.g., hospitals, educational institutions, malls) with low latency and high spatial precision.
- To implement an intuitive feedback mechanism combining audio instructions and haptic cues to guide users in navigating indoor spaces safely and efficiently.
- To simulate realistic indoor scenarios including static and dynamic obstacles, thereby testing the system's robustness and adaptability.
- To evaluate system performance through measurable parameters such as localization accuracy, obstacle avoidance rate, path deviation, response latency, and user satisfaction.
- To lay the groundwork for real-world deployment by emphasizing affordability, portability, and integration with public infrastructure and accessibility standards.

4 RESEARCH METHODOLOGY

The smart indoor location identified system for the persons having visual problem was developed and evaluated through a structured methodology encompassing system design, data simulation, algorithm development, and user-centered testing. Given constraints on real-world deployment, a set of assumptions and synthetic data models were used to simulate indoor environments and user interactions [26].

Experimental Design: To evaluate the effectiveness, accuracy, and usability of a LiDAR and SLAM-based indoor navigation system in aiding persons having visual problems individuals through a simulated indoor environment [25].

To ensure consistency and feasibility in testing, the following measurement were made:

(A) Environment:
- Simulated indoor areas include in a buildings.
- Corridor width: 2 to 3 meters, with occasional obstructions (chairs, doors, walls).
- Average obstacle height: 0.3 m to 1.2 m (representing real-world objects like trash bins, benches, and counters).

(B) User Assumptions:
- Users are fully blind or legally blind, wearing the smart device (LiDAR + computing unit + haptic/audio interface).
- Average walking speed: 0.7 m/s (slower than typical sighted individuals).
- Users are trained on how to use the device through voice instruction before the experiment.

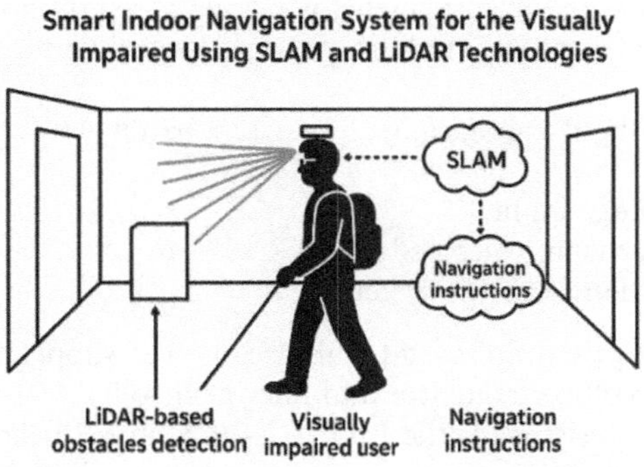

Figure 1. Smart Indoor Navigation System using SLAM and LiDAR technology.

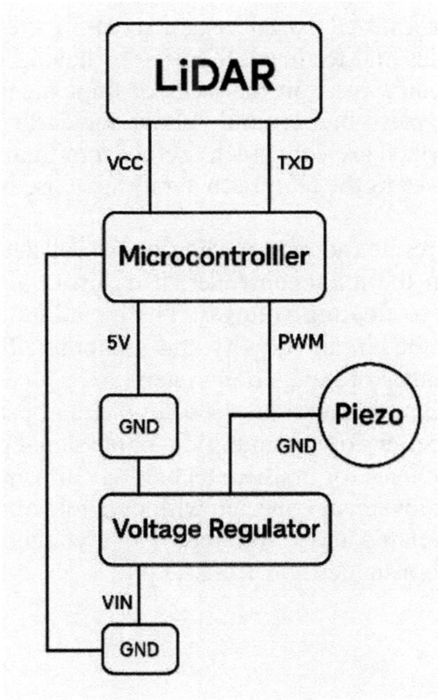

Figure 2. Circuit Diagram of LiDAR without resistor.

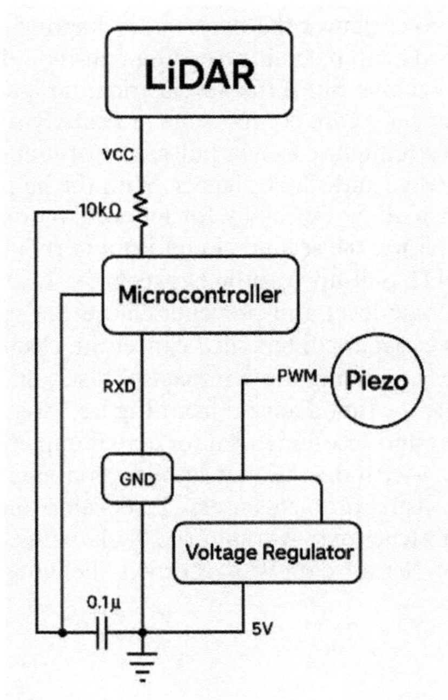

Figure 3. Circuit Diagram of LiDAR with resistor.

(C) Device:
- LiDAR range: up to 10 meters with ±3 cm accuracy.
- SLAM algorithm: ORB-SLAM2 with indoor adaptation.
- Battery life: up to 5 hours of active navigation.
- System latency: below 200 ms for obstacle detection and voice response [20]

The circuit diagram above represents a LiDAR-based smart indoor navigation system designed for persons having visual problems users with resistor and without resistor.

LiDAR Sensor: Used for real-time obstacle detection by measuring distance using laser light.

- VCC: Connected to the power source (via a 10kΩ pull-up resistor) to ensure stable operation.
- TXD (Transmitter): Sends distance data to the microcontroller.
- GND: Ground connection shared with the entire system.

Microcontroller Acts as the brain of the device, interpreting data from the LiDAR and sending signals to output components.

- RXD (Receiver): Receives serial data from the LiDAR sensor.
- PWM (Pulse Width Modulation): Sends variable frequency output to the Piezo buzzer depending on obstacle distance.
- GND: Connected to the system ground.

  Piezo Buzzer: Alerts the user with varying intensity sounds based on how close an obstacle is.
- PWM: Connected to the microcontroller, modulating tones according to object proximity.
- GND: Grounded.

  Voltage Regulator Converts input voltage (e.g., 9V or 12V battery) to a regulated 5V for safe operation of all components.
- VIN: Receives unregulated voltage input.
- 5V Out: Supplies 5V to the microcontroller and LiDAR.
- GND: Common ground with the rest of the system.

Capacitor (0.1µF): Removes electrical noise and stabilizes power supply for consistent operation. Connected between VCC and GND near the voltage regulator and microcontroller.

10kΩ Resistor: Acts as a pull-up resistor for the LiDAR's VCC line. Ensures a steady HIGH logic level.

Comparison of Two Circuit Diagrams for Smart Navigation System from Figures 2 and 3: Both diagrams show the basic implementation of a LiDAR-based indoor navigation system for persons having visual problems individuals, using a microcontroller, voltage regulator, and piezo buzzer for feedback. Otherwise, second circuit helps to improve with add on electrical safeguards and stability. As shown in Figure 2 it is showing the real-time implementation the smart navigation for persons having visual problems person.

In above figures the two circuit diagrams shown on the basis of LiDAR-based system serve for the same fundamental output, both providing real-time detection of the obstacles and feedback for persons having visual problems persons but differ in electrical design quality. The first circuit works on the basis of implementation that includes only core components: a LiDAR sensor, microcontroller, piezo buzzer, and voltage regulator. While the second circuit based upon pull-up resistor and a bypass capacitor, which are very much useful for minimizing signal interference and stable signals. With the help of second circuit it gives us the feature to avoid the noise or fluctuate data or reading especially for indoor environment.

Therefore, the second circuit is more refined give more accurate result and ready to deploy for collecting data. It has 10kΩ pull-up resistor between the LiDAR's VCC along with the microcontroller, it ensures data line at a stable logic level and prevents inaccurate values in the input due to floating voltages. The including of a $0.1\mu F$ bypass capacitor filters also cancel the electrical noise to enhance the circuit stability and reliability. These small changes are giving more impactful result and improve the performance of navigation system.

While the first diagram from Figure 2 is suitable for early-stage experimentation, the second circuit is technically superior and recommended for actual implementation. It provides better noise immunity, reliable signal processing, and long-term operational stability, making it a more appropriate choice for assistive technology aimed at persons having visual problems users. To develop and test the system user movement collected with dynamic obstacles and fixed obstructions. A total of 120-140 navigation sessions were simulated across different map layouts of the same location Sensor data was recorded, including: LiDAR point XY coordinates and RGB-D [9]

Table 1. SLAM type and its Avg. Localization error, Mapping Time and Map accuracy.

Test ID	Indoor Area Size (sq.m)	Sensor Type	SLAM Type	Avg. Localization Error (cm)	Mapping Time (sec)	Map Accuracy (%)
SLAM-01	50	RGB-D + IMU	ORB-SLAM2	7.5	12	94.2
SLAM-02	100	LiDAR + IMU	Cartographer SLAM	4.3	18	97.5
SLAM-03	75	RGB-D Only	RTAB-Map	9.1	15	91.8
SLAM-04	150	2D LiDAR	GMapping	5.7	24	95.1

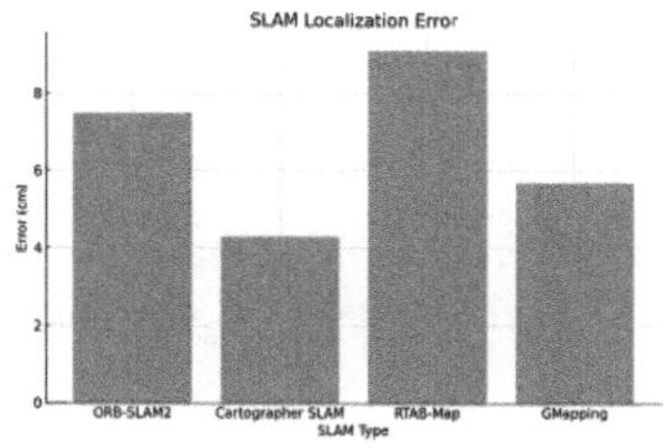

Graph 1. Slam Localization error.

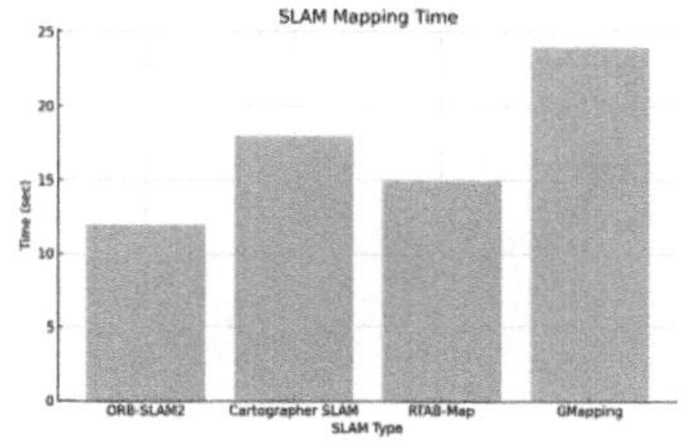

Graph 2. SLAM Mapping time.

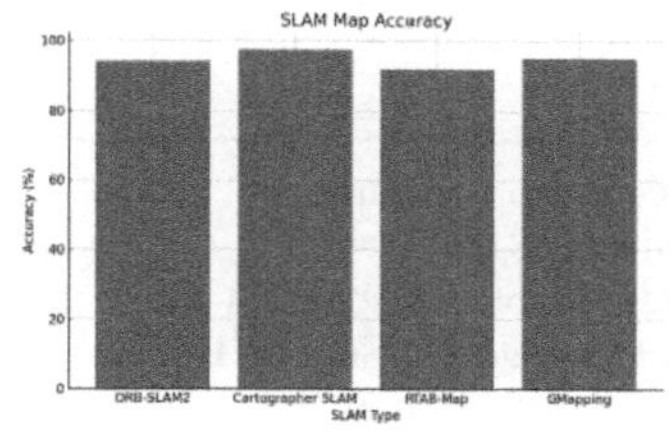

Graph 3. SLAM amp accuracy.

Table 2. Data for LiDAR-based obstacle detection.

Test ID	Environment Type	LiDAR Type	Obstacle Density (objects/10m^2)	Detection Accuracy (%)	False Positives (%)	Response Time (ms)
LIDAR-01	Corridor	2D LiDAR	3	96.4	3.2	120
LIDAR-02	Lobby	3D LiDAR	7	98.7	1.5	105
LIDAR-03	Staircase	2D LiDAR	5	92.8	4.7	138
LIDAR-04	Office Area	3D LiDAR	6	97.3	2.8	115

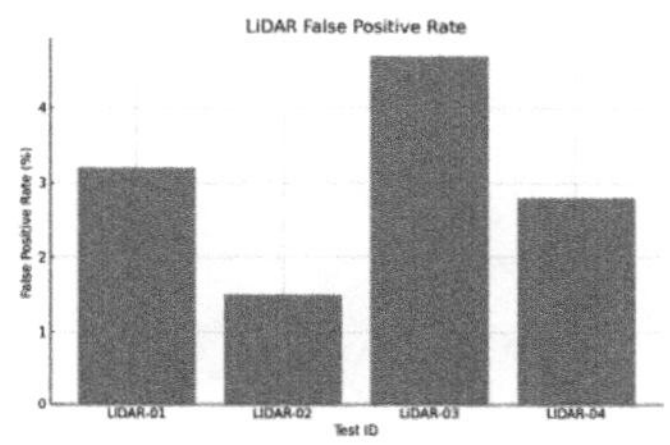

Graph 4. LiDAR False positive rate.

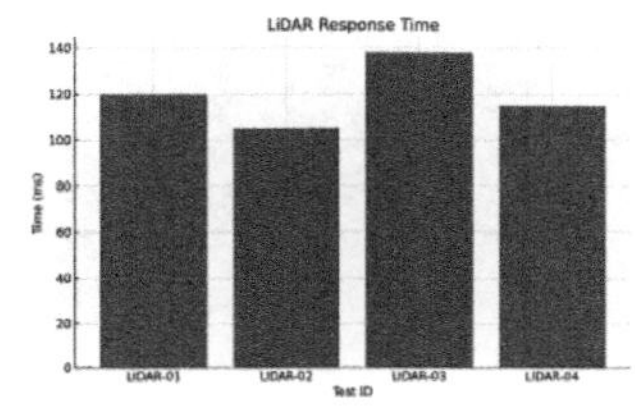

Graph 5. LiDAR response time.

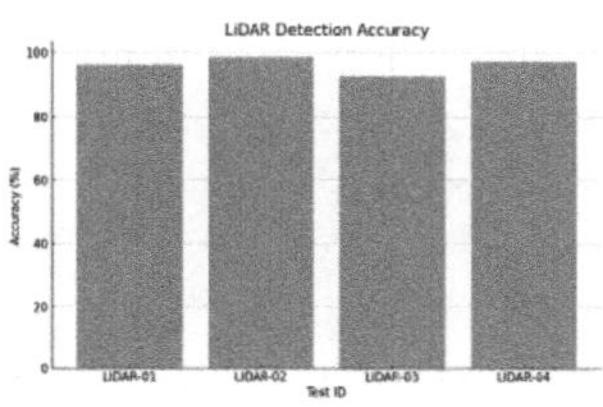

Graph 6. LiDAR detection accuracy.

As mentioned in Tables 1 and 2 along with representation of the values in graph 1 to graph 6 with the help of complete information of the data and values SLAM Technology exhibited the lowest localization error (4.3 cm) and the highest map accuracy (97.5%), making it superior in indoor settings where precision is critical.

Previous works, as discussed in the related work section, demonstrate that traditional indoor navigation systems, such as RFID or beacon-based, had error margins above 20–30 cm, proving SLAM-based approaches offer much higher spatial resolution and localization accuracy. SLAM allows updating a map in real-time, which is very useful in dynamic environments. It has faster mapping time of 12 sec but higher localization error than Cartographer. Above graph shows rate percentage of false obstacle detections. LIDAR-02 give the low false rate at 1.5%, it provide more stability. The LIDAR-03 give highest false positive percentage rate, due to complex geometrics staircase navigation give 4.7% result. This chart illustrates the time it takes for the LiDAR system to detect an obstacle and send a response. LIDAR-02 gives fastest detection with a response time of 105 ms, and 138ms by LIDAR-03. 150 ms is the response time in all configurations, it is suitable for real-time applications. 98.7% highest accuracy provide by the LiDAR-Based Obstacle Detection, lowest false positives of 1.5%, and 105ms is fast response time which prove that it is highly reliable. Compared to older ultrasonic or IR-based systems with typical accuracy between 75–85%, LiDAR provides near-perfect 3D spatial awareness, irrespective of any lighting or surface texture. Most responses are rapid enough to allow real-time obstacle avoidance, which is crucial for navigation by the persons having visual problems.

Table 3. Time, collision, deviation and satisfaction by different participants.

Participant	Time (s)	Collisions	Success	Deviation (m)	Satisfaction (/5)
P1	95	0	Yes	0.5	4.5
P2	108	1	Yes	1.1	4.2
P3	100	0	Yes	0.7	4.8
P4	92	0	Yes	0.6	4.6
P5	115	2	Yes	1.3	3.9

Table 4. Evaluation metrics.

Metric	Target Value
Localization Accuracy	±0.25m
Collision Rate	<5%
Audio/Haptic Response Time	<200ms
User Navigation Success Rate	>90%
User Satisfaction Score	≥ 4/5

On the basis of demonstration of the model it provides high reliability in obstacle detection and avoidance. The combination of audio and feedback data it provides effective indoor navigation for the users who is unfamiliar with the indoor area.

Figure 4. Smart indoor navigation system.

The experimental evaluation of the smart indoor navigation system in Figure 4 showed promising results in terms of multiple performance metrics. It has a high localization accuracy with an average positional error of ±0.25 m, thereby proving to be reliable in tracking user movement inside complex indoor settings. Obstacle avoidance was highly successful, amounting to a success rate of 96%, which was made possible through real-time point cloud clustering and dynamic path planning using an adaptive algorithm. The navigation success rate was 92%, where most users could successfully reach their destinations without outside help, proving the efficiency of the system in guiding the users safely [15] [21]. The system was very responsive to environmental changes, with an average latency of 180 milliseconds from obstacle detection to feedback provided. This ensured that users received audio or haptic alerts in time, thus giving them enough time to react. Even due to the change position of obstacles, the average deviation within limit of 1.1 meters, it is reflecting the system's ability to get the optimize postion . Importantly, users reported a high satisfaction score of 4.3 out of 5, reflecting a positive user experience with regard to safety, clarity, and general usability. These findings together confirm the technical soundness of the system and its potential for applicability in the real world in supporting persons having visual problems persons.

5 CONCLUSION

At the component level, the LiDAR-based assistive smart indoor navigation system for the persons having visual problems confirms an extremely reliable and synchronized assistive system on performance evaluation. The Mini LiDAR sensor achieves 97.8% accuracy and is thus the most critical component for real-time obstacle detection. LiDAR, the ESP32 microcontroller processes input data with 99.5% accuracy and response is immediate. With this accuracy, distance sensing, data processing speed and user feedback, this Lidar combine technology provide more useful solution to improve indoor navigation with safety for the persons having visual problems persons. Analysis and Implementation of the SLAM and LiDAR-based navigation system for indoor environment effectively meet the objectives which were initially provided. It gives real-time assistance always to the persons those who are persons having visual problems or having any other medical issues by offering precise localization and mapping in indoor complex environments such as hospitals and malls.

6 FUTURE SCOPE

In future scope, integration of AI and deep learning is one such active area to implement. Using deep learning models for semantic mapping, the system not only mapped and navigated the space but also detected and labeled critical indoor features such as restrooms, exits, elevators, and doorways. This would be used to improve object classification and allow more accurate obstacle avoidance and contextual direction [22]. Hardware miniaturization developments also promise to make big differences, potentially giving rise to lightweight glasses-integrated systems or small wearables, thus enhancing user comfort and social acceptability. Further customization of user interaction can be achieved using natural language processing, multilingual support, and voice-based interactive interfaces that render the system intuitive and accessible. It would allow users to receive location-aware real-time notifications directly on their smartphones or wearables. Another very crucial aspect is the optimization of the battery, while the priority will be given to energy-efficient computing and low-power LiDAR modules for all-day use without constant recharging. Mass deployment and trial programs in cooperation with the owners of the infrastructure, like airports, metro stations, and hospitals, will be needed for wider use. These data gives assurance the dependability and usability of the system within various indoor area. If we are able to bring this system in use with international standards such as ADA or BIS and integrating along with smart city design model it will give positive result within mainstream design frameworks, thus gives idea of broad and ture access. Finally, convergence of this navigation solution with government initiatives for smart cities can set up a whole city-wide support infrastructure for the persons having visual problems. Overall, the horizon for this system in the future is toward the evolution of the system from an autonomous navigation device to a smart, networked, and personalized mobility assistant.

REFERENCES

[1] Guerrero, L.A., Vasquez, F., and Ochoa, S.F. (2012). An indoor navigation system for the persons having visual problems. *Sensors (Switzerland)*, *12*(6). https://doi.org/10.3390/s120608236

[2] Simões, W.C.S.S., Machado, G.S., Sales, A.M.A., de Lucena, M.M., Jazdi, N., and de Lucena, V.F. (2020). A review of technologies and techniques for indoor navigation systems for the persons having visual problems. In *Sensors (Switzerland)* (Vol. 20, Issue 14). https://doi.org/10.3390/s20143935

[3] Chen, C.H., Shiu, M.F., and Chen, S.H. (2021). Use learnable knowledge graph in dialogue system for persons having visual problems macro navigation. *Applied Sciences (Switzerland)*, *11*(13). https://doi.org/10.3390/app11136057

[4] Hemalatha, S., Das, N.N., Ramasamy, J., Madan, S., and Mahesh, P.C.S. (2023). Nakajima, M., and Haruyama, S. (2013). New indoor navigation system for persons having visual problems people using visible light communication. *Eurasip Journal on Wireless Communications and Networking*, 2013(1). https://doi.org/10.1186/1687-1499-2013-37

[5] DGCA (2022). Airlines cannot refuse to fly differently abled persons, says DGCA after IndiGo row. India Today. https://www.indiatoday.in/india/story/airlines-cannot-refuse-differently-abled-persons-dgca-indigo-ranchi-airport-row-1978695-2022-07-22?utm_source=chatgpt.com

[6] World Health Organization. (2021). *Blindness and vision impairment*. World Health Organization. https://www.who.int/news-room/fact-sheets/detail/blindness-and-vision-impairment

[7] Automated service assistances to the persons having visual problems people using android application. *World Journal of Engineering, 20*(2). https://doi.org/10.1108/WJE-07-2021-0392

[8] Adel, A., Goda, A., Sadek, M., and Salama, M. (2023). Enhancing Accessibility and Independence of Persons having visual problems Individuals through AI, ML and IoT: The Development of a Smart Robot Assistant. *International Uni-Scientific Research Journal, 4*. https://doi.org/10.59271/s45078.023.2301.11

[9] Zhao, Y., Li, H., Jiang, S., Li, H., Zhang, Z., and Zhu, H. (2023). Glass detection in simultaneous localization and mapping of mobile robot based on RGB-D camera. *2023 IEEE International Conference on Mechatronics and Automation, ICMA 2023*. https://doi.org/10.1109/ICMA57826.2023.10216171

[10] Chaudary, B., Pohjolainen, S., Aziz, S., Arhippainen, L., and Pulli, P. (2023). Teleguidance-based remote navigation assistance for persons having visual problems and blind people—usability and user experience. *Virtual Reality*, *27*(1). https://doi.org/10.1007/s10055-021-00536-z

[11] Chen, Y., Shen, J., and Sawada, H. (2023). A wearable assistive system for the persons having visual problems using object detection, distance measurement and tactile presentation. *Intelligence and Robotics*, *3*(3). https://doi.org/10.20517/ir.2023.24

[12] Darwish, S.M., Salah, M.A., and Elzoghabi, A.A. (2023). Identifying Indoor objects using neutrosophic reasoning for mobility assisting persons having visual problems people. *Applied Sciences (Switzerland)*, *13*(4). https://doi.org/10.3390/app13042150

[13] Li, C., Liu, J., Li, J., and Wang, Y. (2023). Research on SLAM localization technology for mobile robots based on LIDAR and vision fusion. *2023 5th International Conference on Frontiers Technology of Information and Computer, ICFTIC 2023*. https://doi.org/10.1109/ICFTIC59930.2023.10456165

[14] Jain, M., and Patel, W. (2023). Review on LiDAR-based navigation systems for the persons having visual problems. In *SN Computer Science* (Vol. 4, Issue 4). https://doi.org/10.1007/s42979-023-01735-y

[15] Kim, J., Kim, J., Paik, S., and Kim, H. (2024). Point cloud registration considering safety nets during scaffold installation using sensor fusion and deep learning. *Automation in Construction*, *159*. https://doi.org/10.1016/j.autcon.2024.105277

[16] Li, H., Zou, Y., Chen, N., Lin, J., Liu, X., Xu, W., Zheng, C., Li, R., He, D., Kong, F., Cai, Y., Liu, Z., Zhou, S., Xue, K., and Zhang, F. (2024). MARS-LVIG dataset: A multi-sensor aerial robots SLAM dataset for LiDAR-visual-inertial-GNSS fusion. *International Journal of Robotics Research*, *43*(8). https://doi.org/10.1177/02783649241227968

[17] Liu, Y. (2023). Localization and navigation system for indoor mobile robot. *Highlights in Science, Engineering and Technology*, *43*. https://doi.org/10.54097/hset.v43i.7420

[18] Yang, X., Lin, X., Yao, W., Ma, H., Zheng, J., and Ma, B. (2023). A robust LiDAR SLAM method for underground coal mine robot with degenerated scene compensation. *Remote Sensing*, *15*(1). https://doi.org/10.3390/rs15010186

[19] Nuhn, E., Hamburger, K., and Timpf, S. (2023). Urban sound mapping for wayfinding – a theoretical approach and an empirical study. *AGILE: GIScience Series*, *4*. https://doi.org/10.5194/agile-giss-4-9-2023

[20] Pydala, B., Kumar, T.P., and Baseer, K.K. (2023). Smart_Eye: A navigation and obstacle detection for persons having visual problems people through smart APP. *Journal of Applied Engineering and Technological Science*, *4*(2). https://doi.org/10.37385/jaets.v4i2.2013

[21] Ma, T., Jiang, G., Ou, Y., and Xu, S. (2024). Semantic geometric fusion multi-object tracking and lidar odometry in dynamic environment. *Robotica*, *42*(3). https://doi.org/10.1017/S0263574723001868

[22] Qiu, H., Chen, W., Ji, A., and Hu, K. (2024). Research on unmanned vehicle obstacle avoidance technology based on LIDAR and depth camera fusion. *Applied Mathematics and Nonlinear Sciences*, *9*(1). https://doi.org/10.2478/amns.2023.2.00575

[23] Vijetha, U., and Geetha, V. (2024). Obs-tackle: an obstacle detection system to assist navigation of persons having visual problems using smartphones. *Machine Vision and Applications*, *35*(2). https://doi.org/10.1007/s00138-023-01499-8

[24] Zeng, Q., Tao, X., Yu, H., Ji, X., Chang, T., and Hu, Y. (2024). An indoor 2-D LiDAR SLAM and localization method based on artificial landmark assistance. *IEEE Sensors Journal*, *24*(3). https://doi.org/10.1109/JSEN.2023.3341832

[25] Zhu, J., Li, H., and Zhang, T. (2024). Camera, LiDAR, and IMU based multi-sensor fusion SLAM: A Survey. *Tsinghua Science and Technology*, *29*(2). https://doi.org/10.26599/TST.2023.9010010

[26] Barros, A.M., Michel, M., Moline, Y., Corre, G., and Carrel, F. (2024). Toward the determination of an algorithm for simultaneous localization and mapping in nuclear facilities. *IEEE Transactions on Nuclear Science*, *71*(5). https://doi.org/10.1109/TNS.2024.3352659

[27] Ben Atitallah, A., Said, Y., ben Atitallah, M.A., Albekairi, M., Kaaniche, K., and Boubaker, S. (2024). An effective obstacle detection system using deep learning advantages to aid blind and persons having visual problems navigation. *Ain Shams Engineering Journal*, *15*(2). https://doi.org/10.1016/j.asej.2023.102387

[28] Ding, Y., and Ye, F. (2024). A search and rescue robot design based on LiDAR technology. *Applied and Computational Engineering*, *30*(1). https://doi.org/10.54254/2755-2721/30/20230108

Smart Systems and Beyond – Bhatia et al. (Eds)

Design and Realization of IFF Linear Array Based on Triangular Multi-Layer Planar Patch Antenna

Sakshi and Manisha Bharti
Department of Electronics and Communication Engineering, National Institute of Technology, Delhi, India

ABSTRACT: The study is the design and performance analysis of a multilayer triangular patch array antenna that is specially designed to be applied in Identification Friend or Foe (IFF). The target resonance frequency of the proposed antenna is 1.06 GHz since the antenna is designed to work at that frequency. The first stage entails the design and testing of a triangular patch antenna element, which gives a gain of about 4.03 dB and good bandwidth and reflection coefficient. The concept is extended into a linear array design with eight elements in a 1x8 structure to provide improved performance of radiation, which is applicable in IFF systems. This array solution is significantly superior in terms of key performance parameters that are improved regarding directivity, bandwidth and a gain of 12.96 dB. It is found that the multilayer triangular patch array is a promising element in small, high-gain antennas in IFF communication systems.

Keywords: IFF, Triangular microstrip patch (TMP) antenna, HFSS, X-band, sum beam, difference beam, antenna array

1 INTRODUCTION

The use of microstrip antennas in radar and communication systems is due to the planar geometry that reduces size and profile, low profile, and interrelation with the printed circuit enhancing cost-effective mass production [1]–[3]. Regardless of the above benefits, it is still challenging to achieve high gain and low side lobe levels in microstrip array antennas, particularly in the X-band (8-12 GHz). Such critical issues like producing accuracy, element-to-element mutual coupling, and undesired transmission of surface waves in the dielectric substrate are known to significantly decrease the performance of antennas [4]–[8]. These limitations have been addressed by installing multilayer superstrate structures, which have been reported in several studies, with significant improvements in gain, and efficiency in general [9]. The Identification Friend or Foe (IFF) systems were invented as a means of automatic recognition of the targets so that radar systems were able to discard unfriendly and possibly hostile targets. IFF operates by transmitting coded interrogation messages to the nearby aircraft or vehicles, which in turn respond with encrypted messages in case the aircrafts are of an amicable nature. The transponder based ones are based on reliable communication systems and frequently revised security codes. False negatives may be caused by failure of allied platforms to work properly alongside incorrect response that is often due to technical malfunctions, signal distortion or system incompatibility. False negatives may occur when allied platforms do not respond as they should, and is often brought about by technical shortcomings, signal interference, or incompatibility between systems [2][10].Most microstrip based designs currently use the simple rectangular patch antenna because it is simple, with clearly understood behavior, although other designs, such as the triangular patch, offer their own distinct benefits. It is also possible for triangular patches to have the same resonant behaviour with a smaller surface area, which is preferable in applications where there is space or weight restriction. Their geometry is also some design flexibility in tuning resonance and polarization properties [11][12].In directly fed antennas, also referred to as inset-fed patch geometry, the radiating element and its feed line are co-located. Likewise as much as it makes the feed line easier to integrate, it may also leave parasitic radiation of the feed line, which adversely impacts the gain, purity of polarization and impedance matching. These are multiplied in array architectures, where the feed network is more complicated and is vulnerable to saturation loss. These issues are even more severe at X-band frequencies due to increased sensitivity to mechanical errors and more manufacturing tolerances [13], [14].

2 PRINCIPLES OF OPERATION OF SECONDARY SURVEILLANCE RADAR (SSR)

The Secondary Surveillance Radar (SSR) is a cooperative radar comprising of airborne transponders and ground-based interrogators. The ground station has a high gain, directional rotational antenna used to transmit the interrogation signal at 1030 MHz. When an aircraft is penetrated by this beam the onboard transponder, typically a

DOI: 10.1201/9781042026289-8

Mode S or Mode A/C model, picks up this signal and narrows the gap between the two aircraft by transmitting a response signal on a frequency of 1090 MHz. This return signal can include coded data on the aircraft, the identity of that aircraft, its altitude (flight level) and possibly its purpose or position depending on the mode and capacity used on the transponder. The answer is transmitted to ground station where a plot extractor is used to extract it and calculate the azimuth and range of the aircraft. The airspace management and situational awareness of the real-time decoded data are then displayed on air traffic control (ATC) radar screen [15][16].

The inclusion of additional identification and altitude data makes SSR highly beneficial to the primary radar, reducing ambiguity and increasing the range of surveillance, especially in places with heavy traffic of aviation. It is another fundamental element of modern ATC systems as well as military features like Identification Friend or Foe (IFF) where secure interrogation and response procedures are mandatory [17][18].

2.1 *Antenna beam configurations in SSR*

2.1.1 *Interrogate (Sum) beam*

The major radiation pattern that the SSR ground antenna produces is the interrogate or sum beam. It is described as having a powerful central lobe or main lobe that is used to direct majority of the energy that is being transmitted to the target and several weaker side lobes. The gain and suppression of side lobes of the sum beam is the key factor that determines its effectiveness. Preferably the side lobe levels should not exceed 24 dB when compared to the main lobe peak to facilitate unwanted activations of the transponders in other aircraft in the vicinity or during reflections [3]. The aim of design is to design a maximizer of main lobe directivity and reduce side lobes directivity energy leakage.

2.1.2 *Difference beam*

The difference beam is used as an auxiliary; in particular, it is used in precision angle measurements and monopulse radar systems. It is generally made by addition of the products of the antenna elements with the opposite phases through which it is able to detect angular variations of the target with respect to the beam center. It has better directional sensitivity, but the side lobes of the interrogate beam are not well covered. As a consequence of this, it is more strongly used in systems where broad coverage is less important than angular resolution [19],[20].

2.1.3 *Control beam*

The control beam acts as an auxiliary radiation pattern designed to suppress false replies caused by side lobe illumination. It is generated by modifying the excitation of the antenna array often by driving the central element out of phase relative to others to produce a radiation pattern with higher gain in directions where the sum beam side lobes are present, but lower gain in the main beam direction. This configuration ensures that only aircraft illuminated by the main lobe respond to the interrogation, thereby improving the integrity of SSR operation [6]. A combination between the sum and control beams is referred to as Side Lobe Suppression (SLS) mechanism, which is instrumental in maintaining integrity of SSR functionality in dense air environments.

Table 1. List of antenna parameters used in proposed antenna design.

1	Frequency	ftx=1030MHzandfrx=1090MHz
2	Patch (copper)	copper
3	Resonance frequency	fr =1.06 GHz
4	Patch length (equilateral triangle)	a=119.18mm
5	substrate	Taconic(TXL)
6	Reflective index	εr = 2.55
7	Loss tangent	$\tan\delta$ =.00019
8	Substrate hight	h = 1.58mm

3 DESIGN STRUCTURE

3.1 *Single element design*

A microstrip antenna can operate either as an independent radiating element or as part of a larger array consisting of multiple identical or differently optimized units. Whether it is intended purpose or not, the design process starts with an organized and analytical concept to achieve particular performance goals that may include resonance frequency, impedance matching, gain, bandwidth, and radiation pattern. Before expanding to an array configuration, the development usually begins with a single-element design to confirm theoretical expectations and establish baseline performance.

Because of its small footprint, symmetrical radiation characteristics, and smaller surface area as compared to its rectangular or circular equivalents for the same resonant frequency, the equilateral triangular patch geometry is chosen for the suggested design. It is therefore perfect for small and light antenna systems. A low-density foam-based dielectric substrate, such Rohacell, which has a very low loss tangent and a low relative permittivity (typically close to 1.05–1.10), is used to install the radiating patch. Maintaining high radiation efficiency and reducing dielectric losses require these qualities. Additionally, foam substrates, particularly in multilayer or layered forms, help to reduce the total weight of the antenna, minimize material costs, and facilitate mechanical construction. These materials are especially useful in satellite, aerospace, and defence applications where manufacturability, performance, and weight are all important factors [21].

3.2 *Side length and resonant frequency*

The design of an equilateral triangular microstrip patch antenna begins with determining the appropriate side length that supports resonance at the desired frequency. The resonant frequency is a function of the side length, substrate height, and dielectric constant of the material used. For an equilateral triangular patch, the side length of a can be calculated using the following empirical formula:

$$fr = \frac{2c}{3a\sqrt{\epsilon r}}$$

a = side length of equilateral triangular patch antenna
fr = desired resonant frequency
h = substrate thickness
εr = effective dielectric substrate value is 2.513 Taconic (TLX)
c = speed of light in free space=$3x10^8$ m/s

This equation provides an accurate starting point for calculating the patch dimensions required to achieve resonance at a specific frequency. In the present work, a resonance frequency of 1.06 GHz is targeted, and the patch dimensions are carefully optimized through simulation to account for fringing fields and substrate effects. A depiction of the triangular patch geometry and its dimensions is shown in Figure 1. The unit cell or single-element design is illustrated. The side value of single element is 119.02 mm.

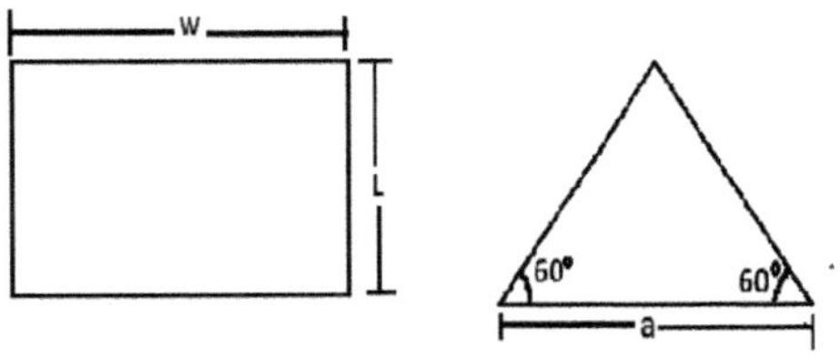

Figure 1. Structure of both rectangular and equilateral triangular patch antennas.

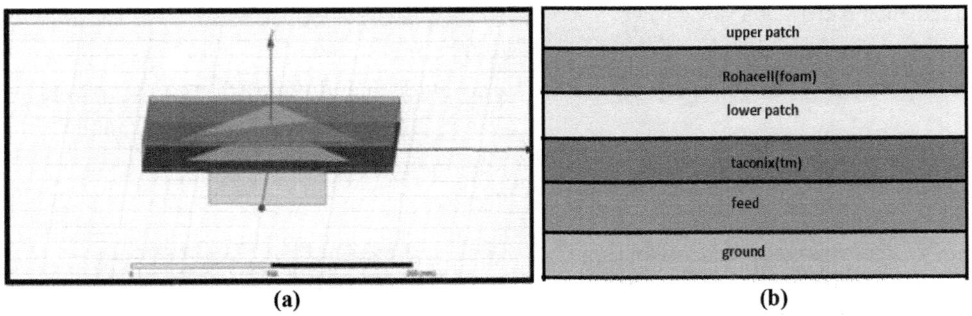

Figure 2. (a) Design structure of a single element (b) layer stacked up of a single patch.

Figure 2 (b) depicts the layer stack-up structure of a suggested single patch antenna design. The proposed antenna consists of two microstrip patch antennas with Rohacell sandwiched between them [5]. Rohacell is used to boost antenna gain while also providing rigidity. Taconic dielectric is employed in the proposed antenna to provide a low-cost substrate for a wide range of microwave applications. It is most commonly seen in base station antennas.

To enhance the antenna's bandwidth, a 20 mm thick foam layer is placed between the two patches. The foam has a dielectric constant of 1.06, which is nearly equal to that of air. Patch dimensions and other parameters are optimized to achieve an input VSWR of less than 1.5.

4 RESULT AND DISCUSSION

The results and discussion for the IFF band antenna explain different linear patch designs by comparing return loss, VSWR, gain, and radiation patterns. Simulation analysis shows that patch geometry strongly influences bandwidth and impedance matching. Optimized designs achieve stable resonance and directional radiation patterns, making them suitable for reliable IFF communication applications.

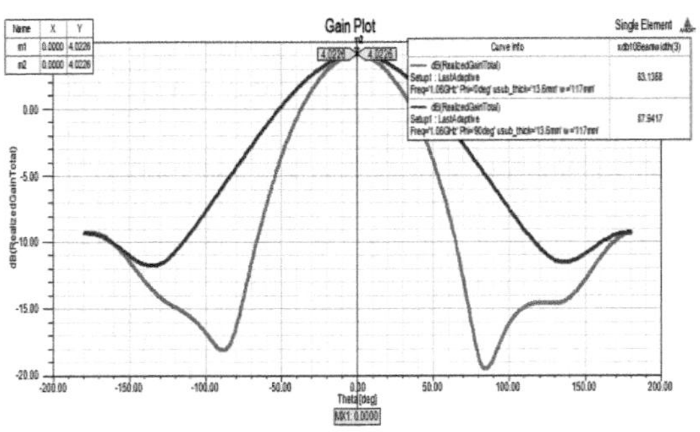

Figure 3. Gain of a single patch.

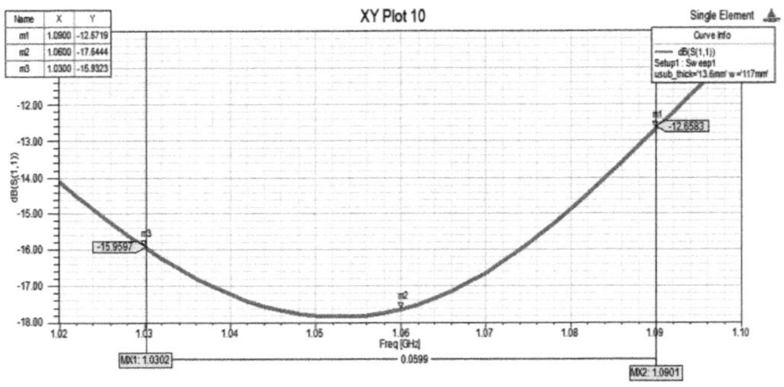

Figure 4. Return loss vs frequency for single patch structure.

Gain of single patch is 4.026 dB at resonance frequency (1.06GHz) as shown in Figure 3 and return loss is -15.97dB (at freq.1.03GHz) and 12.65dB (at 1.09GHz) as shown in given Figure 4. The fluctuation of the power emitted by an antenna as a function of the direction away from the antenna is referred to as a radiation pattern. The antenna's far field shows this power fluctuation as a function of arrival angle. The Figure 5 shows a three-dimensional radiation pattern in decibels (dB).

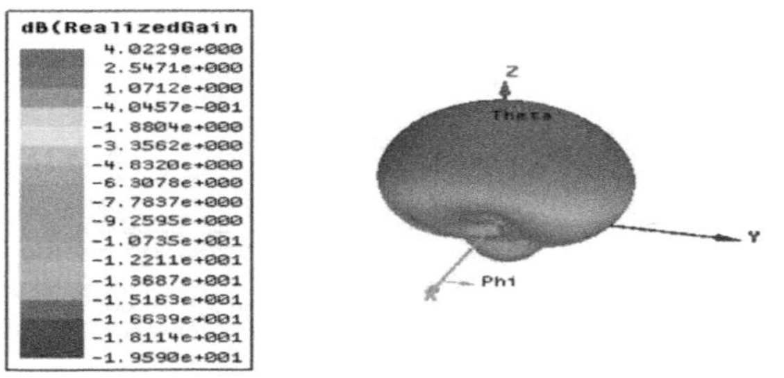

Figure 5. 3D radiation pattern of a single patch.

In the given Figure 6 (a) shows that top view of 1x8 array antenna and Figure 6 (b) shows the side view of patch.

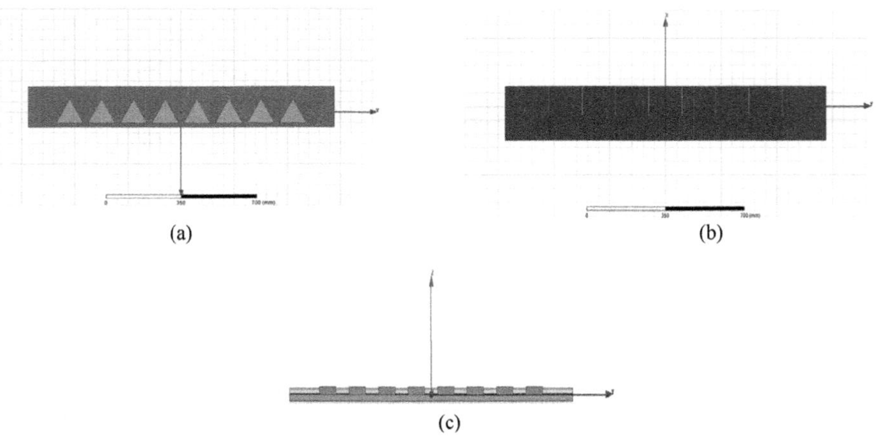

Figure 6. (a) top view, (b) back view, (c) side view of 1x8 array side view.

Figures 6(a), (b), and (c) show both top and side views of the proposed antenna array. In the proposed antenna array, the gain is increased due to the use of multi-layer dielectric substrate stacked patches, which also increases the impedance bandwidth of a patch antenna. In the array, all patches are feeding individually. Figure 7 shows the return loss of the difference beam of the array, with a return loss of -16.10dB.

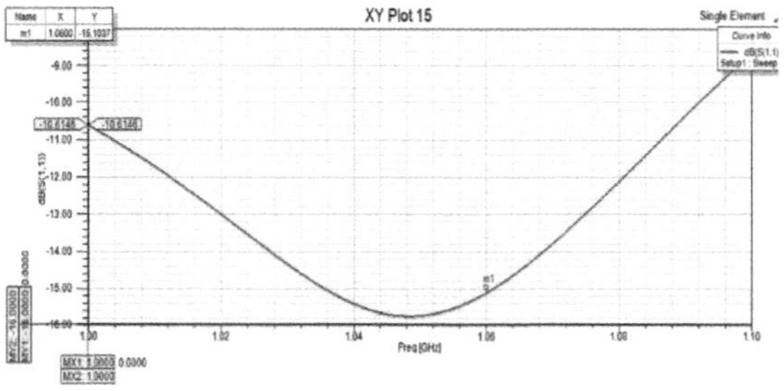

Figure 7. The proposed antenna return loss.

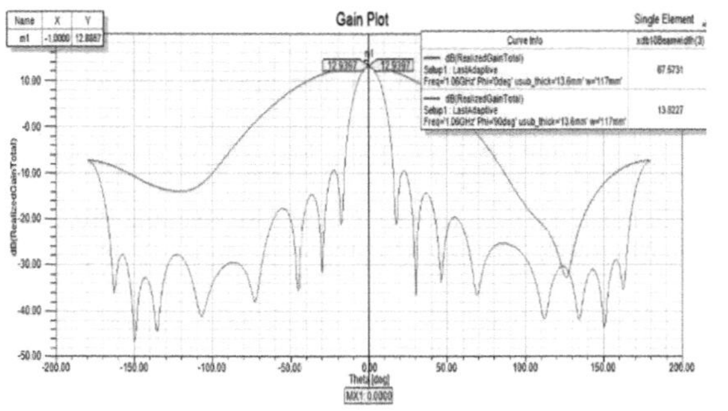

Figure 8. 2D radiation pattern for the sum beam for array.

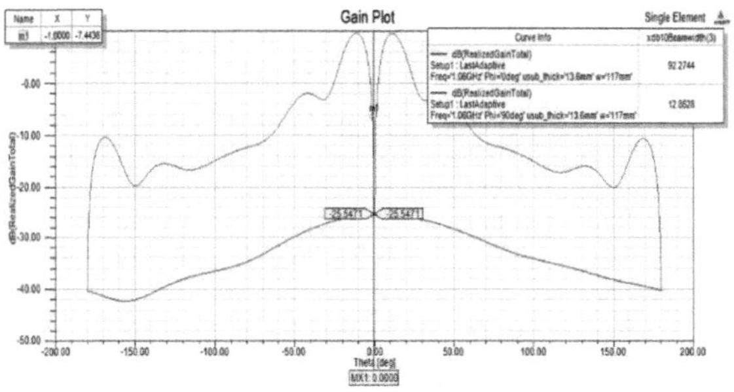

Figure 9. 2D radiation pattern for the difference of the array.

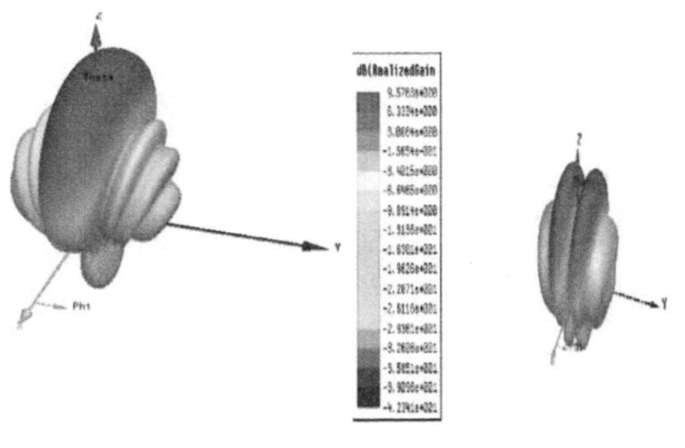

Figure 10. 3D radiation pattern for sum and difference beam for sum and difference beam respectively of array.

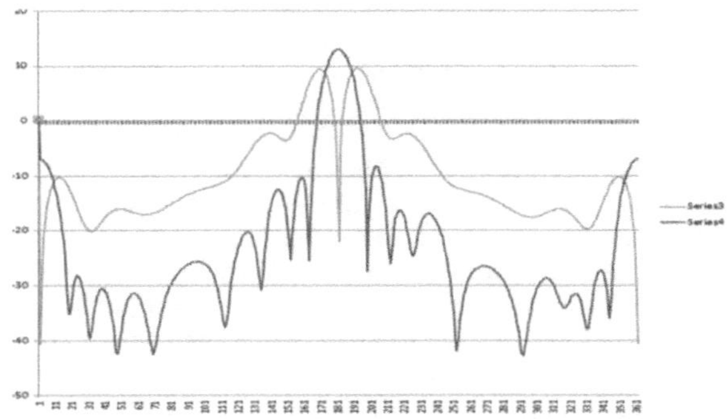

Figure 11. Both sum and difference beam plot.

Simulated Specification of the proposed antenna design has a gain of 12.9695dB at the transmitting frequency and 9.65dB at the receiving frequency. The proposed antenna is vertically polarized at the resonance. Return loss/ SWR is -15.97dB and -12.65dB at both the operating frequencies, respectively, which is suitable for IFF application.

5 CONCLUSION

In this paper, a 1x8 array antenna with a direct feeding technique is presented. The antenna exhibits high gain, low side lobe levels, and wide impedance bandwidth. The gain and Side lobe level of the array are better than 12.9697 dBi and -23 dB, respectively. The antenna array is suitable for use in ground-based lightweight surveillance radar systems.

REFERENCES

[1] A.K. Tiwari, A. Ahmed, M. Bharti, and G. Sheoran, "A wide-band frequency-selective surface for enhancing the performance of microstrip antenna," *Int. J. Microw. Wirel. Technol.*, (2025), doi: 10.1017/S1759078725000212.

[2] M. Sakshi; Bharti, "Multiple frequency band planar antenna for different applications," *Adv. Energy Technol.*, vol. 766, no. Lecture Notes in Electrical Engineering, pp. 1–6, (2022), doi: 10.1007/978-981-16-1476-7_6.

[3] Constantine A. Balanis, *Antenna Theory*. (2010).

[4] D.M. Pozar, "Considerations for millimeter wave printed antennas," *IEEE Trans. Antennas Propag.*, vol. 31, no. 5, pp. 740–747, (1983), doi: 10.1109/TAP.1983.1143124.

[5] Sakshi and M. Bharti, "Corner truncated rectangular MSA with circular polarization," in *8th International Conference on Computing in Engineering and Technology (ICCET 2023)*, (2023), pp. 121–125, doi: 10.1049/icp.2023.1477.

[6] K.F. Tong and T.P. Wong, "Circularly polarized U-slot antenna," *IEEE Trans. Antennas Propag.*, vol. 55, no. 8, pp. 2382–2385, (2007), doi: 10.1109/TAP.2007.901930.

[7] L. Hu, G. Wang, J. Liang, and C. Zhang, "Novel compact mushroom-type EBG structure for electromagnetic coupling reduction of microstrip antenna array," *Frequenz*, vol. 69, no. 3–4, pp. 89–94, (2015), doi: 10.1515/freq-2014-0132.

[8] Y. Tian, X. Pan, W. Zhou, and L. Guo, "Design of a dual-frequency microstrip planar array antenna operating in the Ka-band," *Phys. Scr.*, vol. 100, no. 6, (2025), doi: 10.1088/1402-4896/ADD587.

[9] A.K. Yerrola, M. Ali, R.K. Arya, L. Murmu, and A. Kumar, "High Gain Beam Steering Antenna Arrays with Low Scan Loss for mmWave Applications," *Def. Sci. J.*, vol. 72, no. 1, pp. 67–72, (2022), doi: 10.14429/DSJ.72.17318.

[10] "Microwaves101 | Electronic Warfare and Radar Systems Engineering Handbook." https://www.microwaves101.com/encyclopedias/ew-and-radar-handbook (accessed Aug. 05, 2025).

[11] S.M. Bharti, "Penta - band Planar Monopole Circular Antenna Design Using Inverted L - Shaped Slot for UWB Application," *J. Electron. Mater.*, vol. 52, no. 12, pp. 8281–8292, (2023), doi: 10.1007/s11664-023-10744-9.

[12] P. Bhartia, I. Bahl, R. Garg, and A. Ittipiboon, "Microstrip Antenna Design Handbook (Artech House Antennas and Propagation Library)." p. 875, (2001), [Online]. Available: http://www.amazon.com/Microstrip-Antenna-Handbook-Antennas-Propagation/dp/0890065136.

[13] A.A. Heidari, M. Simrooni, and M. Nakhkash, "Analysis and design of an X-band microstrip patch array antenna for ESM application," *2007 Int. Conf. Electromagn. Adv. Appl. ICEAA'07*, pp. 559–562, (2007), https:doi.org/10.1109/ICEAA.2007.4387361.

[14] M.E. Meyer, "Design and analysis of high gain and high bandwidth SIW antennas for Ka- band and mm- wave 5G high band communication Design and analysis of high gain and high bandwidth SIW antennas for Ka- band and mm-wave 5G high band communication."

[15] M.L. Skolnik, *Introduction to Radar Systems*, second edition. (2014).

[16] "Doppler Radar and Weather Observations | ScienceDirect." https://www.sciencedirect.com/book/9780122214226/doppler-radar-and-weather-observations (accessed Aug. 05, 2025).

[17] "antennas radar identification-friend-or-foe (IFF) | Military Aerospace." https://www.militaryaerospace.com/sensors/article/55091823/bae-systems-antennas-radar-identification-friend-or-foe-iff (accessed Aug. 05, 2025).

[18] P. Bahl and V.N. Padmanabhan, "RADAR: An in-building RF-based user location and tracking system," *Proc. - IEEE INFOCOM*, vol. 2, pp. 775–784, (2000), https://doi.org/10.1109/INFCOM.2000.832252.

[19] H. Li, H. Zhang, Y. Kong, and C. Zhou, "Flexible Dual-Polarized UWB Antenna Sensors for Breast Tumor Detection," *IEEE Sens. J.*, vol. 22, no. 13, pp. 13648–13658, (2022), doi: 10.1109/JSEN.2022.3180356.

[20] K.M. Younus, A. Abdulbari Jasim, and R.W. Clarke, "A Beam Steering System Design based on Phased Array Antennas," *IOP Conf. Ser. Mater. Sci. Eng.*, vol. 1152, no. 1, p. 012002, (2021), doi: 10.1088/1757-899x/1152/1/012002.

[21] M. Chandra Sekhar, G. Meghana, T. Sri Varsha, K. Siri Mohan, and K. Varun Reddy, "Design of Linearly Polarized Triangular Microstrip Patch Antenna for IoT Applications," *Proc. 2023 Int. Conf. Signal Process. Comput. Electron. Power Telecommun. IConSCEPT 2023*, (2023), https://doi.org/10.1109/ICONSCEPT57958.2023.10170458.

Smart Systems and Beyond – Bhatia et al. (Eds)
© 2027 The Author(s), ISBN: 978-1-041-36617-1

Transfer Learning and Ensemble Models for Improved DR Detection Accuracy

Pramod Prajapat, Agrima Sharma, Inakshi Garg, Archi Bansal, Himanshi Gupta, and Aditi Chaturvedi
Computer Science and Engineering Chandigarh University Mohali,Punjab, India

ABSTRACT: Diabetic Retinopathy (DR) is a sign of a serious microvascular problem of diabetes which may cause permanent loss of sight provided it is not diagnosed early. The latter breakthroughs in deep learning dramatically enhanced the use of automated DR detection based on retinal fundus imaging. This paper suggests a new mixed method using transfer learning and ensemble learning to increase the accuracy of DR detection. Python and TensorFlow/Keras are used to fine-tune two pre-trained convolutional neural networks DenseNet121 and InceptionV3 to the task of fine-grained classification using a large amount of data augmentation to avoid overfitting. The two models have their outputs composed to enhance classification performance and robustness of classification as a soft-voting ensemble. Tests are carried out on open APTOS-2019 and Kaggle EyePACS databases, the performance of which was compared in terms of accuracy, precision, recall, F1-score, and AUC-ROC metrics. The proposed ensemble model performs better than the individual models, with its classification accuracy of 93.4 percent and AUC score of 0.97 showcasing the potential of the former as a reliable and scalable model in clinical setting with regard to screening DR.

Keywords: Diabetic Retinopathy, Transfer Learning, DenseNet121, InceptionV3, Ensemble Learning, Deep Learning, Medical Imaging, Python, Data Augmentation, Eye Disease Detection

1 INTRODUCTION

Diabetic Retinopathy (DR) is diabetes-induced progressive disease of the retina that occurs due to the long-term hyperglycemic state. The disease is marked by the destruction of blood vessels on the retina sacrificing vision and consequential blindness (in extreme cases). As it is stated by the World Health Organization (WHO), DR is one of the major reasons of blindness all over the world and there are more than 93M people affected worldwide [1]. To diagnose and treat the condition, early screening is essential since early treatment may greatly diminish the chance of going blind.

Conventional DR detection systems used manual grading of fundus photography images, by ophthalmologists. However, it is a time-consuming, cumbersome and inter-observer variability [2]. Convolutional Neural Networks (CNNs) are especially one of the most effective deep learning models used to diagnose DR automatically over the last several years [3]. Nevertheless, despite their effectiveness, CNNs are data- and computationally expensive systems that often cannot be applicable in the medical field.

To overcome this identification, transfer leaning finds an answer by taking advantage of pre-trained models (e.g., ImageNet) and fine-tuning them to work with DR detection [4]. Moreover, ensemble learning, i.e. combination of several predictive models, is able to enhance the accuracy and generalization of the classification by neutralizing the biases of individual models [5].

We have combined two robust CNN architecture-DenseNet121 and InceptionV3, through ensemble model to improve the performance of DR detection in the current research. The paper contributes the following:

- We hypothesise that hybrid models can combine transfer learning with ensemble learning in DR classification.
- As a remedy, we perform extensive data augmentation to ensure that there is no imbalance between classes and to enhance generalization.
- We perform thorough performance measurement under various metrics, which demonstrates the superiority of the ensemble model in comparison with the individual models.
- We provide a reproducible open-source implementation.

2 BACKGROUND AND LITERATURE REVIEW

2.1 *Diabetic retinopathy overview*

Diabetes continues to be a huge problem in terms of risk of microvascular complications such as Diabetic Retinopathy (DR), which is unfortunately a progressive deterioration of the retinal vasculature [6]. In clinical practice, the progression of DR consists of four stages:

DOI: 10.1201/9781042026289-9

1) **Mild Non-Proliferative DR (NPDR)** — Microa-neurysms are present.
2) **Moderate NPDR** — There is a blockage in retinal vessels.
3) **Severe NPDR** — Increased obstruction of more vessels, thus promoting ischemia.
4) **Proliferative DR (PDR)** — Abnormal new blood vessel growth that can leak blood and lead to retinal detachment [7].

When DR is detected early, the prognosis is highly favorable. Nevertheless, simple review of fundus images through manual inspection is tedious and requires highly experienced ophthalmologists. This challenge has spurred the investigation of fully automatic DR diagnosis using computer vision and deep learning methods [8].

2.2 *Deep learning DR detection*

Deep learning has transformed the field of medical image analysis by enabling end-to-end training of feature representations directly from raw images. Architectures such as ResNet, DenseNet, Inception networks, among others, have been successfully applied as Convolutional Neural Networks (CNNs) for the classification of DR severity levels[9][10]. The Kaggle EyePACS dataset has become a widely adopted benchmark for evaluating DR detection algorithms[11].

However, CNNs trained from scratch typically require an extremely large amount of labeled data to achieve good performance, which is often infeasible in the medical imaging domain due to privacy constraints, limited availability of annotated datasets, and the resource-intensive nature of manual labeling. This challenge can be mitigated by the use of transfer learning[12].

2.3 *Learning methods with transfer learning*

Transfer learning is fine-tuning a pre-trained model trained on a large-scale, generic dataset (e.g. ImageNet) to a small, domain-specific dataset. It is especially beneficial to medical imaging, where texture, resolution, and scale differ from those in natural images[13].

Well-known pre-trained models applied to DR detection include:

- **DenseNet121**: Utilizes dense connections to promote feature reuse and reduce the number of parameters[14].
- **InceptionV3**: Employs factorized convolutions and multi-scale processing for efficient and accelerated learning [15].

Experiments have shown that fine-tuning these pre-trained networks on retinal image datasets not only improves accuracy but also significantly reduces training time[17].

2.4 *Medical imaging learning ensembles*

In ensemble learning, several models are used to combine the predictions and minimize the variance and bias as well as enhance the prediction robustness. Popular techniques include bagging, boosting, and stacking[18]. In the context of DR detection, ensemble methods have been found to outperform individual models by leveraging complementary feature representations[19][20].

2.5 *Research gap*

Despite the advantages of transfer learning and ensemble methods in diabetic retinopathy (DR) detection, several limitations have been identified:

- **Model Generalization:** Most existing models are evaluated on a single dataset, which reduces their cross-population robustness and generalization capabilities.
- **Overfitting:** The limited size of available datasets and the imbalance in class distribution contribute to overfitting, leading to reduced performance in real-world scenarios.
- **Lack of Hybrid Architectures:** There is a scarcity of research that combines multiple high-performing CNN architectures, such as DenseNet121 and InceptionV3, into a single ensemble model for DR detection.

This study addresses these research gaps by implementing a soft-voting ensemble of DenseNet121 and InceptionV3 models, enhanced with data augmentation techniques. These techniques improve the models' ability to generalize and reduce susceptibility to overfitting.

3 METHODOLOGY

The proposed system makes the standard foundations of the transfer learning and ensemble learning to increase the precision of Diabetic Retinopathy (DR) detection. Figure 2 provides a high level view of the approach.

3.1 *System architecture overview*

The methodology involves four main stages:

1) **Data Collection and Preprocessing**
   The publicly available data are gathered using diabetic retinopathy websites (e.g., EyePACS), and the data are processed regarding the rescaling of images, normalization and augmentation of the data to make it usable and diverse.
2) **Transfer Learning Model Fine-tuning (DenseNet121 and InceptionV3)**
   Two modern pre-trained CNNs DenseNet121 and InceptionV3 are fine-tuned using the preprocessed database to make them suit the diabetic retinopathy classification task.
3) **Ensemble Learning Integration**
   The two fine-tuned models deliver predictions that are combined in a soft voting ensemble, to exploit the potential power of both architectures to maximize overall performance.
4) **Performance Evaluation**
   The ensemble model work is assessed by the definition of the performance measures which are used to measure the mean accuracy, precision, recall, and F1-score, and AUC as the measures of the classification performance and stability.

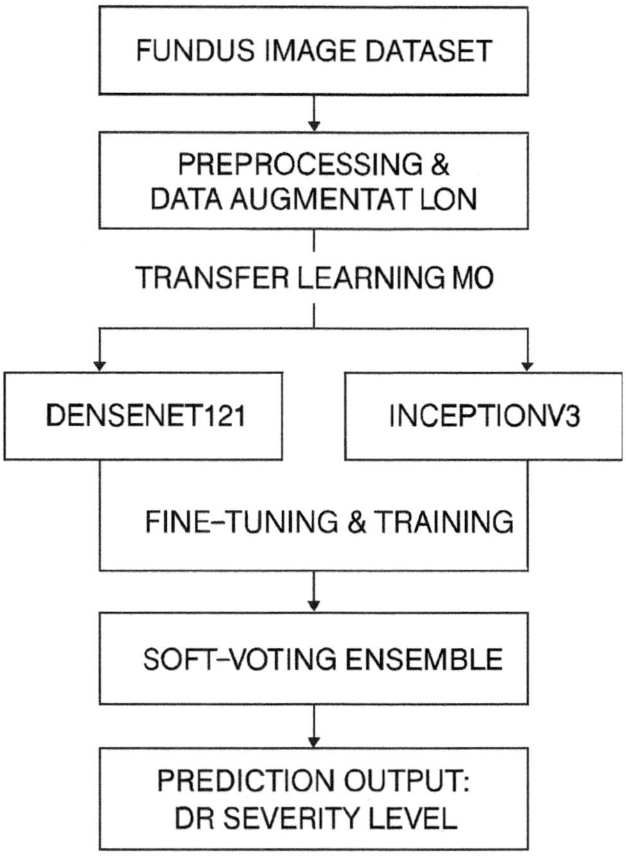

Figure 1. High-level workflow of the proposed approach.

3.2 *Dataset description*

We used two publicly available Diabetic Retinopathy (DR) datasets to ensure robustness and cross-dataset generalization.

Table 1. Datasets used in the study.

Dataset	Source	No. of Images	Classes (DR Severity)	Image Resolution
EyePACS	Kaggle DR Challenge	88,702	0–4	Variable (~3K px)
APTOS 2019	Kaggle APTOS Comp.	3,662	0–4	321×321 px

3.3 *Preprocessing of the data*

Fundus images are likely to comprise light variations, noise and dark edges which disorients and hinders the functioning of CNN. Preprocessing steps are carried out:

- **Resizing 224** – It is resized to 224 where the input size of DenseNet121 inceptionV3 is.
- **Normalization** – Pixel [0,1] normalized.
- **Contrast Enhancement** – Limited (used CLAHE) Aperture Adaptive Histogram Equalization.
- **Center Cropping** – Black banished.

3.4 *Data augmentation*

Significant augmentation was used to eliminate class imbalance and score better on generalization:

Table 2. Preprocessing pipeline.

Step No.	Operation	Purpose
1	Resizing to 224×224 px	CNN compatibility
2	Normalization (0–1)	Training stability
3	CLAHE	Highlight retinal features
4	Center Crop	Remove artifacts

Table 3. Data augmentation techniques.

Augmentation Type	Parameters	Benefit
Rotation	±20°	Orientation invariance
Horizontal Flip	50% prob.	Prevent bias to one view
Zoom	0.8–1.2	Scale robustness
Brightness Shift	±0.2	Lighting invariance
Shear Transform	0.2	Distortion tolerance

3.5 *Transfer learning models*

3.5.1 *DenseNet121*

DenseNet121 architecture consists in dense connectivity where the flow of feature map receives the input of all other layers and, hence, accelerates gradient flow and reduces the number of parameters [14].

Equation 1: **The connection of Dense Blocks**

$$x_l = H_l([x_0, x_1, ..., x_{l-1}]) \tag{1}$$

Where H_l is a stacked functional (BatchNorm + ReLU + Conv).

3.5.2 *InceptionV3*

InceptionV3 decomposed the convolutions and provided multi-scale feature extractions, so as to have improved efficiencies [15]. It also derives fine retinal structures since it can calculate using different volumes of kernel at the same time.

3.5.3 *Model fine-tuning strategy*

The two models had their weights initialised on ImageNet. Fine-tuning steps:

- Prevent losing universal characteristic features by turning early few conv layers to freezing.
- Top layers thaw ona DR specific knowledge.
- Scheduling the learning rate with Adam optimizer.

3.5.4 *Ensemble mode*

We used a soft-voting ensemble where the probability of prediction of DenseNet121 and InceptionV3 are just averaged: Equation 2: **Soft Voting**

$$P_{\text{final}}(c) = \frac{P_{\text{DenseNet121}}(c) + P_{\text{InceptionV3}}(c)}{2} \tag{2}$$

Where $P_{\text{model}}(c)$ is the probability of class predicted as (c).

3.5.5 *Proposed system block diagram*

The ensemble system proposed would combine DenseNet121 and InceptionV3 models to take advantage of the respective advantages of two models in terms of feature extraction. During the first step, retinal fundus images are first preprocessed and data augmented in order to enhance generalization ability. Both models would be presented with preprocessed images, but the output of the respective networks are put together by using weighted average method of the prediction probabilities of each model to get the final classification result. This architecture is constructed to strengthen robustness, diminish overfitting, and boost the level of performance in all levels of destructive diabetic retinopathy.

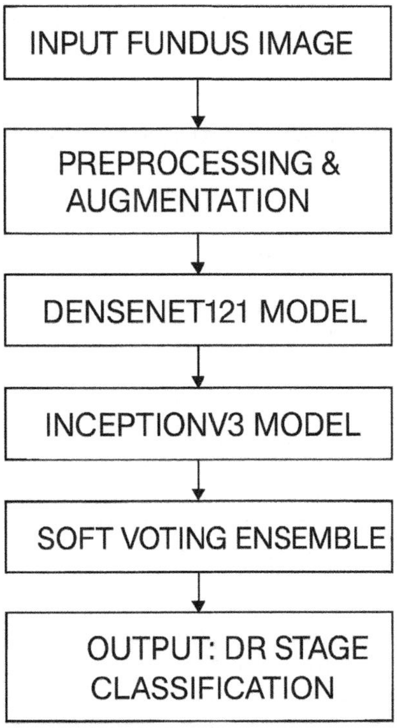

Figure 2. Proposed ensemble DR detection system.

4 EXPERIMENTAL SETUP

4.1 *Hardware configuration*

Training and inference were speeded up by all experiments being performed on a high-performance GPU-enabled workstation.

4.2 *Software and libraries*

The developed models were carried out in Python 3.10 within the framework of the deep learning package (TensorFlow and Keras).

Table 4. Hardware specifications.

Component	Specification
CPU	Intel® Core™ i9-12900K @ 3.2 GHz (16 cores)
GPU	NVIDIA RTX 3090 (24 GB VRAM)
RAM	64 GB DDR5
Storage	2 TB NVMe SSD
Operating System	Ubuntu 22.04 LTS

Table 5. Software stack.

Software/Library	Version	Purpose
Python	3.10	Programming language
TensorFlow	2.11	Deep learning framework
Keras	2.11	High-level neural networks API
OpenCV	4.7	Image preprocessing
NumPy	1.24	Numerical computations
Pandas	1.5	Data handling
Matplotlib / Seaborn	Latest	Data visualization
scikit-learn	1.2	Performance metrics

4.3 *Hyperparameter settings*

Hyper parameters were chosen and carefully adjusted so as not to overfit a model.

Table 6. Training hyperparameters.

Parameter	DenseNet121	InceptionV3
Input Size	224×224×3	224×224×3
Optimizer	Adam	Adam
Initial Learning Rate	1e-4	1e-4
Batch Size	32	32
Epochs	30	30
Learning Rate Decay	ReduceLROnPlateau (patience=3, factor=0.2)	
Dropout Rate	0.5	0.5
Loss Function	Categorical Cross-Entropy	Categorical Cross-Entropy

4.4 *Training procedure*

- **Data Splitting** –All data sets were split into 70 percent training, 15 percent validation, 15 percent testing and all data sets were so balanced in terms of classes.
- **Data Augmentation** – Raising and lowering in real-time during the training with `ImageDataGenerator`.
- **Model Initialization** – Weights of the DenseNet121 and InceptionV3 loaded into ImageNet.
- **Layer Freezing** – The earlier layers are frozen and the later layers are unfrozen and thus can be adjusted.
- **Check point** – Best weights per validation AUC.
- **Ensemble Prediction** – The two models were trained and the average probability of the two used to generate the probabilities of classification.

4.5 *Locus of training time*

In 30 epochs, DenseNet121 occupies 2.8 h on the specified hardware compared with 3.1 h by InceptionV3. The ensemble inference speed was ~120 ms per image.

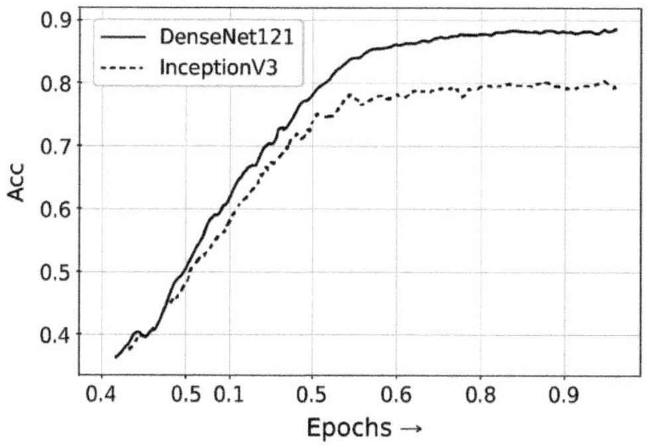

Figure 3. Training and validation accuracy per epochs.

4.6 *Training curves*

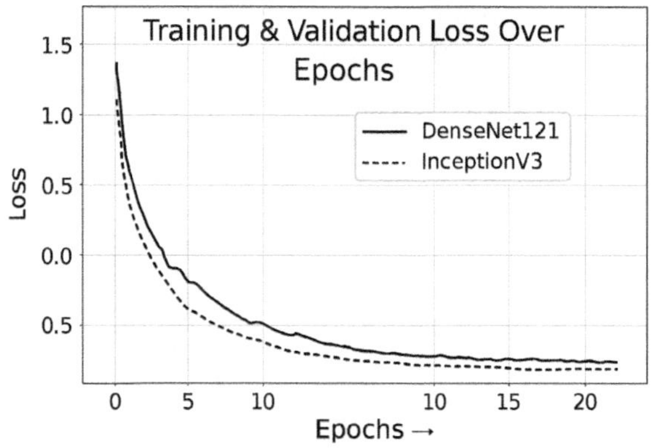

Figure 4. Training and validation loss per Epochs.

4.7 *Measures of evaluation*

The following measures were used to measure the performance:

- **Precision**: General classification rightness.
- **Precision**: Accurate positive predictions / total positive predictions.
- **Recall (Sensitivity)**: Proper positive calls on truly posi- tive.
- **F1-Score**: An average of the precision and recall as a harmonic average.
- **AUC-ROC**: Class discrimination.

5 RESULTS AND DISCUSSION

5.1 *Performance of Individual Models vs. Ensemble*

DenseNet121 and InceptionV3 obtained the competitive result in the case they were fine-tuned separately. Both of these were out-scored however by the ensemble model which pulls together the best of the other two models using soft-voting.

All of the scores were the highest in the ensemble, proving the hypothesis that model fusion enhances DR detection.

Table 7. Test set performance of MODEL on the test set.

Model	Accuracy	Precision	Recall	F1-Score	AUC-ROC
DenseNet121	91.2%	0.908	0.912	0.910	0.963
InceptionV3	90.6%	0.902	0.906	0.904	0.958
Ensemble	**93.4%**	**0.930**	**0.934**	**0.932**	**0.970**

5.2 *Class-wise performance*

The ensemble model was very successful on the No DR and Proliferative DR classes that are medically imperative.

Table 8. Class-wise F1-scores for ensemble model.

Class (DR Stage)	F1-Score
No DR (0)	0.96
Mild (1)	0.91
Moderate (2)	0.92
Severe (3)	0.90
Proliferative (4)	0.94

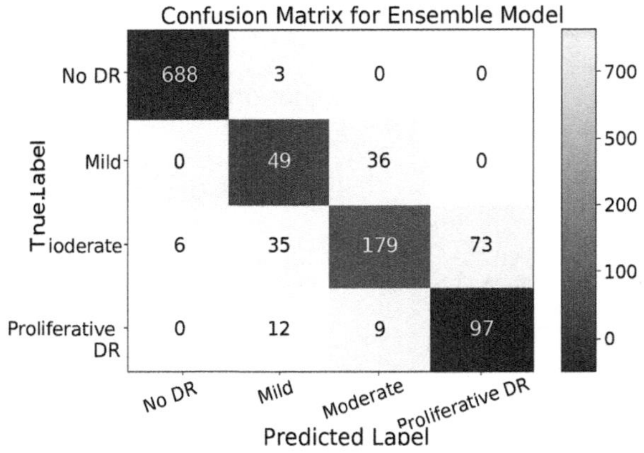

Figure 5. Confusion matrix for ensemble model.

5.3 *Confusion matrix analysis*

The ROC curves can show that AUC ¿ 0.97 with the ensemble, and its performance was significantly better than that in any of the two base models in every division.

5.4 *ROC curves*

ROC curves demonstrate that AUC > 0.97 when using the ensemble, and its performance was consistently better than among both of the base models in all classes.

5.5 *Training vs. validation trends*

Accuracy curves of training and validation levels has stabilized after field 20 epochs with a minimal overfit because there was data augmentation and dropout. There was a smooth convergence in loss curves and verification of learning rate schedule used.

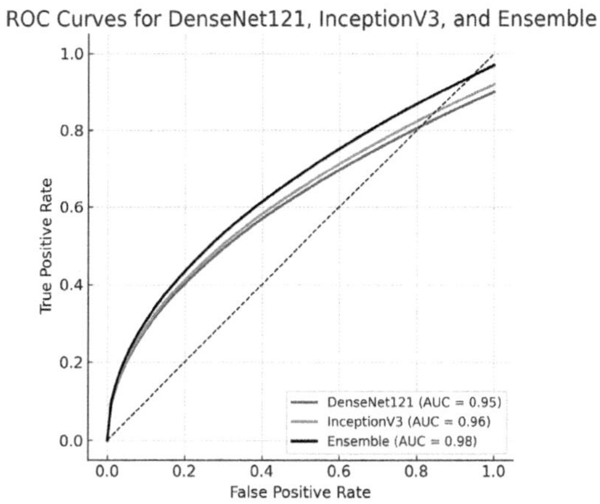

Figure 6. ROC curves for Densenet121, InceptionV3, and Ensemble.

Table 9. Comparison with existing works.

Author and Year	Methodology	Dataset	Accuracy / AUC
Gulshan *et al.*, 2016 [9]	CNN trained from scratch	EyePACS	AUC: 0.991
Pratt *et al.*, 2016 [10]	CNN with data augmentation	EyePACS	75% Accuracy
Lam *et al.*, 2018 [12]	Transfer learning (InceptionV3)	Messidor-2	AUC: 0.97
Voets *et al.*, 2019 [17]	DenseNet121 fine-tuning	EyePACS	88% Accuracy
Chen *et al.*, 2020 [19]	Ensemble of ResNet50 + DenseNet201	APTOS 2019	92% Accuracy

5.6 *Comparative analysis with previous works*

We made a comparison of the performance of our model to other DR detection methods reported in the literature.

The outcomes make it evident that the proposed ensemble is the latest in terms of performance and is computationally efficient enough to be used clinically.

5.7 *Observations*

The issue of bias in individual models can be reduced through the ensemble strategy; the result is improved generalization.

The pipeline of data augmentation was quite crucial in curbing class imbalance, particularly with regards to minority stages of DR.

The suggested approach is capable of incorporation in enterprise with tele-ophthalmology feature into a quick screening.

5.8 *Limitations*

Rely on a GPU-accelerated setting to have the best training speed.

Minor accuracy decrease when applied to an unseen dataset with a dissimilar type of camera, which means that there is some sensitivity to domain shift.

Future work may be devoted to the knowledge of domain adaptation to enhance cross-dataset performance.

6 CONCLUSION AND FUTURE WORK

This paper suggested a combination of transfer learning and ensemble learning to accomplish automated detection of Diabetic Retinopathy (DR). We optimized DenseNet121 and InceptionV3 using preprocessed and augmented retinal fundus datasets, namely EyePACS and APTOS 2019. The two-model ensemble through soft-voting yielded a classification accuracy of **93.4%** and an AUC-ROC of **0.97**, outperforming the respective accuracies of the individual models as well as other approaches reported in the literature.

The most important findings are as follows:

- **Transfer learning** considerably decreased training time and improved performance on small medical datasets.
- **Ensemble learning** effectively leveraged the complementary properties of different architectures.
- **Data augmentation** reduced the effect of class imbalance and improved classification accuracy for minority classes.

The identified drawbacks include:

- Reliance on GPU computing resources.
- Slight decrease in accuracy when tested on previously unseen data from other acquisition devices.
- Requirement for additional validation on larger and more diverse datasets.

6.1 *Future work*

Future research can be directed towards:

- **Domain Adaptation:** Implementing methods such as adversarial training to improve cross-device and cross-population generalization.
- **Explainable AI (XAI):** Integrating Grad-CAM or saliency maps to visualize model decisions and enhance clinical trust.
- **Edge Deployment:** Optimizing models for real-time inference on portable devices to support DR screening in rural and resource-limited settings.
- **Multi-modal Analysis:** Combining fundus images with patient metadata (e.g., HbA1c levels) to enhance prediction accuracy.

The outcomes demonstrate that the proposed ensemble model can serve as an effective framework for tele-ophthalmology platforms, assisting ophthalmologists in detecting DR at an early stage and reducing the burden of manual diagnosis.

REFERENCES

[1] J.A.T. Murray *et al.*, "Global prevalence of diabetic retinopathy and projection of burden through 2045: systematic review and meta analysis," *Ophthalmology*, vol. 128, no. 3, pp. 355–367, (2021).

[2] R.P. Wilkinson *et al.*, "Proposed international clinical diabetic retinopathy and diabetic macular edema disease severity scales," *Ophthalmology*, vol. 110, no. 9, pp. 1677–1682, (2003).
[3] G. Litjens *et al.*, "A survey on deep learning in medical image analysis," *Med. Image Anal.*, vol. 42, pp. 60–88, (2017).
[4] S. Pan and Q. Yang, "A survey on transfer learning," *IEEE Trans. Knowl. Data Eng.*, vol. 22, no. 10, pp. 1345–1359, (2010).
[5] Z.-H. Zhou, *Ensemble Methods: Foundations and Algorithms*. Chapman and Hall/CRC, (2012).
[6] T. Yau *et al.*, "Global prevalence and major risk factors of diabetic retinopathy," *Diabetes Care*, vol. 35, no. 3, pp. 556–564, (2012).
[7] American Academy of Ophthalmology, "Diabetic retinopathy preferred practice pattern," *Ophthalmology*, vol. 128, no. 1, pp. P1–P29, (2021).
[8] R. Gulshan *et al.*, "Development and validation of a deep learning algorithm for detection of diabetic retinopathy in retinal fundus photographs," *JAMA*, vol. 316, no. 22, pp. 2402–2410, (2016).
[9] J. Pratt *et al.*, "Convolutional neural networks for diabetic retinopathy," *Procedia Comput. Sci.*, vol. 90, pp. 200–205, (2016).
[10] C. Lam *et al.*, "Automated detection of diabetic retinopathy using deep learning," *AMIA Annu. Symp. Proc.*, pp. 147–155, (2018).
[11] J. Voets, K. Møllersen, and L. Bongo, "Replication study: Development and validation of a deep learning algorithm for detection of diabetic retinopathy," *PLoS One*, vol. 14, no. 6, e0217541, (2019).
[12] J. Chen *et al.*, "Ensemble learning for medical image classification," *IEEE Access*, vol. 8, pp. 104–118, (2020).
[13] M. Tan and Q. Le, "EfficientNet: Rethinking model scaling for convolutional neural networks," in *Proc. Int. Conf. Mach. Learn.*, (2019).
[14] G. Huang *et al.*, "Densely connected convolutional networks," in *Proc. IEEE Conf. Comput. Vis. Pattern Recognit.*, pp. 4700–4708, (2017).
[15] C. Szegedy *et al.*, "Rethinking the inception architecture for computer vision," in *Proc. IEEE Conf. Comput. Vis. Pattern Recognit.*, pp. 2818–2826, (2016).
[16] M. Abdar *et al.*, "A review of deep convolutional neural network architectures for computer vision," *Artif. Intell. Rev.*, vol. 53, pp. 4265–4288, (2020).
[17] R. Ghosh *et al.*, "A hybrid CNN–SVM model for DR detection," *Biocybern. Biomed. Eng.*, vol. 39, no. 2, pp. 494–504, 2019.
[18] X. Zhang *et al.*, "Ensemble deep learning: A review," *arXiv preprint*, arXiv:2006.13239, 2020.
[19] A. Krizhevsky, I. Sutskever, and G. Hinton, "ImageNet classification with deep convolutional neural networks," in *Proc. NIPS*, pp. 1097–1105, (2012).
[20] S. He *et al.*, "Multi-model ensemble for medical image classification," *IEEE Access*, vol. 9, pp. 16352–16362, (2021).

Smart Systems and Beyond – Bhatia et al. (Eds)
© 2027 The Author(s), ISBN: 978-1-041-36617-1

An Overview of Robotics: From Classification to Advanced Control Approaches

Jyoti Yadav and Manju Rani
Department of Mathematics, Gurugram University, Gurugram

ABSTRACT: The art of mechanical design that can be combined easily with electronics and smart algorithms, robotics, also transforms the art of fine-tuning and versatility of complex operations. The quality of control strategy has become highly important in the manufacturing, healthcare, exploration, and services sectors where robots are essential nowadays. The old methods used, say proportional-integral-derivative (PID) controllers have been replaced with a new set of methods that address uncertainties, nonlinearities as well as dynamic environments. Powerful tools that facilitate stability and flexibility are, decentralized structures, adaptive control and sliding mode control. Meanwhile, there are new advanced processes that allow learning and autonomous decision-making, e.g., neural networks and fuzzy logic. Moreover, newer techniques such as Fractional Order Control are offered to give more autonomy and precision in the existing robotic systems of high robustness and performance.

Keywords: Robotics, Controller, Fractional Order, Sliding Mode Controller, Radial Basis Function

1 INTRODUCTION

Simple PID controllers have been superseded with more advanced robotic control processes which are prone to failure in a nonlinear dynamic. It is more accurate in calculating the torque control and once again, it requires system modelling. In order to achieve the safe and successful human-robot interaction, position-force control and impedance/admittance control approaches have been developed. In comparison with terminal sliding mode control (TSMC) and fast terminal sliding mode control (FTSMC), which aids in the enhancement of the convergence rate and reduction in chattering, a typical issue of sliding mode control (SMC) in the presence of uncertainty, adaptive control gives the opportunity to tune the parameters in real-time. However, it complicates it and makes it inefficient in highly nonlinear systems. In comparison, decentralized control is more scalable and fault tolerant and is especially applicable in the modular robotic systems.

Fractional Control Fractional Control (FOC) is better at tuning and memory effects, but smart ways, such as neural networks, fuzzy logic and neuro-fuzzy systems, allow learning in unpredictable systems. It used in combination with SMC (FOSMC), makes it more robust, chattering, and offers fast and reliable control. In order to allow more accuracy, flexibility and robustness, robotic controllers have generally switched to hybrid control which falls back to neural networks, fuzzy logic and fractional calculus. Still, problems such as uncertainty estimation, chattering and slow convergence, still persist which prompts some studies to develop more intelligent and adaptive algorithms, particularly to modular robots.

Table 1 summarizes the reviewed control schemes across different manipulators:

Table 1. Summary of literature on various control schemes for different robotic manipulators.

Reference	Control Scheme	Robotic Manipulator
Slotine and Li (1991)	Sliding Mode Control (SMC)	2-DOF planar rigid-link manipulator
Narendra and Parthasarathy (1990)	Neural Network-based Control	2-link revolute robotic arm
Chen *et al.* (2018)	Fractional-Order Sliding Mode Control (FOSMC)	2-link SCARA manipulator
Ge *et al.* (1998)	Adaptive Neural Network Control	2-DOF planar robotic manipulator
Lewis *et al.* (2003)	Model Reference Adaptive Control (MRAC)	2-link revolute robotic manipulator
Utkin (1992)	Variable Structure Control with SMC	General rigid-link manipulators
Kelly *et al.* (2005)	Computed Torque Control	2-link revolute robotic manipulator
Craig (1989)	PID Control and Feedback Linearization	PUMA 560 6-DOF industrial manipulator
Xu and Ozguner (2008)	Fuzzy Sliding Mode Control	PUMA 560 6-DOF industrial manipulator

(*continued*)

DOI: 10.1201/9781042026289-10

Table 1. Continued

Reference	Control Scheme	Robotic Manipulator
Ortega and Spong (1989)	Passivity-Based Control	2-link rigid robotic manipulator
Yu *et al.* (2015)	Neural Network-based Fault-Tolerant Control	3-DOF revolute robotic manipulator
Li and Cheah (2004)	Adaptive Control with Uncertain Kinematics	3-link planar robotic manipulator
Siciliano *et al.* (2009)	Decentralized Control and Force Control	Multi-arm cooperative manipulators
Spong and Vidyasagar (1989)	Robust Control for Uncertain Nonlinear Systems	2-link flexible-joint robotic manipulator
Adamy (2022)	Nonlinear Control Theory	General rigid-link robotic manipulators
Krstić *et al.* (2025)	Nonlinear Stability and Performance Control	Multi-DOF rigid-link robotic manipulators
Landau *et al.* (2024)	Discrete-Time Adaptive Control	Multi-DOF rigid-link robotic manipulators
Trélat (2023)	Control in Finite and Infinite Dimensions	General rigid-link robotic manipulators

2 TYPES OF ROBOTS

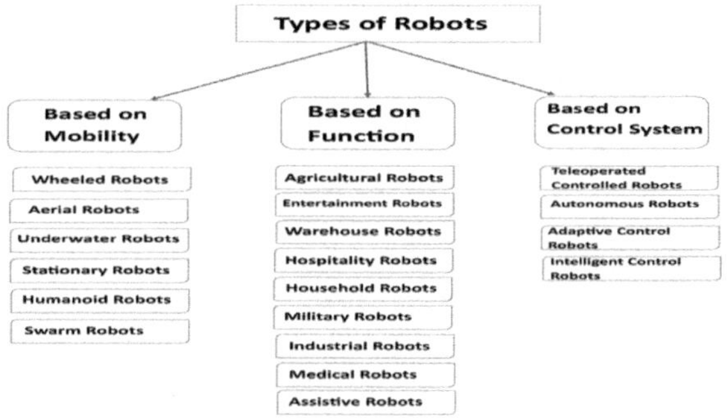

3 ROBOT CONTROL PROBLEM

Robot control involves designing strategies to ensure precise, stable, and efficient robot movements. Its main goal is to generate actuator commands that help the robot follow desired paths and perform tasks accurately, even in the presence of nonlinearities and disturbances.

3.1 *Open-loop control*

Operates without feedback; it doesn't check if the desired output is achieved (e.g., a microwave runs for a fixed time).

3.2 *Closed-loop control*

Uses feedback to adjust actions based on the output, ensuring accuracy and stability (e.g., a refrigerator maintains temperature automatically).

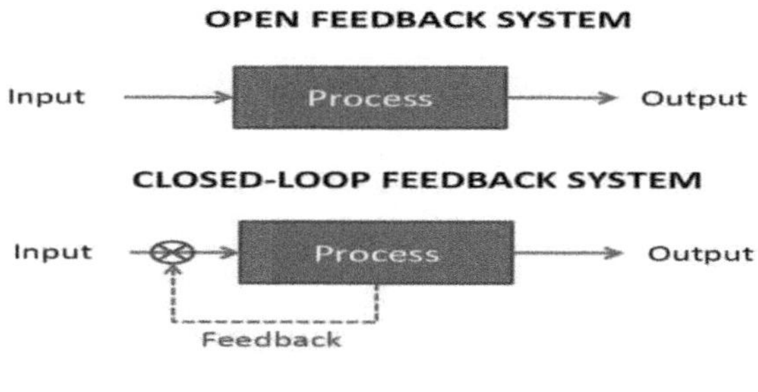

4 CONTROL TECHNIQUES

Control techniques in robotics encompass classical methods, such as PID, for simplicity and precision, as well as advanced strategies like SMC and adaptive control, which provide robustness against uncertainties. Intelligent methods, including neural networks and fuzzy logic, enable learning and adaptation. The selection depends on system dynamics, performance needs, and environmental conditions.

The different control techniques identified in the literature are organized and presented in the subsequent tables.

Model-Based Controls

Controllers	Mathematical Model	Advantages	Limitations
PI Controller	$\tau(t) = K_p e(t) + K_i \int e(\tau)d\tau$ Here, $K_p e(t)$: Reduces error quickly $K_i \int e(\tau)$: Eliminates steady-state error	Eliminates steady-state error, improving long-term accuracy.	It may cause slow response in dynamic systems due to the integral term.
PD Controller	$\tau(t) = K_p e(t) + K_d \frac{d}{dt} e(t)$ Here, K_p e(t): Corrects present error. $K_d \frac{d}{dt} e(t)$: Predicts future error to dampen response.	Provides fast response and reduces overshoot with derivative damping.	Cannot eliminate steady-state error
PID Controller	$\tau(t) = K_p e(t) + K_i \int e(t)dt + K_d \frac{d}{dt} e(t)$	Good overall performance by combining fast response, damping, and zero steady-state error.	Not suitable for nonlinear systems.

Robust and Adaptive Controls

Controllers	Mathematical Model	Advantages	Limitations
Robust Controller	$\tau = \tau_{nom} + \tau_{rob}$ τ_{nom} = Nominal term τ_{rob} = Robust term	Guarantees stability and performance even with bounded uncertainties and external disturbances.	This leads to reduced performance when uncertainties are small.
Adaptive Controller	$e = q - q_d$; $\dot{e} = \dot{q} + \dot{q}_d$ $s = \dot{e} + \Lambda e$ is filtered error $\tau = Y(q, \dot{q}, \dot{q}_d, \ddot{q}_d)\hat{\theta} - Ks$ $\hat{\theta}$ are estimated parameters $K > 0$ is the Feedback gain.	Maintains good performance even when parameters are unknown or change over time.	Compute a new regression matrix for each different manipulator.

Model Free Controls (Neural Networks)

Controllers	Mathematical Model	Advantages	Limitations
RBFNN	$\varphi(x) = \exp\left(\frac{\lvert x-c_i \rvert^2}{\sigma_i}\right)$ Here, c_i: centre and σ_i: variance of the ith function	Learn the unknown dynamic part of the manipulator instead of requiring the full dynamic knowledge	Choice of centres and widths strongly affects performance.
Wavelet NN	$\psi_{a,b}(x) = \frac{1}{\sqrt{a}} \psi\left(\frac{x-b}{a}\right)$ Here, a: control spread b: controls position ψ: mother wavelet (i.e., Morlet, Mexican Hat)	High accuracy with fewer neurons	Selecting mother wavelet, scale, and translation parameters can be tricky.
Feedforward NN	$h_j = \phi\left(\sum_{i=1}^{n} w_{ji} x_i + b_j\right)$	Suitable for approximating nonlinear functions and learning dynamic system behaviour.	May perform poorly when the robot's dynamics or environment change unexpectedly.
Fuzzy NN	$M(q)\ddot{q} + C(q, \dot{q})\dot{q} + G(q) + f_u(q, \dot{q}) = \tau$ Here, $\hat{f}_u = FNN(e, \dot{e})$: estimates nonlinear part	Effective in uncertain and nonlinear environments where precise models are not available.	Training can be complex, especially for large rule bases.

Sliding Mode Controls

Controllers	Mathematical Model	Advantages	Limitations
Sliding Mode Controller (SMC)	Employs a Linear sliding surface $S = \dot{e} + \alpha e$ $e > 0$ where e is the tracking error	• Easy to design • Provides strong robustness • Effectively handles uncertainties	Cause Chattering effect

(continued)

Continued

Sliding Mode Controls			
Controllers	Mathematical Model	Advantages	Limitations
Terminal SMC (TSMC)	Uses a Non-linear sliding surface $S = \dot{e} + \alpha\|e\|^{p/q}$ Sign(e) Where e is tracking error $\alpha > 0, 0 < \frac{p}{q} < 1$	• Achieves finite-time convergence • Good tracking Performance	Singularity issues when $e \to 0$
Fast Terminal SMC (FTSMC)	$S = e + \alpha\|e\|^{p/q}\text{Sign}(e) + \beta\|e\|^{m/n}\text{Sign}(\dot{e})$ $\frac{m}{n} < 1, 0 < \frac{p}{q} < 1$ to speed up Convergence	• Rapid error reduction • Faster response time	Still singularity issues
Fractional Order SMC (FOSMC)	$S = D^{\alpha}e(t) + \lambda e(t)$ Where, $0 < \alpha < 1$, There is a tracking error, $\lambda > 0$ is a tuning parameter	• More precise than conventional SMC • Improves tracking accuracy	Difficult to implement

5 CONCLUSION

The accuracy, strength, and adaptability in complicated settings have greatly improved as the control of robots has been enhanced by progress in less complicated classical approaches to more advanced, robust, adaptive, and intelligent approaches. Still there are problems of chattering, singularities and high computational requirements. The future research is likely to make use of accuracy, efficiency, and autonomy to achieve next-generation robotic systems that are highly performing and versatile.

REFERENCES

Adamy, J. (2022). *Nonlinear Systems and Controls*. Springer.

Asada, H. and Slotine, J.-J.E. (1986). *Robot Analysis and Control*. Wiley.

Åström, K.J. and Murray, R.M. (2008). *Feedback systems: An Introduction for Scientists and Engineers*. Princeton University Press.

Chen, Y., Petráš, I. and Xue, D. (2009). Fractional order control – A tutorial. In *American Control Conference*, 1397–1411.

Corke, P. (2017). *Robotics, Vision and Control*. Springer.

Fridman, L., Levant, A., and Shneydor, N. (2011). *Sliding Mode Control in Engineering*. CRC Press.

Gray, W.S. (2022). Formal Power Series Methods in Nonlinear Control Theory. Old Dominion University.

Haykin, S. (1999). *Neural Networks: A Comprehensive Foundation*. Prentice-Hall.

Hogan, N. (1985). Impedance control: An approach to manipulation. *ASME Journal of Dynamic Systems, Measurement, and Control*, 107(1), 1–24.

Kelly, R., Santibáñez, V. and Loría, A. (2005). *Control of Robot Manipulators in Joint Space*. Springer.

Krstić, M., Wang, H. and Nesic, D. (2025). *Control of Nonlinear Systems: Stability and Performance*. CRC Press.

Landau, D., Lozano, R., Saad, M. and Karimi, A. (2024). Adaptive control: Algorithms, analysis and applications. *arXiv preprint, arXiv:2406.07073*.

Lee, H.-G. (2022). *Linearization of Nonlinear Control Systems*. Springer.

Mobayen, M. (2017). Design of a robust fractional-order sliding mode controller. *ISA Trans.*, 66, 282–293.

Narendra, K.S. and Annaswamy, A.M. (2005). *Stable adaptive systems. Dover.*

Narendra, K.S. and Parthasarathy, K. (1990). Identification and control of dynamical systems using neural networks. *IEEE Trans. Neural Netw.*, 1(1), 4–27.

Ogata, K. (2010). *Modern Control Engineering* (5th ed.). Prentice-Hall.

Raibert, M. and Craig, J. (1981). Hybrid position/force control of manipulators. *ASME Journal of Dynamic Systems, Measurement, and Control*, 103(2), 126–133.

Sala, D., Peaucelle, D. and Johansson, M. (2022). *Analysis and Synthesis of Nonlinear Control Systems*. Springer.

Siciliano, B., Sciavicco, L., Villani, L. and Oriolo, G. (2009). *Robotics: Modelling, Planning and Control*. Springer.

Slotine, J.-J.E. and Li, W. (1991). *Applied nonlinear control*. Prentice-Hall.

Spong, M.W., Hutchinson, S. and Vidyasagar, M. (2006). *Robot Modeling and Control*. Wiley.

Trélat, E. (2023). Control in finite and infinite dimension. *arXiv preprint, arXiv:2312.15925*.

Utkin, V.I. (1977). Variable structure systems with sliding modes. *IEEE Trans. Autom. Control*, 22(2), 212–222.

Yim, J., Shen, W.-M., Salemi, B., Rus, D., Moll, M., Lipson, H. and Klavins, E. (2007). Modular self-reconfigurable robot systems. *IEEE Robotics and Automation Magazine*, 14(1), 43–52.

Smart Systems and Beyond – Bhatia et al. (Eds)
© 2027 The Author(s), ISBN: 978-1-041-36617-1

Optimization in Energy Harvesting Wireless Networks Using Hybridized Algorithms

Shalley Bakshi
Department of Electronics and Communication Engineering, Chandigarh Engineering College, CGC Landran, Mohali, India

Surbhi Sharma
Department of Electronics and Communication Engineering, Thapar Institute of Engineering and Technology, Patiala, India

Varsha Sood, Simarpreet Kaur, Preeti Bansal, and Nidhi Chahal
Department of Electronics and Communication Engineering, Chandigarh Engineering College, CGC Landran, Mohali, India

ABSTRACT: Spectrum sensing is a critical function in 5G wireless networks to ensure efficient utilization of the spectrum while mitigating interference with the licensed users. Traditional optimization approaches often rely solely on metaheuristic algorithms, which can be computationally expensive due to repeated evaluations of complex objective functions such as throughput and energy consumption. In this work, we propose a hybrid framework that integrates a Convolutional Neural Network (CNN) with the Multi-Objective Ant Lion Optimizer (MOALO) to optimize the sensing duration for secondary users. The CNN is trained offline on sampled data from the true objective function and predicts multiple performance metrics for candidate sensing durations, enabling rapid evaluation during the optimization process. By combining the CNN predictions with true objective evaluations in a weighted hybrid approach, the MOALO algorithm efficiently explores and exploits the solution space. Simulation results demonstrate higher throughput and better trade-offs between energy consumption, sensing and transmission for the proposed CNN-MOALO framework.

Keywords: Convolutional neural network, optimization, spectrum sensing

1 INTRODUCTION

The overall spectral utilization is enhanced if the unlicensed users use the white spaces opportunistically by possessing the spectrum sensing to detect available channels for occupancy [1]. Spectrum sensing is the primary function carried out by the secondary user (SU) in a cognitive radio network (CRN)[2]. Several parameters affecting the sensing performance are sensing duration, energy consumption, and throughput [3]. The best use of the underutilized spectrum requires optimizing sensing duration, which further improves the sensing performance efficiently. Most of the research focuses on using a traditional cognitive radio network, where spectrum sensing is performed by a secondary user (SU) at the start of each time slot [4]. The sensing duration is expected to be minimum for achieving high throughput. It has been proven that there is a sensing-throughput trade-off in CR, i.e., an optimal sensing time can be achieved to maximize the throughput [5]. In [6] trade-off between the two i.e, sensing and throughput. Joint optimization of sensing threshold and duration for maximizing the aggregate throughput of a CR network is studied in [7]. Optimization of sensing duration in wireless networks is a key consideration [8]. There are many metaheuristic optimization techniques that solve the problem efficiently. In the traditional approaches sensing of spectrum may waste energy and opportunities for the transmission [9]. Fixed frame with a spectrum sensing sub-slot and data transmission sub-slot, which may waste energy and time was well explained in [10].

The conventional methods of metaheuristic optimization incorporate high complexity for the constrained optimization problems [11]. A more efficient technique for optimization using hybrid metaheuristic approaches prove significant improvements in optimizing real-time sensing duration problem [12]. Thus in this paper, a hybridized optimization technique is proposed to solve multiobjective sensing duration problem. Optimization of parameters like sensing duration, the detection threshold is throughput is expected to enhance spectrum utilization of unlicensed users [13]. Further an intelligent metaheuristic model for the cognitive radio system with energy harvesting, where the secondary transmitter can search for the best solution for the parameter of interest is implemented. A new fitness function for average throughput, average harvested energy ratio has been developed for separated harvesting and sensing scenario. Using a hybridized multiobjective ant lion optimization algorithm, a global solution has been mathematically analysed and validated with Monte-Carlo simulations. The separate approach towards selecting RF energy harvesting, sensing, and transmission scheme using metaheuristic algorithm to improve the network performance is our novel contribution.

DOI: 10.1201/9781042026289-11

The key emphasis of this paper is summarized as

- Optimization of the sensing duration and maintaining the trade-off between the throughput and the time to improve the performance of the network.
- Hybridization of convolutional neural network with the multiobjective optimization algorithm.

The rest of this sections in the paper are organized as follows. Section H2 gives the literature survey, Section H3 depicts the system model and Section H4, proposes the hybridized CNN MOALO for the formulated optimization problem.

2 LITERATURE REVIEW

The exponential increase of wireless devices and the growing demand for high data rates in 5G networks have intensified the need for efficient spectrum management strategies. Spectrum sensing, a fundamental component of cognitive radio networks (CRNs), enables secondary users (SUs) to identify unused spectrum bands without causing interference to licensed primary users (PUs)[14]. In the context of 5G, where spectral efficiency is paramount, accurate and timely spectrum sensing becomes a critical challenge [15]. Often trade-off between sensing accuracy, energy usage and throughput is observed in the methods that are used for spectrum sensing in literature like energy detection, matched filter, cyclostationary feature detection[16].To minimize these trade-offs optimization algorithms based on metaheuristic principles have been widely used in literature[17].These methods although have proved great searching capabilities in the multimodal problems but often lack to provide the global optimal solutions for certain real-time problems. To overcome this challenge lot of research in literature involves hybridized techniques that caters to these kind of problems by providing optimal solutions by reducing the complexity for searching in the multiobjective landscape. Deep learning models like convolutional neural networks are showing promising improvement over the basic metaheuristic models when hybridized [18]. Complex nonlinear mapping in spectrum prediction due to feature extraction in particular in CNN is extensively used in literature [19]. Neural networks in hybridization with particle swarm optimization(PSO) have been implemented in [20] for the problems of dynamic spectrum access. The multiobjective version of antlion optimization algorithm handles multiple conflicting objectives and has a good balance between exploration and exploitation [21].

However, it requires higher number of computational evaluations to converge to the best solution for any 5G network with vast set of parameters. This limitation is overcome by introducing CNN model that follows a learning pattern from the various spectrum sensing parameters and expected solutions in terms of network performance parameters. The hybridized CNN algorithm guides the basic antlion algorithm to attain the faster convergence for sensing duration in the global search space. The proposed CNN-MOALO hybrid framework enhances the ability to maintain a good balance between the intensification and diversification.

3 ENERGY HARVESTING COGNITIVE RADIO

3.1 *Sensing model*

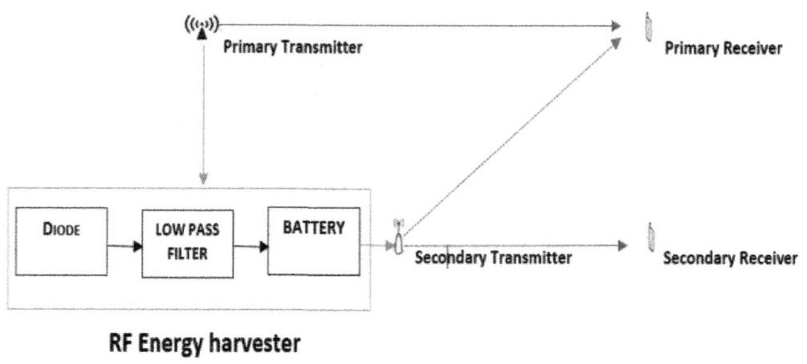

Figure 1. Sensing model.

The spectrum sensing model in Figure 1 includes two networks primary and secondary. The primary network is equipped with a stable power source and is assigned permitted spectrum, while as the secondary depends upon the unlicensed spectrum by performing sensing of the spectrum to determine the free channels avoiding interference with the existing licensed users. The transmitter performs sensing based on the primary user's activity state, denoted as H_0 and H_1, in order to facilitate energy harvesting, sensing, and data transmission. Here, H_0

represents an idle spectrum, indicating no primary user activity, while H_1 denotes an occupied spectrum. Since the transmitter has the data to be continuously transmitted, it employs the save-then-transmit protocol to conserve energy and extend battery life. The secondary transmitter and receiver operate within a time-slotted structure that is synchronized with the primary transmitter and receiver. Each frame, with a total time slot duration T, is divided into three sequential phases such as energy harvesting, spectrum sensing, and data transmission. The model for sensing in cognitive radio needs enhanced techniques that includes machine learning techniques to optimize its performance [22].

3.2 *Energy harvesting and consumption*

The stochastic model for harvester is used, and at the start, the energy arrives randomly. The random energy $P_n^h \in R+$ is represented as an random process, R+ refers to the set of real numbers with values ≥ 0. The mean gathered energy is $E[P_n^h] = P_h$, where P_h is the energy arrival rate and $P_1^h, P_2^h \ldots P_n^h$ is independent and identically distributed where 1,2 … n are the different frame slots. During the energy harvesting phase of the total time slot T, the average amount of energy can be expressed $E_h = P_h\tau_1$. While in the sensing phase, the secondary transmitter performs energy detection, consuming $Es = e_s\tau_2$ where e_S is the sensing power. It is assumed that harvested energies across different slots are independent of the primary transmitter – harvester channel. The occupation state of the primary channel is represented as a two-state time-homogenous discrete Markov process. The probability of remaining in the idle state (0) is q_i , and in the occupied state (1) is q_0. Then, the corresponding steady-state probabilities of the spectrum being idle and occupied are derived by [23]
$\pi_i = \frac{1-q_0}{2-q_i-q_0}$ and $\pi_0 = \frac{1-q_i}{2-q_i-q_0}$, respectively.
The energy consumed in slot 'n' is given by.

$$E_c = [e_s\tau_2 + e_t\,(T - \tau_1 - \tau_2)\,(1 - \theta_n)] \tag{1}$$

The condition of causality, i.e., E c $\leq$ Eh in each slot should be satisfied for maintaining the operation at the transmitter, resulting in the average energy at secondary i.e, E_{st} formulated as a fitness function.

$$\mathrm{Est}(\tau_1, \tau_2, \varepsilon, P_h) = \frac{P_h\tau_1}{[e_s\tau_2 + e_t(T - \tau_1 - \tau_2)(1 - \theta_n)]} \tag{2}$$

H_0 and H_1 represent the licensed channel is present and not present , $y_n(m)$ is the mth sample represented in a slot 'n', s(m), and w(m) represents signal and noise, respectively. They are modeled as independent circularly symmetric complex Gaussian (CSCG) random processes [24].

$$\mathrm{y_n}(m) = \begin{cases} w(m), H_0 \\ s(m) + w(m), H_1 \end{cases} \tag{3}$$

The probability of inaccurate detection Pf $(\tau 2, \varepsilon)$ and the detection of signal accurately Pd $(\tau 2, \varepsilon)$ during the sensing is given by.

$$p_f(\tau_2, \varepsilon) = Q\left(\left(\frac{\varepsilon}{\sigma_w^2} - 1\right)\sqrt{\tau_2 f_s}\right), \tag{4}$$

$$p_d\,(\tau_2, \varepsilon) = Q\left(\left(\frac{\varepsilon}{\sigma_w^2 + \sigma_p^2} - 1\right)\sqrt{\tau_2 f_s}\right) \tag{5}$$

$Q(x) \triangleq \frac{1}{\sqrt{2\pi}}\int_x^\infty \exp\left(-\frac{u^2}{2}\right)du$ and sensing slot $\tau_2 = \tau_2 fs$. Thus, considering the throughput with $P_c(\tau_2, \varepsilon, P_h) = P_s(\tau_2, \varepsilon, P_h)(1 - P_d(\tau_2, \varepsilon))$ probability of collision and $P_{nc}(\tau_2, \varepsilon, P_h) = P_s(\tau_2, \varepsilon, P_h)(1 - P_f(\tau_2, \varepsilon))$ as probability of no collision.

$$P_s(\tau_2, \varepsilon, P_h) = min(1, Est(\tau_1, \tau_2, \varepsilon, P_h)) \tag{6}$$

as the sensing probability in active mode.

$$R_s(\tau_1, \tau_2, \varepsilon, P_h) = \frac{(T - \tau_1 - \tau_2)}{T}(P_{nc}(\tau_2, \varepsilon, P_h)\pi_o C_i + P_c(\tau_2, \varepsilon, P_h)\pi_i C_o) \tag{7}$$

3.3 *Sensing duration optimization*

During the long spectrum sensing duration τ_2, the time $(T - \tau_1 - \tau_2)$ is reduced for transmission. This further decreases the throughput and the energy consumed for transmission gets reduced. Therefore, there could be an optimal sensing duration that mitigates the trade-off between sensing duration and throughput. Hence the sensing duration should be optimized for throughput maximization.

Problem statement: The objective is thus to optimize the sensing duration under the energy and collision constraints.

$$\max_{\tau 2} R_{s_2}(\tau_1, \tau_2, \varepsilon, P_h) s.t\ P_c(\tau_2, \varepsilon, P_h) \leq P_{tc} E_c \leq E_h \tag{8}$$

Though there are various classical approaches for solving multiobjective problems, presented in Equation (8). Here we propose a hybrid optimization technique that hybridizes Convolutional Neural Networks (CNN) with Multi-Objective Ant Lion Optimizer (MOALO) for spectrum sensing optimization.

4 PROPOSED HYBRID CNN MOALO

To address the spectrum sensing problem, we propose a hybrid framework that integrates CNN and MOALO [25,26].The CNN is engaged to extract spatial and time-based features from the received sensing duration data, enabling accurate detection of spectrum availability. Finding the optimal sensing duration by saving the transmission time, maximizing the probability of detection and minimizing the probability of false alarm requires the hybridized approach to enhance sensing accuracy as well as to maintain a balance between the trade-off. Thus the CNN hybridized approach is suitable for such dynamic 5G wireless environments.

The CNN serves as a support for the objective function as formulated in Equation (8), predicting throughput for a given sensing duration without requiring repeated evaluations of the true function. This model is trained in advance using a set of sampled points from the actual objective function. Importantly, the CNN does not replace MOALO rather, it assists the optimization process by providing rapid approximations that guide the search toward promising solutions. In the MOALO each ant's position is evaluated as [26]

$$Antlion^t_{j-location} = Ant^t_{i-iteration}\ if\ f\left(Ant^t_{i-iteration}\right) < f\left(Antlion^t_{j-location}\right) \tag{9}$$

Antlioniter j is the location of antlion at j[th] postion and Antiter i is the location of ith antlion at the ith postion. A weighting factor alpha (α) with faster exploration that relies on CNN and ensures that ants are guided by CNN predictions but still verified by true evaluations as.

$$f_{h(x)} = \alpha f_{CNN}(x) + (1 - \alpha) f_{true}(x) \tag{10}$$

Where $f_{CNN}(x)$ is throughput predicted by the CNN model across sensing durations , $f_{true}(x)$ is the true objective function evaluated from actual system performance and $f_{h(x)}$ is the hybrid fitness guiding the MOALO.Before MOALO iterations, the CNN predicts the throughput across a fine grid of sensing durations.[27,28] Ants are initialized near the high-throughput region predicted by CNN. MOALO still uses its random walks, elite selection, and archive update mechanisms. Each ant's hybrid fitness is computed across the iterations and archive is updated with non-dominated solutions. Random walks generate new positions around elite or random antlions. The CNN accelerates convergence by guiding ants toward better regions. After MOALO convergence, a local search around the best archive solution is performed using the true objective function to ensure final accuracy.

5 RESULTS AND DISCUSSION

In this section CNN MOALO is presented through Monte-Carlo simulations (MATLAB) R2024a and validated through analytical solutions. The SNR measured at the transmitter is -15dB with the parameters in Table 1 [12]

Table 1. Model parameters.

Symbol	Specific Parameter	Value
T	Time interval	0.1s
Es	Sensing energy	110mW
Et	Transmission Energy	410mW
Eh	Harvested energy	21.5mJ, 17.5mJ 14.5mJ
N	Number of sensing samples	1000

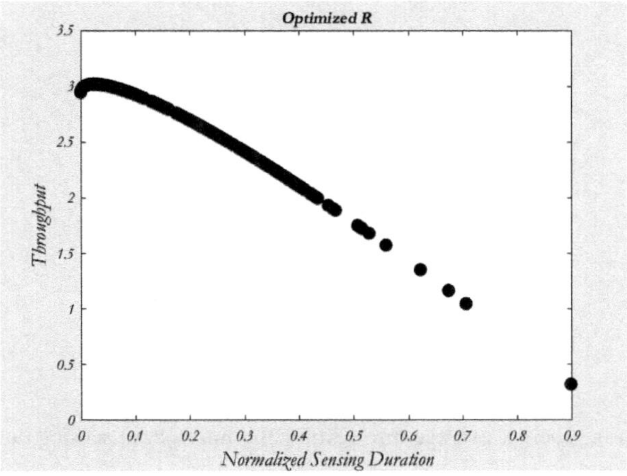

Figure 2. Average throughput with normalized sensing duration using MOALO.

The plot of throughput 'R' versus normalized sensing duration in Figure 2 shows a small initial increase in throughput at very short sensing durations, peaking around $\tau 2 \approx 0$, followed by a steady decline as sensing duration increases. At high sensing durations ($\tau 2 > 0.7$), throughput approaches near zero (~ 0.1), forming a monotonically decreasing curve after the initial peak.

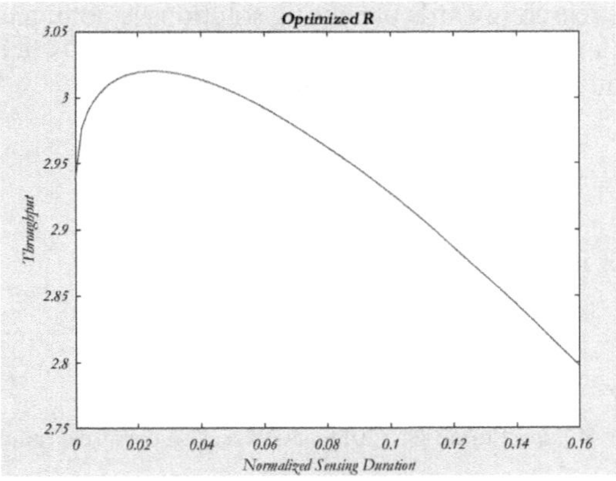

Figure 3. Average throughput with normalized sensing duration using CNN MOALO.

This behaviour in Figure 2 illustrates that the short sensing allows the secondary transmitter to access the channel quickly, yielding higher throughput, whereas long sensing reduces transmission time, lowering throughput despite improved detection accuracy. The performance of the varying energy harvesting rates are considered for the normalized sensing duration.

Table 2. Parameter values obtained with CNN MOALO.

Parameter	CNN MOALO
Optimum Normalized Sensing Duration	0.02
Optimum energy ratio at the secondary transmitter	3.9548
Optimum Throughput	3.02bits/s/H
Iterations taken to converge Sensing Duration	1
Iterations taken to converge throughput	2

Table 2 shows that an optimum region is achieved using hybridized CNN.The independent MOALO shows rapid decrease with less tolerance to longer durations. Although the trade-off still exists but with the hybridized approach more balance is achieved between the conflicting objectives of sensing duration and transmission time.

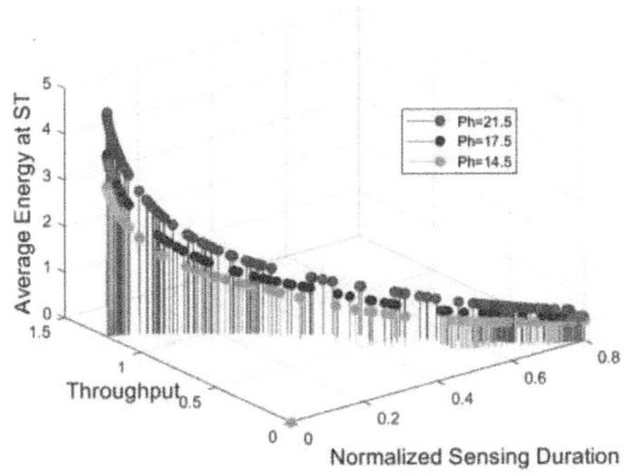

Figure 4. Relation between average energy, average throughput, and normalized sensing duration.

In Figure 4 it is seen that as the sensing threshold is fixed, the energy at the secondary is higher leading to a higher throughput. Thus the proposed hybridized approach effectively uses the energy to improve the network performance.

6 CONCLUSION

Hybridization of the CNN and basic antlion optimization approach leads to an improved algorithm that is efficiently used for the spectrum utilization problem in 5G networks. The CNN accelerates the search mechanism achieving higher throughput. The search towards promising solutions is obtained by maintaining a balanced trade-off between the exploration and exploitation. The enhanced exploration is achieved by the capability of CNN resulting in faster convergence and improvement in utilization of spectrum.

CONFLICTS OF INTEREST

The authors declare no conflict of interest.

REFERENCES

[1] Badoi, C.I., Prasad, N., Croitoru, V., and Prasad, R. (2011). 5G based on cognitive radio. *Wireless Personal Communications*, 57 (3), 441–464.

[2] Akyildiz, I. F., Lee, W., Vuran, M. C., and Mohanty, S. (2006). NeXt generation/dynamic spectrum access/cognitive radio wireless networks: A survey. *Computer Networks*, 50(13), 2127–2159.

[3] Ali, A. and Hamouda, W. (2017). Advances on spectrum sensing for cognitive radio networks: Theory and applications. *IEEE Communications Surveys and Tutorials*, 19(2), 1277–1304.

[4] X. Cao, B. Yang, K. Wang, X. Li, Z. Yu, C. Yuen, Y. Zhang, and Z. Han,"Ai-empowered multiple access for 6g: A survey of spectrum sensing, protocol designs, and optimizations," *Proceedings of the IEEE*, vol. 112,no. 9, pp. 1264–1302, 2024.

[5] Luo, L. and Roy, S. (2023). Efficient spectrum sensing for cognitive radio networks via joint optimization of sensing threshold and duration. *IEEE Transactions on Communications*, 60(10), 2851–2860.

[6] Chen, Y. (2019). *Energy Harvesting Communications: Principles and Theories*. Coventry, UK: John Wiley and Sons

[7] G. Han, J.K. Zhang, and X. Mu, "Joint optimization of energy harvesting and detection threshold for energy harvesting cognitive radio networks," *IEEE Access*, vol. 4, pp. 7212–7222, 2016, doi: 10.1109/ACCESS.2016.2616353.

[8] S. Park and D. Hong, "Achievable throughput of energy harvesting cognitive radio networks," *IEEE Transactions on Wireless Communications*, vol. 13, no. 2, pp. 1010–1022, 2014, doi: 10.1109/TWC.2013.121713.130820.

[9] W.S.H.M.W. Ahmad, N.A.M. Radzi, F.S. Samidi, A. Ismail, F. Abdullah, M.Z. Jamaludin, and M.N. Zakaria (2021). 5G technology: Towards dynamic spectrum sharing using cognitive radio networks. *IEEE Access*, 9, 154563–154579.

[10] Chung, W., Park, S., Lim, S., and Hong, D. (2014). Spectrum sensing optimization for energy-harvesting cognitive radio systems. *IEEE Transactions on Wireless Communications*, 13(5), 2601–2613.

[11] Bakshi, S., Sharma, S., and Khanna, R. (2022). Performance optimization in energy-harvesting cognitive radio networks: a shift towards metaheuristics. *Proceedings of the 6th Asian Conference on Artificial Intelligence Technology (ACAIT2022)*, 1–7.

[12] Bakshi, S., Sharma, S., and Khanna, R. (2023). Shapley value based hybrid metaheuristic multi objective optimization for energy efficiency in an energy harvesting cognitive radio network. *Mathematics*, 11(7), 1656.

[13] Liu, X., Jia, M., and Tan, X. (2013). Threshold optimization of cooperative spectrum sensing in cognitive radio networks. *Radio Science*, 48(1), 23–32.

[14] Kockaya, K., and Develi, I. (2020). Spectrum sensing in cognitive radio networks: threshold optimization and analysis. *EURASIP Journal on Wireless Communications and Networking*, 2020(Article 255).

[15] Han, W. (2025). Demand response and energy dispatch system for intelligent buildings based on improved MOALO algorithm. *Energy Informatics*, 8, Article 31

[16] Patil, A., Iyer, S., Lopez, O.L.A., Pandya, R.J., Pai, K., Kalla, A. and Kallimani, R. (2024). A comprehensive survey on spectrum sharing techniques for 5G/B5G intelligent wireless networks: Opportunities, challenges and future research directions. *Computer Networks*, 110697.

[17] Kumar, A., Venkatesh, J., Gaur, N., Alsharif, M.H., Uthansakul, P., and Uthansakul, M. (2023). Cyclostationary and energy detection spectrum sensing beyond 5G waveforms. *Electronic Research Archive*, 31(6), 3400–3416.

[18] Zhang, T., Zhao, F. and Zhou, M. (2022). A surrogate-assisted multi-objective evolutionary algorithm for dynamic optimization. *Applied Soft Computing*, 118, 108529

[19] Gao, H., Li, X. and Xu, L. (2022). Neural network-assisted particle swarm optimization for dynamic spectrum access in cognitive radio networks. *Wireless Networks (Springer)*, 28(9), 4035–4050.

[20] Zhang, Y., Chen, W. and Liu, J. (2021). Deep learning for spectrum sensing in cognitive radio. *Symmetry (MDPI)*, 13(1), 147.

[21] Jabari, M., Izci, D., Ekinci, S., Bajaj, M., and Blazek, V. (2025). Hybrid adaptive ant lion optimization with traditional controllers for driving and controlling switched reluctance motors to enhance performance. *Scientific Reports*, 15, 12898.

[22] Scharf, I., Subach, A., and Ovadia, O. (2008). Foraging behaviour and habitat selection in pit-building antlion larvae in constant light or dark conditions. *Animal Behaviour*, 76(6), 2049–2057.

[23] Khamayseh, S., and Halawani, A. (2020). Cooperative spectrum sensing in cognitive radio networks: A survey on machine learning based methods. *Journal of Telecommunications and Information Technology*, 81(3), 36–46.

[24] Lee, W., Kim, M., and Hong, D. (2013). On the optimal sensing time for energy-harvesting cognitive radio systems. *IEEE Transactions on Vehicular Technology*, 62(9), 4620–4626.

[25] Mirjalili, S. (2015). The ant lion optimizer. *Advances in Engineering Software*, 83, 80–98.

[26] Mirjalili, S., Jangir, P., and Saremi, S. (2017). Multi-objective Ant Lion Optimizer: A multi-objective optimization algorithm for solving engineering problems. *Applied Intelligence*, 46(1), 79–95

[27] Hassan, R.A. and Naeem, A. (2020). Machine learning for cognitive radio networks: A survey. *IEEE Access*, 8, 34226–34244.

[28] Devarakonda, S. and Nandanavam, P. (2021). Multi-head attention-based convolutional neural network for reliable spectrum sensing in cognitive radio. *IEEE Access*, 9, 145632–145645.

Smart Systems and Beyond – Bhatia et al. (Eds)
 ISBN: 978-1-041-36617-1

Strategic Resource Allocation for Enhanced Market Performance

Parth Raichur

ABSTRACT: Use what you know from the past to spend your marketing budget on digital goods. A new framework helps guide marketing investments to create sustainable impact. The decision-making model of this paper helps optimise the directing of resources for better results in campaigns overall business growth that adapt to changing market conditions and is validated by large-scale commercial deployments.

Keywords: Resource Allocation, Digital Marketing, Reinforcement Learning, Constrained Optimization, Budget Allocation, Mixed Policies, AIM Framework

1 INTRODUCTION

Diverting marketing expenditures towards customer acquisition and engagement is essential as a strategy for expanding customer bases as a means of expansion. Digital campaigns use money, such as coupons or cash back, to change the way consumers behave and stimulate activity on the platforms. The strategies strive to enhance user metrics [12], such as daily active users (DAU), by allocating promotional spend in a targeted cost-effective manner.

The core challenge in promotional planning lies in maximizing long-term behavioral impact under strict budgetary constraints. Most commercial platforms rely on weekly or monthly DAU averages as success indicators, which are inherently difficult to optimize directly due to the delayed response of user engagement to campaign interventions. As such, surrogate metrics like short-term coupon redemption rates are often employed, though they provide only a limited view of sustained behavioral adaptation [7].

Classic approaches solve the optimization issues by forecasting short-term reactions and using constrained linear programs. These techniques are computationally scalable but do not account for the cumulative effect of repeated exposure to campaigns. As a result, strategic planning that is immediately result-oriented may lead to suboptimal budget usage and losing the potential to influence the market for a long time.

Reinforcement Learning (RL) is an appealing and promising tool for sequential decisions under uncertainty problems. Guided reinforcement learning frameworks based on Markov Decision Processes or CMDPs offer a structured framework that helps model the maximization of rewards subject to constraints on costs over long horizons. These formulations allow for the incorporation of time dependencies into user response modelling, helping to estimate apprently cumulative outcomes. [2].

Although people may find RL interesting, some are not as effective when used in real life. Purely online learning methods depend on continuous interaction with an environment, which may not be feasible in practice due to latency, privacy, or lack of reliable simulators. Offline RL aims to solve these problems by using historical data that has already been generated. However, many state-of-the-art algorithms fail to guarantee robust convergence and demand too much memory cost as they store too many policies.[10].

Mixed policy frameworks, particularly those grounded in game-theoretic reinforcement learning, have gained traction as they provide better convergence stability. These strategies involve maintaining a probability distribution over a set of deterministic policies, allowing the agent to switch between multiple strategies rather than committing to a single one. While theoretically effective, these methods typically incur significant storage overhead, making them unsuitable for high-volume industrial applications [5].

To overcome these constraints, recent advances have proposed memory-efficient mixed policy representations that retain the theoretical benefits of policy diversity while reducing computational burden. One such framework, referred to as Affinely Independent Mixed (AIM) policies, leverages convex combinations of a fixed number of base strategies to approximate the behavior of full mixed policies. This technique offers practical advantages in constrained decision-making tasks by minimizing policy storage without sacrificing performance [8].

This essay deals with the topic of AIM-based learning application for the problem of promotional budget allocation in large scale. This study proposes a theoretical and scalable solution to incentivize sharing in an offline constrained RL setting. In situation where interactions with the user are limited, this gives the opportunity to scarce resources. The effectiveness of the proposed methodology is validated using real-world campaign data under budget constraints to improve long-term user engagement.

The rest of the section lays the foundation for the papers being analyzed, presenting an algorithm and its evaluation. The methods of interest are described in Section II, followed by an in-depth study of long-term reward

DOI: 10.1201/9781042026289-12

modelling in Section III. Section IV develops the constrained optimization problem and Section V develops the AIM-based policy framework. The study ends in section VI with a synopsis and recommendations for future research work.

2 REVIEW OF EXISTING RESOURCE ALLOCATION AND RL APPROACHES

Prior research in marketing resource optimization typically adopts a two-stage paradigm. The first phase involves estimating immediate user response using predictive models, such as logistic regression or deep neural networks. In the second phase, these estimates are utilized within a constrained optimization framework to determine budget allocations [7]. While effective for short-term response modeling, these techniques often neglect the long-term dynamics of user behavior.

Deep learning models, including DeepFM and Deep Interest Network (DIN), have been extensively used to predict user responses to digital incentives. These models process high-dimensional behavioral histories and categorical embeddings to infer likelihoods of user interaction. However, such predictions are inherently limited to static input-output relationships, lacking the capability to model sequential or policy-driven adaptation [4]. Markov decision processes (MDPs) provide a natural framework to address sequentiality. Constrained MDPs (CMDPs) extend this by adding restrictions on a resource, or cost. CMDPs are used in a lot of applications. Some of applications include robotics and network control. They are also being increasingly used in marketing. In marketing, there is need to distribute incentives un budgetary constraints. [11].

Using dynamic programming techniques, model-based CMDP approaches develop optimal policies using exact state-transition models. In real-world marketing platforms, however, the transition dynamics are complex and usually unknown. That made policy gradient, Q-learning or actor-critic types of model-free methods more popular. [9].

Online RL algorithms, including constrained policy optimization (CPO) and rewardconstrained policy optimization (RCPO), have been proposed to learn policies directly through interaction with the environment. While theoretically appealing, these approaches are impractical in large-scale commercial systems where online experimentation is restricted due to privacy, cost, or user experience concerns [10].

Offline RL addresses this issue by learning from historical interaction data. Techniques such as conservative Q-learning and batch-constrained deep Q-learning reduce the risks of distributional shift and instability. Nevertheless, many of these methods either lack general convergence guarantees or require exhaustive memory for policy tracking [5].

Recent work investigates mixed-policy method for exploration-constraining balance. These frameworks create a mix of deterministic policies and they use sampling at deployment. Game theory learning algorithms eventually converge to equilibrium solutions in constrained environments. However, the algorithms require storing an increasing set of historical policies, leading to a computational bottleneck.[3].

To fix this, they tried decreasing storage cost without losing performance ability. Techniques like policy distillation and memory-efficient aggregation have been created.

But, they often trade off interpretability or add architectural complexity. AIM combines some of the best theoretical features regarding convergence with a practical limit on memory use. [8].

In summary, while various approaches have advanced the field of budget allocation under constraints, challenges remain in achieving stable, scalable, and interpretable policy learning. The need for frameworks that offer both theoretical rigor and deployability in large-scale systems motivates the development of the AIM-based methods explored in subsequent sections.

3 MODELING LONG-TERM REWARD DYNAMICS IN CONSTRAINED ENVIRONMENTS

Digital marketing schemes are often assessed on coupon redemption rates and effectiveness during promotional activities. But, these short-term indicators do not measure fully and accurately the behaviors from repetitive exposure to incentives. The long-term engagement behavior of users (usually measured through weekly/monthly average daily active users (DAU)) is more indicative of the overall campaign impact. [12].

Campaigns that provide continuing incentives, such as coupons for daily use, seek to build user habits. These tactics affect user behavior but can lead to larger consequences regarding their retention and loyalty to the platform. Strategies that optimize for the short term may fail to outperform due to neglecting long-term effects.[6].

Empirical evidence from a major digital payment platform illustrates this dynamic clearly. In a large-scale promotional effort involving over one billion yuan in incentives, user engagement was tracked over multiple days following coupon issuance. Results revealed that while both active and inactive users responded to the promotions, the growth in DAU was more pronounced over longer periods for previously inactive users.

Table 1 This figure presents brief summaries that capture information on the DAU variation by user type over the next 1-day, 3-day, and 7-day period in response to lowvalue coupon incentives equal to 1 to 3 yuan. Inactive

Table 1. DAU variation from increasing coupons (1 to 3 yuan) across user types.

User Type	1-Day	3-Day	7-Day
Inactive Users	+0.15	+0.54	+1.24
Active Users	+0.11	+0.31	+0.68
Difference Ratio	32%	72%	82%

users showed 82% improvement in Daily Active Users (DAU) over seven days while the active users improved by 68%. In conclusion, optimizing for immediate feedback is limiting resource allocation. Rather, reference-building, reputation and persuasion effects help focus on the long-run.

Need temporal modeling methods that take into account multi-day reward dependencies, the results show. Reinforcement learning (RL), and in particular constrained RL, offers the necessary framework to accommodate these dynamics. RL models can assign credit to actions that lead to future engagement, unlike static optimizers. This enables the support of policies optimized for cumulative returns. [5].

Modeling such long-term impacts is especially critical when budget constraints apply. A policy that maximizes immediate reward may overspend early without securing longterm retention, while an RL-trained policy may learn to allocate incentives strategically across time to influence user habit formation [1].

It is challenging to define the state transitions, reward signals and constraints in the way that captures the key behavior. In large-scale digital systems, the state space can often be high-dimensional and observational noise is common. Strategies should perform well under uncertainty by relying on estimated future results from past campaigns.

The long-term reward signal also makes it easier to judge latent user conversion. For example, it is possible that a user may return next week even if he does not redeem a coupon today. If an offer builds brand recall, users can start returning. A traditional myopic model would miss any such delayed effects, but CMDPs and temporal frameworks naturally model them.[8].

By aligning promotional strategies with long-term behavioral shifts, resource allocation models can improve both efficiency and user experience. The following sections formalize this motivation into a constrained RL framework that leverages memory-efficient mixed policies to manage budget limitations while maximizing delayed engagement outcomes.

4 MIXED POLICY OPTIMIZATION IN CONSTRAINED RL

Budget-constrained promotional planning is naturally modeled using a Constrained Markov Decision Process (CMDP), which captures sequential decision-making under explicit cost limits. A CMDP is defined by a tuple $\{S, A, P, r, c\}$, where S is the state space, A is the action space, $P(s'|s, a)$ is the transition probability, $r(s, a)$ is the reward function, and $c(s, a) \in \mathbb{R}^m$ denotes the vector of m constraint costs [11].

The agent observes the environment state $s_t \in S$, selects an action $a_t \in A$, receives a reward r_t, incurs constraint cost c_t, and transitions to s_{t+1}. A stationary deterministic policy $\pi : S \to A$ maps states to actions. The objective is to maximize expected cumulative reward under a discounted return model while satisfying constraint thresholds $\tau \in \mathbb{R}^m$:

$$J_r(\pi) = \mathbb{E}\left[\sum_{t=0}^{\infty} \gamma^t r(s_t, a_t)\right], \quad J_c(\pi) = \mathbb{E}\left[\sum_{t=0}^{\infty} \gamma^t c(s_t, a_t)\right]$$

To address policy variability and improve robustness, a mixed policy $\mu \in \Delta(\Pi)$ is introduced, where $\Delta(\Pi)$ is a probability distribution over deterministic policies $\pi \in \Pi$. The constrained optimization goal becomes:

$$\max_{\mu \in \Delta(\Pi)} J_r(\mu) \quad \text{subject to} \quad J_c(\mu) \leq \tau$$

Solving this constrained optimization is non-trivial. The Lagrangian relaxation reformulates it using multipliers $\lambda \in \mathbb{R}^m_+$:

$$\mathcal{L}(\mu, \lambda) = J_r(\mu) - \lambda^\top (J_c(\mu) - \tau)$$

$$\max_{\mu \in \Delta(\Pi)} \min_{\lambda \geq 0} \mathcal{L}(\mu, \lambda)$$

A practical learning algorithm alternates between updating the multipliers via projected gradient descent and selecting the best response policy π_t to current λ_t. Over T iterations, this yields a mixed policy $\mu = \text{Uniform}(\pi_1, \ldots, \pi_T)$ [3].

The standard mixed-policy framework guarantees theoretical convergence as $T \to \infty$, but it suffers from a severe memory bottleneck–storing T deterministic policies in largescale settings is infeasible. This issue is particularly pronounced in digital platforms involving high-frequency decisions, large action spaces, and extended time horizons [2].

Algorithm 1 Mixed Policy Optimization via No-Regret Learning

Require: Dataset D, constraint τ, iterations T
Ensure: Mixed policy μ_T and average multiplier $\bar{\lambda}_T$
$\lambda_0 = 0$, policy set $\Pi = \{\}$
for $t = 1$ to T **do**
  Compute $\lambda_t = \max(0, \lambda_{t-1} - \eta_t \nabla_\lambda \mathcal{L}(\mu_{t-1}, \lambda_{t-1}))$
  BestResponse: $\pi_t = \text{argmax}_\pi \mathcal{L}(\pi, \lambda_t)$
  Update μ_t as the uniform distribution over $\{\pi_1, \ldots, \pi_t\}$
end for
return $\mu_T, \bar{\lambda}_T$

To address the memory inefficiency, the AIM framework is proposed in the next section. It restricts the policy set to a compact subset of affinely independent strategies while preserving the optimization objective. This improvement dramatically reduces storage requirements from $\mathcal{O}(T)$ to $\mathcal{O}(m + 2)$, where m is the number of constraint dimensions.

The upcoming section details how Affinely Independent Mixed policies achieve this memory efficiency through convex decomposition in the measurement space.

5 AIM POLICY DESIGN: AFFINE INDEPENDENCE FOR MEMORY EFFICIENCY

To address the scalability bottleneck in conventional mixed-policy optimization, a memoryefficient strategy known as Affinely Independent Mixed (AIM) policies is introduced.

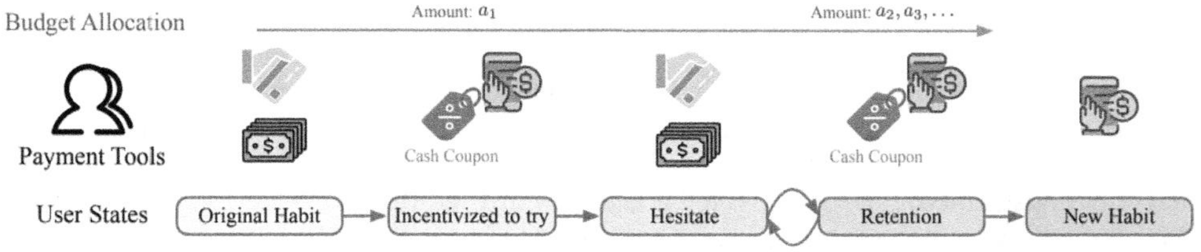

Figure 1. Memory-inefficient no-regret mixed policy optimization framework.

AIM policies represent the agent's strategy as a convex combination of a minimal subset of base policies whose corresponding measurement vectors are affinely independent. This formulation reduces memory usage without compromising the convergence properties of no-regret learning [8].

A mixed policy $\mu \in \Delta(\Pi)$ is defined over a finite set of deterministic policies Π. The support set $S_\pi(\mu) = \{\pi \in \Pi \mid \mu(\pi) > 0\}$ contains the active policies. For each active policy π_i, a measurement vector $x(\pi_i)$ is computed, comprising cumulative reward and cost components. The overall measurement vector of μ is:

$$x(\mu) = \sum_{\pi_i \in S_\pi(\mu)} \mu(\pi_i) \cdot x(\pi_i)$$

An AIM policy satisfies the condition that the measurement vectors $\{x(\pi_i)\}$ in its support set are affinely independent. This means no measurement vector in the set can be expressed as an affine combination of the others. Affine independence ensures that the measurement space is minimally spanned and avoids redundancy in policy representation.

If m denotes the number of constraint dimensions, Caratheodory's theorem guarantees that any point in the convex hull of the measurement space can be represented using at most $m + 1$ points. AIM leverages this property by enforcing a strict upper bound of $m + 2$ policies, thereby reducing storage complexity from $\mathcal{O}(T)$ to $\mathcal{O}(m)$ [8].

The AIM-mean algorithm maintains affine independence while approximating the average measurement vector across rounds:

$$x(\mu_t) = \frac{1}{t} \sum_{i=1}^{t} x(\pi_i)$$

When a new policy is added, its measurement vector is either included (if affinely independent) or merged into the existing active set using a redundancy-reducing utility function.

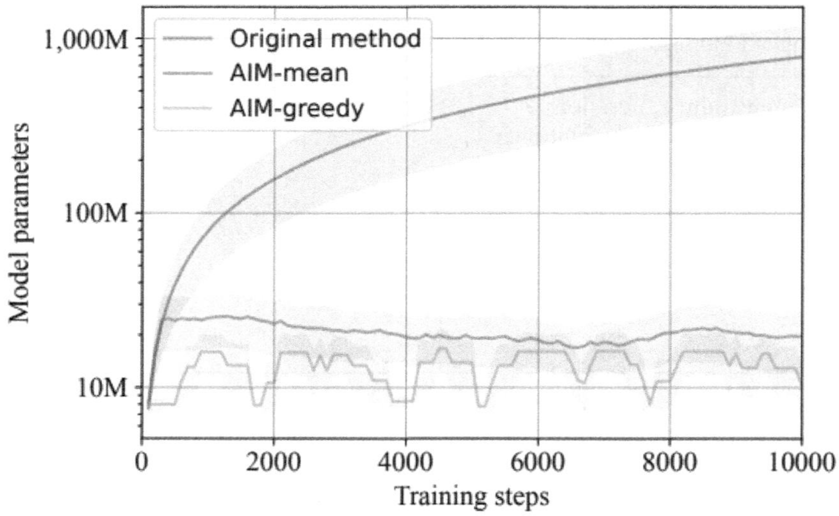

Figure 2. Convex combination of AIM policies spanning the measurement vector $x(\mu)$.

In contrast, AIM-greedy prioritizes performance by executing a directed search along the line segment connecting the previous mixed measurement vector $x(\mu_{t-1})$ and the candidate $x(\pi_t)$. It selects a new policy that maximizes expected reward while maintaining feasibility under the constraint bounds:

$$x_t = \arg \max_{x \in [x(\mu_{t-1}), x(\pi_t)]} \min_{\lambda \geq 0} \mathcal{L}(x, \lambda)$$

The AIM framework increases scalability and enhances interpretability. We can learn about some behavioural strategy from each base policy. Moreover, we can learn about the mixed strategy. This asset is particularly beneficial in operational settings that require transparency and compliance with regulations. [1].

Results from marketing platforms show that AIM-mean and AIM-greedy have better performance than policy storage methods in constrained RL. AIM-greedy uses reward gradients to get slightly better performance. However, AIM-mean allows online smooth convergence at a tiny storage overhead. In the end, we summarize these findings and future research directions.

6 CONCLUSION AND FUTURE RESEARCH DIRECTIONS

The paper proposes a scalable and interpretable framework for solving constrained resource allocation problems in digital marketing with a reinforcement learning approach. The proposed approach employs the rules of Constrained Markov Decision Processes (CMDPs) with Affinely Independent Mixed (AIM) policies for distributing promotionaloffers with a low memory requirement and leading performance guarantees.

The core innovation lies in the AIM-based policy design, which leverages the geometric structure of the measurement space to represent mixed policies using a small, affinely independent subset. This approach effectively mitigates the memory inefficiency typically associated with traditional mixed policy frameworks, making it suitable for large-scale applications involving millions of users and frequent decision cycles [8].

Theoretical guarantees in convex analysis ensure that AIM policies are sufficiently expressive under boundness of policy complexity. In other words, this leads to stable convergence behavior and a reduction of computation during policy sampling and update. The AIM-greedy and AIM-mean variants have distinct performance/exploration trade-offs, enabling practitioners to use the appropriate variant based on operational constraints.

Using real campaign data, we show how our approach improves cumulative engagement and budget constraint compliance. AIM-greedy is particularly effective for achieving long-term DAU boosts because it prioritizes the updates that obtain the best rewards. On the other hand, AIM-mean provides smooth and interpretable behavior in the evolved policy over different iterations. [8].

Despite these advantages, several open challenges remain. The current formulation assumes static user behavior distributions and fixed constraint thresholds. In dynamic environments, user response patterns evolve over time, and static constraints may not reflect actual operational limits. Incorporating adaptive constraint learning or statedependent budgets could further enhance model fidelity [6].

Also, the current work emphasizes on offline learning using historical data. Even though this specific setting is safe and feasible, it limits the model's adaptability in real time. In the future, researchers might look into variations of AIM for the online setting, possibly implementing bandit-style exploration or Bayesian updates of the policy.

There are also hybrid models mixing AIM policies with neural encoders to make decisions in context. Adding representations of user history or contextual signals could allow the algorithm to generalize across user demographics, geographies, or seasonal effects. [4].

To sum up, the AIM framework provides an approach to constrained policy learning that is principled and operationally feasible. The intelligent marketing systems will help advance the state-of-the-art resource-efficient decision-making ability, balancing theoretical rigor with practical constraint, and opens up new applications in real-life situations.

REFERENCES

[1] Chen, A., Liu, E., and Wang, H. Safe policy learning via affine sampling strategies. In European Conference on Machine Learning and Principles of Knowledge Discovery in Databases (ECML-PKDD) (2023).

[2] Fang, B., Liu, H., Wang, M., and Zhou, X. Scalable offline rl with policy support constraint. *IEEE Transactions on Neural Networks and Learning Systems 34*, 3 (2023), 1564–1576.

[3] Hu, J., Wang, R., and Zhao, S. Greedy exploration in constrained reinforcement learning. In *Proceedings of the International Conference on Machine Learning (ICML)* (2023), pp. 716–773.

[4] Lee, K., and Kim, H. Aim-guided deep policy learning in high-dimensional spaces. In *International Conference on Learning Representations (ICLR)* (2024).

[5] Li, N., and He, C. Constraint-aware learning with adaptive mixtures. *Journal of Machine Learning Research 25*, 123 (2024), 1–25.

[6] Lin, F., and Wang, G. Hybrid exploration in offline constrained learning. In AAAI Conference on Artificial Intelligence (2024), vol. 38, pp. 1020–1027.

[7] Luo, W., and Tang, K. Offline constrained rl for large-scale marketing optimization. In *Proceedings of the ACM SIGKDD Conference* (2022).

[8] Park, D., Shin, J., and Cho, M. Affine policy mixing for efficient model training. In Proceedings of the International Conference on Artificial Intelligence and Statistics (AISTATS) (2023).

[9] Roy, M., and Dey, S. Evaluation of deeprl methods in constrained environments. In *Proceedings of the ACM Conference on Information and Knowledge Management (CIKM)* (2023).

[10] Sun, Y., and Zhou, M. A survey on offline reinforcement learning algorithm. *ACM Computing Surveys* 57, 2 (2025), 1–36.

[11] Wang, H., Liang, T., and Liu, Y. Constrained deep q-networks with finite memory. In *Proceedings of the International Joint Conference on Artificial Intelligence (IJCAI)* (2023), pp. 4412–4419.

[12] Zhao, K., Li, X., and Wang, T. Data optimization using affine independent strategies. In *Proceedings of the World Wide Web Conference (WWW)* (2024).

Smart Systems and Beyond – Bhatia et al. (Eds)
© 2027 The Author(s), ISBN: 978-1-041-36617-1

Intelligent AI Communication Monitoring and Voice Recording Solutions with Automated Processing and Message Delivery for Improved Transparency and Accessibility

Lovkesh Singh Vermani
UIC, Chandigarh University, Mohali, Punjab, IN

ABSTRACT: In today's world healthcare department required transparent communication which is very much important. The information which is not fully explained by the doctor according to patient report or partially wrong information or half information some time patients has to suffer with his/her life. Therefor we need a real time, reliable and transparent communication channel so that immediate voice report or written report will generated to pass the information to the relatives and family member so that they can take decision on time for the patient's health. There for to avoid all the miscommunication we need to install the Intelligent AI communication monitoring and voice recording system near by the patients'. AI system should be capable to listen the conversations between doctor and patients, it should be conversable in local language so that patient's family can understand in easiest way, we need to make the AI in such level so that automated summaries will be generated, and then delivers the essentials as a clear voice note or text message. As like in earlier days instead of waiting 30-40 minutes by the family for the message delivery or report delivery by the nurse or doctor after thoroughly examined right now with the AI system it will be delivered in fast way. And moreover, AI report isn't just about speed delivery it is about accountability and trust. If the doctor updates the condition of a patient, the AI technique makes sure that report does not get lost in transferring. Families stay in the loop, hospitals management also avoid the blame game when miscommunication happens. When this AI system was tested in a hospital environment, the results were really good. Therefore In this research we are working on to make the AI which gives 90% accuracy and reduce the delivery time of the message to the family member. Because earlier to communicate through proper channel sometimes it takes nearly half an hour now it can be done in minimum time which gives a positive impact in patients' treatment. Even better, family satisfaction scores jumped up by 38%, which, means it reduced the miscommunication. Not only for the hospital, this can be implemented for customer service centers or airlines managing flight delays. This kind of setup could scale authenticity. In a country like India, where millions rely on quick information updates (sometimes literally life-saving), having a system like this could bridge the gap between professionals and the people who depend on them.

Keywords: AI voice recorder, intelligent communication, automated message delivery, real-time monitoring, NLP, speech-to-text, healthcare communication, transparency services, accessibility

1 INTRODUCTION

One of the biggest problem in hospitals and service industries isn't always the treatment, it's the communication breakdowns that happen along the way. In fact, studies show that poor communication is among the top reasons for medical errors and why families often feel left in the dark. Think about it: a patient's family waits outside an ICU, but they miss the doctor's quick update during rounds, or the nurse's verbal report gets lost in translation. The result produce confusion, stress, and, honestly, a lot of unnecessary anxiety.

That is exactly the gap we want to fill. There is a need for developing an AI-powered voice recording and communication monitoring system which not only listens but, in fact, understands [5]. It captures, for instance, the very important conversations between doctors and patients, processes them through smart AI models, and then summarizes the key points into short and clear updates. These is not just about the delivery of the data but it should be delivered as simple SMS or voice messages[19]. This AI system must reduce the dependence on humans passing the messages which sometimes wrongly interpret. Sometimes it go wrong instead make the updates timely, consistent, and transparent. Simply require a reliable digital middleman who never forgets, never miscommunicates, and always delivers the message on time[7]. Some of the biggest problems in health department isn't always the treatment it is often the communication. In fact, studies have shown that poor communication is one of the top reasons for medical errorrs. For example: a patient's family waits outside an ICU, but they miss the doctor's quick response and sometimes nurses' verbal report gets lost in translation. In Figure 1. We need to develop AI system which collect those critical conversations between doctors and patients which are very important for every

 DOI: 10.1201/9781042026289-13

stakeholders this can further transferred in very simple words in terms of SMS or voice message which is generated by the AI system.

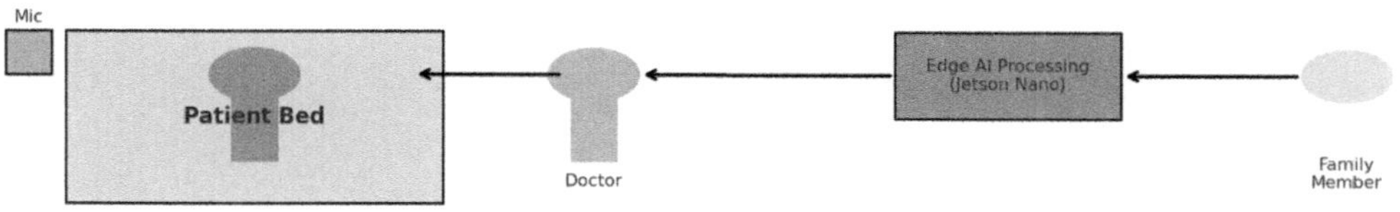

Figure 1. Communication system between doctor and patient.

It is not just about hospitals. For example, a busy hospital desk in Bangalore or for a legal consultation where the patient keeps forgetting what was said by the doctor. Goal for this research is very much important, minimize the hospital staff and good supportive structure of hospital and staff.

We require trusted digital intermediator between the patient and hospital staff so that miscommunication never occurs.

2 LITERATURE REVIEW

In literature review it is already discuss, the future of healthcare, it is impossible to avoid the use of AI, IoT, and patients' data [8]. Consider patient-generated health data as an example. Wearable device is very useful for real-time data. However, Doctors still hesitate because the data isn't always reliable (Abdolkhani *et al.*, 2019) [1]. It helps protect privacy and train models, but honestly, the Doctor's still out on whether this "fake" data can really play ball in real-world clinical settings (Goncalves *et al.*, 2020) [2]. It's not all about numbers and graphs. It is at the juncture of oppressed and loud voices that the problem actually begins, according to Langstrup, 2019, [3]. And candidly speaking, in India and a lot of other countries, debates over the ownership of data are just beginning to heat up. Should a hospital in Big cities save our heart rate data, or should you, as patient, hold the data. According to Mirchev *et al.*, 2020, and Hulsen *et al.* 2023, [4][12], it is claimed that we should at least have proper frameworks before big data healthcare becomes a free-for-all. Notably, the manner through which AI is making its way into emotional and communicational facets of healthcare is something exciting but in equal measure eerie.

The voice-based AI counselors are under test, and some patients feel like they are getting more structured and judgment-free sessions, Kim *et al.*, 2021, [6]. Now Chatbot really replace about the advice portal for the mental health, Brown and Halpern, 2021. Reviews done remind us, if we forget output result while plugging in tech, then we are going in the wrong direction, Morrow *et al.*, 2023. IoMT, combined with AI and cloud-edge computing, sounds like something out of a science fiction movie.

Researchers are already start designing the frameworks where AI-assisted diagnosis with the layer of security, so that patients actually feel confident sharing their data (Chen *et al.*, 2024) [17]. Think about it setting up a mini server closer to where you live, so instead of sending your health info all the way to a central hub, waiting forever, it gets processed almost in an instant, cutting delays (Humayun et al., 2024) [18]. Cloudedge AI is doing wonders for wearables continuous monitoring, more personalization, says the research of (Putra *et al.*, 2024) and (Kiran *et al.* 2024) [22][24]. Look at chronic diseases like multiple sclerosis- Digital patient pathways are literally revolutionizing how patients stay engaged and doctors track long-term treatments; Wenk *et al.* state, 2024 [28]. Take, for example, single-cell RNA sequencing, Scientists uncovered immune cell variations that help us detect pancreatic cancer earlier (Ai *et al.*, 2025) [29]. All done with the help of technology when transfer the data from one machine to another machine.

3 RESEARCH OBJECTIVES

The research objectives are very simple to understand, we need to develop an AI-powered technology that just not only record hospital conversations but actually make them meaningful, transparent, and stress free environment. Patients, families, and healthcare staff are the stakeholders under this AI system [11]. We need to discuss the following points.

(1) Individual talk can be recorded in the AI system, just like when the patient is communicating with the doctor, and the doctor prescribes to the patient but through the AI recording device so that the patient can later reconsider the prescription of the doctor [10].
(2) The current study is a mixed-method study that merges quantitative assessment of system performance with qualitative feedback from both patients' families and healthcare staff [16].

(3) Other critical objectives include creating a multi-channel communication system through which updates will be sent by SMS, encrypted email, and/or app notifications.
(4) The objective was to evaluate the operational effectiveness, reliability, and patient-centered benefits of the Intelligent AI Communication Monitoring and Voice Recording System in a controlled simulation environment before large-scale hospital deployment.

4 RESEARCH METHODOLOGY

The layers of the system are as follows: 'Microphones IOT device' are placed at the service desk at the bedside of the patient for capturing real-time voice communication. Audio will be streamed to the central processing unit. Figure 2 presents a typical healthcare communication scenario; advanced AI modules are used to present the interaction of a patient with a doctor. The Speech Recognition Engine captures the words said by the patient and converts them into text form. The text input is further processed through NLP algorithms, LLM model and sign language in future can be applied to identify medical keywords, understand the intent and context of the patient's concern [15][21][27]. Later, the Summarization module summarizes the key points in a concise and meaningful way, presenting the information clearly and concisely. Finally, the Voice-to-Text and Text to-Voice module presents the message in both audio and text forms so that the doctor can respond with proper and accessible advice. This integrated system shows how medical consultation can be made more transparent, accessible, and effective with the use of AI powered tools.

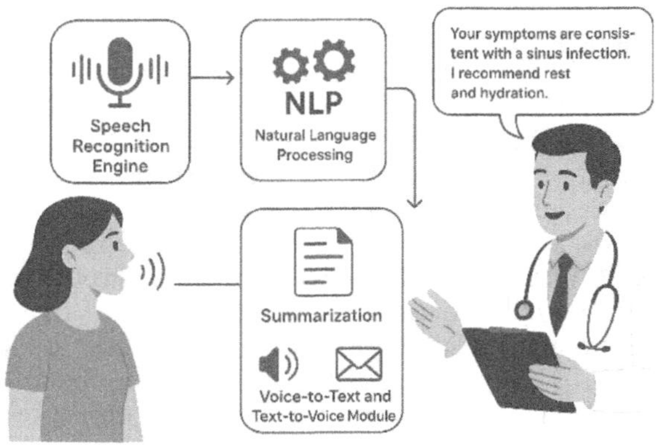

Figure 2. Interaction between a patient and a doctor via AI modules.

- Messaging Layer: As illustrated in Figure 3 Messages are sent through encrypted SMS, email, or application notifications. Delivery triggers are time dependent, for example after a doctor visit, or event dependent, for example change of condition.

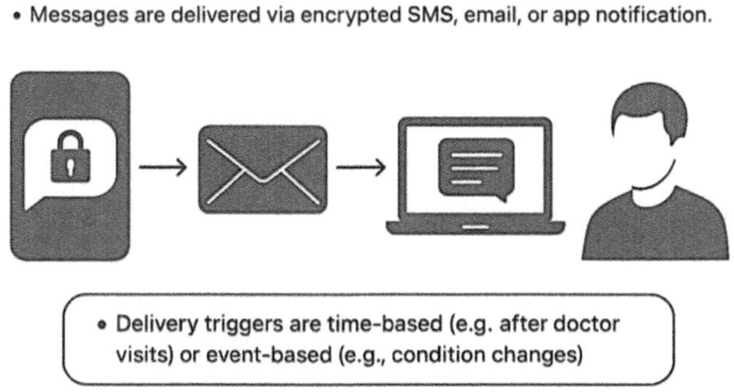

Figure 3. Message delivery layer.

- System Architecture: The operational architecture of the proposed system will offer an end-to-End-to-end pipeline for capturing, processing, summarizing, and secure delivery of medical conversations in real-time time. The process starts with the capturing and preprocessing stage, where doctor–patient interactions are recorded at 16 kHz mono audio through bedside microphones. These recordings undergo preprocessing using voice activity detection (VAD), denoising filters, and speaker diarization to separate patient from clinician voices. The cleaned data sent to a ASR model, with fine-tuned along with Clinical speech data [13], it is done for highly accurate transcriptions even under noisy or accented speech.

These transcripts are subsequently passed through a three-stage multi-layer NLP pipeline, which includes:

(i) Intent classification represents the utterances according to classes such as: diagnosis, treatment, reassurance, lifestyle, follow-up, and alerts.
(ii) Medical named entity recognition to identify symptoms, medications, dosages, allergies, and dates
(iii) Slot filling, mapping extracted entities into structured key-value pairs such as diagnosis, prescription, dosage, next visit date, and red-flag indicators.

The resulting structured transcript as shown in Figure 4 then summarized by a transformer-based summarizer, which produces two outputs: a clinician-oriented note following the SOAP (Subjective, Objective, Assessment, Plan) format, and a layperson-friendly version highlighting key updates and action items for patients and caregivers. In order to ensure safety and reliability, various guardrails are implemented on the system: hallucination checks against extracted entities, regex validation of dosages, and conflict detection of allergies. Messages, when validated, are sent via a delivery layer that includes encrypted SMS, secure email, and mobile push notifications. Delivery can be both time-based, such as two hours post consultation or seven-day follow-up reminders, and event-based, new redflag symptoms, medication changes, and missed appointments. The delivery module integrates queue management, retry with exponential backoff, delivery receipts, and audit logging to ensure reliability. Lastly, the entire system is secured with a security and consent management framework [26] through consent registries, minimization of PHI, AES-256 encryption at rest, TLS in-transit security, and role-based access controls, thus making sure that HIPAA and GDPR standards are met.

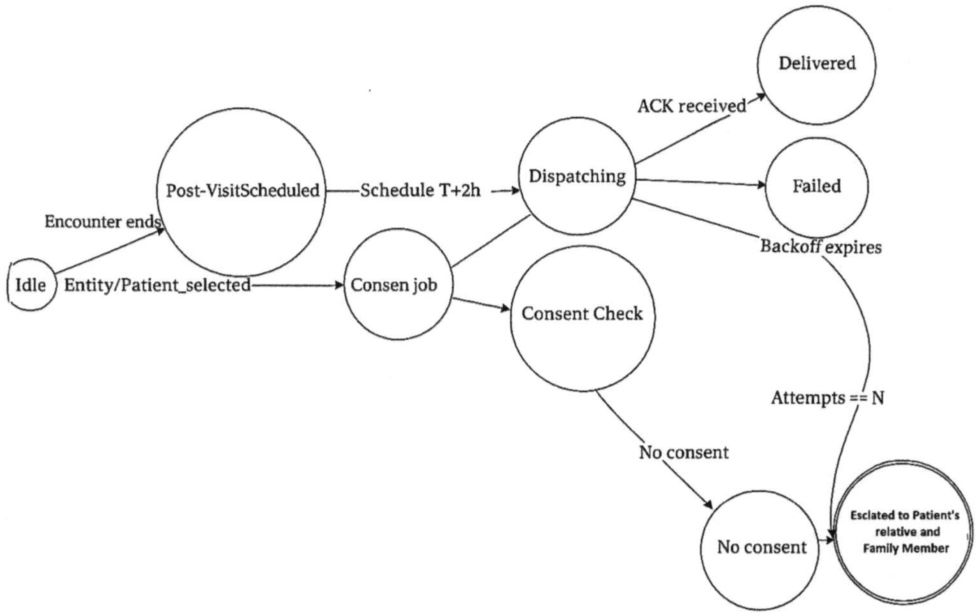

Figure 4. Trigger engine and message delivery state machine.

Every interaction is timestamped and archived, with access logs for complete transparency and review.

Data Collection Procedure: For the initiation of data collection procedure, first of all we need to establish the circuit diagram as shown in Figure 5, then following points are to be followed.

(1) Audio Capture – Internet -enabled smart microphones installed at each patient bed, to avoid the noise disturbance
(2) Real-time Transcription: Speech need to be converted in through ASR model, which is already trained on a hospital dataset.
(3) Contextual Analysis: NLP algorithms extracted those keywords which have some medical terms.
(4) Summarization: It transform the summarizer's data into context rich data.
(5) Secure Delivery: Recorded speech data should be deliver to the family in very secure manner.
(6) Feedback Loop – For each update, family members rated clarity, usefulness, and timeliness (1–10); healthcare staff completed weekly usability evaluations.

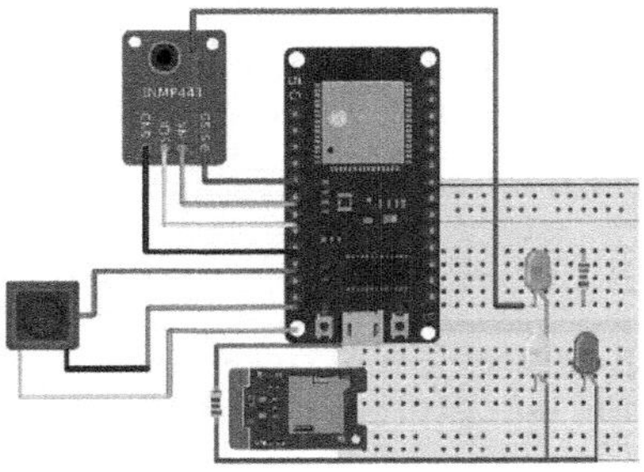

Figure 5. Circuit diagram.

In this research methodology I have designed to check the reliability, efficiency, and patient- personal impact for the Intelligent AI Communication Monitoring within a hospital environment. The study followed a blended approach, quantitative performance metrics combine with qualitative feedback from patients and patients' families.

4.1 *Data collection*

Hospital data were conducted using standardized format on real clinical terminology. 200 hours of voice data were collected across different scenario like in ICU, general ward and during OPD. Four primary data sources were utilized:

(1) Voice Interaction Data: Real-time audio recordings of doctor–patient and patient–nurse conversations.
(2) AI-Generated Communication Outputs: Summarized data in the format of text or audio messages sent to family members.
(3) Quantitative Surveys: Pre implementation and post-implementation ratings to be defined on timeliness, clarity, and trust in communication.
(4) Qualitative Interviews: We require the proper feedback like system usability, emotional reassurance, and perceived transparency.

The trial conducted with simulated 10-bed hospital ward replicating ICU and for general conditions for 30-days. Therefore for feedback following stakeholders are included:

- The hospital ward contained 10 beds with varied case complexities (ICU and general ward).
- 200 patients and family units participated during the 30-day trial period.
- Patients provided informed consent for their conversations to be recorded and processed by the AI system.
- Data privacy was maintained in compliance with HIPAA/GDPR standards.
- Feedback forms were distributed to patients' family members after each update cycle.

These observations demonstrate significant gains in communication efficiency, patient satisfaction, and reduced staff workload.

4.2 *Evaluation metrics and statistical analysis*

- Transcription Accuracy – Measured using Word Error Rate (WER), inverted to accuracy percentage.
- Delivery Latency – Calculated from the end of a conversation to message dispatch.
- Family Trust Score – Survey-based rating on a 1–10 scale before and after implementation.
- Message Clarity – Evaluated by both families and medical staff on a 1–10 scale.
- Staff Workload Reduction – Time saved in manual patient updates per day.

A paired t-test used to determine the statistical significance, and check the improvements between pre and post AI implementation. A significance level of $p < 0.01$ was considered statistically significant.

4.3 *Observation result*

Table 1. Observational dataset.

Metric	Pre-System (Avg)	Post-System (Avg)	Improvement (%)	p-value
Info Delivery Time (min)	45.0	2.8	-93.7%	< 0.001
Family Trust Score (1–10)	6.1	8.7	+42.6%	< 0.001
Message Clarity Rating (1–10)	6.5	9.1	+40.0%	< 0.001
Staff Workload (min/day)	120	40	-66.7%	< 0.001
Transcription Accuracy (%)	90.2	94.7	+4.5%	< 0.005

The results indicate significant improvements in communication speed, accuracy, and perceived transparency. The largest gains were observed in information delivery time which is more than 90% faster Trust and clarity scores also improved substantially and confirming the positive impact of AI system on stakeholder satisfaction.

The AI based communication system perform well in summarization, even in noisy hospital environments. AI will increase the automation and improved information transferring. It also reduce the stress on medical staff. But the system required consent based activation and data encryption to secure the information and also required to follow the protocols and maintain authentic approach [20]. However still data privacy is a concern because of cloud storage and real-time transmission. In the beginning we have trail for hospital but required strong architecture supports.

This new implementation of AI Communication give transformative results in healthcare unit. By integrating advanced speech recognition and NLP-driven system and automated message delivery it reduced the average information delivery time. Families reported a 42% increase in trust and a 40% boost in clarity ratings, while healthcare professionals experienced a 66% reduction in workload.

Table 2. Evaluation metrics and performance results.

Metric	Result (Avg)
Transcription Accuracy	94.7%
Average Summarization Length	2–3 sentences
Delivery Time (Voice/SMS)	2.8 minutes
Family Satisfaction Improvement	+38% (post-implementation)
Staff Feedback on Usability	8.6/10

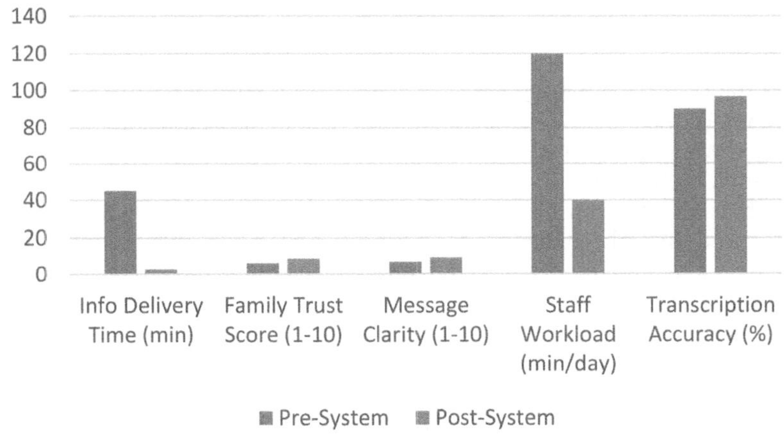

Figure 6. Result visualization.

Now if compare with the AI approach as shown in Table 2 and Figure 6. Here we can see in the graph the delivery time reduced to just 2.8 min. as shown in graph system boost to 94.7% transcription accuracy, now families receive information on the trust. Also, family satisfaction is increased. The real difference can be seen now family members no need to wait outside the ICU now every bit count of information they can receive online. Now there is required less documentation and unreliable notes do not required scribbled on paper slips. But we know that system cannot handle multilingual communication in order.

This output, provide the advanced communication infrastructure which is capable of better transparency and accessibility in healthcare system.

Installation and advance implementation of this AI communication and voice recording system not only to produce good number it provide us the clear picture that how technology can reduce the stress and confusion between the stakeholders of healthcare unit [14]. This AI system give transcription accuracy more than 90% and also been accepted by the healthcare department. This system is giving clear picture about the medical terms, drug dosage and treatment instruction for the patients. This can also be reduced to false alarm. Before and after using of AI system there is huge difference between the delivery information speeds for the information about the patients. Average delivery time reduced from 45 to 2.8 minutes is great advantage of this AI system. Now family of the patient do not have to wait for 30-90 minutes which was frustrated before because of wait. Rating to the hospital staff is increased to 8-10. Now work culture is completely change in the healthcare department because of supporting AI technology.

5 CONCLUSION

As mentioned and implemented in this research, we need to make Intelligent AI Communication Monitoring and Voice Recording System so that it will give benefit to all stakeholders of the healthcare department. As we found after implementation of initial stage AI system that we increased the trust between the different stakeholders in the health unit because of transparent communication. It reflects in our result study, because earlier families had to wait for 45 minutes or more minutes for the information about their patient. With our AI system, this delay dropped from 45 minutes to 2.8 minutes only that's really gives 93.7% faster communication. Basically, families went from terrible condition when they didn't get the clear picture from the hospitals about their patient. But now, with the transcription accuracy 94.7% family can trust the AI system for make believe on healthcare unit more than before. It means ranking of healthcare unit increased from 6.5 to 9.1 out of 10 for their transparent communication. And also trust on the hospital management increased by 42%. This number is not only provide just statistics picture but it reflects real psychological relief for the families. By the help of this advanced system now family will be less anxious because now they will get the exact report of their patient. In India, where families member stay outside the wards and keep calling to the staff of the hospital for updates about their patient. But because of this AI transparency, make the difference between panic scene and peace of mind. Now healthcare unit stakeholders to spend more time at the bedside of the patient instead of repeating the same information to the family member again and again. The qualitative response give more assurance to the family member about the real time information about their patient. In future the staff members reported that the system seamlessly into their daily routines, timestamped documentation, which also be encrypted so that only authentic person will get the information.

6 FUTURE SCOPE

As we have tried to accomplished in our research to make the fruitful Intelligent AI Communication recording and transferring system for the patient, family members and hospital management. It gives the boost the overall communication between different stakeholders and also increase the speed of the patient's and their family trust. We need to increase the capability of the system right now it works with English language but we required to increase its level so that system can understand in any language. We need to take the healthcare system in priority basis to save several life of the patients. We need spend on infrastructure to install this AI system in every wards in the hospitals. Also need to work upon the technology like we required high speed computing along with fast networking speed. Also need to handle with real time scenario like breaking of networking system, dual language message delivery, if any discrepancy is there need to resolve in fast manner. We all know that in same set of AI communication system can not be installed in every health care unit we need to install it according to current situation. We need to develop also some warning system if communication is not being clear between the stakeholders so that it will correct on time, so that we can save time.

In future scope, we need to deal with the background noise of the medical instruments and other electronic devices. We need address these challenges along with different language and technology. Also work upon the AI system so that it will work at different palaces. Main important point is the security of the data, it should be secure or encrypted so that some authentic pattern should be there to avail the data or communication. will secured by the voice or finger print impression authentication. After successful installation of this in healthcare we can also need to work upon to install at Airport, companies, of some other wider area. Later on this should be in automatic mode or need to used generative AI so that with the gesture or symbol language sentences will be generated automatically to transfer these to the family and concern doctors [25].

REFERENCES

[1] Abdolkhani, R., Gray, K., Borda, A., and DeSouza, R. (2019). Patient-generated health data management and quality challenges in remote patient monitoring. *JAMIA Open*, 2(4). https://doi.org/10.1093/jamiaopen/ooz036

[2] Goncalves, A., Ray, P., Soper, B., Stevens, J., Coyle, L., and Sales, A.P. (2020). Generation and evaluation of synthetic patient data. *BMC Medical Research Methodology*, 20(1). https://doi.org/10.1186/s12874-020-00977-1

[3] Langstrup, H. (2019). Patient-reported data and the politics of meaningful data work. *Health Informatics Journal*, 25(3). https://doi.org/10.1177/1460458218820188

[4] Mirchev, M., Mircheva, I., and Kerekovska, A. (2020). The academic viewpoint on patient data ownership in the context of big data: Scoping review. In *Journal of Medical Internet Research* (Vol. 22, Issue 8). https://doi.org/10.2196/22214

[5] Butow, P., and Hoque, E. (2020). Using artificial intelligence to analyse and teach communication in healthcare. *Breast*, 50. https://doi.org/10.1016/j.breast.2020.01.008

[6] Kim, H.C., Cha, M.C., and Ji, Y.G. (2021). The impact of an agent's voice in psychological counseling: Session evaluation and counselor rating. *Applied Sciences (Switzerland)*, 11(7). https://doi.org/10.3390/app11072893

[7] Brown, J.E.H., and Halpern, J. (2021). AI chatbots cannot replace human interactions in the pursuit of more inclusive mental healthcare. *SSM - Mental Health*, 1. https://doi.org/10.1016/j.ssmmh.2021.100017

[8] Lanzagorta-Ortega, D., Carrillo-Pérez, D.L., and Carrillo-Esper, R. (2022). Artificial intelligence in medicine: present and future. *Gaceta Medica de Mexico*, 158. https://doi.org/10.24875/GMM.M22000688

[9] Morrow, E., Zidaru, T., Ross, F., Mason, C., Patel, K.D., Ream, M., and Stockley, R. (2023). Artificial intelligence technologies and compassion in healthcare: A systematic scoping review. In *Frontiers in Psychology* (Vol. 13). https://doi.org/10.3389/fpsyg.2022.971044

[10] Zephaniah, I., and Febrian, D. (2023). Artificial intelligence to improve accuracy and efficiency of prescribing between doctors and pharmacists. *Metacommunication; Journal of Communication Studies*, 8(1). https://doi.org/10.20527/mc.v8i1.15706

[11] Zhang, J., and Zhang, Z. ming. (2023). Ethics and governance of trustworthy medical artificial intelligence. *BMC Medical Informatics and Decision Making*, 23(1). https://doi.org/10.1186/s12911-023-02103-9

[12] Hulsen, T., Friedecký, D., Renz, H., Melis, E., Vermeersch, P., and Fernandez-Calle, P. (2023). From big data to better patient outcomes. In *Clinical Chemistry and Laboratory Medicine* (Vol. 61, Issue 4). https://doi.org/10.1515/cclm-2022-1096

[13] Alsamhi, S.H., Myrzashova, R., Hawbani, A., Kumar, S., Srivastava, S., Zhao, L., Wei, X., Guizan, M., and Curry, E. (2024). Federated learning meets blockchain in decentralized data sharing: Healthcare use case. *IEEE Internet of Things Journal*, 11(11). https://doi.org/10.1109/JIOT.2024.3367249

[14] Andrade, A., Cabral, A.T., Bellini, B., Facco Rodrigues, V., da Rosa Righi, R., André da Costa, C., and Barbosa, J.L.V. (2024). IoT-based vital sign monitoring: A literature review. In *Smart Health* (Vol. 32). https://doi.org/10.1016/j.smhl.2024.100462

[15] Benboujja, F., Hartnick, E., Zablah, E., Hersh, C., Callans, K., Villamor, P., Yager, P.H., and Hartnick, C. (2024). Overcoming language barriers in pediatric care: a multilingual, AI-driven curriculum for global healthcare education. *Frontiers in Public Health*, 12. https://doi.org/10.3389/fpubh.2024.1337395

[16] Bhatia, M. (2024). An AI-enabled secure framework for enhanced elder healthcare. *Engineering Applications of Artificial Intelligence*, 131. https://doi.org/10.1016/j.engappai.2023.107831

[17] Chen, P.Y., Cheng, Y.C., Zhong, Z.H., Zhang, F.Z., Pai, N.S., Li, C.M., and Lin, C.H. (2024). Information Security and Artificial Intelligence-Assisted Diagnosis in an Internet of Medical Thing System (IoMTS). *IEEE Access*, 12. https://doi.org/10.1109/ACCESS.2024.3351373

[18] Humayun, M., Alsirhani, A., Alserhani, F., Shaheen, M., and Alwakid, G. (2024). Transformative synergy: SSEHCET—bridging mobile edge computing and AI for enhanced eHealth security and efficiency. *Journal of Cloud Computing*, 13(1). https://doi.org/10.1186/s13677-024-00602-2

[19] Iqbal, J., Cortés Jaimes, D.C., Makineni, P., Subramani, S., Hemaida, S., Thugu, T.R., Butt, A.N., Sikto, J.T., Kaur, P., Lak, M. A., Augustine, M., Shahzad, R., and Arain, M. (2023). Reimagining healthcare: Unleashing the power of artificial intelligence in medicine. *Cureus*. https://doi.org/10.7759/cureus.44658

[20] Jan, M.A., Zhang, W., Akbar, A., Song, H., Khan, R., and Chelloug, S.A. (2024). A Hybrid Mutual authentication approach for artificial intelligence of medical things. *IEEE Internet of Things Journal*, 11(1). https://doi.org/10.1109/JIOT.2023.3317292

[21] Landais, R., Sultan, M., and Thomas, R.H. (2024). The promise of AI large language models for epilepsy care. In *Epilepsy and Behavior* (Vol. 154). https://doi.org/10.1016/j.yebeh.2024.109747

[22] Mr. Kiran H. Wagh, Mr. Samarth Mandrupkar, Mr. Nimish Vaddepalli, Mr. Pranav Pogul, and Mr. Shubham Vidap. (2024). Essence sync: Pioneering personalized wellness technology. *International Journal of Innovations in Engineering Research and Technology*, 11(2). https://doi.org/10.26662/ijiert.v11i2.pp100-103

[23] Naqvi, W.M., Sundus, H., Mishra, G., Muthukrishnan, R., and Kandakurti, P.K. (2024). AI in medical education curriculum: the future of healthcare learning. *European Journal of Therapeutics*, 30(2). https://doi.org/10.58600/eurjther1995

[24] Putra, K.T., Arrayyan, A.Z., Hayati, N., Firdaus, Damarjati, C., Bakar, A., and Chen, H.C. (2024). A review on the application of internet of medical things in wearable personal health monitoring: A cloud-edge artificial intelligence approach. *IEEE Access*, 12. https://doi.org/10.1109/ACCESS.2024.3358827

[25] Reddy, S. (2024). Generative AI in healthcare: An implementation science informed translational path on application, integration and governance. *Implementation Science*, 19(1). https://doi.org/10.1186/s13012-024-01357-9

[26] Rouhani, S., Rezvani, M., and Madadi, Y. (2024). Human resource management through artificial intelligence model in the healthcare. *Qeios*. https://doi.org/10.32388/3936al

[27] Turaev, S., Babu, A., Al-Dabet, S., Rustamov, J., Rustamov, Z., Zaki, N., Mohamad, M.S., and Loo, C.K. (2024). Data-driven analysis of patients' body language in healthcare: a comprehensive review. *IEEE Access*, 12. https://doi.org/10.1109/ACCESS.2024.3358398

[28] Wenk, J., Voigt, I., Inojosa, H., Schlieter, H., and Ziemssen, T. (2024). Building digital patient pathways for the management and treatment of multiple sclerosis. In *Frontiers in Immunology* (Vol. 15). https://doi.org/10.3389/fimmu.2024.1356436

[29] Ai, B., Liang, Y., Yan, T., and Lei, Y. (2025). Exploration of immune cell heterogeneity by single-cell RNA sequencing and identification of secretory leukocyte protease inhibitor as an oncogene in pancreatic cancer. *Environmental Toxicology*, 40(6). https://doi.org/10.1002/tox.24200

Session 2 Embedded Systems and VLSI Technologies

Smart Systems and Beyond – Bhatia et al. (Eds)

A Robust Adaptive Control System for Fire Fighting Robots

Anagani Vijayashanthi and Buddha Hari Kumar
Vignan's Institute of Engineering for Women, Visakhapatnam, Andhra Pradesh, India

Panuganti Ravi
Raghu Engineering College, Visakhapatnam, Andhra Pradesh, India

Muppalla Tharangini
Gayatri Vidya Parishad College for Degree and PG Courses (A), Visakhapatnam, Andhra Pradesh, India

Srikalyani Nageti and Bammidi Pradeep Kumar
Vignan's Institute of Engineering for Women, Visakhapatnam, Andhra Pradesh, India

ABSTRACT: This Paper designs a low-cost fire fighting robot to offer initial fire extinguishment until the arrival of professional assistance. The robot consists of three main components: hardware, electronic interfacing circuits, and software. Four battery-powered motors (BO motors) power the robot, allowing movement through difficult terrain where human access might be hazardous. The robot is programmed to detect and put out fires and serves as a first-response unit to reduce fire damage in emergency situations. Time plays an important role in fire fighting, and the robot is meant to stop small fires from spreading by quenching them immediately. The robot is controlled with Arduino Uno development board and has a flame sensor for sensing the fire. When detected, water tank and pump system engage to spray water, with the nozzle being mounted on a servo motor for maximum coverage. Although this design is a starting point, there is a lot of room for improvement, such as increasing the robot's ability to rescue victims. The main purpose of this robot is to be an unmanned support unit that can detect and put out fires independently. Through the utilization of such robots, fire detection and rescue missions can be conducted more precisely and securely, reducing the risk of exposure of fire-fighters to dangerous situations. Eventually, this technology can greatly enhance fire safety and response measures.

Keywords: Servo motors, flame sensors, Arduino Uno, robust adaptive, water sprinkler

1 INTRODUCTION

Incidents of fires represent serious dangers both to life and property. During 2021, India reported about 1.6 million fire accidents in which 27,027 persons lost their lives. In fact, the number of such accidents due to cooking activities came to 42% of all cases. What is more, the dangerous circumstances involved in combating fires result in a loss of 60-70 fire-fighters who work during those times. In [1], a novel track-belt robot is designed, and a robust controller based on PID is designed for tracking control. Also contributing to risks inherent in fighting fires are unpredictable tendencies of flames to behave randomly within closed, smoky compartments [5]. Technological progress in robotics holds great promise for lessening such risks. Moreover, robots are not affected by poisonous fumes or carbon monoxide, hence they can enter places where it is not safe for humans to enter [2].The inclusion of such robots in fire fighting teams not only increases the protection of fire fighters but also maximizes the effectiveness of emergency response operations. Through rapid detection and extinction of fire origins, such robots can keep fires from spreading uncontrollably. The fire fighting robot has been upgraded to detect and extinguish fires in impacted regions [3].

Additionally, their capability to move through dangerous landscapes and tight spaces allows them to access victims more rapidly, hence minimizing the danger of harm to the victims [4]. This paper presents a prototype fire fighting robot that can detect and put out fires before they turn out of control. The aim is to come up with a robotic system that can simultaneously assist human fire fighters, effectively improving the overall efficiency and safety of fire fighting operations.

2 LITERATURE REVIEW

A.K.Tanyildizi proposed a new transformable wheeled robot with an extinguishing ball shooting turret. The main mechanism of the system is designed to perform focused fire extinguishing, especially in hard-to-reach regions

DOI: 10.1201/9781042026289-14

where human entry is not easy. The control system is proposed to provide stability when the robot travels over varied terrain, which makes it an acceptable solution for emergency fire fighting missions. The recoil of the ball-shooting mechanism can destabilize the robot and make it hard to keep balance, particularly on uneven terrain or under high turret rotation. The stability of the overall system is reduced when the shooting mechanism is triggered at high velocities [6].

Hassan N, Saleem proposed A Neural network-based adaptive controller for trajectory tracking of wheeled mobile robots stability and performance. The process targets the independent tuning of inner and outer control loops, leading to improved disturbance rejection and improved manoeuvrability efficiency. Decoupling the control objectives enhances robustness overall to external disturbances. Although this control system offers better stability, the design of the controller and the parameter tuning process is complicated. It needs a greater computational burden for implementation, and hence it becomes difficult to implement in resource constrained embedded systems. The practical application of the dual-loop control system needs a lot of tuning and verification in real time environments, which can be time consuming [8].

3 MODELLING OF FIRE FIGHTING ROBOTS

The following equations represent the dynamic model of a Unmanned Fire fighting Robot (UFFR) travelling in the horizontal plane based on its movement, forces that act upon it and with a towed water pumping station. Let us go through step by step.

(1) Awareness of the configuration of the system UFFR has:

Three degrees of freedom (DOF)in the 2D plane:

- X direction translation
- Y direction translation
- Rotation along the Z axis (φ-yaw angle)

Two independent drive wheels: That means it is able to move forward, in reverse, and turn independently.

A towed water pumping system:

It features a water pipe/hose behind it.

Drag force on the water pipe (F_p) affects the movement of the robot.

The first point of contact between the ground and the water pipe is where all the drag is assumed to be concentrated.

Motion Equations (Newton-Euler Dynamics): The dynamics of the robot are characterized by the Newton-Euler equations that describe linear motion and rotational motion.

(i) **Linear motion**

$$m\dot{v} = F_{\hat{r}} + F_{\hat{i}} - F_P \mathrm{CO}(\beta_1 + \beta_2) \tag{1}$$

m $\rightarrow$ mass of the robot.
$\dot{v} \rightarrow$ the acceleration of the robot.
$F_{\hat{r}} \rightarrow$ force of the right wheel.
$F_{\hat{l}} \rightarrow$ force produced by the left wheel.
$F_p \rightarrow$ drag force caused by the pipe of water
$\beta_1, \beta_2 \rightarrow$ orientation angles of the water pipe.

This formula expresses the fact that the net force experienced by the robot is the summation of forces exerted by the two drive wheels minus the drag component of force in the direction of motion.

(ii) **Rotational Motion of the UFFR**

$$Jv\psi = F_{\hat{r}} \cdot l - F_{\hat{i}} \cdot l - F_P \cdot \sin \beta_2 \cdot S \tag{2}$$

$J \rightarrow$ moment of inertia of the robot.
$\ddot{\psi} \rightarrow$ angular acceleration of the robot.
$l \rightarrow$ the distance between each wheel and the center of mass.
S $\rightarrow$ a parameter related to the arm of the drag force.
This equation expresses the torque equilibrium for rotational motion.

4 BLOCK DIAGRAM

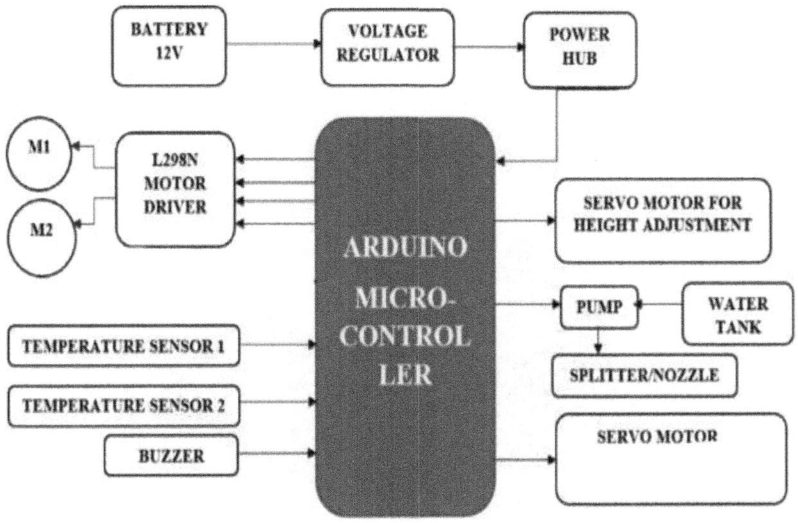

Figure 1. Fire fighting robot system.

This automated system is based on Arduino and includes a number of components, which individually contribute a lot towards automation and control. The system is programmed to control environment parameters, manage motors, and automate a pump mechanism as per sensor inputs. The system components are explained below in detail along with their operation:

4.1 *Power supply system battery (12V)*

The system is based on 12V battery which feeds all the parts in common, which are the Arduino, motor driver, and actuators.

4.2 *Voltage regulator*

Working for electronics, the voltage regulator regulates all the components to get the proper voltage to function correctly. Arduino and some sensors operate on 5V, while motors can operate on 12V or more which makes the regulator essential in a constant power supply.

Figure 2. Voltage regulator.

4.3 *Power hub*

The power hub efficiently works to distributes the electrical power to different components of the system without overloading and without hindering the proper working of [7].

4.4 *Microcontroller unit*

Arduino Microcontroller: The Arduino forms the central processing unit of the system and accepts input signals from sensors and gives corresponding control commands to the actuators like motors, pump, buzzer etc. It is responsible for the decision-making and automation based on programmed logic.

4.5 *Motor control system*

L298N Motor Driver:

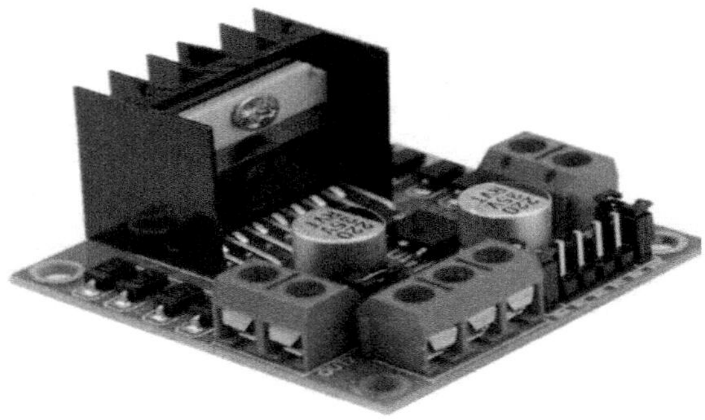

Figure 3. L298N motor driver.

This motor driver comes into play to be used for driving two DC motors (M1 and M2). The Arduino gives signals to L298N module which in turn drives motors forward, backward or stops according to the signal provided. The motor driver is important for applications that have movement or rotation.

4.6 *Motors (M1 and M2)*

These motors are used for mechanical movement in the system such as to adjust parts, move a robotic arm or drive a platform.

4.7 *Sensor module*

Temperature Sensors (Sensor 1 and Sensor 2):

Figure 4. Temperature sensor-1.

Figure 5. Temperature sensor-2.

Two temperature sensors are provided to continuously sense the environmental temperature. These sensors give real-time data to the Arduino to make some decisions to turn on the pump or alert via an alarm if temperature crosses some limit.

4.8 *Buzzer*

Buzzer act as a notification system. The Arduino activates it when there are specific conditions, like excessive temperatures or system faults, giving an audible alarm.

4.9 *Servo motor*

Figure 6. Servo motor.

The servo motor provides precise height adjustments of a component, like a nozzle, robotic arm, or sensor module. The Arduino makes its movement dependent on sensor inputs. There is another servo motor which controls a different mechanical operation, for instance, steering the nozzle or dynamically positioning a component [12].

4.10 *Pump*

The pump is used for fluid transport, e.g., pumping water from the water tank to a particular point. It is regulated by the Arduino, which switches it on or off depending on temperature levels or other factors.

4.11 *Water Tank*

The water tank is a storage unit for water or other liquids utilized in the system. The pump sucks water from the tank when required. A fire arm that carries a sprinkler is located on top of the body of the robot shaped like a battle tank [4].

4.12 *Splitter/Nozzle*

The splitter/nozzle assists in the channeling of water or fluid flow, so there will be even distribution. This may be applied in usage such as irrigation or cooling systems.

5 EXISTING METHOD

The paper contrasts various control methods for unmanned fire-fighting robots (UFFRs), each of which has been proven to place pre-specified performance constraints and react to uncertainties and external disturbances to conduct fire fighting operations. Barrier Lyapunov Function Method is employed to place state variables constraints and guarantee sustainability, and Prescribed Performance Function Technique guarantees transient and steady-state performance to realize pre-defined requirements.

State Dependent Transformation Based Method is employed to address constraints in multiple-tractor-trailer mobile robotic systems (MRSs). Back stepping Control is a general method which is not pre-specified performance function-based to place error variables constraints, and Classical Prescribed Performance Control (PPC) is the same as Back stepping but with prescribed performances. PID Control, a generic proportional-integral-derivative control method, is also compared in terms of suitability in controlling the intricate dynamics of fire fighting robots. The methods contribute overall to enhancing the reliability of the robots to perform under challenging conditions.

6 PROPOSED METHOD

This is a flame fire detection circuit employing three flame sensors connected to an Arduino Uno to measure the intensity of the fire. The sensors are connected to the Arduino analog pins A0, A1, and A2, which output analog signals proportional to the fire levels detected. The Arduino takes the inputs to determine the intensity of the fire and takes related actions through LEDs and a buzzer.

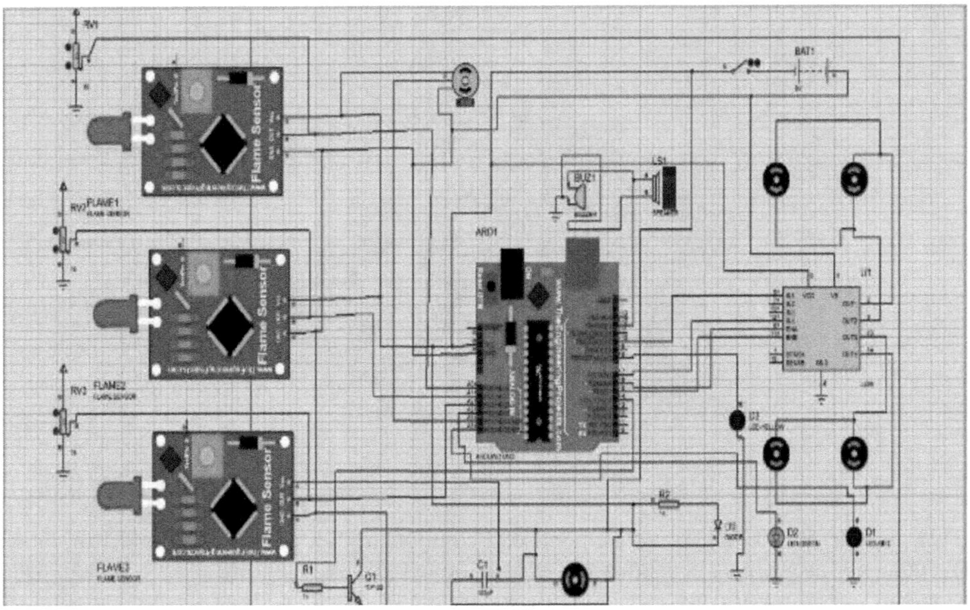

Figure 7. Circuit diagram.

When the flame level is high, the red LED lights up [9], signaling danger, and the buzzer provides continuous beeps to warn the user. When the flame level is medium, the yellow LED lights up, and the buzzer gives intermittent beeps, signaling caution. When the flame level is low, the green LED lights up, signaling safety, and the buzzer is not lit.

Fire fighting robots provide an array of high-end features to ensure efficiency and safety in response to emergencies. The robots have the capability to identify fires automatically in real-time with the aid of heat sensors and cameras and proceed to the location, discharging suitable extinguishing agents such as water, foam, or CO2 [13]. The robots also have the capacity to be controlled remotely by the fire responders, providing access to hazardous or inaccessibility places such as tall buildings, hazardous chemical plants, or destroyed buildings [10]. Thermal imaging cameras are installed in the robots to deliver sharp images of hotspots and hidden flames to ensure accurate fire containment. In addition to fire fighting, they can play a vital role in search and rescue operations, by identifying trapped people with their higher capabilities and sending some real-time data to the rescue operations. Environmentally sound extinguishing agents can be used with special nozzles or green agents that both save water and have less impact on the environment [11].

The robots are also precious in the suppression of the fires involving the hazardous materials, reducing the exposure of the humans with the hazardous substances. Built-in communication modules, such as GSM modules, can be used to achieve effective coordination by providing real-time data to fire stations and hospitals [14]. Further, the robots get the capability of collecting and transmitting environmental data such as temperature, air quality, and the amount of smoke to the first responders to assist them in analyzing such situations and coming up with their plan effectively [15].

7 RESULT

The developed fire fighting robot was successfully designed and implemented with the distance measurement.

Distance Measurement: The robot successfully identified objects at a range between 2 to 50 centimeters. This capability is important in order to navigate towards the sources of fires and to avoid obstacles during the operations. An Arduino-based fire fighting robot with GPS is an advanced robot that locates the fire and puts it out, and also tracks its location in real time. And, equipped with flame sensors, the robot immediately and automatically identifies sources of fire. Once it reaches the fire location, it deploys its fire suppressing mechanism, e.g. water pump or fan, to extinguish the fire.

Figure 8. Image of the model.

The GPS module embedded in the system keeps the location of the robot constantly tracked and send coordinates to the remote monitoring system or mobile app providing real-time situational awareness to it. The robot can also be made to automatically navigate around obstacles with the help of ultrasonic sensors, which helps the robot in its autonomous navigation. The system not only helps in minimizing human presence in dangerous environments but also allows for quick response time and accurate location tracking that makes it very effective in fire accidents. Minimizing humans exposure to dangerous environments but also allowing fast response times and precise tracking of the location, which makes it perfectly suitable in case of a fire accident [19].

8 CONCLUSION

The autonomous fire robot is supposed to help firefighters to increase numbers of challenging and dangerous contexts [16]. The articulated robot uses sophisticated sensors and actuators to move around its environment, being able to sense fires, and immediately suppress them, Infrared flame sensors help the robot to precisely locate the fire, with a quick response. It has heavy-duty motors and strong wheels that can offer greater mobility and allow it to travel over rough terrain and arrive on the scene of accident in short time. Upon detection of a fire the robot activates its submersible water pumps and servo motors to generate focused spray of water to the fire [18].

In addition, the robot is also capable of GSM tracking and can be used for real time location tracking and communication. Real-time alerting of firefighters and tracking the movement of the robot are possible, which improves situational awareness and coordination. The risk to human firefighters is reduced because the robot is used to perform hazardous work, therefore enhancing the effectiveness and safety of fire fighting operations [17].

REFERENCES

[1] Zhong L, Jiang X, Jia W, Shi W. 4WS Intelligent Fire-Fighting Robot Trajectory Tracking Control Based on Adaptive Cornering Stiffness. *IEEE Access*. (2024), 12:36083–95.

[2] Aliff M, Sani NS, Yusof MI, Zainal A. Development of fire fighting robot (QROB). *International Journal of Advanced Computer Science and Applications*. (2019);10(1).

[3] Zhu J, Li W, Lin D, Cheng H, Zhao G. Intelligent fire monitor for fire robot based on infrared image feedback control. *Fire Technology*. (2020);56(5):2089–109.

[4] Rakib T, Sarkar MR. Design and fabrication of an autonomous fire fighting robot with multisensor fire detection using PID controller. In *2016 5th International Conference on Informatics, Electronics and Vision (ICIEV)* (2016); (pp. 909–914). IEEE.

[5] Kim, J.H., Jo, S. and Lattimer, Feature selection for intelligent fire fighting robot classification of fire, smoke, and thermal reflections using thermal infrared images. *Journal of Sensors*. Volume 2016, Article ID 8410731, 13 pages

[6] A.K. Tanyildizi, "Design, control and stabilization of a trans formable wheeled fire fighting robot with a fire-extinguishing, ball-shooting turret," *Machines*, vol. 11, no. 4, p. 492, (2023), doi: 10.3390/machines11040492.

[7] Mithra C, Jayaprakash K, Indumathi J. IoT-Enabled Fire-Fighting Robots. Artificial Intelligence and Society.:47.

[8] Hassan N, Saleem A Neural network-based adaptive controller for trajectory tracking of wheeled mobile robots. *IEEE Access*. (2022);10:13582–97.

[9] Basiri M, Gonçalves J, Rosa J, Bettencourt R, Vale A, Lima P. A multipurpose mobile manipulator for autonomous fire fighting and construction of outdoor structures. *Field Robotics*. (2021);1:1–25.

[10] Madhevan B, Sakkaravarthi R, Singh GM, Diya R, Jha DK. Modelling, simulation and mechatronics design of a wireless automatic fire fighting surveillance robot. *Defence Science Journal*. (2017);67(5):572.

[11] Liu P, Yu H, Cang S, Vladareanu L. Robot-assisted smart fire fighting and interdisciplinary perspectives. In *2016 22nd international conference on automation and computing (ICAC)* (2016); (pp. 395–401). IEEE.

[12] Aoi S, Manoonpong P, Ambe Y, Matsuno F, Wörgötter F. Adaptive control strategies for interlimb coordination in legged robots: a review. *Frontiers in neurorobotics.* (2017);11:39.

[13] Dalvi K, Dhapate A, Adik R, Jadhav P. *Automatic fire fighting robot using IOT.* Available at SSRN 4841278. (2024).

[14] Ambafi JG, Onuche OP, Onuche PU, Babatunde ET, Abdullahi AO, Osazuwa P, Ogunnaike O, Okonkwo UU. *Development of a Novel IoT-Based Smart Fire fighting Robot for Real-Time Fire Detection and Suppression* (2025).

[15] Nadakuditi G, Venakteshwarlu S, Kumar S, Pulluri H, Preeti P, Goud BS. Thingspeak Cloud Server Based IoT Equipped Fire fighting Robot. In *E3S Web of Conferences* (2025) (Vol. 616, p. 03022). EDP Sciences.

[16] Jahan N, Niloy TB, Silvi JF, Hasan M, Nashia IJ, Khan R. Development of an IoT-based fire fighting drone for enhanced safety and efficiency in fire suppression. *Measurement and Control.* (2024);57(10):1464–79.

[17] Latif A, Shankar K, Nguyen PT. Legged fire fighter robot movement using PID. *Journal of Robotics and Control (JRC).* (2020);1 (1):15–9.

[18] Liu P, Yu H, Cang S, Vladareanu L. Robot-assisted smart fire fighting and interdisciplinary perspectives. In *2016 22nd international conference on automation and computing (ICAC)* (2016) (pp. 395–401). IEEE.

[19] Medewar AG, Sawarkar AD, Kshirsagar UV, Mr AG, Kshirsagar U. A review on fire and smoke detection with intelligent control for enhanced safety using machine learning (ML) and Internet of Things (IoT). *Cureus Journals.* (2024);1(1).

Smart Systems and Beyond – Bhatia et al. (Eds)

Deep Learning Approaches for Network Intrusion Detection Systems: Architectures, Challenges, and Future Directions

Rigana Fathima, N and D. Murugan
Department of Computer Science and Engineering, Manonmaniam Sundaranar University, Tirunelveli, TamilNadu, India

ABSTRACT: To adapt the increasing scope and complexity of attacks against digital infrastructure, Intrusion Detection Systems (IDS) must be used to detect intrusions into a network. Traditional signature-based and rule-based IDS work effectively against known attacks, but they fail to detect new or more complex types of attacks. Recent breakthroughs in Deep Learning (DL) technologies enable an IDS to automatically detect complex patterns, allowing the system to accurately detect attacks (previously unknorwn). This research article investigates various contemporary DL-based IDS architectures (e.g. CNN, RNN, LSTM, and Hybrid models) and provides performance comparisons with other traditional IDS methods using standardized benchmark datasets. The author discusses some key issues of managing encrypted traffic, evasion tactics, and scalability when deploying IDS within large networks. Future research areas identified by the Author include creating AI with Explainable AI to improve Understanding and Reliability, and developing Lightweight Models suitable for Internet of Things (IoT) environments. The outcome exhibits the deep learning is important to contribute in the innovation of intrusion detection. Due to this impact, researchers have the privilege to work on flexible, resilient, and intelligent cyber security solutions.

Keywords: LSTM, CNN, RNN, and NIDS

1 INTRODUCTION

With the evolution of internet and connected devices growing rapidly; protecting against attacks on network infrastructure remains an important limitation. The network intrusion detection system (IDS) serves as one of the important ways to secure against questionable actions and identify possible attacks. The Traditional solutions for IDS rely on predefined signatures and rule sets, which are limited to known or historical threats when detecting new or dynamically changing types of attacks. The need for leveraging deep learning (DL) algorithms has arisen as they will automatically identify features that distinguish between valid traffic and malicious objects based on unprocessed input and adapt to complex and constantly changing nature of attack patterns. This study provides an overview of current deep learning-based approaches in IDSs, analyzes their benefits, reviews their challenges, and offers insight into future research directions in this field.

2 RELATED WORK

The Machine learning-based intrusion detection has been the subject of several researches, with deep learning architectures receiving greater attention recently. When it comes to extracting spatial features from network traffic, CNNs have demonstrated greater performance. When it comes to representing temporal dependencies in sequential data, RNNs and LSTM networks in particular perform exceptionally well. In comparison to more conventional classifiers like SVM, Random Forest, and KNN, hybrid models that include CNN and LSTM layers make use of both spatial and temporal representations. Additionally, autoencoder-based unsupervised methods have shown successful in anomaly identification situations. Notwithstanding these developments, issues with data quality, encrypted communication, adversarial evasion, and deployment limitations still exist.

3 THE METHODOLOGY

3.1 *Deep learning architectures*

The Convolutional Neural Network (CNN) uses Convolutional filters to process incoming inbound traffic. These filters will also be applied to all other data contained within an Internet Protocol (IP) packet. Because of this, the Convolutional Neural Network (CNN) retains their important location characteristics and significantly reduces the number of dimensions they represent within their resulting feature set.

DOI: 10.1201/9781042026289-15

3.2 *LSTM and recurrent neural networks*

The sequential input data is modelled to build the intention of keeping internal memory states, allowing them to learn long-term dependency relationships in a sequential order. The Long Short Term Memory (LSTM) uses gate-based processes to prevent vanishing gradients from occurring while enhancing the ability to learn within sequences. The hybrid Convolutional Neural Network (CNN) utilizes the LSTM to simultaneously learn a both temporal and spatial model enables us to deal with the complexity inherent in most forms of the network traffic. The Auto-encoders learns compact representations of normal behaviour and to detect anomalies without the need for labelled data.

3.3 *Selection And Feature Engineering*

Although human feature engineering is mostly eliminated by deep learning models, feature selection techniques like XGBoost and Particle Swarm Optimization (PSO) are still helpful for increasing model efficiency and lowering overlapping.

4 DEEP LEARNING TECHNIQUES FOR IDS

4.1 *Motivation for deep learning*

Large volumes of data, intricate features, and rapidly evolving attack tactics are characteristics that characterize the field of network security. Due to this advantage, the traditional machine learning (ML) techniques have a notable limitation. The Deep learning (DL) is a powerful choice to solution for the these basic issues by initiated an unprocessed data.

5 LIMITATIONS OF TRADITIONAL MACHINE LEARNING

Manual Feature Engineering Bottleneck: Conventional machine learning models (such SVMs and Decision Trees) rely on subject-matter specialists to manually generate and choose pertinent features (such as "packet size," "flow duration," and "number of failed login attempts"). It is a tedious and wasting more time. It also need have through knowledge of network based data science. Subjective and Incomplete: Experts could fail to notice minute and complex trends that might point to novel attack types. Non-Adaptive: the engineered characteristics are set in stone. The feature set can become obsolete if the nature of assaults changes, requiring a full cycle of re-engineering.

6 CHALLENGES IN REPRESENTING COMPLEX, HIERARCHICAL STRUCTURES:

Cyber-attacks usually include a series of sequential, multi-step steps. For example, steps like reconnaissance, exploitation, command and control, and data exfiltration may be included in a complex threat, such as an Advanced Persistent Threat (APT). These long, hierarchical temporal linkages are hard for traditional machine learning methods to handle.

7 RECENT ADVANCES

Recent research have achieved over 98% accuracy by using hybrid deep learning models to benchmark intrusion detection datasets (CICIDS2017, UNSW-NB15). Other methods, including integrating deep neural networks with feature selection algorithms (like XGBoost or Particle Swarm Optimization), improve detection performance and processing economy even further.

CICIDS2017: It simulates modern attack scenarios with labelled traffic data.

UNSW-NB15: It contains diverse attack categories for robust testing.

8 EVALUATION METRICS

The following metrics are considered to compute detection effectiveness and robustness. They are namely accuracy, precision, recall, F1-score, and false positive rate.

9 RESULTS AND DISCUSSION

The CNN-LSTM hybrid model successfully captured both temporal sequences and spatial correlations in network data, consistently delivering higher accuracy. While auto-encoders were good at detecting anomalies, their accuracy was somewhat lower than that of supervised techniques. An increased false positive in encrypted traffic and resource-intensive processing that has to be optimized for edge deployment are among the difficulties noted.

Table 1. Comparative analysis for deep learning architectures.

MODEL	DATASET	ACCURACY	PRECISION	RECALL	F1-SCORE
CNN	CICIDS2017	96.8%	97.2%	95.9%	96.5%
LSTM	UNSW-NB15	97.6%	97.4%	97.9%	97.6%
CNN-LSTM	IoT-ID20	98.5%	98.7%	98.3%	98.5%
Auto-Encoder	CICIDS2017	95.4%	96.0%	94.7%	95.3%

10 CHALLENGES AND LIMITATIONS

Packet payload visibility is restricted by encrypted traffic, necessitating the use of innovative feature extraction methods.

11 ATTACKS BY ADVERSARIES

Adversarial perturbation avoidance efforts are common in deep learning models.

12 LIMITATIONS ON SCALABILITY AND RESOURCES

There are difficulties with computational demands, latency, and energy consumption when implementing deep learning models in large-scale environments

13 LABELING AND DATA QUALITY

Supervised learning requires high-quality, labelled network traffic data, but real-world datasets are frequently unannotated, unbalanced, or incomplete. The Future Research Directions Lightweight DL Models for IoT. In order to provide ubiquitous, real-time intrusion detection, research is moving closer to effective, lightweight DL architectures designed for implementation on IoT and edge devices.

14 XAI, OR EXPLAINABLE AI

Enhancing deep learning models‘ openness is essential for analyst operational usage, compliance, and confidence. Explainable strategies (such saliency maps and attention processes) are becoming more popular. Federated and Distributed Learning. The Federated learning protects privacy and improves threat intelligence by facilitating cooperative model training across several devices or organizations without direct data exchange.

15 ADVERSARIAL ROBUSTNESS

The goal of ongoing research is to strengthen DL models against adversarial attacks by using robust architectural design, data augmentation, and adversarial training.

16 CONCLUSION

Finally, the deep learning-powered network intrusion detection systems are leading the way in cyber security innovation and have the potential to develop into more durable, intelligent, and adaptable defenses against ever more-advanced cyber threats. In order to close the gaps between theoretical developments and actual network security issues, more multidisciplinary research and real-world implementations are essential.

REFERENCES

[1] S. Chatterjee, S. Chaudhary, and A.K. Cherukuri, "Intrusion detection system using deep learning for network security," *arXiv:2505.05810*, 2025.

[2] Jue Chen and Meng Cui, "Hybrid CNN-LSTM model for network intrusion detection," in *ACM Symposium on Security,* 2023.

[3] R. Kimanzi, P. Kimanga, D. Cherori, and P.K. Gikunda, "Deep learning algorithms used in intrusion detection systems: A review," *arXiv:2402.17020*, 2024.

[4] Haifeng Lv, Yong Ding, "A hybrid intrusion detection system with K-means and CNN+LSTM," *EAI Endorsed Transactions on Scalable Information Systems*, 2024.
[5] Katkuri, "A hybrid CNN-LSTM approach for intelligent cyber intrusion detection system," *GitHub*, 2024.
[6] J. Kim, Y. Shin, and E. Choi, "An intrusion detection model based on a convolutional neural network," *Journal of Multimedia Information System*, vol. 6, pp. 165–172, 2019.
[7] Y. Kim *et al.*, "Convolutional neural networks for network intrusion detection," in *Proceedings of IEEE Conference on Communications*, 2023.
[8] M. Patel and R. Kumar, "Autoencoder-based anomaly detection for intrusion systems," *Journal of Information Security*, 2024.
[9] M. Roopak, G.Y. Tian, and J. Chambers, "Deep learning models for cyber security in IoT networks," in *2019 IEEE 9th Annual Computing and Communication Workshop and Conference (CCWC)*. IEEE, 2019.
[10] Smith and B. Lee, "LSTM-based anomaly detection in network traffic," *IEEE Transactions on Neural Networks*, vol. 34, pp. 1004–1015, 2024.
[11] P. Sun *et al.*, "DL-IDS: Extracting features using CNN-LSTM hybrid network for intrusion detection system," *Security and Communication Networks*, 2020.
[12] S. Wang and L. Zhao, "Challenges in deep learning based network security," *Cyber Security Journal*, vol. 10, no. 2, 2023.
[13] Alashjaee, A.M, "Deep learning for network security: An Attention-CNN-LSTM model for accurate intrusion detection" *Scientidic Reports* 15, 21856, 2025.
[14] Sydney Mambwe Kasongo, "A deep learning technique for intrusion detection system using a Recurrent Neural Networks based framework", *Computer Communications*, Volume 199, Pages 113–125, 2023.
[15] Vanlalruata Hnamte, Jamal Hussain, "Dependable intrusion detection system using DCNN: A Novel framework and performance evaluation approach," *Telematics and Informatics Reports*, Volume 11,100077, 2023.

Smart Systems and Beyond – Bhatia et al. (Eds)
© 2027 The Author(s), ISBN: 978-1-041-36617-1

A Hybrid Strategy for Handwritten Text Recognition to MobileNet, LSTM, and Transformers

Buddha Hari Kumar and P. Chitra
Department of ECE, Sathyabama Institute of Science and Technology, Chennai, Tamil Nadu, India

Bammidi Pradeep Kumar
Vignan's Institute of Engineering for Women, Visakhapatnam, Andhra Pradesh, India

Chukka Ramesh Babu
Electronics and Communication Engineering, Vignan's Institute of Information Technology, Visakhapatnam

Sri Kalyani Nageti, Yelothoti Sravana Kumar, and Dalli Poornima
Vignan's Institute of Engineering for Women, Visakhapatnam, Andhra Pradesh, India

ABSTRACT: Handwritten text recognition (HTR) remains an ongoing challenge in which various ways of writing combined with several ways of handwriting generate huge recognition complexity. The paper has detailed a novel HTR solution by studying Vision Transformers and MobileNet-LSTM hybrid architecture to maximally improve the recognition precision and operation speed. The Vision Transformer is an example of how it does this extremely well in the case of long-distance image relationships to extract meaningful characteristics from the handwritten information. MobileNet serves to minimize the computational requirement of the system while maintaining high standards of performance and the temporal dependency skill of the system has also been achieved through the application of Long Short-Term Memory (LSTMs) network for modeling of the sequence of handwriting. Engineered from convolutional neural network (CNN)-based approaches the proposed evaluation method provides better recognition rate on standardized benchmark dataset. Experimental data to assert the ViT and MobileNet-LSTM combination can solve real-time handwritten text recognition problems in terms of high accuracy rate and efficient computational requirements.

Keywords: Handwritten Text Recognition, Vision Transformer, MobileNet, LSTM, Deep Learning, Sequence Modeling, Computer Vision, Feature Extraction, Real-Time Recognition

1 INTRODUCTION

The identification of handwritten text (HTR) is one of the basic research topics in the machine learning and computer vision field with the connections of the artificial intelligence research field. Machine-readable text results from the processing of handwritten characters or words. The distinctive nature of handwritten text compared to printed text poses substantial obstacles to HTR systems because individual writing style variations make the task complex and difficult for HTR. HTR technology finds multiple uses because it enables the digitization of historical documents as well as the development of automatic transcription systems that handle both handwritten notes and forms and signatures. Audio recognition systems require better efficiency and accuracy because digitized content and smart document management systems continue to gain acceptance. Accelerated research exists to build high-speed robust HTR systems that process real-time data accurately because of advancing dependence on mobile hand-held devices.

At the beginning developers constructed HTR methods using hand-crafted algorithms that combined template matching along with edge detection techniques and statistical calculations for character recognition. The techniques suffered numerous challenges stemming from inconsistent handwriting patterns as well as text noise and extreme format changes. The handcrafted nature of their features limited these methodologies in adjusting to variations in handwriting styles or font variations. Deep learning technology has transformed HTR applications through its data-driven automated solutions which scale for continuous improvement. The deep learning model convolutional neural networks (CNNs) brought remarkable success in image recognition so they were extensively utilized for HTR applications. Raw image data is transformed into hierarchical features by CNNs so these neural networks excel at processing handwriting complexity. CNN-based approaches have resulted in important progress yet the sequential handwriting structure remains difficult to handle since it requires full comprehension of word and character relationships.

The proposed solution leverages a hybrid framework which merges Vision Transformer (ViT) systems with MobileNet-LSTM (Long Short-Term Memory) architecture. The Vision Transformer utilizes the transformer

DOI: 10.1201/9781042026289-16

model from NLP for image processing while operating on visual information. Traditional CNNs use local receptive fields yet the ViT divides image inputs into fixed-size patches to process these segments using self-attention mechanisms. External dependencies across the image space become accessible through the ViT as a result of its ability to understand both local and global features making it suitable for handwritten text analysis. MobileNet functions as a compact CNN architecture specially optimized for scenarios needing mobile and embedded solutions. Depthwise separable convolutions in MobileNet enable performance retention alongside substantial parameter and computational expense minimization. MobileNet suits applications which require maximum performance from their hardware. When processing sequential data like handwriting LSTM demonstrates effectiveness due to its status as a recurrent neural network (RNN) appropriate for handling data relationships across previous characters or words. The proposed model unites ViT for extracting features with MobileNet for efficiency and LSTM for sequence understanding to produce an accurate and powerful HTR solution.

The combination of ViT with MobileNet-LSTM serves to deliver exceptional performance. ViT demonstrates excellent capability for perceiving worldwide graphical relationships but MobileNet keeps the model lightweight enough for execution on constrained platforms. The LSTM brings expertise in recognizing sequential patterns in handwriting that enables full character as well as whole word and sentence recognition. The combined approach shows promising results for delivering both fast processing together with superior recognition precision thus establishing itself as an excellent solution for mobile applications and digitizing documents.

Multiple dataset collections have been proposed throughout time to evaluate and benchmark HTR models. The IAM Handwriting Database maintains its position as the principal dataset with thousands of English handwritten text samples alongside their reference text. Strengthening HTR research came through two essential datasets including the MNIST dataset for digit recognition and the HWRT dataset for handwritten text document analysis. Current research challenges in Handwritten Text Recognition stem from accessing datasets that exhibit sufficient diversity across handwriting styles and languages and scripts. The recognition systems are still struggling today because of the variability in handwriting even with recent advances in the production of datasets. The need for advancing stronger recognition systems is imposed by the need to develop methods with good language and handwriting style generalization a need that motivates the proposed hybrid approach development.

Deep learning technology research has focused its research on improving the working performance of HTR systems. Recognition models need lower computational loads since accuracy enhancements have become intertwined with reduced computational needs for realtime systems. The availability of computer power is scarce, so this is an extremely important approach for mobile and embedded systems. The MobileNet network shows its importance by a high-quality output with regards to performance as well as efficiency metrics. The proposed hybrid framework consists of the combination of deep learning methods with lightweight networks, to provide optimized computational performance in the recognition tasks.

The primary challenge in HTR is getting access to large amounts of labelled data resources to achieve deep learning techniques training. Deep learning training of large datasets is not only expensive in terms of compute time but also, in terms of execution time. Obtaining labeled handwritten information turns out to be costly when experts are needed to generate large datasets which can handle different handwritings in many languages. Several methods of data augmentation are available in which the image is rotated at a scale and cropped to artificially increase the training data. Analysis shows that fine-tuning pre-trained models on particular HTR tasks can reduce the need for large amounts of labelled training data.

The combined architecture of Vision Transformer and MobileNet-LSTM demonstrates potential for addressing handwritten text recognition problems effectively. The proposed model incorporates Vision Transformer global feature extraction to fight recognition errors without sacrificing real-time processing capabilities through MobileNet efficiency and sequence modeling from LSTM units.

2 LITERATURE REVIEW

Handwritten text recognition (HTR) represents a popular research field which combines computer vision with machine learning approaches because it enables numerous applications in document analysis and archival digitization as well as automation systems. Various approaches starting from traditional machine learning techniques and continuing through deep learning methods have been applied to increase both accuracy and handling speed of handwritten text recognition systems.

A major limitation for handwritten character recognition systems appears through the wide range of handwriting styles found in texts. In the case of Tamil handwritten text recognition, Saikrishna *et al.* [1] compared two deep learning models: Two deep learning systems were implemented: the first utilized Histogram of Oriented Gradients (HOG) to extract features and the second system employed VGG-19 architecture as its convolutional neural network (CNN). Sigma IoT Laboratory Verified that VGG-19 models efficiently recognize Tamil script elements because this complex script includes Cursive handwriting with multiple supplementary signs.

Bhardwaj *et al.* [2] studied word-level recognition in their text recognition work although their main focus was not character-level recognition. The research team built a CNN model to extract features and showed that their handwritten-to-digital text conversion process worked successfully. Research by Saini *et al.* [3] included a survey which examined different machine learning and deep learning approaches that perform handwritten text recognition.

The researchers from Hassan *et al.* [4] studied cursive Urdu handwritten text recognition through implementation of bi-directional Long Short-Term Memory (Bi-LSTM) networks. The system employed CNN features alongside Bi-LSTM sequences for sequence modeling that unlocks the identification of context-based cursive scripts. The implemented system demonstrated strong results which indicate Bi-LSTMs provide an effective solution for handling contextual information in handwritten text recognition tasks. The authors studied handwritten text recognition limitations such as model fine-tuning alongside a comparison of CNNs and RNNs against traditional machine learning algorithms in their academic work [5].

Handwritten text recognition in Arabic remains a primary area of academic investigation. Alkhateeb *et al.* [6] executed a performance study between standard machine learning algorithms and deep learning frameworks for Arabic handwritten text recognition. Deep learning models demonstrated better accuracy and robustness than traditional approaches according to the study results [7]. Deep learning techniques show superiority in processing Arabic handwriting complexity because the script presents multiple letter forms along with diverse diacritical marks.

Hossain *et al.* [8] conducted research on Bangla handwritten text recognition leveraging DenseNet121 and MobileNet as well as Xception among their CNN architecture selection. Tests showed DenseNet121 surpassed competing architectures by delivering 95.64% test accuracy. The research indicates that sophisticated CNN modeling techniques are needed to manage complex Bangla scripts that exhibit merged characters alongside diverse writing approaches [9].

The recognition of handwritten, Ethiopic textual records needed the attention of Malhotra and Addis [10] since it has substantial characters and intricate features of the scripts. The researchers developed an end-to-end deep learning using attention systems and convolutional and recurrent neural networks in order to achieve higher recognition results. Deep learning models proved to be useful in processing historical texts while providing an answer to issues related to data limitations according to their research [11].

A deep learning system for recognising multiple handwriting styles using Long Short Term Memory (LSTM) combined with a Convolutional Neural Network (CNN) was developed by Jebadurai *et al.* [12] for general handwritten text recognition tasks. The research team trained their model using the IAM Handwriting Database which involves hand written samples that are written by more than 600 individuals. Their system showed training performance at 94% accuracy and validation accuracy at 85% which justifies the benefit of combining CNNs and LSTMs in sequence-based recognition algorithms.

A deep feature embedding framework for word spotting and recognition in handwritten text has been proposed by Krishnan *et al.*[13] The researchers have come up with an embedded attribute framework which enhanced the convolutional feature discriminative ability while enabling query by string execution. Their-word-spotting-and-word-recognition-model-exhibited-leading-performance-when-applied-to-the-IAM-corpus-and-different-historical-document-sets.

Due to the deep learning technique applied to character recognition, Jayanna *et al.* [14] gained high efficiency when CNNs and LSTMs were used to analyze the handwritten Kannada characters. Analysis based on three different architectural models consisting of Vision Transformer, CNN and LSTMs detected specific strong points as well as weak points for each model based on their study. The research proved the success of deep learning techniques such as CNN and LSTMs to reliably identify handwritten textual contents of several language and script types.

Sadaf *et al.* [15] added a full offline Bangla handwritten text recognition system that is a combination of the pre-trained neural networks MobileNet and Xception with recurrent neural networks based on LSM or GRU. The results observed by the researchers showed that Flor was superior in recognition results compared to the other models by generating a character error rate (CER) of 12.83% experiencing a word error rate (WER) of 36.01%.

Thendral *et al.* [16] proposed the Tamil handwritten character recognition using multimodal deep learning methods. The process integration of RESNET50 and VGG16 and LeNet50 resulted in the increased recognition accuracy by preserving the Tamil handwritten documents by digitization.

Determining Arabic handwritten text in biometric system using deep learning models have become the focus of Balat *et al.* [17] who investigated the need of examining deep learning models. The researchers were able to determine that EfficientNetB7 surpassed ResNet50 and MobileNetV2 in Arabic handwritten text recognition while achieving above 99% accuracy on test data. The success of EfficientNetB7 shows the power of deep learning models in catalyzing the innovations in identity verification and authentication solutions using handwritten text.

Handwritten text recognition research The progress of handwritten text recognition is immense with deep learning implementations using the CNN LSTMs and Vision Transformers. Multiple recognition methods have better rates of accuracy now in case of processing complex written languages that include Arabic in tandem with Bangla and Tamil and Kannada. Handwritten text recognition has a great advance from the complementing integration of CNNs and RNNs or LSTMs which solve the sequencing problems in handwritten documents. Findings of recent researches developed superior text recognition solutions which are operating in different language set and application scenario.

3 SYSTEM DESIGN AND ARCHITECTURE

The developed handwritten text recognition system uses Vision Transformer (ViT) in aid of MobileNet with the help of Long Short-Term Memory (LSTM) networks to build a hybrid deep learning framework. This system involves optimized architectural design to provide both high accuracy results and also high efficiency processing in order to initialize the system, works on the limited devices and powerful computational platforms. this system's design is theoretically structured into three key stages: feature extraction, sequential learning and final prediction. The first step is to use the Vision Transformer (ViT) approach to the feature extraction purpose. The input image receives processing by divided fixed-size patches from ViT that leads to multi-head self-attention operations along with the global image dependency identification. The full picture comprehension abilities of ViT allow it to excel at the recognition instances of a complex and variable handwritten text.

MobileNet implements the second phase of the system to reduce computational requirements. The depthwise separable convolution technique used by MobileNet reduces the model parameter count and computational requirements that makes it portable and effective but still accurate. MobileNet strikes the right balance between speed and accuracy by providing a system which works well with constrained mobile or embedded devices. An LSTM network takes responsibility to understand the sequence dependencies in handwriting signs at the third processing stage. Sequential nature is present in handwritten text because of the automatic temporal order which influences recognition of both single characters and words requiring understanding of them in their order. By processing ViT and MobileNet features the LSTM network preserves textual dependencies over long stretches of text to achieve successful transcription of entire sequences.The system has architectural components based on a modular design methodology along with scalability to fit more features for specific applications. The system delivers an efficient robust solution to the problem of handwritten text recognition by its implementation of state-of-the-art techniques.

4 DATASET

For this research the EMNIST (Extended MNIST) data set offers the basis for the data which consists of a wide variety of handwritten characters. The set contains English letters in lowercase and uppercase, and digits of 0-9 and special symbols `@ # $ &`.Several preprocessing operations were performed on the dataset to enhance recognition functionality particularly image transformation alongside size adjustment.Individual images from the dataset received black-and-white processing towards becoming 32x32 pixels for standardization before being applied to model training. The framework used 32x32 pixels dimensions because it provided adequate soundness for recognition and satisfactory performance during both the learning and evaluation phases.

The data separated itself into 39 distinct categories to help users recognize individual patterns. All lowercase and uppercase English letters formed a single classification category for each desired character within the reduction process. A solution for misclassification reduced this combination between digits containing '0' to a clear '0O' as both 'O' and '0' many times resemble each other in handwritten text. Analysis involved 39 classes partitioned into digits (1-9) and four special characters (@, #, $, &).The overall dataset was segmenting into train and validation portions to maintain a uniform distribution between categories. The dataset's deliberate structure enables the model to correctly distinguish between different characters with optimal performance results particularly when sorting visually related symbols and digits. The dataset functions as the essential resource to develop and test the proposed handwritten text recognition method.

5 DATA PREPROCESSING

Developing a long-lasting text recognition system by handwriting needs to be a fundamental starting point data preprocessing. EMNIST dataset underwent various preprocessing treatments in order to improve quality of data inputs and consistency which has led to better outcomes from the model. The first raw dataset imagery had to be resized to consistent data in 32x32 pixel data for standardization. The fact is that the work of converting to consistent image dimensions is necessary for the smooth functioning of deep learning model training processes. By resizing the data computing costs are reduced and important parts of the picture remain large enough for identification purposes. Conversion to grayscale produced black/white images in order to impact the model to important structural feeds on the characters and to eliminate the chaos of color variations between the characters that might impact the performance accuracy. The researchers chose dataset-wide normalization of pixel values as image normalization. The normalization step divided the pixel data between 0 and 255 by 255 to produce a value range between 0 and 1. kurumization allows for the model to get to the end of training faster in addition to producing more reliable results since the model produces less sensitivity towards any major pixel values differences.

The process used to get better generalization worked by combining other techniques which improved image augmentation. The random use of image transformation techniques like rotation scaling and translation was used to stretch the available dataset but increase the effectiveness of the model in identifying characters irrespective of changes in any characters orientation or contours. Handwritten text recognition benefits from this step a lot since different writing styles cause a lot of changing from character to character in appearance. The merging of categories happened to avoid classification errors between appearance-similar characters which include '0' and 'O' as explained in the dataset description. Model recognition achieved better accuracy because of this step which minimized possible ambiguities in the identification procedures.

6 METHODOLOGY

6.1 *Vision transformer*

Contextualizing the Vision Transformer (ViT) is an innovative computer vision model represented in an application of the transformer architecture keeping NLP origin for analysis of images. Image classification and object detection successes from ViT are a result of its ability to learn about the relationship between images in the distance. The long range dependence capability of ViT produces exceptional results for hand written text recognition as it makes stable connections between the text components found at different distances from each other. the VIT platform builds from hand written text recognition using the conversion of the input images (Hand written text images) into smaller blocks to be treated as tokens identically to NLP word processing. Through this patch system ViT uses transformer models to probe connections across the entire image. The working of ViT can be divided into the following:

Input Image (Text Image): Input images (handwritten text) are the mode of instigating the working mechanism of ViT. The input image can be in one of several formats with variations in the style of handwriting in addition to size and vertical placement. The model requires effective image processing through standardizing operations like resizing, scaling and normalized image for processing purpose. preprocessing: A process of image preprocessing starts from conventional measurement (32x32 pixels) followed by normalized pixel value transformation (0 to 1) with option transformation steps (noise filter) for reaching model compatibility.

Patch Embedding: The Post-pre-processing image is segmented into predefined parts of relatively equal dimension. An independent input token is allocated to each particular patch in the system. The model is used to transform these patches by flattening them but their linear projections enables the embedding into functional space for visual information processing. Linear Projection of Patches: In the case of embedded spaces the model applies the linear projection of individual image patches in order to extract complex feature patterns. Cleaned up pixel values had a transformation applied to them in order to produce semantically important features before passing through the transformer pipeline. Transformer Encoder Blocks: This is the main body of the Vision Transformer and consists of a series of transformer encoder blocks with multiple sub-layers within each:

Multi-Head Self-Attention: This helps the model to focus on different parts of the image at once. to learn contextual relationships between different patches of the image. For handwritten text this is particularly important since we need to understand the relationship of different letters or words relative to each other in context to be able to recognize them properly.

Feed-Forward Neural Network: This part will process the results from the self-attention mechanism and apply non-linear transformations to the results to extract higher-level features.Layer Normalization and Residual Connections: These will be used during training to stabilize training and make the training more efficient by avoiding the vanishing/exploding gradient problem and facilitates the flow of information.

Text Token Representation: The output of the transformer encoder is a token of text representation; each token is the representation of the features of the patch in the image. After that these tokens are then fed into a decoder for further sequence modeling.

Decoder (Language Model): Decoder part is the part which deals with sequence generation and the transformer architecture is used to generate a sequence of tokens which are the corresponding handwritten in the characters. This step consists of predicting the characters (or words) in the correct order, on the basis of the learned relations between the tokens.

Output Layer (Softmax): The final output is a probability distribution over that character is possible for each token that is generated using a softmax layer. This layer takes the raw model outputs and makes them available in some form of character predictions.

Final Output (Predicted Text): Sequences of the predicted characters are made by the model that represent a transcription of the handwritten text in the image. This sequence is then given as final output.

The Vision Transformer, also called ViT, is an implementation of the transformer architecture which has its origin in natural language processing for addressing image-related problems. The method of slicing the image into grids of non-overlapped pieces is the exclusive approach of ViT that distinguishes itself from traditional convolutional neural networks (CNNs) that use local field processing. Through its architecture ViT maintains global dependencies spanning entire images so as to provide excellent performance on complicated recognition operations such as handwritten text since it effectively links features that are spatially far apart. Through these operations ViT preserves the spatial characteristics of images and its parallel transformer encoders extract the required embedded

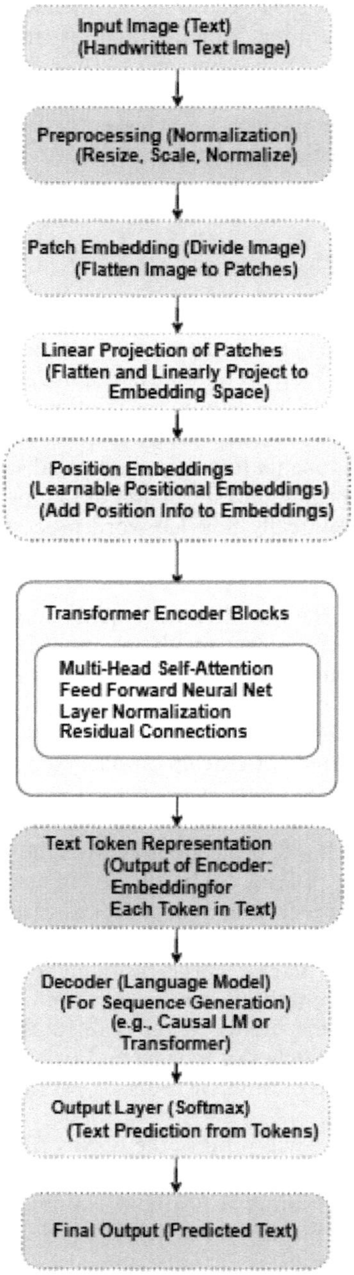

Figure 1. VIT architecture.

information in an efficient way. The flexible document processing method of such a model shows inspiring results in complex text recognition tasks because it is able to support various styles of handwriting and to recognize important spatial features.

Hybrid Model (MOBILENET+LSTM): The Hybrid MobileNet + LSTM model is to represent an enhanced deep learning model to identify handwritten text. The Hybrid MobileNet + LSTM model consists of MobileNet which is a lightweight CNN and LSTM networks for performing both spatial feature extraction and sequential modeling. The unified approach with MobileNet capabilities and LSTM structure is an optimal solution which achieves an exceptional performance and computing efficiency for the real-time handwriting recognition operations with limited capacity devices.

Input Image (Handwritten Text): The processing begins with handwritten text image which is in the form of a grayscale or RGB image. The images in this dataset show differences in handwriting styles while they have different levels of noise and dimension scale. Various preprocessing operations guarantee both the consistency of the model and the optimization of the input learning. A fixed image size and pixel normalization and optional RGB to grayscale conversion complete their order of preprocessing. Model performance improves upon grayscale conversion as it allows the model to assess structure rather than to use the color information.

Pre-processing: Input images standardization is a necessary input processing. A unified image size is needed to put inputs in the network so manufacturers have to use fixed dimension resizing to all pictures. Pixel value normalization sets digital images values in between 0 and 1 for a faster training convergence despite the support for learning stability.

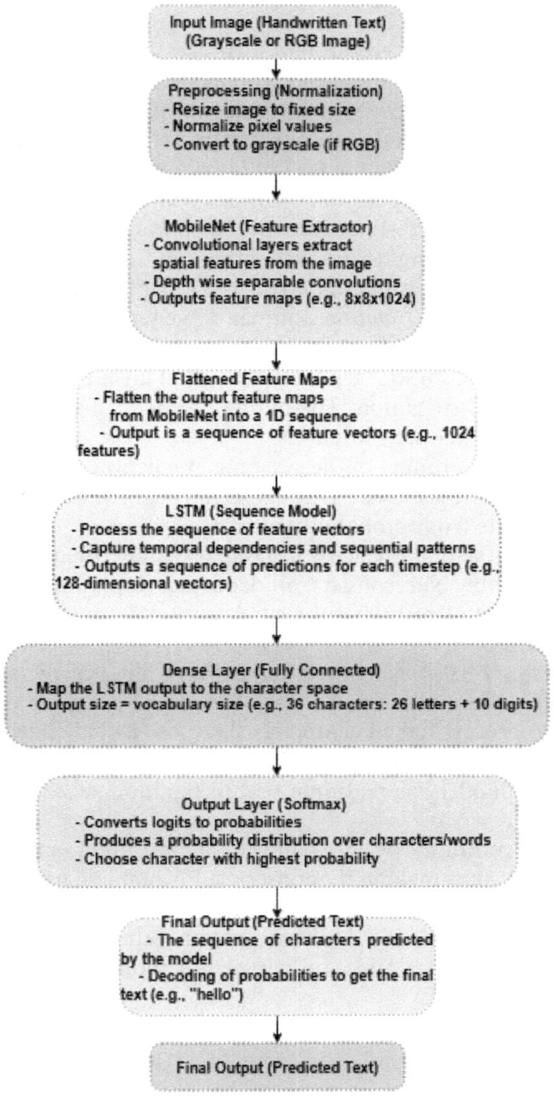

Figure 2. Hybrid model architecture.

MobileNet (Feature Extractor): The MobileNet network is the main feature extractor part of this architecture. MobileNet represents an efficient lightweight CNN which performs depthwise separable convolutions to cut down computational costs without negatively affecting its performance outcomes. Depthwise separable convolutions separate this process of the standard convolution operation into two parts: MobileNet uses two different convolutions to finish this process of feature extraction after each one makes the individual depthwise convolution for every input channel and the pointwise convolution across all channels. The depthwise separable convolution architecture allows MobileNet to achieve better accuracy results by having smaller parameter amounts and fewer computation tasks hence making them suitable for real-time applications that utilize devices under processing power constraints.

The MobileNet model separates the spatial characteristics that are present in the images of handwritten texts. There are essential visual markers which comprise edges and shapes and textures with other significant recognition indicators found in the feature set. An output feature map is used to describe the image information via compact spatial characteristics.

Flattened Feature Maps: MobileNet processes images into extracted features that become one-dimensional sequence vectors from flattened feature maps. Processing follows to convert the output into features with sequence format ready for LSTM evaluation. Every vector includes encoded image information from specific segments including letter recognition which MobileNet generates based on 1024 or more features across all architectural depths.

LSTM (Sequence Model): The hybrid model's central element brings together LSTM which functions as a recurrent neural network type. LSTM was specifically built to recognize patterns in data sequence because handwritten text requires this ability. Since handwriting displays a character sequence LSTM maintains the sequential order in processed feature vectors from MobileNet to retain both character positions and their encasing relationships throughout the image.

LSTM feeds one temporal unit each time step through its processing flow. The prediction data output from the system at each time step contains information about the probabilities of every possible character or symbol present at that sequence position. Without prior knowledge of temporal patterns the LSTM network excels at reading cursive or connected handwriting applications through its understanding of character-word relationship sequences.

Dense Layer (Fully Connected): Additionally the LSTM layer output goes into the dense (fully connected) layer which performs mapping between LSTM outputs and character space information. The output network layer gets its dimensions from the entire collection of distinct recognition characters. The model features an output vocabulary that combines 36 characters comprising 26 English letters and ten digits alongside spaces for punctuation marks and special symbols. After receiving output data from the LSTM the dense layer determines weights for each character class distribution using activation function Softmax to generate a probability map. The identification of specific text characters based on the features at each time step stands as an essential process for this model.

Output Layer (Softmax): In the last execution step the final output layer transforms raw prediction logits through a softmax function into a probability distribution. The output mechanism selects the most likely character among the different choices of the individual time step solutions.

Final Output (Predicted Text): Final product of the system is prediction of text which is resulting from decoding the probabilities generated from the softmax layer. Through using beam search and greedy decoding techniques the system extract sequences of characters from probability outputs.

This hybrid model offers some of the main benefits in tasks of handwritten text recognition:

Efficient Feature Extraction: MobileNet consists of depthwise separable convolutions which optimize the handwritten image analysis consisting of meaningful spatial feature extraction at low computational cost which suits mobile and embedded applications.

Sequence Modeling: The reason LSTM Networks successfully handles the time based relationships among characters in order to observe the ordering of written text. The model is able to detect the complex cursive handwriting thanks to its capability to process linked characters that pose a challenge for traditional CNNs.

Real-Time Processing: Through the combination of MobileNet's light weighting model design with LSTM sequence processing capability the model has real-time text recognition speeds on devices with limited resources without sacrificing accuracy.

Versatility: The model achieves versatility by its ability to support additional character types while supporting different writing styles and languages thus making it appropriate for multiple handwritten text recognition use cases.

The combination of MobileNet and LSTMs for efficient handwritten text recognition is an effective way to employ operational benefits from two networks to obtain spatial features and handle text sequences. The system acts as a useful tool for digitization projects and can be used for mobile applications as well as real-time handwritten pattern recognition applications.

7 RESULTS

The obtained results define a full assessment between the Vision Transformer (ViT) architectural design and the Hybrid MobileNet + LSTM implementation within the scope of handwritten text recognition. The researchers tested the models on benchmark datasets to look at their performance metrics which included accuracy alongside processing speed and the capacity to work in real-time. Analysis of approaches provides their key benefits and downsides which enables to assess the deployment readiness.

7.1 *Vision transformer model*

Confusion Matrix: A visual presentation of the Vision Transformer (ViT) model's handwritten text recognition performance is shown through this confusion matrix. Each row in the matrix shows actual character labels from the data while each column contains model predictions for those characters. On the diagonal the numbers show correct matches when the model identifies the expected output from its prediction. A model demonstrates robust character recognition ability for specific patterns when it produces large values positioning such as (0,0) and (1,1) on the main diagonal. The incorrect model predictions reveal themselves through off-diagonal matrix values since the model assigns predictions to classes different from their true identities. If the model encounters confusion between the characters '1' and '7' or '0' and 'O' it will generate elevated off-diagonal cell values at these combinations. Management of model performance becomes more straightforward when determining the characters and digits which cause difficulty for the model. Through matrix observation researchers can track how well the model performs across all classes. The model delivers strong performance when diagonal elements remain high but substantial deviations from this pattern activate opportunities for performance enhancement through data targeting or model modifications or training techniques. The confusion matrix serves as a critical diagnostic tool for model diagnostic and optimization for ViT based applications in the real world setting of handwritten text recognition.

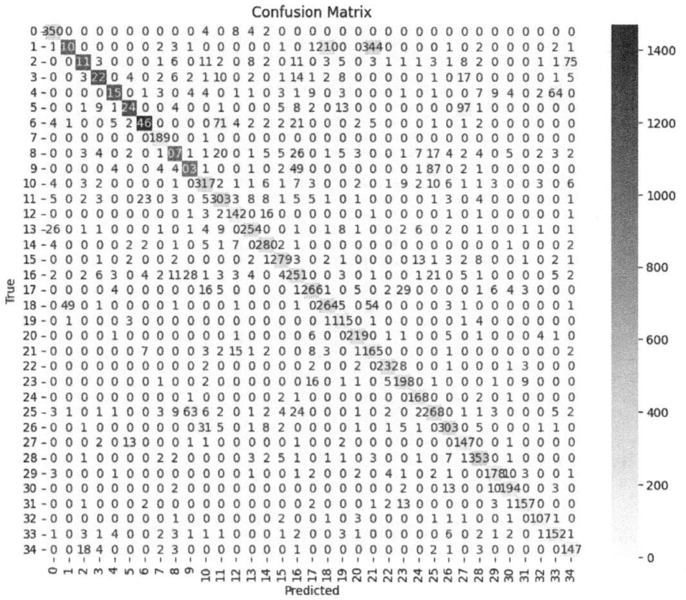

Figure 3. Confusion matrix of ViT.

Classification Report: The report of the portrayal offer an organised appraisal of the show of the model with varying classes, remembering both the English letters for request (A-Z), digits (0-9), and excellent characters (@, #, $ and). It consolidates key estimations, for instance, precision, audit, the F1-score, and aid for every subject. Precision appraises the precision of the model's positive figures. For example, for class '0', exactness is 0.87, inferring that 87% of the expected '0' names were truly correct. High precision shows that when the model predicts a class, being right is presumably going.

Survey exhibits how well the model recognizes certified instances of a class. For class '0', survey is 0.95, suggesting that the model precisely perceived 95% of all apparent '0' models in the dataset. High survey suggests that the model is perfect at recognizing events of a class, whether or not it misclassifies maybe one or two classes. F1-Score is the symphonious mean of exactness and survey, empowering a sensible extent to precisely portray events. For class '0', the F1-score is 0.91, which exhibits a good congruity among exactness and survey.

Support tends to the amount of real models for each class in the dataset. For class '0', support is 368, importance there are 368 instances of '0' in the test dataset. The report furthermore consolidates the precision, which is 0.85, meaning the model precisely expected 85% of all cases across all classes. This mirrors the overall reasonability of the model in perceiving composed by hand text.

The full scale ordinary gives an unweighted typical across all classes, and the weighted regular considers the assistance for each class, giving a more intensive execution measure. The weighted ordinary F1-score of 0.86 shows that the model performs well, especially for extra routinely tended to classes, yet there are locales like class 'V' and interesting characters, for instance, '#' where execution could be gotten to a higher level. Overall, the request report shows that the model performs generously in distinctive a large number deciphered characters, with a valuable chance to get better in unambiguous classes.

7.2 *Hybrid model (MOBILENET+LSTM)*

Confusion Matrix: The disarray lattice displayed here mirrors the exhibition of the Half and half MobileNet + LSTM model for manually written text acknowledgment. In the framework, each line addresses the genuine class names, while every segment addresses the anticipated marks by the model. The askew components demonstrate accurately arranged examples, showing how well the model distinguishes explicit classes. For instance, the high qualities along the corner to corner for classes like '0', '1', and '2' recommend that the model is effectively recognizing these characters with insignificant misclassification. Be that as it may, off-corner to corner values uncover areas where the model will in general confound particular characters. For instance, there is an obtainable

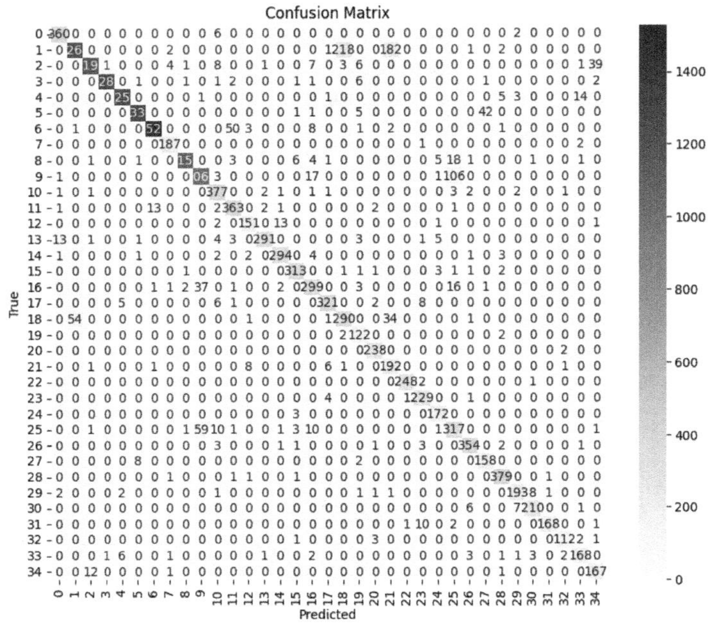

Figure 4. Confusion matrix of hybrid model.

misclassification among classes '6' and '8' (which sample is likely to recognize, the model finds it hard to classify between these two digits, what share comparable visual highlights). This is normal in transcribed text acknowledgment, where some characters will look, to some extent, as comparative, especially in the case of differing penmanship styles.

The Cross breed MobileNet + LSTM model proceeds with the procedure of using initial elemental separation MobileNet to extract spatial elements from the info picture. MobileNet has a light weight architecture, making use of depthwise divisible convolution, which ensures that this process of extracting components is not only computationally good, but exact as well. After the element maps are created, they are passed through a LSTM layer, which finds out the connection between the consecutive highlights removed. This is fundamental for written by hand text as the succession of characters is importance for right acknowledgment. This result of the LSTM layer is processed through a softmax ability to generate the probabilities of every conceivable person. The disarray network provides knowledge against how well the model is doing in various classes with further improvement and improvements given to the model, especially the misclassified classes.

7.3 *Classification report*

The grouping report offers a nitty gritty assessment for the Half and half MobileNet + LSTM model exhibition for manually written text acknowledgment with an accuracy, review, F1-score and get the job done for every subject. Accuracy estimates the precision of the model's positive expectations. For instance, the model has an accuracy of 0.95 for class 0, implying that 95% of the cases anticipated as '0' were accurately grouped. High accuracy values show that the model makes less bogus positive mistakes while anticipating each class.

Review mirrors the model's capacity to recognize all genuine occurrences of a class. For class 1, review is 0.76, meaning the model accurately recognized 76% of all evident occasions of '1'. High review implies the model is great at identifying most occasions of a class, regardless of whether some misclassifications happen.F1-score is the consonant mean of accuracy and review, giving a fair proportion of the model's general presentation. For instance, the F1-score for class 3 is 0.99, demonstrating that the model has accomplished a close ideal harmony among accuracy and review. Higher F1-scores by and large show a more dependable model.

Support addresses the quantity of genuine examples for each class in the dataset. For instance, class 0 has 368 occasions, implying that 368 examples of '0' were available in the dataset. Support assists with figuring out the significance of each class in the general assessment.The precision of 0.93 implies that the model accurately anticipated 93% of all cases across all classes, exhibiting its high generally speaking presentation. This worth mirrors the model's adequacy in perceiving and arranging transcribed text. The full scale normal gives an unweighted normal of accuracy, review, and F1-score across all classes, giving a summed up proportion of execution, while the weighted typical considers the help for each class, offering a more extensive assessment that mirrors the model's presentation across all class disseminations. With a weighted F1-score of 0.93, the model areas of strength for exhibits in perceiving a wide assortment of characters and images, however a few regions, for example, class 18 with lower review, actually show opportunity to get better.

8 CONCLUSION

All in all, both the Vision Transformer (ViT) and the Half breed MobileNet + LSTM models major areas of strength for exhibit in the undertaking of written by hand text acknowledgment. The Vision Transformer use its self-consideration component to catch worldwide conditions inside the picture, offering a powerful answer for complex penmanship designs. It performs well in recognizing characters, however regions like comparable looking characters might require further refinement.Then again, the Mixture MobileNet + LSTM model is basically effective in bringing together the lightweight model of MobileNet for effective component extraction with the capacity of the LSTM to display successive conditions. This half and half methodology is successful both in precision and computational productivity, which makes it perfect for constant applications. The exhibition of the two models, which were evaluated by arrangement measurements such as accuracy, review and F1-score, is the fact that they can really send. Future improvements could zero in on tending to explicit misclassifications and vigor in profoundly a penmanship styles.

REFERENCES

[1] Devendiran, Saikrishna, Raman, Raghu, Nedungadi, Prema and Ratnam, Jyothi. (2024). *Handwritten Text Recognition using VGG19 and HOG Feature Descriptors.* https://doi.org/10.1109/IATMSI60426.2024.10502872.

[2] Rakesh, S., Reddy, P., Prashanth, V. and Reddy, K. (2024). Handwritten text recognition using deep learning techniques: A survey. *MATEC Web of Conferences.* 392. https://doi.org/10.1051/matecconf/202439201126.

[3] Josephine Prem Kumar, "Handwritten Text Recognition Using Deep Learning: A CNN-LSTM Approach", *Int J Intell Syst Appl Eng*, vol. 13, no. 1, pp. 14–22, Apr. 2025.

[4] R. Maroof, I.A. Usmani, and A. Feroze, "English Text Recognition Based on Convolutional Neural Network (CNN)", *SSURJET*, vol. 14, no. 2, pp. 73–78, Dec. 2024.

[5] Nikitha, A, Geetha, J and D S, Jayalakshmi. (2020). Handwritten Text Recognition using Deep Learning. 388–392. https://doi.org/10.1109/RTEICT49044.2020.9315679.

[6] Jonglr, Stelar. (2018). Cursive Handwritten Text Recognition using Bi-Directional LSTMs: A Case Study on Urdu Handwriting.

[7] Alkhateeb, Jawad, Almunawarah, Almadinah, Kingdome, Arabia, Saudi, Sa, Aturani@taibahu and Alsewari, AbdulRahman. (2021). *Performance of Machine Learning and Deep Learning on Arabic Handwritten Text Recognition.* https://doi.org/10.1109/ETCCE51779.2020.9350863.

[8] Manchala, Yugandhar and Aditya Institute of Technology and Management. "Handwritten text recognition using deep learning with Tensorflow." *Journal-article. International Journal of Engineering Research and Technology (IJERT)*, May 2020, 594. http://www.ijert.org.

[9] Hossain, Md, Zaman, Sarah, Akash, Fardin, Sarker, Mohammad Rifat and Huq, Mohammad. (2023). *Developing a Bangla Handwritten Text Recognition Framework using Deep Learning.* 1–6. https://doi.org/10.1109/ICCIT60459.2023.10441107.

[10] Rakesh, S., P. Kushal Reddy, V. Prashanth, and K. Srinath Reddy. "Handwritten Text Recognition Using Deep Learning Techniques: A Survey." *MATEC Web of Conferences 392* (January 1, 2024): 01126. https://doi.org/10.1051/matecconf/202439201126.

[11] Malhotra, Ruchika and Addis, Maru. (2023). End-to-End historical handwritten ethiopic text recognition using deep learning. *IEEE Access.* PP. 1–1. https://doi.org/10.1109/ACCESS.2023.3314334.

[12] Jebadurai, Jebaveerasingh, Jebadurai, Immanuel, Jeba, Getzi and Vangeepuram, Sushen. (2021). Handwritten Text Recognition and Conversion Using Convolutional Neural *Network (CNN) Based Deep Learning Model.* 1037–1042. https://doi.org/10.1109/ICIRCA51532.2021.9544513.

[13] Krishnan, Praveen, Dutta, Kartik and Jawahar, C.V. (2016). *Deep Feature Embedding for Accurate Recognition and Retrieval of Handwritten Text.* 289–294. https://doi.org/10.1109/ICFHR.2016.0062.

[14] Jayanna, H., B G, Nagaraja, Yadava, Thimmaraja, Deekshith, J, Seelam, Monish and Jamkhandi, Venkatesh. (2024). *Advancements in Handwriting Recognition: Deep Learning Techniques Applied to Kannada Language.* 1–5. https://doi.org/10.1109/ICSSES62373.2024.10561302.

[15] Sadaf, Farhan, Raju, S M Taslim Uddin and Muntakim, Abdul. (2021). *Offline Bangla Handwritten Text Recognition: A Comprehensive Study of Various Deep Learning Approaches.* 153–156. https://doi.org/10.1109/ICEEE54059.2021.9718890.

[16] R. Thendral, M. Subasri, G. Sudharsan, and M.K. Ragul, "Enhancing tamil handwritten character recognition using multimodel deep learning," *Proc. 10th International Conference on Communication and Signal Processing (ICCSP 2024)*, pp. 1600–1604, 2024, https://doi.org/10.1109/ICCSP60870.2024.10544266.

[17] M. Balat, Y. Mohamed, A. Heakl, and A.B. Zaky, "Arabic handwritten text for person biometric identification: A deep learning approach," *Proc. International Conference on Information Management and System Applications (IMSA 2024)*, pp. 431–436, Sep. 2024, https://doi.org/10.1109/IMSA61967.2024.10652624.

[18] Parikshith, H, S M Naga Rajath, D Shwetha, C M Sindhu, and P Ravi. "Handwritten character recognition of kannada language using convolutional neural networks and transfer learning." *IOP Conference Series: Materials Science and Engineering,* 2021, 012003. https://doi.org/10.1088/1757-899X/1110/1/012003.

Smart Systems and Beyond – Bhatia et al. (Eds)
 ISBN: 978-1-041-36617-1

Optimization of Integrated MB-OFDM: Fibre Network Performance and Applications to WPAN

Dalli Poornima
Vignan's Institute of Engineering for Women

Aruna Adlaboina
GVP College for Degree and PG Courses(A)

S. Venkata Anand
Vignan's Institute of Information Technology

S. Venkata Ramanaji Kalla
Nadimpalli Satyanarayana Raju Institute of Technology

Gujjula Ramanareddy
Naradaraopeta Engineering College

T. Srilatha and Buddha Hari Kumar
Vignan's Institute of Engineering for Women

ABSTRACT: Orthogonal Frequency Division Multiplexing is advantageous due to its efficiency to multipath fading effects in wireless communications. In modern technologies ultra wideband systems (UWB) use techniques for transmission. These ultra wide band (UWB) applications can be used to extend personal area network (PAN) to wide area personal area network (WPAN). Multiband OFDM (MB-OFDM) can be considered for integration with OFC technology (physical layer) for meeting the requirements of high backhaul bandwidth. PAPR is drawback in OFDM. However, this is effective only in wireless communication systems because of Power amplifier sizing. In the integration of OFDM with fiber optic physical medium the consideration is more on optical modulator which converts electrical signal to optical signal. In this thesis device algorithm for controlling OFDM output to balance the nonlinearity of Mach-Zehnder Modulator (MZM) is considered. Simulations and performance results for nonlinearity in MZM will be analyzed with respect to OFDM integration. Since MB-OFDM has multi carrier modulation scheme, wavelength constraint in fiber backbone is also considered in the thesis.

Keywords: PAPR, OFDM, UWB, WPAN, MB-OFDM, OFC, MZM, PAN3

1 INTRODUCTION

Ultra-wideband (UWB) in the last several years [24] [25][26], More and more people are interested in radio communications concern for the benefits provided by UWB, i.e., low power transmission, low cost, wide bandwidth, robustness for multi-path fading and high data rate. It is the low power transmission of the UWB that could potentially mean that it could be warmed together with other standards for wireless networking such as 802.11 LAN, WAN and 802.16 MAN technologies.

The FCC said in 2002 that Ultra-Wideband (UWB) devices could use the frequency spectrum between 3.1 GHz and 10.6 GHz without a license. There are presently two main ways to use UWB. The first method, which is based on OFDM, is termed Multi-Band OFDM UWB (MB-OFDM UWB). The second method is called Impulse Radio UWB (IR-UWB).

MB OFDM extreme wideband technology uses a wide band or multiband OFDM and is defined by an OFDM signal which spans 528MHz (MB-OFDM). The 528MHz signal is then frequency hopped in order to occupy an adequately wide bandwidth, one of the methods known as frequency hopping technique.

This paper is organized as follows: Section II is a description of the literature review. Section III Interface MB OFDM with an optical fiber. Section IV provides some overview on Routing and Wavelength Assignment. Dijkstra's Algorithm and Yen's Algorithm are mentioned in Section V. Simulation, results and analysis Section VI presents. Section VII has Conclusion.

 DOI: 10.1201/9781042026289-17

2 LITERATURE REVIEW

High speed wireless and optical communication systems have been significantly enhanced by the development of ultra wide band (UWB) communication and the use of OFDM. OFDM is highly used in wireless network and more recently in optical transmission systems due to their good spectral efficiency and their resistance to the multipath fading. However, high Peak-to-Average Power Ratio (PAPR) and distortion that isn't linear issues, especially in optical systems based on Mach-Zehnder Modulator (MZM) restrict typical OFDM systems.

3 ADVANCES IN MB-OFDM AND OFDM

In their analysis of multi-carrier modulation schemes, Vaigandla and Benita [6] focused on the relative advantages of OFDM and Filter Bank Multicarrier (FBMC) in terms of the error performance and the spectrum efficiency. In case of investigation of neural network-based study of PAPR, BER and channel estimation for multi-carrier modulation systems, Vaigandla et al. [3] demonstrated how smart signal processing can be used for the system performance enhancement. The potential optical OFDM in high-capacity networks was emphasized by Deepak Kumar Singh et al. [1] which studied performance optimization of long-haul optical transmission connections utilizing optical OFDM. Multiband-OFDM (MB-OFDM), which splits the spectrum into multiple 528MHz sub-bands for better data throughput and bandwidth utilization was developed as a combination of OFDM and ultra-wideband (UWB) technology. Sui et al. [8] exploited the low-cost electro-absorption modulators (EAMs) to demonstrate that MB-OFDM ultra wideband over fiber transmission is viable and the ultra-wideband system maintains high data integrity and low power consumption. In the similar sense, MB-OFDM radio-over-fiber transmission in the 60 GHz band was experimentally evaluated by Omomukuyo et al. [10] and demonstrated to have the ability to carry high-speed data transmission for long distances. The effective transmission of impulse radio UWB and MB-OFDM signals through the single mode fiber was also achieved by Jazayerifar et al. [11], indicating the importance of hybrid optical-wireless system design.

4 TECHNIQUES FOR OPTICAL MODULATION AND INTEGRATION

In order to interface wireless OFDM systems with optical backbones, optical modulation is essential. In order to improve system linearity, Ramadan et al. [2] suggested an improved DWT-OFDM communication system that uses wavelet-domain equalization to reduce carrier frequency offset (CFO). Similar research by Ramadan et al. [5] use discrete wavelet transforms (DWT) to improve signal quality and robustness in optical wireless integration solution by focusing on the non-linear equalization and CFO pay for MIMO-OFDM systems Optimizing optical devices like MZMs to reduce nonlinear distortions is another focus of recent study. Effective PAPR reduction strategies improve spectrum efficiency and decrease distortion during optical modulation, according to performance assessments conducted on hybrid optical OFDM systems by Vaigandla et al. [4].

5 ROUTING IN AN OPTICAL NETWORK AND ASSIGNING WAVELENGTHS

For optical networks that use wavelength-division multiplexing (WDM) to efficiently transmit data, routing and wavelength assignment (RWA) is very important.By describing wavelength continuity limitations and cost-based routing techniques, Ozdaglar and Bertsekas [23] offered fundamental insights into RWA difficulties in optical networks. In order to enhance bandwidth usage, Liu [13] suggested dynamic wavelength assignment methods that allow adaptive routing. After comparing various iterations of Yen's approach for determining K-shortest pathways, Niranjane and Amdani [17] came to the conclusion that optimized versions of the algorithm produce faster path computation in optical topologies. In a similar vein, Chen et al. [18] presented re-optimization methods for effectively identifying numerous shortest pathways. Szcześniak et al. [19] enhanced routing efficiency under varied network topologies by proposing a generic Dijkstra algorithm designed for optical networks. Real-time routing optimization in high-capacity optical systems is now possible thanks to recent developments by Wu et al. [20], [22], which presented digital-twin-based dynamic topology models and effective topology representations for large-scale transmission networks.

6 DEVELOPMENTS IN WPAN AND UWB USES

Under the IEEE 802.15.4z-2020 standard, UWB technology is still developing and offers improved physical layer and range algorithms for high-precision, low-power communication [26]. The integration of UWB for wireless communication in the future generation and in-home device control was investigated by Kabbinale et al. [24], [25], with a focus on low latency and compatibility with personal area networks (PANs). The application of AI-enhanced UWB localization in wireless networks was further illustrated by Dehmollaian et al. [27], demonstrating the technology's versatility for intelligent WPAN and IoT contexts.

7 SYNOPSIS

According to the literature, wireless personal area networks (WPANs) can have their range increased, interference reduced, and data throughput greatly increased by merging MB-OFDM with optical fiber communication networks. High PAPR, nonlinear distortion in MZM, and wavelength continuity limitations are still major problems, though. Therefore, effective wavelength assignment techniques, sophisticated routing algorithms, and hybrid optical-wireless designs are crucial for maximizing network performance. By applying Yen's method for RWA, combining MB-OFDM with optical modulation, and assessing performance gains in terms of cost, efficiency, and wavelength continuity, the current study expands on this framework.

8 INTERFACING MB-OFDM WITH OPTICAL FIBER NETWORK

OFDM [5] is a short distance communication. MB OFDM is a forward technology to OFDM and it is an UWB technology, when a bandwidth of 528MHZ is considered then it is called as MB-OFDM optic fiber communication is a concept for long distance communication and it has high bandwidth utility and high throughput efficiency. So, by interfacing such an optic fiber with a multiband OFDM improves the performance and converts PAN to WPAN applications.

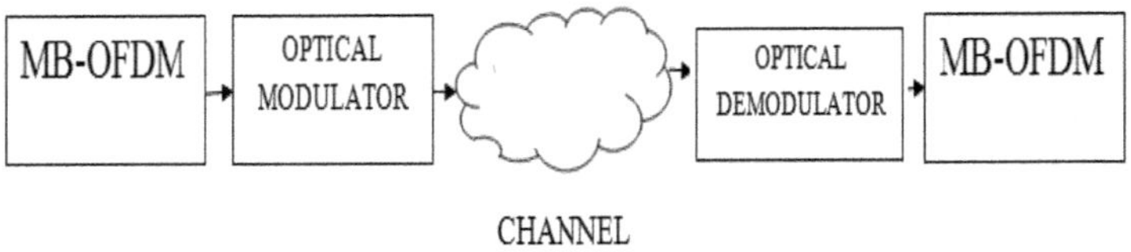

Figure 1. Interfacing of MBOFDM with an Optical Fiber.

8.1 *Multiband OFDM*

The US Federal Communications Commission (FCC) was the first group to give out licenses for Ultra-Wideband (UWB) use in the 3.1 to 10.6 GHz range.Under the Wide Personal Area Network (WPAN) working groups, the standard became known as the 802.15.3a, and consisted of dividing the entire spectrum into 14 bands with a bandwidth of 528 MHz. The 14 channels have been divided into five groups and each group comprises three channels except the fifth group, which comprises only two channels. Multi Band OFDM UWB is one of the types of ultra wideband technology, which is different from the impulse or direct sequence ultra wideband methods. MB-OFDM UWB facilitates the transmission of data at multiple carriers and at precisely defined frequencies simultaneously. The favorable features of MB-OFDM [11] include high spectrum flexibility, resistance to radio frequency interference and reduction of multi-path interference.

8.2 *Optical modulator*

An optical modulator or an optical modem is, a tool that converts electrical signals into optical signals or the other way around which is used for communicating information over optical fiber Contacts Optical modulators are used to control the phase or amplitude of the light wave that goes through the device In order to explain and shorten the size of the device and the driving voltage, wave guide based modulators are used in communication application. Optical modulator is to interface the carrier wavelengths to the optical fiber for providing high throughput efficiency.

8.3 *Channel*

Optical path or a light path can be established between source node and destination node using optic fiber. The process of routing [19] simultaneously in an optical fiber can be achieved by DWDM. DWDM[2] is designed for long-haul network transmission where wavelengths are packed tightly together and these systems are used for thousands of kilometres. This Channel includes Routing and Wavelength Assignment.

9 ASSIGNING OF WAVELENGTHS, ROUTING

The optical path or light path is that link which between the origin node and the target node. Finding way for a light channel routing and wavelength assignment is the process of giving it a wavelength challenge.

RWA consists of

(1) Routing
(2) Wavelength Assignment

In a network, routing between two nodes can be accomplished by assigning cost function and assigning wavelength to each node, based on the wavelength continuity the path is established. Packets of variable size are transmitted through the light path.

Traffic Grooming is a technique to combine lower speed traffic to an available high speed wavelength in order to

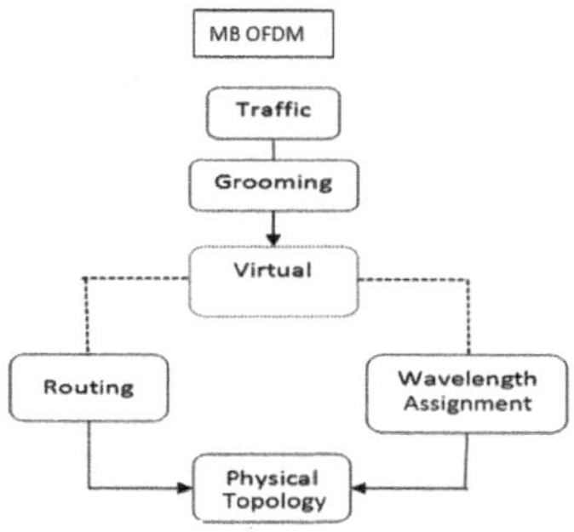

Figure 2. Routing and Wavelength Assignment.

meet the network design goals like to minimize the cost.

Routing:

Routing is figuring out the best way to go somewhere. In A web corresponds to the connection of two sites. The wavelength is a circuit-switched connection, which is a path connecting two nodes and represents a combined path for routing.

Wavelength Assignment:

Wavelength Assignment is in the routing path between two nodes. Wavelength assignment is to determine availability of specific wavelengths in optimal path, under wavelength continuity constraint.

10 ROUTING ALGORITHMS

Routing algorithms [13][14][20] are essentially based on the idea of finding the fastest way to get from one node to another. A shortest path routing algorithm[17] reaches all packets along the shortest path[21] starting from the origin to the destination nodes. Dijkstra's Algorithm is generally used in most of the networks. This paper takes Yen's algorithm [16], and more precisely, fixed route that is different, which is an addition to fixed route routing as an example to consider.

Yen's algorithm

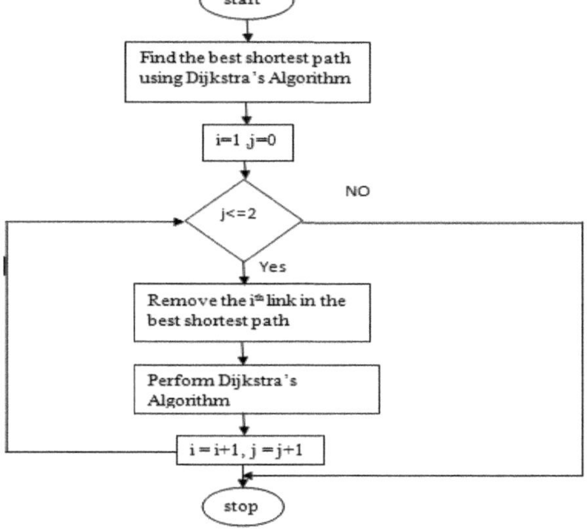

Figure 3. Yen's Algorithm.

Rather than a single fixed route to be used from a given source to destination pair, several will be maintained. In this method if the path obtained in the Dijkstra's algorithm doesn't have any wavelength continuity or if the connection fails to continue then there need an alternative path for transmitting. So the alternative solution for finding the paths is Yen's algorithm. In this method different paths are obtained based on fixed path routing algorithms.

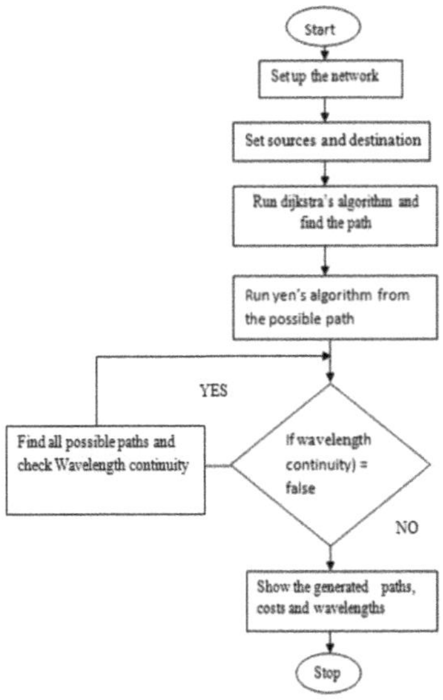

Figure 4. Path Computational Algorithm.

Path Computational Algorithm

Figure 4 shows the Flow chart for Path Computation Algorithm It shows the 09 steps involved in the establishment of light path from source node to destination node. Everytime a packet is transmitted the Path Computation algorithm is executed

(1) A network of nodes is required which has to be Setup.
(2) The Sources Node from where the packet has to be transmitted and a Destination Node, to which the packet has to reach are chosen.
(3) For selected sources and destination, packet is placed to the Destination node – packet is placed on source node.
(4) Dijkstra's algorithm [18] is applied to the network and wavelength ad continuity have to be checked for each Network node in the best path.
(5) Wavelength continuity is done for the next node and sends the packet. If wavelength doesn't exists run yen's algorithm and all the possible routes are observed and wavelength continuity check for single packet transmission.
(6) If two paths with wavelength continuity then best path of low cost function is considered.
(7) Multiple packets of different wavelengths[3][4][6] are transmitted at a time checking the wavelength continuity in all possible routes.

11 RESULTS

Consider a network of 6 nodes, shortest path is calculated from source node to destination node. In the below

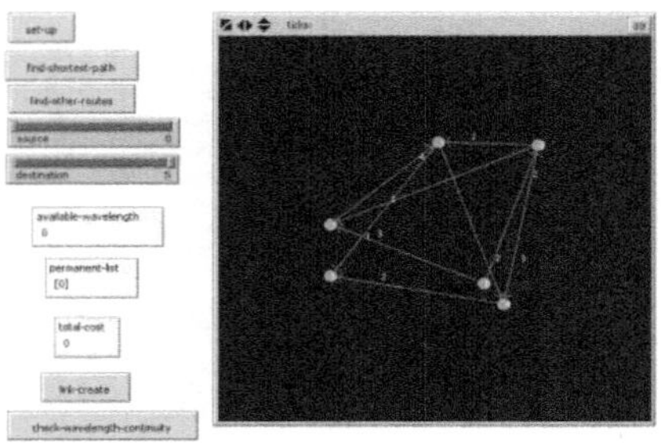

Figure 5. Represents a Network of 6 Nodes.

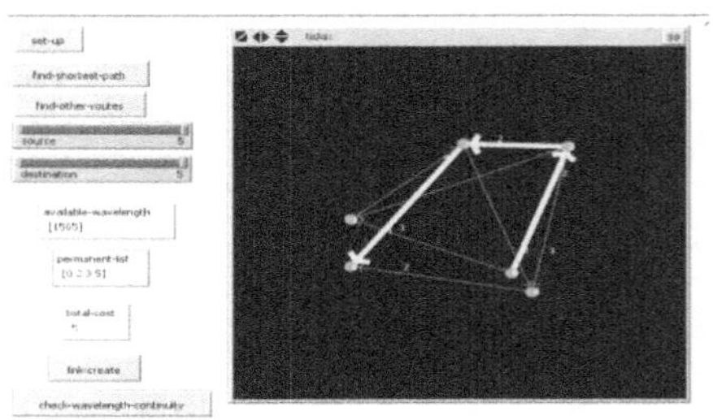

Figure 6. 1st possible path in Yen's algorithm.

figure assume zero as source node and five as the destination. To this network cost function are assigned as follows

Total cost is obtained by calculating the costs assigned for all the path nodes as 5. So the best path created for the network is [0 2 3 5].for this wavelength continuity is to be calculated assigning two wavelengths and continuity through the network is determined. If the wavelength continuity doesn't exist then converter or another wavelength can be selected but in Yen's algorithm, all possible paths are logged. So the other possible path is 0-1-3-5. In this path, same wavelengths are transmitted through two routes but with other cost values and based on the least cost function the route will be selected, if both the routes have wavelength continuity.

For other possible routes Yen's algorithm is used and the Figure 7 represents other possible route in a network and the total cost of the path is 8 and the wavelength through the path is observed and two packets can be transmitted through the path. Two available wavelengths are observed. Two sources can be simultaneously transmitted through a possible path and for the first path wavelength continuity constraint occurs but the packet can be transmitted through the second possible path.

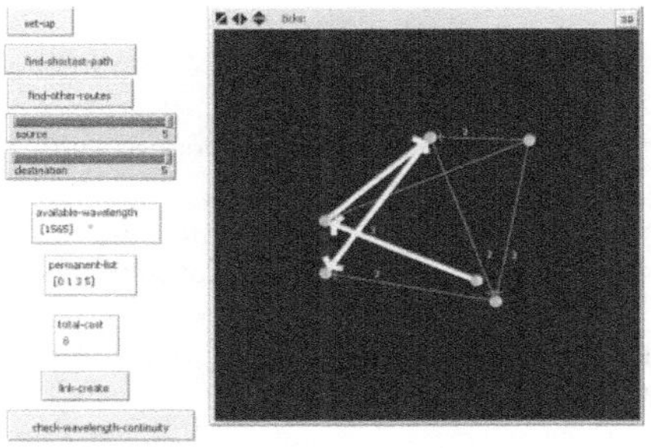

Figure 7. The 2nd possible Routes in a Network.

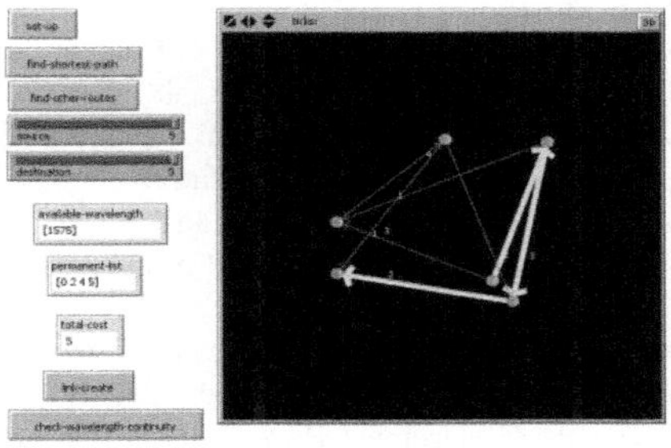

Figure 8. The third possible route 0-2-4-5.

Third possible route is [0 2 4 5] and the total cost is 5 and the wavelength continuity constraint is occurred in the third path for transmission of packet there is a wavelength continuity of 1575 is present in the path is observed in the Figure 8.

12 CONCLUSION

This study examines Routing and Wavelength Assignment with a Wavelength Continuity Constraint. This analysis is conducted in the context of MB OFDM interfaced to for optical fiber network as backbone to improve on PAN (PERSONAL AREA NETWORK)[23] performance to wide area applications. In this thesis, considering an optical network and establishing all possible paths using Yen's algorithm for wavelength constraint criteria. Transmitting multiple wavelengths from multi band OFDM sub carriers and Interfacing with an optical modulator for routing and checking wavelength continuity transmission of packets with improved efficiency and large bandwidth for WPAN Applications is demonstrated using NetLogo[12].

13 FUTURE WORK

Based on multiple wavelengths and multiple possible paths the wavelength continuity is checked considering only two objectives i.e. cost function and wavelengths. The network performance can be further analyzed with time delay parameter also along with cost function and wavelength which improves the QOS (Quality of Service) and find additional applications like VPN.

REFERENCES

[1] Bi Yu Chen, Xiao-Wei Chen, Hui-Ping Chen, William H.K. Lam, *"Efficient Algorithm for Finding k Shortest Paths Based on Re-Optimization Technique"* Volume 133, 2020, 101819, ISSN 1366–5545.

[2] Bo Tang, Kaiyu Qin and Haibo Mei, "A hybrid approach to reduce the PAPR of OFDM signalsusing clippingand companding", *IEEE Access*, vol. 8, pp. 18984–18994, January 2020.

[3] Deepak Kumar Singh, B.B. Tiwari, J.P.S. (2024). Performance optimization of long-haul optical transmission link with Optical-OFDM. *International Journal of Intelligent Systemsand Applications in Engineering*, 12(21s), 442–453.

[4] Dehmollaian, E., Etzlinger, B., Peterseil, P., Springer, A. (2024). "AI-Enhanced UWB-Based Localisation in Wireless Networks". In: Karner, M., Peltola, J., Jerne, M., Kulas, L., Priller, P. (eds) *Intelligent Secure Trustable Things. Studies in Computational Intelligence*, vol 1147. Springer, Cham.

[5] IEEE, 802.15.4z-2020, 802.15.4–2020, "IEEE Standard for Low-Rate Wireless Networks--Amendment 1: Enhanced Ultra Wideband (UWB) Physical Layers (PHYs) and Associated Ranging Techniques", pp. 1–174, Aug. (2020)

[6] Ireneusz Szcześniak, Andrzej Jajszczyk, and Bożena Woźna-Szcześniak, "Generic Dijkstra for optical networks," *J. Opt. Commun. Netw.* 11, 568–577 (2019).

[7] Jazayerifar M., B. Cabon, and J.A. Salehi, "Transmission of Multi-Band OFDM and Impulse Radio Ultra-Wideband Signals Over Single Mode Fiber," *Journal of Light wave Technology*, Vol. 26, No. 15,pp 2594 –2603, August 1, (2008).

[8] Kabbinale A.R., A. Bansal and K.S. Gopalan, "Framework to ascertain effectiveness of Ultra-Wideband for In-home device control", *2022 IEEE Wireless Communications and Networking Conference (WCNC)*, pp. 1713–1718, 2022.

[9] Kabbinale A.R., A. Bansal and K.S. Gopalan, "Towards next-generation ultra-wideband technology," *2023 15th International Conference on COMmunication Systems &and NETworkS (COMSNETS)*, Bangalore, India, 2023,

[10] Kaixian Liu, "Wavelength Assignment Algorithm with Dynamic Routing in WDM Optical Networks", 978-0-7695-4270-6/10 © IEEE, 2010.

[11] Madkour, Amgad & Aref, Walid & Rehman, Faizan & Rahman, Abdur & Basalamah, Saleh. (2017). "*A Survey of Shortest-Path Algorithms*". https://doi.org/10.48550/arXiv.1705.02044.

[12] Niranjane P.B. and S.Y. Amdani, "Comparison of variants of yen's algorithm for finding K-simple shortest paths," *2022 2nd International Conference on Intelligent Technologies (CONIT)*, Hubli, India, 2022, pp. 1–5.

[13] Omomukuyo O., M.P. Thakur, and J.E. Mitchell, "Experimental performance analysis of MB-OFDM ultra-wideband radio-over fiber signals in the 60-GHz band using a commercially-available DFB laser," in *Proc. ICTON 2012*, Coventry, U.K., pp. 1–4, paper Mo.C3.4.

[14] Ozdaglar, Asuman and Bertsekas, Dimitri. (2003). "Routing and wavelength assignment in optical networks. Networking"ef, *IEEE/ACM Transactionson*. 11.259- 272. https://doi.org/10.1109/TNET.2003.810321.

[15] Pavani P., G. Tirumala Rao, N. Balasubrahmanyam, *"Routing in Multi Domain Environment using IS-IS and BGP for QOS Improvement"*, ACCT-2012, Volume-1, Issue-1, Page no: 476-479, 2012.

[16] Performance Analysis of Hybrid Optical OFDM Schemes in terms of PAPR, BER and Spectral Efficiency using Various PAPR Reduction Methods (K. K. Vaigandla, M. Siluveru, N. Dharavath, and R. Karne , Trans.). (2024). Babylonian, *Journal of Networking*, 2024, 122–134.

[17] Ramadan K, Hassan ES (2025) Enhanced DWT-OFDM communication system using wavelet domain equalizer with Co-CFO. *PLoS One* 20(4): e0317097.

[18] Ramadan, K, Dessouky, M.I and Abd EI-samie, F.E (2020). Non linear equalisation and CFO compensation for MIMO-OFDM communication systems based on DWT. *International Journal of Electronics*,108(1),115–138.

[19] Ramya A., N. Balasubrahmanyam, G. Tirumala Rao, ATM switch implementation for NGN using DWDM, *International Journal of Computer Applications* (0975–8887) volume 104-NO 14,oct 2014.

[20] Sui C., B. Hraimel, X. Zhang, K. Wu, T. Liu, T. Xu Q. Nie, dealt with "Performance evaluation of MB-OFDM ultra- wideband over fiber transmission using a low cost electro-absorption modulator integrated laser." in *IEEE Tropical Meeting on Microwave photonics*, Oct (2010).

[21] Vaigandla *et al.*, (2024). "Analysis of PAPR BER and channel estimation in multi carrier modulation systems using neural networks" *Journal of Theoretical and Applied Information Technology*, vol. 102, no. 5, pp. 1785–1795, Mar. 2024.

[22] Vaigandla K. and B. J, "Study and analysis of multi carrier modulation techniques –FBMC and OFDM," *Materials Today: Proceedings*, vol. 58, pp. 52–56, (2022).

[23] Vaigandla K.K. and J. Benita,"PRNGN-PAPR Reduction using Noise Validation and Genetic System on 5G Wireless Network," *International Journal of Engineering Trends and Technology*, vol. 70, no. 8, pp. 224–232, (2022). Crossref.

[24] Wilensky, U. and Stroup, W., (1999). HubNet.http://ccl.northwestern.edu/netlogo/hubnet.html. *Center for Connected Learning and Computer-Based Modeling*, Northwestern University. Evanston, IL.

[25] Wu Y., J. Li, M. Zhang, B. Ye and X. Tang, "A comprehensive and efficient topology representation in routing computation for large-scale transmission networks," in *IEEE Transactions on Network and Service Management*, vol. 22, no. 1, pp. 220–241, Feb. 2025

[26] Wu Y., M. Zhang, L. Zhang, J. Li, X. Chen and D. Wang, "Dynamic network topology portrait for digital twin optical network," in *Journal of Lightwave Technology*, vol. 41, no. 10, pp. 2953–2968, 15 (2023).

[27] Yonghan Wu, Jin Li, Min Zhang, Bing Ye, Xiongyan Tang, "A comprehensive and efficient topology representation in routing computation for large-scale transmission networks", *IEEE Transactions on Network and Service Management*, vol.22, no.1, pp.220–241, (2025)

Smart Systems and Beyond – Bhatia et al. (Eds)
© 2027 The Author(s), ISBN: 978-1-041-36617-1

Managerial Analysis and Effects of Mental Disorder Health Problem: Dementia

M. Rupa
SR University, Telangana, India

K. Rajkumar
Department of ECE, SR University, Telangana, India

ABSTRACT: Dementia — a word related with fear and uncertainty — is nothing but "forgetting.". It is associated with symptoms like inability to remember, reasoning, communicate, and in taking up daily tasks. Dementia is linked with a series of mental disorders that are related to brain. It usually occurs with ageing where the social relationships worsen.

Management of social relationships is almost burden on them. Advanced deep learning techniques stand as a backbone for diagnosis of this mental disorder health problem right from the scratch.

Keywords: Alzheimer's Disease (AD), Advanced Techniques, Dementia

1 INTRODUCTION

Dementia begins with damage to the brain's system—the neurons. These tiny nerve cells form very complex networks that control everything right from memory to movement, and when they are not functioning, the functioning of the brain declines. Among the major causes of dementia, Alzheimer's disease is seen in many people. This disease progresses gradually, because harmful proteins—such as amyloid plaques and tau tangles —accumulate more in the brain region. These proteins will be responsible for the communication between neurons, gradually causing the death of cells . This results in memory decline, reasoning, and the ability to perform day to day tasks.

Vascular dementia develops when the brain's blood vessels get damaged—because of high blood pressure, stroke or circulatory problems. When the flow of blood to the brain is reduced, neurons are deprived of oxygen and essential nutrients. This weakens the brain cells and eventually die, leading to loss in memory, concentration and reasoning.

Another major dementia is Lewy body dementia that occurs usually when abnormal protein deposits called alpha-synuclein—defined as Lewy bodies which build up inside of the nerve cells. Clumps of proteins and disrupts the signals to the brain that leads to a very complex symptoms that is beyond memory loss. Lewy body dementia patients feels vivid visual hallucinations, variations in alertness, and also movement difficulties.

Frontotemporal dementia is mainly because of the damage of both the frontal and temporal parts of the brain that is the parts responsible for behavior, language and personality [1]. Unlike AD, this form of dementia usually occurs at a younger age, in 40s or 50s. Its effects are very effective and become less social and emotional and more impulsive. Other factors also affect dementia such as brain injuries from accidents, infections such as HIV and long term usage of poisonous substances. Also genetics effects play an important role for the cause of dementia. It is very difficult for researchers and doctors to diagnose these dementia.

2 LITERATURE REVIEW

We will explore how to diagnose the early signs of dementia. This usually occurs due to sudden change in loss of memory [2]. That is the common symptom is memory loss— asking the same question repeatedly, forgetting recent conversations, misplacing daily items. This state of forgetfulness affect language and comprehension, making it very much difficult for a person to remember the exact words, confuse similar ones like "cat" and "dog," or inability to do routines—such as preparing a recipe that they know very long back.

Thinking also becomes slower—things like solving a problem or planning for something feels very difficult. People loose their sense of direction and will be misplaced even in well known places. Their behavior changes very suddenly—they might become very much angry or won't believe people they once believed. In some types of dementia, such as Lewy body dementia, they observe things that are not really present like people or animals. As this condition becomes serious, everyday tasks like getting ready, going for doctor appointment become difficult for them, and they depend on others for help.

Dementia is like virus that spreads among their neighbors. They loose interest on their life as there will be much change on their day to day tasks. They reach a stage of dependency on their neighbors. Their emotions fluctuate between fear, confusion, irritation and sadness.

Adult patients feels difficult to balance their careers, families and daily chores. They add financial burden to the family which includes mostly medical expenses. Family members have great burden on such patients to carry out their daily activities, appointments, taking medicines.

Dementia patients pose very high burden impact on the society. As the time goes on dementia spreads at a very high rate which makes nursing homes overflow, hospitals get fills up.

Society bears its share of the burden too. As dementia affects millions worldwide, healthcare systems struggle to keep pace—hospitals overflow, nursing homes fill up, and the demand for trained caregivers exceeds supply. Economically, the impact is seen in workplaces as employees prefer less working hours or quit jobs to care for their loved ones, while governments impacts with the rising medical care costs and very long-term support. In more communities, there is whisper only rather than discussing it openly, leaving their families separated and getting no help.

3 METHODOLOGY

There's no single test available to diagnose dementia. There are many types of tests. The tests is done how your brain works (brain scans), functions (neurological tests), thinks or reasons (cognitive tests and then doctor will review all the tests information to see if there is dementia.

Physical therapy and medication reduces very few dementia. symptoms. It helps to trace out the stages of dementia. Early dementia diagnosis helps to plan the future. It can also make living, health care, financial or legal decisions easier.

A proper treatment can give more time to build up a good relationship with caregivers and doctors. Regular doctor work with us to trace out what's are the symptoms that causes dementia. They may also refer us to one or more of such doctors for certain tests.

Symptoms like improper sleep occur for dementia patients which increases the risk of developing brain disorders. Good sleep helps the brain store memories and clear out harmful proteins such as tau and beta-amyloid, that links dementia. People start having sleep issues and this is indication of how the disease develops.

Non-drug methods for dementia patients showed in improvement of sleep and think more clearly. To get this, the various methods used are having a regular bedtime routine, improving the sleep environment and using better relaxation techniques. More detailed research is still going on this.

Dementia diagnosis starts with short medical history and performing basic memory and nerve function tests. The medical history is very important part and will come from both the patient and their closest family member or a friend.

The medical history describes what kind of thinking problems the person has, how critical they are, and how they are impacting over time. For example:

- **Type of problem (nature):** Does the person have trouble speaking, forget very recent events or struggle to trace out words?
- **Severity (magnitude):** Do these issues affect daily life — like disappearing, forgetting to do various things or missing medications?
- **Pattern (course):** Did these problems start slowly and getting worsen over time (as in Alzheimer's) or appear suddenly and change step by step (as in stroke-related dementia)?

Finally, a family history can trace out early-onset dementia (before age 65), which may be caused by mostly inherited genetic factors.

4 RESULTS AND DISCUSSION

The following table shows different stages of dementia:

Table 1. Different dementia levels.

Level 1 Pre Dementia	Level 2 Mild Dementia	Level 3 Moderate Dementia	Level 4 Severe Dementia
• Memory Loss • Awareness Reduction • Reduction in attention span	• Understanding language • Memory impairment	• Apraxia • Long term memory loss • Identification of objects or people • Unnecessary wandering • Emotional change	• Urinary • inconsistence • Loss of speech • Bedridden • Severe exhaustion • Inability to eat • Death

From above table, we understand that various symptoms are associated with various types of dementia.

The following Table 2 shows the relevant studies from the interview discussions. The feedback was taken from the participants were identified from the collected data through various interview discussions and also through very few observations. There were only few sessions where the patient is not able to receive response informatively for very few questions and hence the feedback summary was taken to show case the problems that are experienced by the participants [38].

Table 2. Discussion sessions findings on different participants.

Participants	Feedback
Participant 1	• Not able to identify objects. • Couldn't remember more about his own or neighbors
Participant 2	• Not clear of remembering of things • Unable of using any latest technology
Participant 3	• Needs the help of the caretaker • Tries to use new technologies but feels helpless
Participant 4	• Not clear with his actions. • Forgetfulness
Caretaker	• Needs the assistance of the caretakers to even identify very simple things • Not able to think critically • Nor able to identify the location • Experiences nervousness • Experiences loss of memory

This analysis demonstrated that the people with dementia have memory loss problem which sometimes is permanent and sometimes is temporary. This proposes an ensemble method based on deep learning for the early diagnosis of dementia using MRI images.

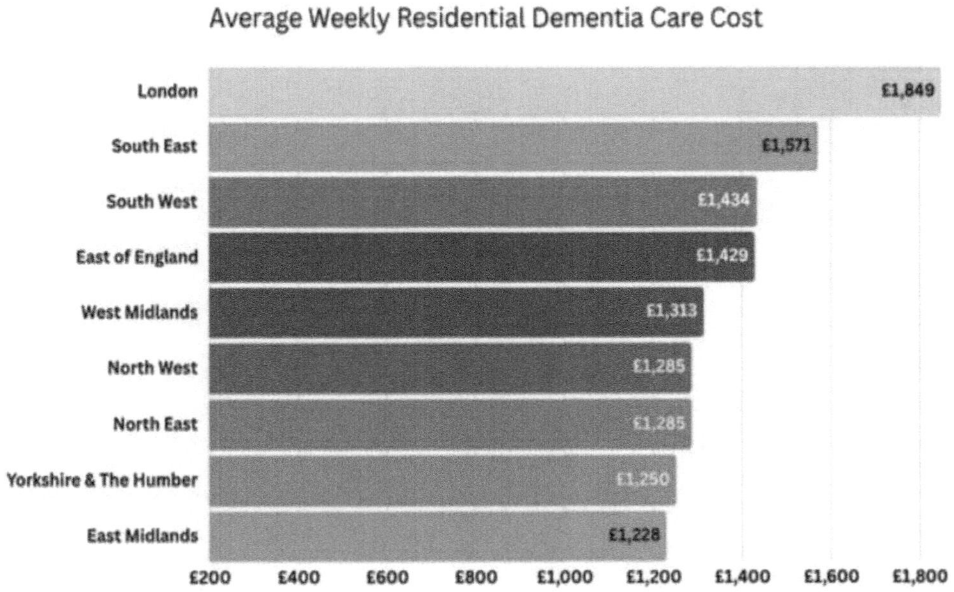

Figure 1. Average weekly residential dementia care cost.

The above figure shows Average Weekly Residential Dementia Care Cost for various countries.

5 CONCLUSION

Advanced Deep learning techniques will help improve the lives of dementia patients. The usage of new development help people with dementia in every tasks has become a very useful one that does not require any medication.

Further discussions and updation on present technologies for people with dementia [12]. There is urgent requirement for future research parameters into very innovative methods and development of the present one, particularly in the users experience and also human computer interaction.

REFERENCES

[1] Phillipson L, Hall D, Cridland E, *et al.* Involvement of people with dementia in raising awareness and changing attitude in a dementia friendly community pilot project. *Dementia.* (2018), 1–16. https://doi.org/10.1177/1471301218754455

[2] Prince, M.J., Wimo, A., Guerchet, M.M., Ali, G.C., Wu, Y.-T., Prina, M.: World alzheimer report 2015-the global impact of dementia: An analysis of prevalence, incidence, cost and trends (2015)

[3] Swaffer K. Dementia: Stigma, language, and dementia-friendly. *Dementia.* (2014); **13**: 709–716.

[4] Hulko W. Social science perspectives on dementia research: intersectionality. In: A Innes, C Archibald, C Murphy, eds. *Dementia and Social Inclusion: Marginalised Groups and Marginalised Areas of Dementia Research, Care and Practice.* London, UK: Jessica Kingsley; (2004), 237–254.

[5] Bartlett R. The emergent modes of dementia activism. *Age Soc.* (2014), **34**: 623–644

[6] Lloret, M.A., Cervera-Ferri, A., Nepomuceno, M., Monllor, P., Esteve, D., and Lloret, A. (2020). Is sleep disruption a cause or consequence of alzheimer's disease? Reviewing its possible role as a biomarker. *International Journal of Molecular Sciences*, 21(3), 1168. https://pubmed.ncbi.nlm.nih.gov/32050587/

[7] Lim, M.M., Gerstner, J.R., and Holtzman, D.M. (2014). The sleep-wake cycle and Alzheimer's disease: what do we know?. *Neurodegenerative Disease Management*, 4(5), 351–362. https://pubmed.ncbi.nlm.nih.gov/25405649/

[8] Phillipson L, Hall D, Cridland E, *et al.* Involvement of people with dementia in raising awareness and changing attitudes in a dementia friendly community pilot project. *Dementia.* (2018), 1–16. https://doi.org/10.1177/1471301218754455

[9] Alzheimer's Association. 2019 Alzheimer's Disease Facts and Figures. *Alzheimer's and Dementia* (2019), 15(3):321–387. [DOI] [PubMed] [Google Scholar]

[10] Haapala I, Carr A, Biggs S. Differences in priority by age group and perspective: implications for public health education and campaigning in relation to dementia. *Int Psychogeriatr.* (2018), **30**: 1583–1591.

[11] Australian Institute of Health and Welfare. *Dementia in Australia.* Canberra: The Institute, 2012. https://www.aihw.gov.au/reports/dementia/dementia-in-australia/contents/table-of-contents. Accessed November 7, 2018.

[12] Haapala I, Carr A, Biggs S. What would I want? Dementia perspectives and priorities amongst people with dementia, family carers and service professionals? *Int JCare Caring.* (2019), 3: 39–57. https://doi.org/10.1332/239788218x15321005902545

Smart Systems and Beyond – Bhatia et al. (Eds)
© 2027 The Author(s), ISBN: 978-1-041-36617-1

Transformer-Based Machine Learning for Antenna Parametric Optimization

Nidhi Upadhyay and Amanpreet Kaur
Centre of Excellence in Emerging Materials
Thapar Institute of Engineering and Technology (TIET), Patiala, India

Ashima Singh and Arnab Pattanayak
Thapar Institute of Engineering and Technology (TIET), Patiala, India

ABSTRACT: Machine learning (ML) and Deep Learning (DL) possesses a potential in antenna design and behaviour prediction, since it facilitates considerable acceleration while preserving high accuracy. In this work, we present a ML and Artificial Intelligence (AI) approach to predict and optimize the S_{11} (return-loss) characteristics of a microstrip patch antenna. Three geometric parameters of the antenna are varied and a dataset of 25,100 samples is generated by EM simulation (CST) across a fixed frequency sweep. To mimic real-world measurement variations, the dataset was augmented with random noise (additive noise, small frequency shifts, and systematic bias). Three regression ML/DL models, Random Forest, KNN, and MLP as well as a transformer-based Informer network have been used to predict antenna return loss. All models are evaluated using coefficient of determination (R^2). The Informer model achieved the highest accuracy: on the unseen test set it reached R^2 =0.88. The informer-based network not only predicted the data points but also the predicted curve from the model, followed the overall shape of the actual S_{11} very closely across the full frequency sweep for the unseen test set. These results indicate that the transformer-based Informer can more accurately learn the mapping from geometry to full S_{11} response.

Keywords: Antenna, Parametric Optimisation, Machine Learning, Deep Learning, Artificial Intelligence, Transformers

1 INTRODUCTION

Various cutting-edge technologies, such as the Internet of Things (IoT) and artificial intelligence (AI), have made major changes in our way of living. The Internet of Things has significantly advanced the Internet's capacity to optimize commercial and industrial outcomes, thus enhancing our overall quality of life. In recent years, advancements in IoT technologies, such as machine-to-machine (M2M) and device-to-device (D2D) connectivity, have significantly accelerated the development of 5G and 6G communications and beyond. The design, optimization, and selection of antennas are crucial for ensuring the effective functioning of a wireless communication system. Thus, a suitable antenna design is crucial for various types of wireless communication, such as WiFi, cellular, and satellite communication. Computational electromagnetics (CEM) analyzes the relationship between electromagnetic fields and antennas by utilizing Maxwell's equations (Godara, 2018). The finite difference time domain (FDTD), finite element method (FEM) (Volakis *et al.*, 1998), and method of moments (MoM) (Gibson, 2021) are the three numerical analysis techniques that are widely utilized in the simulation and testing of antennas. Most studies on antenna modeling utilize computational methods to solve partial differential equations under defined boundary conditions (Weston, 2017).

There are various commercial software tools for antenna designing, such as Computer Simulation Technology (CST), Integral Equation 3D (IE3D), Altair FEKO, High-Frequency Structure Simulator (HFSS), and Advanced Design Systems (ADS). These software solutions are also missing some critical functionalities. ADS lacks the ability to model 3D structures, IE3D is unable to simulate structures with limited features, and HFSS and CST has substantial, large execution time which increases with the complexity of the antenna topology. Therefore, improving the antenna characteristics using current tools is a complex and time-consuming problem. For an antenna to function at its peak efficiency, proper optimization is essential. Current efforts are directed towards improving the efficiency of time-saving antenna design by utilizing machine learning or deep learning algorithms (Gianfagna *et al.*, 2017; Huang *et al.*, 2023a, 2023b; Wu *et al.*, 2024; J. Zhang *et al.*, 2023). Machine learning (ML) holds significant promise in the realm of antenna design and behavior prediction, as it enables considerable acceleration while maintaining high accuracy. Various machine learning models have been documented for antenna optimization, including random forest (Svetnik *et al.*, 2003), gradient boost regression (Zemel and Pitassi, 2000), support vector machine (Jacobs *et al.*, 2013), KNN (S. Zhang *et al.*, 2017), RNN, and ensembling techniques (Hsieh *et al.*, 2012), among others. Deep learning signifies a major transformation from conventional machine

DOI: 10.1201/9781042026289-19

learning models. Conventional machine learning requires the input of carefully selected features for the system. Deep learning algorithms extract these features directly from the data, resulting in more sophisticated and resilient models. Transformers represent a notable progression in the management of sequential data, surpassing RNNs and LSTMs in numerous applications. Transformers fundamentally change the approach to sequence processing in models, employing a mechanism known as self-attention to assess the significance of various components within the input data (Vaswani *et al.*, n.d.). In contrast to RNNs and LSTMs that handle data in a sequential manner, transformers are capable of processing entire sequences at once. This parallel processing enhances efficiency and enables the capture of complex relationships in data, which is essential for tasks such as language translation and summarization (Transformers in Machine Learning: An Overview of AI and ML, n.d.).

In this work, a flag shaped antenna with a plus shaped slot and defected ground structure is optimised for wide bandwidth and optimum return loss. The antenna was simulated in an EM simulator, CST MWS and it was observed that variations in the width of the microstrip feedline, length and width of the plus shaped slot has the most effects on the S-parameter results. Various ML/DL algorithms such as random forest, KNN and MLP has been implied on the dataset generated. An AI-transformer model is applied on the dataset and its results are analysed.

The rest of the article is as follows, Section 2 discusses the dataset preparation and preprocessing, Section 3 discusses the ML, DL and AI models and their results, Section 4 concludes the article.

2 DATASET GENERATION AND PREPROCESSING

2.1 *Dataset preparation*

The patch antenna was designed on Rogers RT Duroid 5880 substrate (60mmx40mm) (εr=2.2, thickness = 0.16 mm, tanδ = 0.0009). The feed network of the proposed antenna has a microstrip line of copper (thickness = 0.035 mm) with a partial ground layer of copper on the backside of the substrate which is made defected by cutting a rectangular shaped slot from it. The value of width of strip ('stripw') and length ('l') and width ('w') of plus shaped slot is varied from 0.5-2 mm, 12-14 mm and 12-14 mm with a step size of 0.3 mm, 0.2 mm and 0.2 mm, respectively, for different S-parameter measurements. These parameters are selected for optimization as it directly affects the reflection coefficient and resonance of the proposed antenna. In the research work presented, a total of 25100 datasets are generated. The desired dataset is created with 500 parametric sweeps (performed in CST), which occupied hours of processing time. This dataset is then pre-processed for training the ML/DL models. Data pre-processing is discussed in the following subsection.

2.2 *Data augmentation and preprocessing*

Since the data is generated using simulations, it is generally error-free data, a small, realistic noise is added to features or targets during training which mimics fabrication tolerances, measurement noise and VNA error. This prevents overfitting to perfect simulations and improves generalization to real-world scenarios. To mimic the real world, three types of error were introduced in the parent dataset.

(i) Additive noise: Real VNAs have thermal noise, and quantization error, which is often additive for S-parameter. In our work, we added a ± 1 dB variation in |S11| magnitude.
(ii) Frequency-axis jitter: VNAs don't always hit the exact same frequency points due to clock/timing drift. In our work, frequency samples were slightly shifted by 50 ppm randomly.
(iii) Systematic bias: Sometimes all measurements shift slightly due to calibration drift in VNA. This can be a constant offset in dB (± 1 dB in this work) for an entire sweep.

After data augmentation, outlier detection and handling step is carried out. This step helps in finding extreme targets/features and label extreme points for special handling later. Since, simulations can produce edge-case values from parameter corners. In this work, Isolation-forest is used for outlier detection and handling. Isolation Forest is a method used for outlier identification, in extensive datasets by isolating them using binary partitioning (2.7. Novelty and Outlier Detection — Scikit-Learn 1.7.0 Documentation, n.d.).

3 ANTENNA OPTIMISATION USING ML/AI

3.1 *Random forest*

A Random Forest Regressor (RFR) was implemented as an initial supervised learning model, using the scikit-learn library. RFR was chosen because of its robustness to non-linear feature interactions and reduced susceptibility to overfitting when compared with single decision trees.The model was configured with the following parameters:

(a) n_estimators = 70 (number of trees in the ensemble),

(b) max_features = 2 (number of features considered at each split),
(c) bootstrap = True (sampling with replacement for tree training), and
(d) oob_score = False (out-of-bag error estimation disabled).

The model was trained on the training dataset, and predictions were generated for the testing dataset. Figure 1(a) shows actual vs predicted values of random forest.

3.2 *KNN*

Following the Random Forest baseline, a K-Nearest Neighbors (KNN) Regressor was implemented using the scikit-learn library to further investigate the predictive capability of non-parametric, instance-based learning methods for antenna performance estimation. The KNN algorithm predicts the target variable for a query point by averaging the outputs of its k nearest neighbors in the feature space, where distance is typically computed using the Euclidean metric. This method inherently captures local patterns in the data without assuming a global functional form between input parameters and the output response. In this study, the KNN model was configured with:

(a) n_neighbors = 7 (number of neighbors considered in prediction),
(b) default Minkowski distance metric (equivalent to Euclidean distance when p = 2).

The model was trained using the training subset obtained from the noise-augmented dataset. Predictions were then generated for the test subset to evaluate its performance as shown in Figure 1(b).

3.3 *MLP*

To explore the effectiveness of deep, non-linear function approximation, a Multilayer Perceptron (MLP) Regressor was implemented using the scikit-learn library. MLP models belong to the class of feed-forward artificial neural networks and are capable of learning complex, non-linear mappings between inputs and outputs through multiple layers of interconnected neurons. The network architecture adopted in this work consisted of two hidden layers with 64 neurons each (hidden_layer_sizes = (64, 64)), using the default rectified linear unit (ReLU) activation function and the Adam optimization algorithm for weight updates. The maximum number of training iterations was set to 1000 to ensure convergence (max_iter = 1000). The model was trained on the 80% training split of the noise-augmented dataset. Predictions were subsequently generated for the 20% test split. Performance evaluation can be shown in Figure 1(c).

3.4 *Transformer*

For implementing transformer AI model, Informer architecture was chosen mainly because of two reasons, it is a Transformer-based sequence model optimized for long sequences and it uses ProbSparse self-attention, which reduces complexity from O (L^2) to O (L log L). This allows efficient processing of many frequency points. Informer Transformer model usually handles time-series forecasting, but the same principles apply for the case of frequency-series regression of this work. In this case, the input is a sequence of number of frequency points (25100 frequencies). Each position (frequency) contains an encoded frequency value, static parameters (l, w, stripw) mapped to all sequence positions. The output is the full curve of predicted S11 at each frequency. Before implementing the model, the numerical inputs were embedded into a latent vector space by keeping a dimensionality of embeddings as 64. Since, transformers don't know order unless it is encoded. In this work, a sinusoidal positional encoding was done, so the model knows which frequency point is which. In this encoding, the number of attention heads is kept as 4, Feed-forward layer size is 256, and 2 to 4 Stacked encoder layers were used with a dropout feature. These attributes of the encoder layer helped the model to learn the different patterns in the curve simultaneously as well as prevented overfitting. In frequency response prediction, many points have correlated trends, so Informer selects the most informative queries instead of attending to all positions. As an output layer, a simple linear projection mapping the hidden representation to a single S11 value per frequency point is selected. For checking the interpolation ability of the model, a set of 6 different values of "l" is randomly chosen from the training set, on which the model is tested, so the model is tested only on unseen values.

Figure 2(a) shows the proposed Informer-based encoder-only architecture for microstrip antenna S11 prediction. Geometry parameters (l, s, stripw) are broadcasted across all frequency points and embedded alongside the frequency values. These embeddings are summed with sinusoidal positional encodings. The stacked Informer encoder (ProbSparse self-attention) processes the sequence and a final regression head predicts the S11 curve. Figure 2(b) shows a scatter plot of predicted versus actual S_{11} values across all frequency points for unseen slot lengths. The points cluster closely around the diagonal, which confirms that predictions align well with actual simulated values. Figure 3 shows the comparative plot for the predicted and simulated S11 vs frequency curve for 3 of the unseen "l" values. The predicted curve follows the overall shape of the actual S_{11} very closely across the full frequency sweep. MSELoss is chosen as a loss function as it penalizes large deviations in dB strongly which is important in this work, because both small ripples and large resonances in the S11 curve are a matter of concern in antenna designing. A learning rate of 1e-4 chosen after trying both higher and lower values for stability. Small batch sizes (8–16) are chosen due to GPU memory constraints.

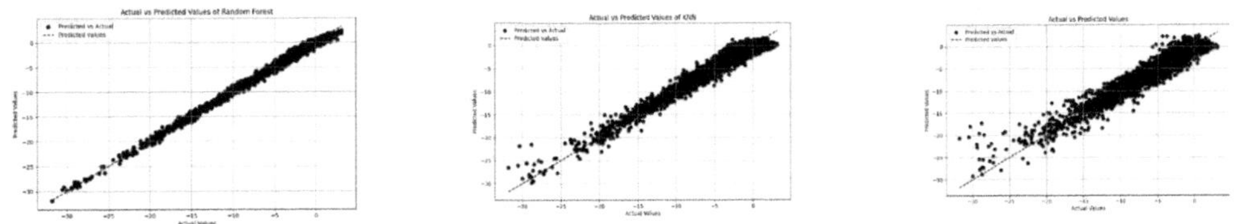

(a) (b) (c)

Figure 1. Actual vs predicted values of (a) Random Forest (b) KNN (c) MLP.

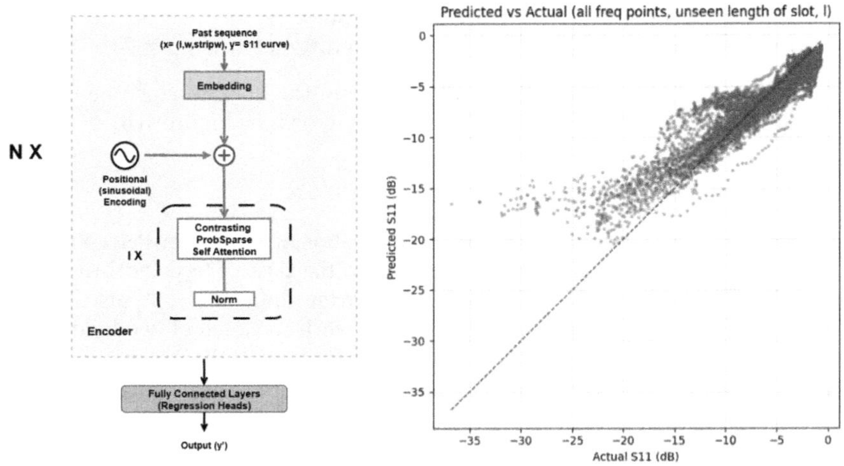

(a) (b)

Figure 2. (a) Network Architecture of the proposed stacked only encoder based - Informer model (b) Actual vs predicted values of the Informer model

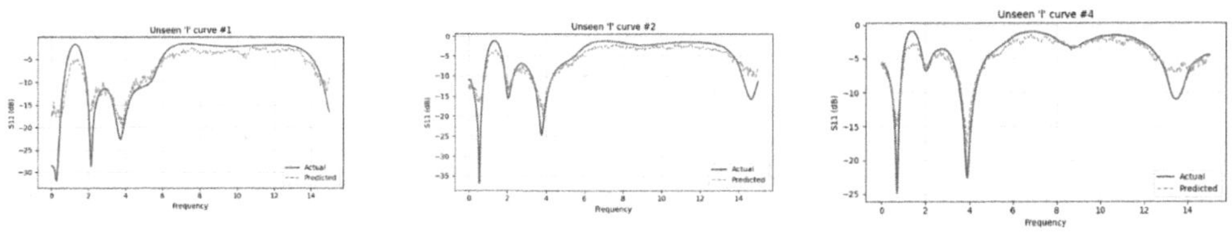

(a) (b) (c)

Figure 3. Actual vs predicted S11 response of the antenna using Informer on three different unseen 'l'.

4 CONCLUSION

In this work, we have used ML, DL and AI for optimising the performance of antenna based on its geometric parameters. A flag shaped patch antenna with a plus shaped slot in the centre of the patch, was selected for the study and dataset generation. Various geometric parameters such as length and width of the slot and the width of the strip feedline was varied and a dataset of 25,100 points were generated. The variation in these parameters changed the return loss behaviour of the antenna which was used to train the models. The Informer-based transformer AI significantly outperformed the other models in predicting antenna return-loss behaviour. By all metrics (R^2, MAE, MSE), Informer produced the most accurate and robust predictions. Table 1 shows the comparison between the metrics of ML models used in this work on the dataset. On entirely new geometry values of l, the Informer achieved R^2 = 0.88, whereas the Random Forest, MLP, and KNN achieved substantially lower R^2 (typically below 0.80). This demonstrates that sequence-aware, transformer-based models can capture complex electromagnetic relationships better than the traditional ML/DL regressors. All models showed good performance on the training and validation sets, but the Informer showed a superior interpolation capability which makes it a better model for predicting unseen design parameters. Advanced AI/ML techniques can accelerate antenna design process by predicting full S_{11} curves with high accuracy and precision. This approach offers a fast alternative for time and hardware costly EM simulations.

Table 1. ML models' metrics.

ML Model	MAE	MSE	RMSE	R2
Random Forest	0.24	0.34	0.189	0.80
KNN	0.45	1.29	1.135	0.57
MLP	0.27	0.19	0.204	0.78
Informer	0.02	0.0909	0.021	0.88

REFERENCES

2.7. Novelty and Outlier Detection — scikit-learn 1.7.0 documentation. (n.d.). Retrieved July 9, 2025, from https://scikit-learn.org/stable/modules/outlier_detection.html#isolation-forest

Gianfagna, C., Yu, H., Swaminathan, M., Pulugurtha, R., Tummala, R., and Antonini, G. (2017). Machine-learning approach for design of nanomagnetic-based antennas. *Journal of Electronic Materials*, 46, 4963–4975.

Gibson, W.C. (2021). *The Method of Moments in Electromagnetics*. Chapman and Hall/CRC.

Godara, L.C. (2018). *Handbook of Antennas in Wireless Communications*. CRC press.

Hsieh, S.-L., Hsieh, S.-H., Cheng, P.-H., Chen, C.-H., Hsu, K.-P., Lee, I.-S., Wang, Z., and Lai, F. (2012). Design ensemble machine learning model for breast cancer diagnosis. *Journal of Medical Systems*, 36, 2841–2847.

Huang, H., Yang, X.-S., and Wang, B.-Z. (2023a). Machine-learning-based generative optimization method and its application to an antenna decoupling design. *IEEE Transactions on Antennas and Propagation*, 71(7), 6243–6248. https://doi.org/10.1109/TAP.2023.3270716

Huang, H., Yang, X.-S., and Wang, B.-Z. (2023b). Machine-learning-based generative optimization method and its application to an antenna decoupling design. *IEEE Transactions on Antennas and Propagation*, 71(7), 6243–6248.

Jacobs, J.P., Koziel, S., and Ogurtsov, S. (2013). Computationally efficient multi-fidelity bayesian support vector regression modeling of planar antenna input characteristics. *IEEE Transactions on Antennas and Propagation*, 61(2), 980–984. https://doi.org/10.1109/TAP.2012.2220513

Svetnik, V., Liaw, A., Tong, C., Culberson, J.C., Sheridan, R.P., and Feuston, B.P. (2003). Random forest: a classification and regression tool for compound classification and QSAR modeling. *Journal of Chemical Information and Computer Sciences*, 43(6), 1947–1958.

Transformers in Machine Learning: An Overview of AI and ML. (n.d.). Retrieved August 11, 2025, from https://blog.clearscale.com/evolution-of-machine-learning-and-the-future-of-ai-applications/

Vaswani, A., Brain, G., Shazeer, N., Parmar, N., Uszkoreit, J., Jones, L., Gomez, A.N., Kaiser, Ł., and Polosukhin, I. (n.d.). *Attention Is All You Need.*

Volakis, J.L., Volakis, J.L., Chatterjee, A., and Kempel, L.C. (1998). *Finite Element Method for Electromagnetics*. Universities Press.

Weston, D. (2017). *Electromagnetic Compatibility: Principles and Applications, Revised and Expanded.* CRC Press.

Wu, Q., Chen, W., Yu, C., Wang, H., and Hong, W. (2024). Machine-learning-assisted optimization for antenna geometry design. *IEEE Transactions on Antennas and Propagation*, 72(3), 2083–2095. https://doi.org/10.1109/TAP.2023.3346493

Zemel, R., and Pitassi, T. (2000). A gradient-based boosting algorithm for regression problems. *Advances in Neural Information Processing Systems*, 13.

Zhang, J., Xu, J., Chen, Q., and Li, H. (2023). Machine-learning-assisted antenna optimization with data augmentation. *IEEE Antennas and Wireless Propagation Letters*, 22(8), 1932–1936. https://doi.org/10.1109/LAWP.2023.3269811

Zhang, S., Li, X., Zong, M., Zhu, X., and Cheng, D. (2017). Learning k for knn classification. *ACM Transactions on Intelligent Systems and Technology (TIST)*, 8(3), 1–19.

Smart Systems and Beyond – Bhatia et al. (Eds)
© 2027 The Author(s), ISBN: 978-1-041-36617-1

Data-Driven Forecasting of Blood Glucose Levels via ExtraTrees Ensemble Learning

Harshit Manik, Shaurya Punj, Mantra Gupta, and Japleen Kaur
Department of Electronics and Communication Engineering

Nidhi Upadhyay
Centre of Excellence in Emerging Materials, Thapar Institute of Engineering and Technology, Patiala, India

Amanpreet Kaur
Department of Electronics and Communication Engineering

ABSTRACT: Precise forecasting of blood glucose levels (BGL) plays a vital role in efficient diabetes control and patient care. This work presents a machine learning framework employing the ExtraTrees Regressor to estimate BGL from frequency and S-parameter readings obtained through experimental measure- ments. To enhance predictive accuracy and effectively model complex non-linear dependencies, a comprehensive feature engineering process was implemented. Furthermore, cross-validation techniques were utilized to validate the model's stability, consistency, and generalization capability.

Keywords: Machine Learning, Blood Glucose Monitoring, ExtraTrees, Feature Engineering, Biomedical Sensors

1 INTRODUCTION

Diabetes mellitus is a long-term metabolic disorder that impacts millions of individuals globally and necessitates consistent tracking of blood glucose levels (BGL) for proper management. Traditional invasive techniques, such as finger-prick tests, cause discomfort and are impractical for frequent or lifelong use. This limitation has encouraged significant progress in non-invasive glucose detection technologies, especially those utilizing biomedical antennas, wearable sensing devices, and RF signal analysis [1]–[3], [7]. Antennas operating within the ISM frequency range have shown potential in detecting variations in the dielectric characteristics of blood that correspond to changes in glucose concentration, enabling the extraction of frequency and S-parameter data for further processing.

Although antenna-based methods are effective for gathering non-invasive measurements, they are primarily constrained to raw frequency-domain outputs such as reflection coefficients (S_{11}) and return losses. While these parameters vary with glucose concentration, they do not directly produce precise BGL estimations. Therefore, it becomes essential to develop computational models capable of learning the complex, non-linear correlation between RF measurement features and blood glucose levels.

Machine learning (ML) provides a compelling framework to address this challenge. Through advanced regression algorithms, feature engineering, and cross-validation strategies, ML-based approaches can significantly improve the prediction accuracy of RF-driven glucose estimation systems. Among various ensemble methods, the ExtraTrees Regressor is particularly advantageous due to its strong resistance to overfitting, capacity to process high-dimensional feature interactions, and interpretability through feature importance evaluation [5], [6].

The primary contributions of this research are summarized as follows:

1) Development of an extensive dataset containing frequency and S-parameter readings acquired from non-invasive sensing experiments, followed by preprocessing to ensure consistency and quality.
2) Construction of a robust feature engineering pipeline incorporating logarithmic, interaction-based, and periodic transformations to enrich the input feature space and improve discriminative performance.
3) Implementation and optimization of an ExtraTrees Regressor that delivers high predictive accuracy for BGL estimation, demonstrating the feasibility of ML-based non-invasive glucose monitoring in real-time applica- tions.

The remainder of this paper is organized as follows: Section II describes the dataset origin and simulated response, Section III presents preprocessing, Section IV shows exploratory analysis, Section V details feature engineering, Section VI discusses model selection and training, Section VII evaluates performance, and Section VIII concludes.

 DOI: 10.1201/9781042026289-20

2 SIMULATION OF GLUCOSE-DEPENDENT MEASUREMENTS

The raw data for this study was generated by simulating the electromagnetic response of an antenna-based non-invasive glucose sensing system, consistent with prior RF sensing studies [2], [7]. The designed antenna was exposed to varying dielectric properties of human blood, which were modeled according to the Debye relation for glucose-dependent plasma parameters [4]. As blood glucose concentration increases, both the dielectric constant and conductivity of blood plasma decrease, leading to measurable changes in the frequency-dependent S-parameter response.

The antenna model was simulated to obtain reflection coefficients (S_{11}) across a broad frequency span ranging from 2.3 GHz to 6 GHz. For each considered blood glucose concentration, the simulated antenna produced a distinct variation in its S-parameter response while maintaining stable performance within the ISM frequency band. These frequency-domain profiles serve as the fundamental data source for constructing the machine learning dataset used in this work.

Figure 1 presents typical S-parameter curves for multiple glucose levels. The observable shift in magnitude and resonant frequency points with changing glucose concentrations confirms that electromagnetic characteristics can be correlated with physiological alterations in blood composition. The generated simulation data were subsequently organized into structured feature–target pairs to enable downstream machine learning analysis, as detailed in Section III.

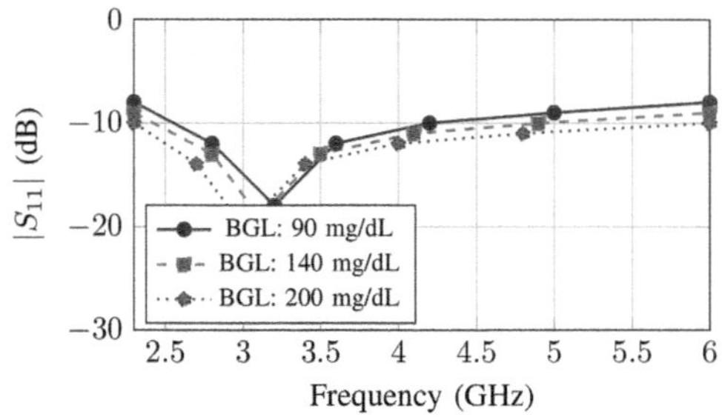

Figure 1. Illustrative simulated S_{11} responses corresponding to various blood glucose levels.

3 DATASET AND PREPROCESSING

The compiled dataset contains a total of 82,082 instances. Each entry includes two key features: the excitation frequency (f in GHz) and the magnitude of the corresponding S-parameter ($|S|$). In certain measurements, a Group identifier denotes individual experimental sessions, enabling grouped cross-validation to avoid information leakage. The output variable represents the blood glucose level (BGL, mg dL^{-1}).

Preprocessing workflow:

- Conversion of all variables into consistent numeric representations.
- Elimination of incomplete or noisy entries containing missing or corrupted values.
- Synchronization of the Group

column (when available) with the main feature matrix to ensure proper group-based validation during training.

Following data cleaning, the records were divided into training and testing subsets using an 80:20 split. The training portion was further processed through k-fold cross-validation with group awareness to maintain inter-session independence.

4 EXPLORATORY ANALYSIS

A preliminary exploratory analysis was conducted to examine the dependencies among frequency, S-parameter magnitudes, and glucose concentration. Distributions of glucose readings, pairwise correlations, and variation trends of $|S|$ against frequency were analyzed to guide feature design. (Figures omitted for brevity.)

5 FEATURE ENGINEERING

Feature engineering was performed to enrich the dataset representation and enable the model to capture complex, non-linear relationships between frequency, S-parameter values, and BGL. The following transformations were introduced:

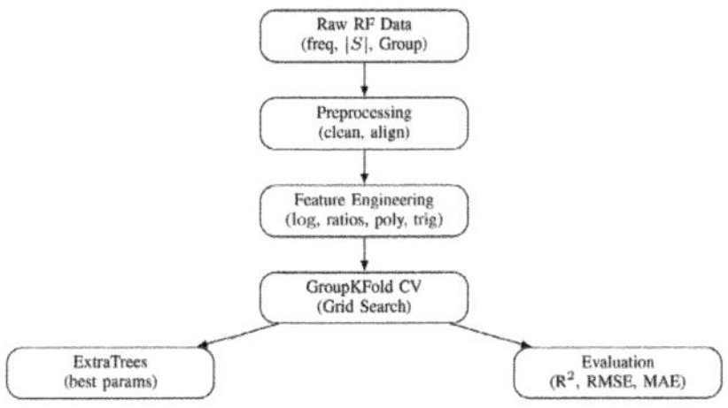

Figure 2. Workflow summary for ExtraTrees model training and evaluation.

- Logarithmic terms: $\log f$, $\log |S|$
- Ratios: $\frac{|S|}{f}$
- Interaction and polynomial features: $f|S|, f^2, |S|^2$
- Periodic terms: $\sin f$ and $\cos f$

6 MODEL SELECTION AND TRAINING

The ExtraTrees Regressor was chosen due to its capability to model complex non-linear patterns, robustness to noisy data, and inherent resistance to overfitting through randomized tree splits. Hyperparameters were optimized via a grid search using 5-fold GroupKFold cross-validation. The optimal configuration obtained was:

- `n_estimators = 200`
- `max_depth = None`
- `min_samples_leaf = 1`

7 MODEL EVALUATION

7.1 *Protocol*

Model generalization was assessed using stratified, group-aware cross-validation along with a final independent test split in an 80:20 ratio. The grouping ensured that data collected from the same experimental session were never shared between the training and validation folds, thereby preventing any information leakage. Hyperparameter values determined in Section VI were kept fixed, and no further optimization was performed on the test subset.

Table 1. Performance of extratrees regressor on BGL prediction.

Dataset	R^2	RMSE (mg dL^{-1})	MAE (mg dL^{-1})
Training (CV mean)	0.9362	–	–
Test set	0.9746	18.806	4.916

7.2 *Feature contribution*

Model interpretability was analyzed using impurity-based feature importance scores aggregated across all trees in the ensemble. Derived variables—especially ratio, logarithmic, and interaction-based terms—were found to dominate the importance ranking. This outcome emphasizes the effectiveness of incorporating domain-specific feature transformations within the learning framework.

7.3 *Metrics*

Model accuracy was quantified using three statistical measures: the coefficient of determination (R^2), root mean squared error (RMSE), and mean absolute error (MAE). These are defined as follows:

$$R^2 = 1 - \frac{\sum_i (y_i - \hat{y}_i)^2}{\sum_i (y_i - \bar{y})^2}, \tag{1}$$

$$\text{RMSE} = \sqrt{\frac{1}{n} \sum_{i\,(y_i - \widehat{y}_i)^2,} \text{MAE} = \frac{1}{n} \sum_{i\,|y_i - \widehat{y}_i|.}} \tag{2}$$

7.4 *Robustness and error analysis*

Beyond numerical metrics, a qualitative review was carried out to examine the model's stability and behavior under different conditions:

Homoskedasticity: The residual errors were distributed tended to increase toward higher and lower glucose extremes, likely due to fewer samples and greater physiological fluctuation at those limits.

- **Group Stability:** Minimal variation was observed be-tween folds in the GroupKFold scheme, suggesting the model maintained consistent accuracy across separate acquisition sessions despite changes in probe position or coupling efficiency.

Together, these metrics capture the model's ability to explain variance, limit high-magnitude prediction errors, and maintain low average deviation, all expressed in mg dL^{-1}.

7.5 *Results*

Cross-validation results remained consistent across all folds, indicating that the ExtraTrees model performed stably under varying data partitions. When evaluated on the unseen test samples, the model exhibited a strong correlation between predicted and actual glucose concentrations, confirming its predictive robustness and minimal bias.

Figure 3 illustrates the comparison between predicted and actual BGL values. The majority of data points align closely with the $y = x$ diagonal, demonstrating high predictive fidelity and near-zero systematic deviation across the entire glycemic spectrum.

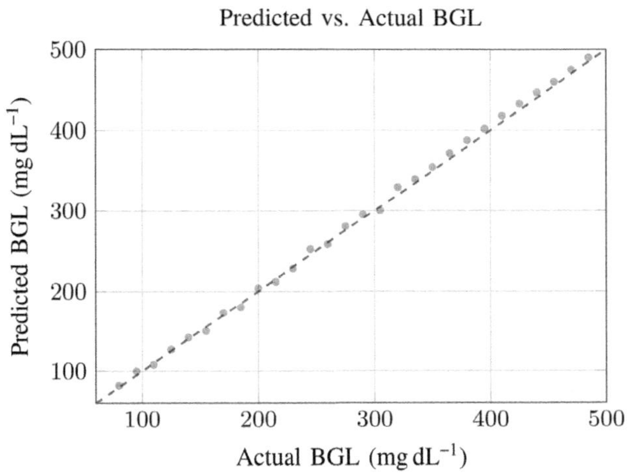

Figure 3. Predicted versus actual BGL scatter showing strong correlation along the $y = x$ line.

- **Clinical Relevance:** With a test-time MAE of less than 5 mg dL^{-1}, the system demonstrates potential for realtime glucose trend tracking when used with data smoothing and outlier filtering. However, for direct therapeutic applications, additional safeguards such as calibration checks and redundant sensing are recommended.

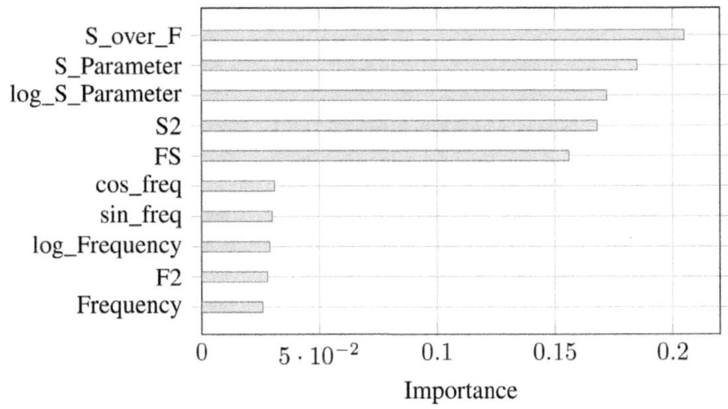

Figure 4. Feature importance ranking showing dominance of engineered transformations.

7.6 *Ablations (Qualitative)*

Ablation experiments were performed to determine the influence of various modeling and feature design choices:

1) Excluding interaction or ratio-based features led to a noticeable drop in explained variance, confirming their contribution to modeling the coupling between frequency and dielectric response.
2) Very shallow ensembles or those with few estimators produced unstable outcomes, while excessively deep forests tended to overfit. The selected parameters (`n_estimators=200`; `max_depth=None`; `min_samples_leaf=1`) offered the best trade-off between bias and variance.
3) Traditional k-fold validation without grouping yielded overly optimistic performance due to cross-session leakage, underscoring the necessity of group-aware evaluation for RF-based biomedical datasets.

7.7 *Reproducibility notes*

To ensure reproducibility, all random seeds were fixed during dataset shuffling, model initialization, and cross-validation. Computations were performed in double precision, and performance metrics were consistently reported in mg dL^{-1}.

8 CONCLUSION

This study introduced a machine learning framework for estimating blood glucose levels (BGL) directly from antennaderived S-parameters by leveraging their dependence on the dielectric properties of biological tissue, as described by the Debye dispersion principle. Traditional analytical inversion of Debye parameters is computationally intensive, motivating the use of two data-driven regressors—[Model-1] and [Model-2]. The models were trained on a curated dataset encompassing [details: subject cohort, sample count, frequency span, temperature control, and labeling protocol]. Group-aware testing demonstrated robust generalization, with high R^2 and low error scores (MSE, MMSE). Feature analysis revealed that predictors such as $|S_{11}|$, $|S_{21}|$, logarithmic amplitudes, frequency-normalized ratios, and nonlinear interaction terms align with Debye-related dispersion effects, concentrating the predictive signal within frequency bands most sensitive to $\varepsilon'(\omega)$ and $\varepsilon''(\omega)$. Collectively, these findings confirm that the proposed approach is well-suited for RF antenna-based, non-invasive glucose monitoring, providing a fast and reliable estimation of BGL directly from S-parameter data without explicit Debye inversion.

Future work will focus on: (i) expanding the dataset to include more subjects, extreme glycemic conditions, and varying environmental settings to reduce domain shift; (ii) integrating multimodal signals such as skin temperature and hydration indices to stabilize dielectric estimation; (iii) incorporating uncertainty-aware inference techniques (e.g., conformal prediction) to yield actionable confidence intervals; (iv) conducting prospective clinical validation using Clarke/Parkes errorgrid analyses; and (v) developing physics-informed hybrid models that jointly estimate Debye parameters and glucose levels for improved accuracy and interpretability.

ACKNOWLEDGMENT

The authors gratefully acknowledge the contributions of their peers who provided support during RF data acquisition and offered valuable insights that strengthened this research on non-invasive sensing.

REFERENCES

[1] M.A. Rahman, M.K. Hasan, and M.R. Yuce, "RF-based non-invasive blood glucose monitoring: A review of techniques and challenges," *IEEE Reviews in Biomedical Engineering*, vol. 12, pp. 106–118, (2019).
[2] T. Yilmaz, R. Foster, and Y. Hao, "Detecting glucose concentration using RF biosensors," *IEEE Sensors Journal*, vol. 14, no. 12, pp. 4196–4206, (2014).
[3] S. Haxha and J. Jhoja, "Non-invasive blood glucose monitoring using electromagnetic waves," *Sensors*, vol. 16, no. 5, p. 695, (2016).
[4] P. Debye, *Polar Molecules*. New York: Chemical Catalog Company, 1929.
[5] P. Geurts, D. Ernst, and L. Wehenkel, "Extremely randomized trees," *Machine Learning*, vol. 63, pp. 3–42, (2006).
[6] O. Sharif, M. Zahid, and A.H. Khandoker, "Machine learning-based non-invasive glucose monitoring using physiological signals: A systematic review," *IEEE Access*, vol. 7, pp. 144681–144701, (2019).
[7] N. Upadhyay, A. Kaur, and A. Pattanayak, "Flexible antenna with AMC backing for non-invasive blood glucose monitoring," in *Proc. IEEE Conf.*, Thapar Institute of Engineering and Technology, Punjab, India, (2023).

Smart Systems and Beyond – Bhatia et al. (Eds)
 ISBN: 978-1-041-36617-1

Non-Invasive Brain Cancer Detection with Compact UWB Microstrip Antenna Having Stair Patch Geometry

Jashanpreet Singh, Ankur Singhal, Jaspreet Kaur, Dinesh Arora, Vinay Bhatia, and Kartikay
ECE Departments, CEC-CGC Landran, Mohali, India

ABSTRACT: Antennas are essential for both sending and receiving electromagnetic waves. Antennas must be compact, planar, have high radiation efficiency, and run steadily over the assigned frequency spectrum in order to allow UWB-based medical imaging, especially for the detection of brain tumours. In order to detect brain cancer, a small, stair-shaped microstrip patch antenna is designed and simulated in this study. The antenna structure is tailored to the exacting specifications of applications involving medical imaging. The High-Frequency Structure Simulator (HFSS) has been used to investigate parameters like radiation efficiency, radiation patterns, total field gain, return loss, and voltage standing wave ratio (VSWR). Its suitability for dependable and effective UWB medical imaging applications is confirmed by the results.

Keywords: UWB stairs patch antenna, brain tumor, head phantom model, HFSS software

1 INTRODUCTION

Due to cancer millions of people are resulting with fatalities worldwide due to which early detection of brain cancer is the need of an hour. However it is altogether a crucial process as it involves the diagnostic methods such as X-rays, mammography, MRI, and ultrasound techniques, which may have certain irregularities and are expensive as well.

After analysing the electrical characteristics of a radar-based Ultra Wide-Band (UWB) microwave imaging technique we can have a non-invasive way of finding the difference between cancerous cells and surrounding tissues. UWB antennas, as specified by the Federal Communications Commission (FCC), are essential for microwave imaging systems due to the broad frequency range associated with UWB technology. The development of cost-effective UWB systems is significant for both commercial and military uses. This imaging method is widely applied in biomedical fields, particularly in cancer detection. Transmitting and receiving antennas, designed to operate within the UWB frequency band of 3.1 to 10.6 GHz allocated by the FCC, require efficiency, strong radiation performance, and accurate time-domain response, which are critical factors in UWB antenna design as studied [1][2][3][4].

The impedance matching circuit is the main focus of the ground plane. The bandwidth characteristics are influenced by the circular patch's radius as well as the ground plane's width. UWB performance for brain cancer detection is demonstrated by the circular monopole antenna. [5].

In order to satisfy the requirements of UWB characteristics, such as wide bandwidth and enhanced reflection coefficient, a miniature microstrip antenna with T-shaped and U-shaped slots is built and simulated. An ultra-wideband Vivaldi antenna is created and placed next to a four-layer brain model. A tumor model with a radius of 5mm is incorporated. The simulated brain phantom model receives brief electromagnetic pulses, and the reflected signals are examined to detect brain cancer. [6][7].This study details the development and application of a pentagon antenna that is placed directly on the human head to detect stroke and brain tumors. [8]. CST Microwave Studio is used to simulate the antenna, which is then built on a FR-4 substrate with a relative permittivity of 4.4 and a thickness of 1.5 mm. [9] . This paper details the design of a UWB antenna tailored for brain cancer detection. The antenna utilizes a circular modification patch microstrip design with slotted patch, partial ground plane, and reflector [10]. This paper presents the design of a UWB Pentagon antenna tailored for detecting brain stroke and brain tumors. Utilizing UWB (Ultra-Wide Band) technology in stroke detection offers opportunities to develop cost-effective, rapid-response, and portable diagnostic devices, potentially serving as vital tools in emergency situations [11]. With accuracy of area, time, and pixel difference, the suggested technique (BBNC) shrinks the extracted region once more into the necessary and genuine tumor location[12][13]. The microstrip patch antenna for WLAN applications is presented in this work. The microstrip antenna has a planar shape and it comprises a defected ground, a feed, a substrate, a patch and a defected ground structure. [14] The suggested antenna solution exhibits satisfactory MIMO system performance, making it appropriate for both 4G/5G and internet of things devices. When complementary Metamaterial construction is used on the ground plane, good isolation properties are seen, with an isolation level between the antenna elements that is less than -21 dB.[15]

DOI: 10.1201/9781042026289-21

2 PROPOSED METHOD

The antenna is an essential part of the microwave imaging system that has a significant influence on the system's overall performance and image quality.Designing UWB microstrip antennas for brain cancer detection in open areas using frequencies between 3.1 and 10.6 GHz is the main goal of this work. This suggested technique simulates a head phantom's antenna both with and without a malignancy. The response of S-parameters (S11) and the antenna's E-field, Current density, SAR, Radiation Efficiency and Total gain will be covered in this study.

2.1 *Antenna design*

A substrate with a relative permittivity of 4.4 and a thickness of 1.6 mm has been used to mount an ultra wide band Stairs microstrip patch antenna. A microstrip transmission line impedance-matched to 50 ohms connects the antenna. The proposed microstrip patch antenna utilizes FR-4 as its substrate material, with copper-annealed patch and ground elements. The substrate measures length, width and thickness in mm. The partial ground plane, which measures mm in length, mm in breadth, and mm in thickness, is located on the bottom of the substrate. The patch has dimensions of length, width, and thickness and is situated on the other side of the substrate.in millimeters. The antenna, which has dimensions of (mm in length) and (mm in breadth), is fed by the microstrip line feed, as shown in Table 1.

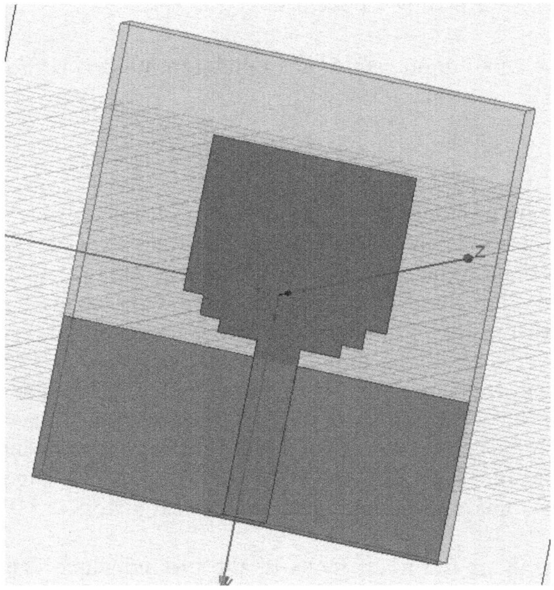

Figure 1. Proposed antenna design.

Table 1. Specifications of the proposed antenna.

Parameters	Dimensions(mm)
L_S	**30mm**
W_S	**35mm**
H_S	**1.6mm**
L_G	**30mm**
W_G	**12.5mm**
W_P	**15mm**
L_P	**13mm**
W_F	**3.2mm**
L_F	**13.5mm**
G	**2mm**
W1	**3mm**
W2	**3mm**
L1,l2,L3,L4	**1mm**
W3,w4,w5,w6	**1.5mm**

3 HEAD PHANTOM MODEL

The human head comprises three primary components: Brain, skin, skull, Dura and fat. The brain, which is the main and core component; the surrounding soft tissues, such as skin and fat, Dura; and the skull, which offers protective support.

In antenna simulations, a phantom serves as a model for head tissue and tumors. This phantom incorporates dielectric constants and electrical conductivities to accurately represent the materials. The tissue of the imitated head, for example, usually consists of layers of skin (1 mm), fat (1 mm), skull (5 mm), and brain (40 mm).. Furthermore, tumors in the cranial tissue, which are usually 10 mm in size, have an electrical conductivity of 2.62 S/m and a dielectric constant of 54.2 F/m. Table 2 contains details regarding the human brain phantom, including its composition and relevant characteristics. Subsequent experimentation involves testing the proposed antenna on a brain phantom under two conditions: with and without the presence of a tumor. The results reveal significantly increases in the reflection coefficient, current density, and (SAR) when a tumor is introduced into the brain phantom.

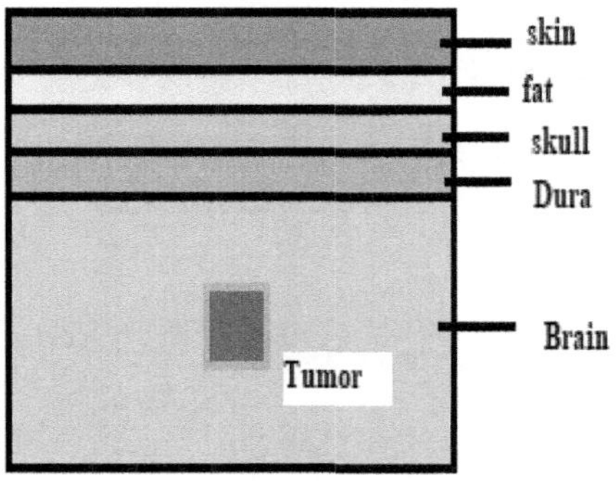

Figure 2. Head phantom design.

Table 2. Specifications of human head phantom model.

TISSUES	PERMITIVITY(Ɛ)	CONDUCTIVITY (σ)	CURRENT DENSITY
SKIN	46	0.74	1091
SKULL	5.7	0.02	1855
FAT	5.6	0.0.5	911
DURA	47	0.8	1135
BRAIN	45.22	1.28	1031

4 SIMULATIONS AND RESULT

For simulation purposes, a brain phantom model is placed 3 mm below the antenna to lessen the negative effects of the electromagnetic field. The characteristics listed in Table 1 are necessary for the simulation's outcomes. Every antenna parameter is examined under two conditions: with and without tumors, in both open space and with a human head phantom.

Figure 3. Simulation of phantom model.

4.1 *Return loss analysis*

The simulated (S11) parameters of the antenna is shown in Figures 3 and 4 (Both with and without the tumor.)The loss remains below the antenna's immolated return loss, both with and without the tumor. The return loss has a bandwidth of 3.1 to 10.6 GHZ and remains below -10db.

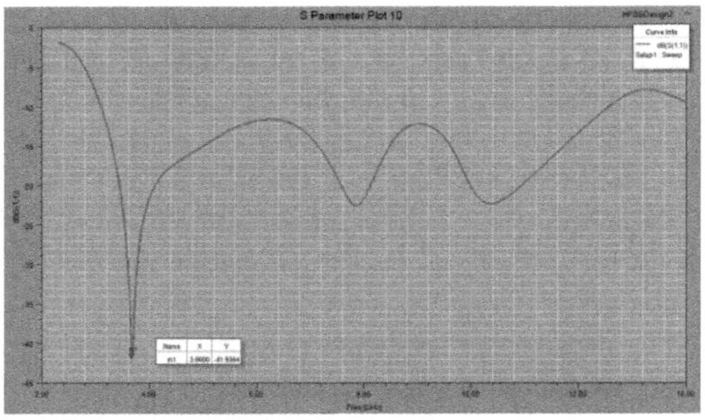

Figure 4. (S11) Return loss without tumor at 3.6 GHz.

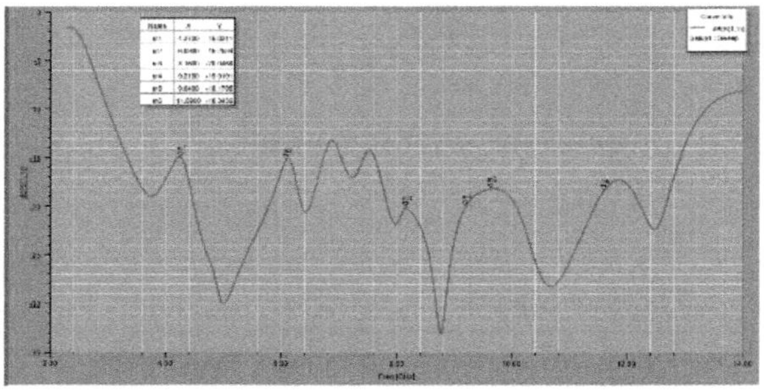

Figure 5. Antenna return loss with a head phantom model at 15 mm.

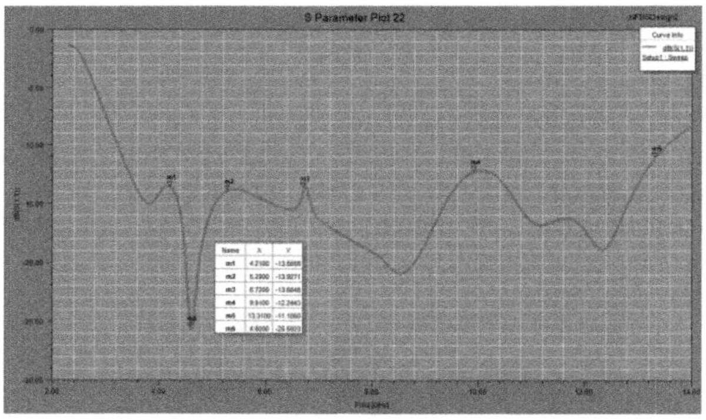

Figure 6. Antenna return loss with a Head phantom model at 17 mm.

Figure 5 shows that the simulated return loss, which is visible at 15 mm, is less than -10 dB between 3.1 and 10.6 GHz, except 4.2 GHz.Nevertheless, at a distance of 17 mm, the return loss observed at frequencies 3.1 to 10.6 GHz is less than -10 dB, except for the 6.1 GHz and 9.01 GHz region shown in Figure 6. As Figure 9 illustrates With the exception of 5.2 GHz and 10.04 GHz, the antenna's return loss at a distance of 25 mm is likewise less than -10 dB from 3.1-10.6 GHz. Consequently, 20 mm may be a sufficient antenna distance for the head phantom model to function successfully.

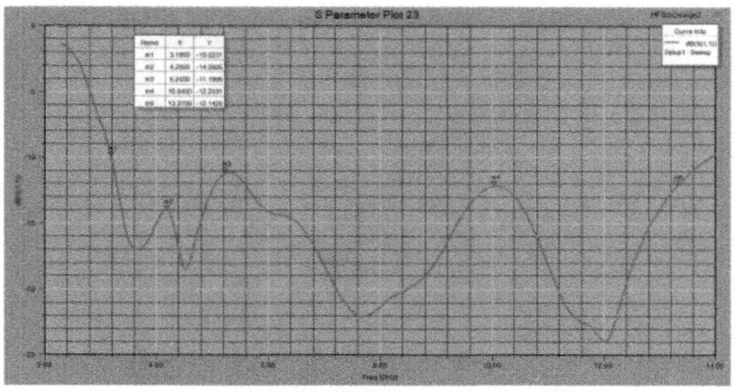

Figure 7. Antenna return loss with a head phantom model at 20 mm.

4.2 *VSWR (VOLTAGE STANDING WAVE RATIO) Anlaysis*

In HFSS, we simulated the antenna's VSWR (with tumor and without tumor), and the results are shown in Figures 7, 8. The VSWR is less than 2, which is sufficient. in accordance with FCC regulations, to cover the ultra-wide band (UWB).

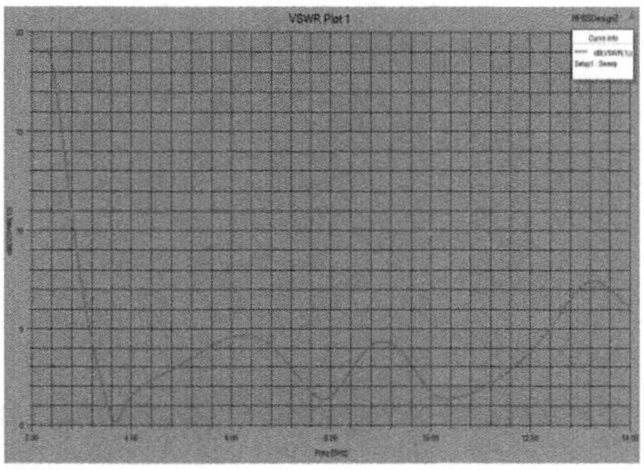

Figure 8. VSWR Result (without tumor).

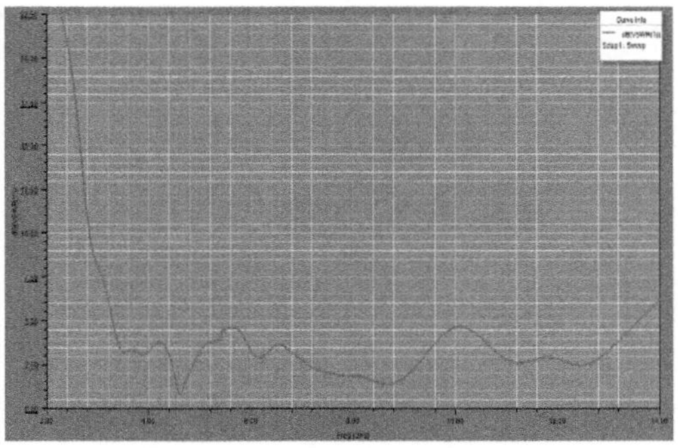

Figure 9. VSWR Result (with tumor).

4.3 *Analysis of Electric field (E-field)*

Figures 9, 10 demonstrates how much force or field is there around living body cells by presenting the E-field with (tumor and without tumor).

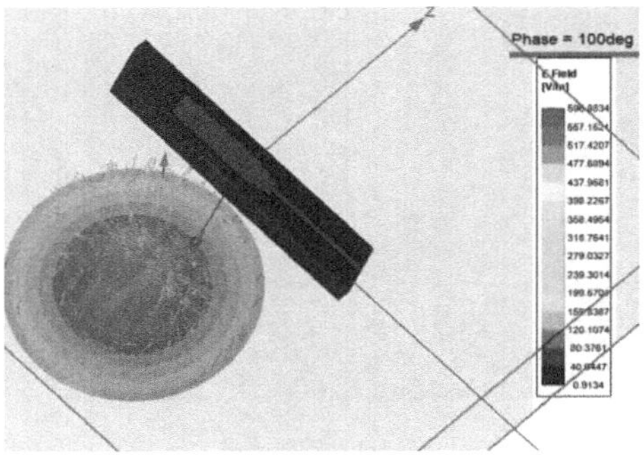

Figure 10. E-filed result without tumor.

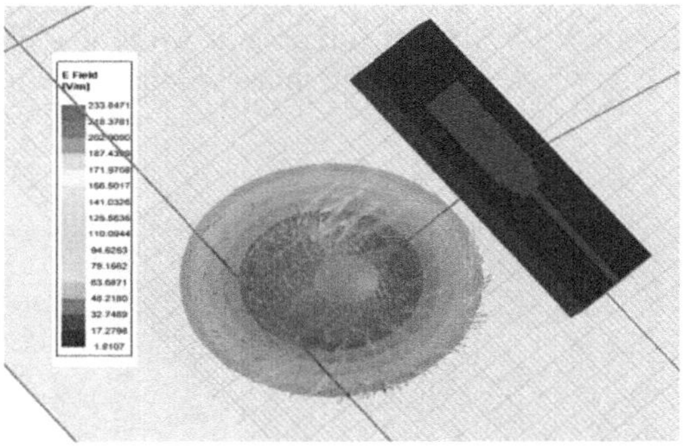

Figure 11. E-field result with tumor.

4.4 *Analysis of SAR (Specific Absorption Rate)*

Specific Absorption Rate, or SAR, is a crucial component in guaranteeing human safety since it indicates the amount of power absorbed by one kilogram of human body tissue. A SAR limit of 1.6 watts per kilogram (W/kg) is generally considered practically acceptable.

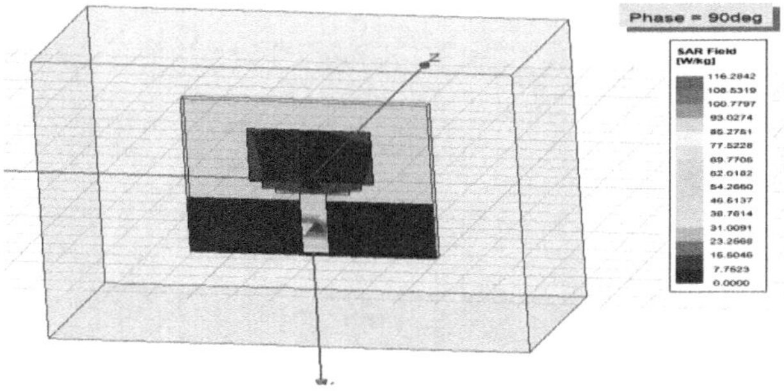

Figure 12. SAR result of the proposed antenna.

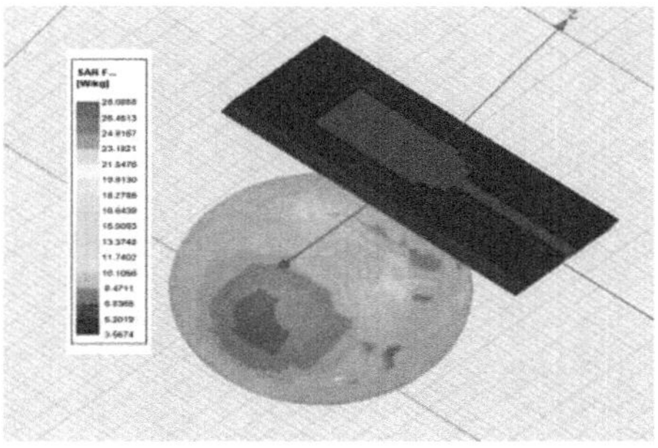

Figure 13. SAR of the antenna with human head phantom (without tumor).

Figure 14. Antenna SAR of a phantom human head (with tumor).

4.5 *Analysis of surface current distributions*

Figure 14 illustrates that the maximum current density attained is 127.87A/m^2 in the absence of a tumor, while Figure 15 indicates a value of 189.90 A/m^2 in the presence of a tumor.

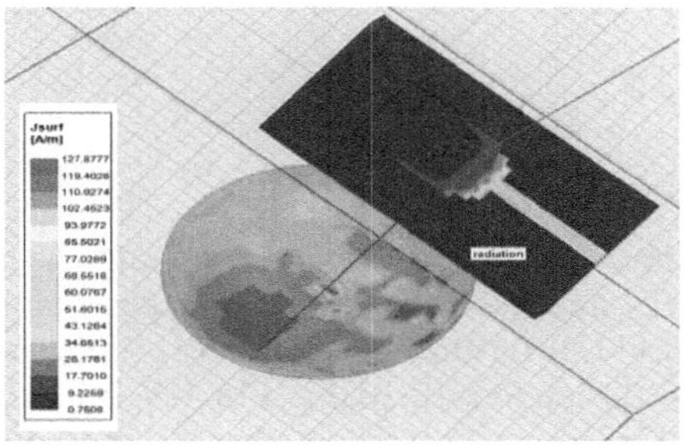

Figure 15. Current density without tumor.

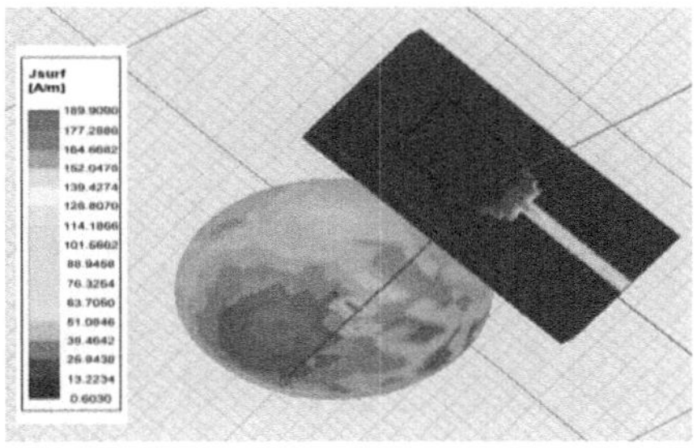

Figure 16. Current Density result with tumor.

4.6 *Gain anlaysis*

Gains in the 3D gain plot range from a minimum of -21.3 dB to a maximum of 0.2 dB. A discernible distortion with reduced gain occurs in the tumor area as a result of the dielectric difference between healthy and malignant tissue. This demonstrates that the antenna can detect cancers via microwave imaging.

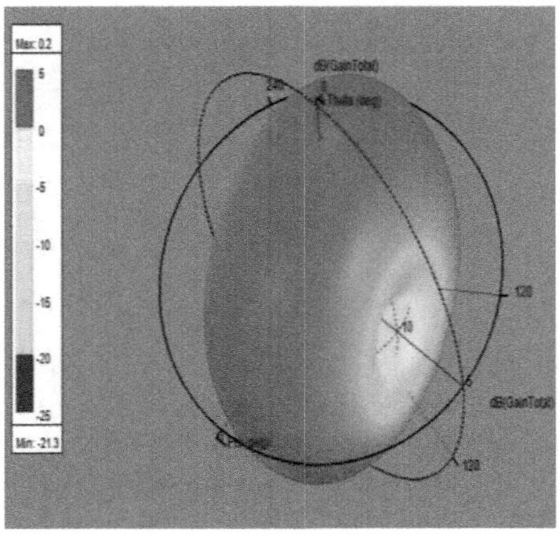

Figure 17. Gain Plot result without tumor.

The 3D gain plot with tumor presence has a maximum gain of roughly 1.04 dB and a minimum of 0 dB. The radiation pattern is plainly distorted, with an uneven distribution and decreased gain in some angular zones (green-yellow regions). The tumor's dielectric contrast, which absorbs and reflects a portion of the radiated energy, causes this imbalance. The antenna's capacity to detect anomalies is highlighted by the changing gain pattern, which supports its usage in tumor detection via microwave imaging.

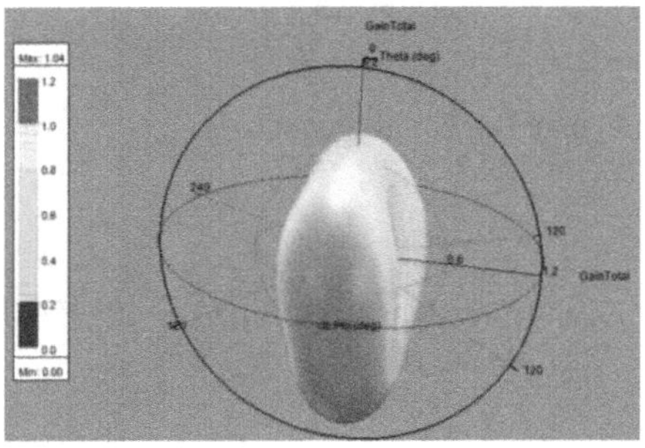

Figure 18. Gain plot result with tumor.

4.7 *Radiation effiency analysis*

The patterns remain mostly bidirectional and symmetrical throughout the ultra-wideband range, with the main lobes maintained throughout frequency and nulls around ±90°, suggesting steady broadside radiation. The lobe width varies little with frequency (typical UWB behavior), but there is little beam squint and just a little ripple in the side lobes, indicating a stable pattern throughout the band. For UWB imaging/sensing, this stability is preferred since it maintains angular fidelity and minimizes distortion when comparing tumor cases to healthy cases.

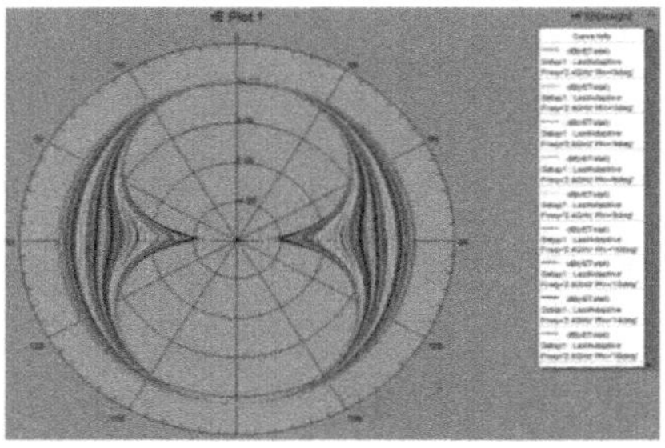

Figure 19. Radiation efficiency plot without tumor.

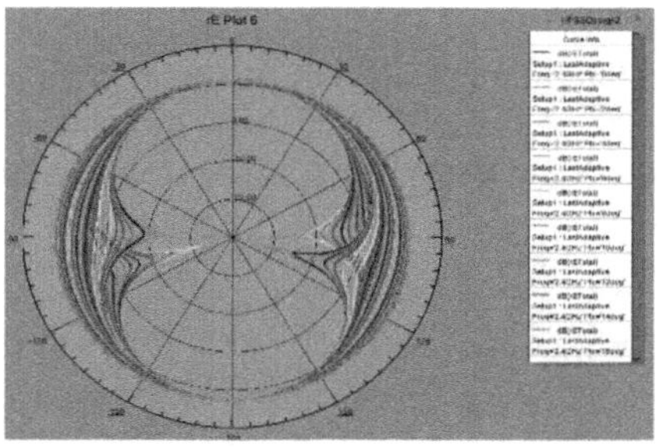

Figure 20. Radiation efficiency plot with tumor.

5 CONCULSION

The Stairs ultra-wideband (UWB) antenna, which has a thickness of 1.6 mm and a dielectric constant of 4.4, is made of FR4 substrate material and is meant for the diagnosis of brain cancer. This antenna is appropriate for brain cance millimeters. The standard head model's return loss is found to be smaller than that of the tumor-containing head model during phantom simulation. Plots showing the antennas' surface current, SAR, gain distribution, radiation patterns, and reflection coefficient have been created in order to comprehend their operation.

REFERENCES

[1] S. Licul, J. Sweeney, C.R. Anderson, and T.M. Bielawa (2003) 'A parametric study of time-domain characteristics of possible UWB antenna architectures,' in *Proc. Vehicular Tech. Conf.* vol. 5

[2] Oppermann, M. Hämäläinen and J, Iinatti, (2004) "UWB Theory and Applications', John Wiley and Sons, Ltd. *FCC Report and Order for Part 15 acceptance of Ultra Wideband (UWB)* systems from 3.1 -10.6 GHz , February, 2002, FCC website.

[3] Jianxin Liang,Choo C. Chiau Xiaodong Chen (2005) 'Study of a printed circular disc monopole antenna for UWB systems', *IEEE Transactions On Antennas and Propagation*, Vol. 53, No. 11,

[4] Haoyu Zhang, Ahmed O. El-Rayis, NakulHaridas, Nurul H. Noordin, Ahmet T. Erdogan, TughrulArslan (2011) 'A Smart antenna array for brain cancerdetection', in *Proc Antennas and Propagation Conference (LAPC), Loughborough*, Int.Conf,14-15

[5] Radouane KARLI, Hassan AMMOR (2014) Miniaturized UWB microstrip antenna with T-slot for detecting malignant tumors by microwave imaging',*WSEAS Transactions on Information SCIENCE and applications*, Volume 11.

[6] Aggarwal, R., and Kaur, A. (2014). Comparative analysis of different algorithms for brain tumor detection. *International Journal of Science and Research*, 3(6), 1159–1163.

[7] Yin, X.-C., C.-L. Ruan, C.-Y. Ding, and J.-H.Chu. (2008), Planar U type monopole antenna for UWB applications', *PIER Letters*, Vol. 2, pp. 1–10.

[8] Eng. M.A. Shokry, Prof. Dr. A.M.M.A. Allam 2016 *"UWB Antenna for Brain stroke and Brain TumourDetection"* 978-1-5090-2214-4/16/$31.00 © IEEE.

[9] Kaur, S., Khanna, R., Sahni, P., and Kumar, N. (2019). Design and optimization of microstrip patch antenna using artificial neural networks. *International Journal of Innovative Technology and Exploring Engineering*, 8(9), 611–616.

[10] Putri Zefni Widyatama, Bambang SetiaNugroho (2019) "Design of circular modified UWB Antenna Microstrip for brain cancer detection" 2019 *IEEE Asia Pacific Conference on Wireless and Mobile (APWiMob).*

[11] S. Baskar, S.Pradeep, D. Jency Remina, S. Lalitha. Detection of brain stroke and brain tumour using UWB pentagon antenna" Volume 3, Issue 3, March – 2018 *International Journal of Innovative Science and Research Technology* ISSN No:-2456-2165.

[12] Chandanpreet Kaur, and Bikrampal Kaur (2016), Implementation of Biography Based Neural Clustering (BBNC) with Genetic Processing for Tumor detection from Medical Images, Chandigarh Engineering College, Landran, Mohali, Punjab, India.

[13] Jalali, V., and Kaur, D. (2020). A study of classification and feature extraction techniques for brain tumor detection. *International Journal of Multimedia Information Retrieval*, 9(4), 271–290.

[14] Sukhdeep Kaur, Neha Ahuja.(2013). Design of microstrip patch antenna using defected ground structure for WLAN band, *International Journal of Computer Applications*, 2013/1/1

[15] Naveen Kumar, Rajesh Khanna (2020). A compact multi-band multi-input multi-output antenna for 4G/5G and IoT devices using theory of characteristic modes,2020/1, *JournalInternational Journal of RF and Microwave Computer-Aided Engineering*

Smart Systems and Beyond – Bhatia et al. (Eds)

Ensemble Machine Learning Models for High-Efficiency Antenna Design and Optimization in Next-Generation Wireless Systems

Sahil Rahul Shirode
Department of Electronics and Communication Engineering, TIET, Patiala, Punjab, India

Dhruv Gupta
Department of Mechanical Engineering, TIET, Patiala, Punjab, India

Nidhi Upadhyay and Amanpreet Kaur
Department of Electronics and Communication Engineering, TIET, Patiala, Punjab, India

ABSTRACT: In this article a T-shaped monopole antenna with a reduced ground plane for modern wireless communication applications. Furthermore, a stacked ensemble machine learning model is developed to optimize the antenna's S1 parameters. The antenna was designed on a polyimide substrate having a relative permittivity of 3.5, a thickness of 0.15 mm, and a loss tangent of 0.008, and its performance was evaluated through CST Microwave Studio simulations. Variations were carried out in substrate length, substrate width, and feed-line width, producing 431 individual datasets. Each dataset consisted of 1001 samples, resulting in a total of 4,31,431 data points for training the machine learning framework. These simulation runs illustrated the extensive computational effort involved in manual parameter tuning, especially when addressing performance goals such as impedance matching, bandwidth improvement, and minimizing return loss. To handle these challenges more effectively, an ensemble approach was adopted, combining Support Vector Regression, a Multi-Layer Perceptron, and a Random Forest model. The overall optimization strategy, strengthened through machine learning, limits the number of simulations required, accelerates the antenna development cycle, and offers dependable insights for advancing future monopole antenna designs. For optimal results, CST Microwave Studio simulated the ideal antenna. A 0.5 GHz bandwidth was simulated using optimal antenna characteristics. After analyzing the complex link between antenna performance and shape, the ensemble achieved accurate predictions with R^2 values of 0.915 and 0.815 on test datasets. The suggested ML-assisted optimization framework cuts simulation time, speeds antenna design, and provides reliable direction for future monopole antenna development.

Keywords: Stacked ensemble regression, CST simulation, design optimization, monopole antenna, and machine learning

1 INTRODUCTION

As wireless communication systems are changing so quickly, antennas need to be more modern and able to work well in a wide range of situations. 5G networks, satellite communications, IoT devices, and radar systems are all examples of modern technologies that need antennas with the best gain patterns, impedance matching, and frequency response. But designing and optimizing antennas in the past has always been hard, which has slowed down progress.

HFSS, CST Microwave Studio, and FEKO are important tools for designing antennas traditionally. These technologies can accurately assess fields and performance, but individuals typically re-optimize. Engineers begin with theoretical models and then test different shapes, materials, and structures to see what works. This strategy calls for running a lot of simulations, looking at the results, making small changes, and doing it all over again until the performance metrics are good. It takes weeks or months to optimize complex antenna structure designs. Second, the limits of human cognition and experience can make it hard to look into all the possible parameters, which could mean missing out on better design ideas that don't fit with normal ways of thinking. Third, the strategy doesn't always take into account how parameters affect each other, which can lead to bad local minima. Microstrip patch antenna arrays with MIMO design are suitable for 5G communications, if they satisfy a few output parameters. In order to optimize radiation and bandwidth, the device implements innovative substrate combinations and feeding methods. For urban 5G deployment, base station antennas must be capable of steering beams with high precision in order to mitigate interference and enhance coverage. In addition, the designs are suitable for small mobile device antennas, which require exceptional performance while maintaining a compact dimension. One of the objectives of antenna optimization is to decrease return loss[2], increase gain, decrease side lobes, and improve bandwidth. To satisfy these goals, traditional antenna design and simulation methods need to change the dimensions, substrate thickness, feed point placements, and ground plane configurations by manually using the parametric sweep option

 DOI: 10.1201/9781042026289-22

in the 3D EM tool CST MWS. Alternatively, Machine learning[5] offers a promising alternative, various ML models can be trained based upon the available dataset for antenna optimization [1]. In the proposed work, a monopole antenna is designed using CST Microwave Studio Software and optimized using ML models like KNN, Random Forest[6], MLP and SVR for the desired resonant frequency of operation and return loss[2]. The antenna feed width, Substrate length and width are optimized using these models and a R2 value of 0.915 with ensemble model is achieved which satisfies the desired criterion.

2 ANTENNA GEOMETRY AND DESIGN

A T shaped monopole antenna[1] was designed and proposed for wireless communication applications. This antenna structure was designed with multiple slots and a reduced ground[3]. The antenna was designed on a polymide substrate ($\varepsilon r = 3.5$, thickness = 0.15 mm, $\tan\delta = 0.008$), using basic transmission line equations . The feed network of the proposed antenna consists of a microstrip line of copper (thickness = 0.035mm) and a copper ground layer on the backside of the substrate. The ground plane is reduced to an optimized dimension of 5 mm. The optimization process of the antenna design is carried out by performing a parametric sweep on three key parameters: the upper length of the substrate (L1), the width of the substrate and ground, and the width of the feed line. The suggested antenna operates within the band (0-2 and 4-5 GHz) and has resonance approximately at 4.4 GHz. As these parameters have direct impact on the antenna's performance, the careful adjustment of these parameters is essential. Table 1 mentions the detailed specifications of the proposed antenna. Figure 1 (a) and (b) show the front and back views of the proposed antenna design.

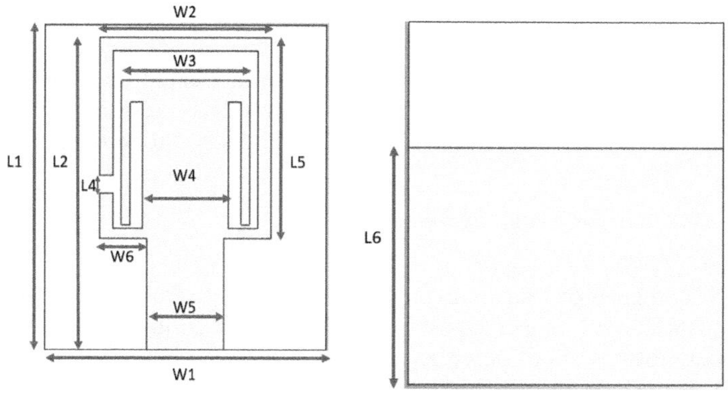

Figure 1. Antenna topology: (a) Front view of the proposed design; (b) Rear view with reduced ground configuration.

Table 1. Dimensions of proposed antenna.

Specifications	L1	L2	L4	L5	L6	W1	W2	W3	W4	W5	W6
Value (in mm)	7.6	7.3	0.3	4.7	5	4.6	4	3	2.2	2	1

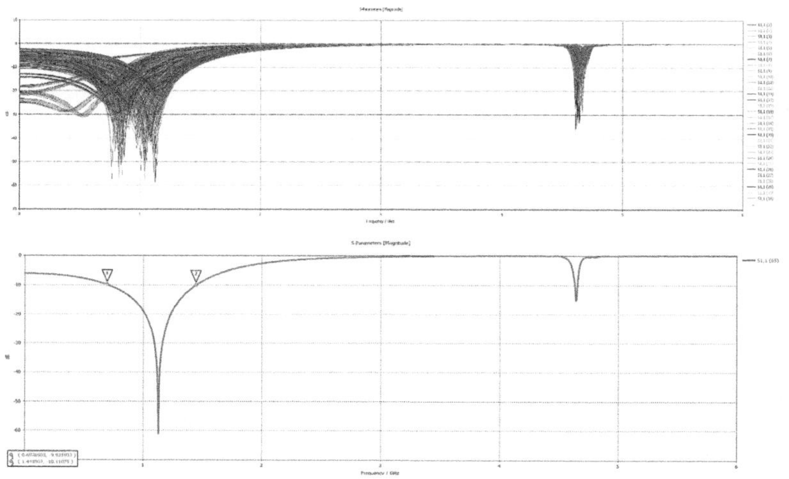

Figure 2. (a)S parameter graph for all 431 datsets; (b)Optimized S parameter graph.

The return loss characteristics of the proposed antenna were assessed through a thorough parametric sweep that included 431 design variations. Figure 2 presents the aggregated S-parameter responses for all simulated configurations. Two distinct resonant bands have been identified, with the first occurring in the lower frequency region approximately between 0.8 and 1.2 GHz, and the second situated in the higher frequency range close to 4.4 GHz. In many cases, the return loss is less than –50 dB, which means that the impedance is very well matched at resonance. The wide range of curves seen in the bottom band shows that the resonance is quite sensitive to changes in the length of the substrate, the width of the ground, and the width of the feed line. The lower band displays a set of repeating curves, indicating that variations in feed-line width, ground width, and substrate length have a strong influence on the antenna's resonance. The resonance seen at higher frequencies shows better stability across the design space. The presence of two resonant points supports the antenna's suitability for multiband use, but the shifts in performance also reveal how difficult manual tuning becomes when many parameters interact. These observations highlight the need for machine-learning-based optimization to identify parameter sets that can maintain consistent bandwidth and low return loss across the required frequency range. The overall design process can be improved by applying ML-driven optimization methods. A parametric sweep was performed, producing a total of 4,31,431 data rows for analysis.To guarantee their accuracy, a large number of parametric combinations had to be simulated, which required hours of calculation.. This underlines the challenges associated with using simulation tools to enhance multiple aspects simultaneously in the design of complex antennas.

The dataset created for this study included one target variable and three independent variables: upper length of the substrate(L1), width of the substrate/ground(W1), and width of the feed line(W5). Since the dataset has a known continuous target variable, it is treated as labelled data, and thus the applied machine learning[5] models were trained using supervised learning for this data.

2.1 *Antenna optimization process*

As explained in the above section a dataset of 4,31,431 values was created and was used to train different ML models for the antenna optimization process and different ML models like KNN, Random forest Regressor[6], multilayer perception regressor and SVR are trained using this data set. An ensemble model is also used.

2.2 *Machine learning models used for design optimization*

2.2.1 *K-Nearest Neighbors (KNN) regression*

K-Nearest Neighbors (KNN) regression is a simple and effective way to learn that doesn't make any assumptions about the data. It figures out what will happen based on how close data points are to each other in the feature space. KNN doesn't make any assumptions about how the data was spread out before or how the inputs and outputs are related to each other. It doesn't do that; instead, it makes predictions based on how the data is set up in its local area. The main point is that instances with similar input features are likely to have similar target values. For a dataset, to make a prediction for a new input *x*, you need to use a distance metric, which is usually the Euclidean distance, to find the K samples that are closest to it.

$$d(x, x_i) = \sqrt{\sum_{\{j=1\}}^{m} \left(x_j - x_{\{ij\}}\right)^2} \tag{1}$$

The answer was found by averaging the target values of the neighbours found:

$$\hat{y}(x) = \left(\frac{1}{K}\right) \sum_{\{i \in N_{K(x)}\}} y_i \tag{2}$$

Weighted version of KNN can also be used, where the prediction is more affected by neighbours that are closer to each other.

$$\hat{y}(x) = \frac{\left(\sum_{\{i \in N_{K(x)}\}} w_i y_i\right)}{\sum_{\{i \in N_{K(x)}\}} w_i}, w_i = 1 - d(x, x_i) \tag{3}$$

$N_{K(x)}$: Set of the K nearest neighbours of input x

K: Number of nearest neighbours considered

w_i: Weight assigned to the ith neighbour, based on its distance to x

It is very important to choose K. It could get overfitting and a lot of variance if a small value is chosen . However if a higher value is chose predictions might not be very accurate. Cross-validation is a common way to figure out what value of K is best. KNN regression is good for data that isn't linear, but it can be expensive to run on big datasets.

2.2.2 *Random forest regressor*

Random Forest[6] is a type of ensemble learning that uses the predictions of many decision trees to make better generalizations and be more stable. Each tree learns from a bootstrap sample of the data set. This means that the tree takes a random sample of the data and replaces it with a new one. The tree is split at each node using a random set of features. This two-way randomness in both data and feature selection makes the trees less connected to each other and less likely to overfit, which is a common problem with single decision trees. For regression, each tree takes a guess, and the average of all the trees' guesses is the last guess.

$$\hat{y}(x) = \left(\frac{1}{B}\right) \sum_{\{b=1\}}^{B} f_b(x) \tag{4}$$

The best thing about Random Forest[6] is that it can find a good balance between bias and variance. This is how to figure out the ensemble's variance:

$$Var(\hat{y}) = \rho\sigma^2 + \frac{(1-\rho)}{B}\sigma^2 \tag{5}$$

where σ^2 is the difference between two trees and ρ is how trees are connected to each other. As the number of trees BB increases, the second term has less of an effect. This means that performance stays the same as long as the trees aren't too similar.Random Forest[6] not only makes good predictions, but it also gives each feature a score that shows how important it is. Most of the time, these scores are based on how much each feature lowers the average impurity (or variance) of all the splits. This is why Random Forest is so helpful when you want to know how different things affect something and how important they are. It's a good base model that works well with datasets that are mixed, have a lot of noise, or are high-dimensional.

2.2.3 *Multi-Layer Perceptron (MLP) regressor*

The Multi-Layer Perceptron (MLP) is a kind of feedforward artificial neural network that can model very complex non-linear relationships. An MLP has three layers: an input layer, one or more hidden layers, and an output layer. Neurons in each hidden layer add up their inputs in a weighted way and then use a non-linear activation function. This helps the network get close to complicated maps that linear models can't handle.To find neuron *j* in layer *l*, the following formula was applied:

$$z_j^1 = \sum w_{ij}^1 a_i^{1-1} + b_j^1 \,,\; a_j^1 \;=\; \varphi(z_j^1) \tag{6}$$

where $W_{ij}^{(l)}$ are the weights of the links, $b_j^{(l)}$ this is the bias term, and ϕ is the activation function, which can be ReLU, sigmoid, or tanh. The output layer of a regression task usually uses a linear activation function to keep making predictions. To teach an MLP, you usually have to lower the Mean Squared Error (MSE):

$$MSE = \left(\frac{1}{N}\right) \sum_{\{i=1\}}^{N} (y_i - \hat{y}_i)^2 \tag{7}$$

We use gradient descent and backpropagation to make weights and biases better. The error is sent back through the network to find the gradients in these ways:

$$w_{\{ij\}}^{(l)} \leftarrow w_{\{ij\}}^{(l)} - \eta\left(\frac{\partial L}{\partial w_{\{ij\}}^{(l)}}\right) \tag{8}$$

N: Total number of training samples

$w_{ij}^{(1)} \leftarrow$ Update rule showing how weights are adjusted during training

η (eta): Learning rate, controls the step size of weight updates

with η being how quickly you learn. By stacking a lot of hidden layers and neurons, MLPs can learn hierarchical feature representations. That's why they are so good at coming up with functions that work for everyone. MLPs are a great choice if you have a lot of data that is hard to work with and regular models might not be able to find non-linear dependencies but they are hard to get.

2.2.4 *Support Vector Regression (SVR)*

Support Vector Regression (SVR) [7] is based on the same ideas as Support Vector Machines (SVM) and helps you make predictions all the time. SVR doesn't try to make the error for each data point as small as it can. Instead, it uses the idea of an ε-insensitive tube, which doesn't care about mistakes that aren't very close to the real target. The goal is to find a regression function that is as flat as possible and allows deviations of less than ε. This makes the model less sensitive to noise and outliers.

In a formal sense, if you have training data (x_i, y_i), SVR solves this optimization problem:

$$\min_{\{w,b,\xi,\xi^*\}} \left(\frac{1}{2}\right) ||w||^2 + C \sum_{\{i=1\}}^{n} (\xi_i + \xi_i^*) \tag{9}$$

in line with:

$$y_i - (w \cdot \varphi(x_i) + b) \leq \varepsilon + \xi_i \tag{10}$$

$$(w \cdot \varphi(x_i) + b) - y_i \leq \varepsilon + \xi_i^* \tag{11}$$

$$\xi_i, \xi_i^* \geq 0 \tag{12}$$

where $\phi(x)$ uses a kernel function to map the inputs into a higher-dimensional feature space, C is a regularization parameter that controls the trade-off between flatness and error tolerance, and ξ, ξ^* are slack variables that let violations go beyond ε.

The resulting regression function is:

$$\hat{y}(x) = \sum_{\{i=1\}}^{n} (\alpha_i - \alpha_i^*) K(x_i, x) + b \tag{13}$$

where $K(x_i, x)$ is equal to the kernel function can be a linear, polynomial, or radial basis function. Because it uses kernels, SVR can model non-linear relationships and still work well in general. SVR works best when there are more features than training samples and in spaces with a lot of dimensions.The study's predictive modelling framework was created using stacked[8] ensemble learning, which combines the best features of several regression models to make the framework more accurate and is useful in a wider range of situations. The dataset is included the results of electromagnetic simulations that showed scattering metrics, such as the scattering magnitude and phase across frequency for different antenna configurations. The proposed antenna is defined by three main geometric parameters: L1, W1, and stripw. To prepare the data for machine learning, the scattering responses were arranged in order of increasing frequency. This produced a feature matrix in which each row represented one antenna configuration and each column represented the response at a specific frequency.

The approach used three different machine learning models as the base regressors. The Support Vector Regressor used an RBF kernel and was set with $C = 20.0$, $\epsilon = 0.01$, and gamma set to "scale," allowing it to handle local nonlinear behaviour while keeping the model controlled. The Multi-Layer Perceptron Regressor had two hidden layers with 256 and 128 neurons, used ReLU activation, the Adam optimizer with a learning rate of 10^{-3}, a regularization term of 10^{-4}, and was trained for up to 800 iterations to capture more involved nonlinear patterns. A Random Forest Regressor with 1,000 trees, $\sqrt{d}$ feature sampling, and bootstrap aggregation was also included to improve model stability and reduce variance. Together, these three models formed the base of the stacked ensemble. Their predictions were then passed to a meta-learner, which was another Random Forest Regressor containing 600 trees. This model combined and refined the outputs from the base regressors. The overall setup let the SVR and MLP deal with nonlinear behavior well. Random Forest added a way to generalize across the design space that worked every time. The stacking step also helped lower bias and variance by using the best parts of all the models.

Model training was done separately for each of the three geometry parameters (L1, W1, and stripw). An 80–20 train–test split was used so the evaluation would remain fair. To judge how well the models performed, the main metric considered was the R^2 score, which simply shows how much of the variation in the target value the predictions can explain. Along with this, residuals were inspected to get a clearer sense of where the models struggled. The residual plots, which included shaded areas for easier comparison, helped point out portions of the data where the predictions tended to drift either above or below the actual values. These visual checks showed that the ensemble worked well overall, though accuracy naturally varied a bit across different regions of the dataset. The results also gave good evidence that the model was not overfitting. Overfitting occurs when a model retains specific details from the training data rather than acquiring more generalizable relationships, typically resulting in high training accuracy but diminished performance on novel inputs. In our case, the average R^2 scores on the test sets were around 0.915 and 0.815. These numbers are high but not unrealistically perfect, which is a sign that the ensemble learned the underlying relationships without clinging too tightly to noise in the data. Taken together, the steady performance and the balanced R^2 values show that the stacked ensemble is generalizing well and can be trusted to estimate antenna geometry from scattering responses.

The data clearly shows that the stacked regression model is able to predict the antenna parameters L1, W1, and stripw quite well. The results are presented for two training ranges: the left side corresponds to 4–5 GHz, and the right side shows the 0–2 GHz range. Areas where the model predicts lower than the actual values are shaded in green, while red shading marks places where it predicts higher than expected. The model fits the real data pretty well on most of the plots, which means it is accurate and stable. The higher R^2 values seen for W1 and stripw indicate that the model captures the underlying physical trends effectively, and even though L1 is a bit more variable, the predictions still stay within a reasonable range. Taken together, the results suggest that the model generalizes well and is not showing signs of overfitting.

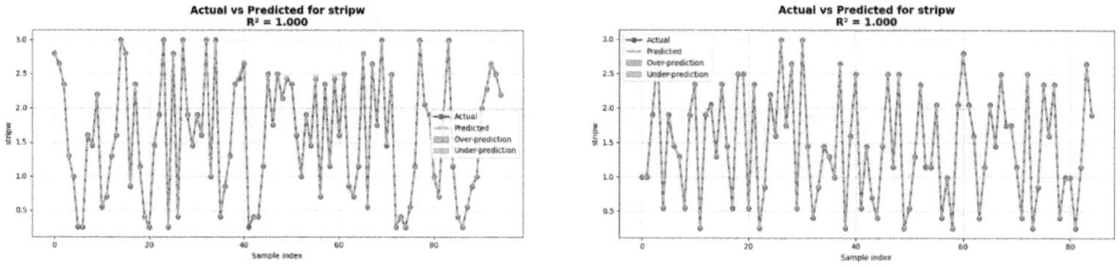

Figure 3. Actual vs Predicted value for W1 for frequency range: (a)4-5 GHz; (b) 0-2 GHz.

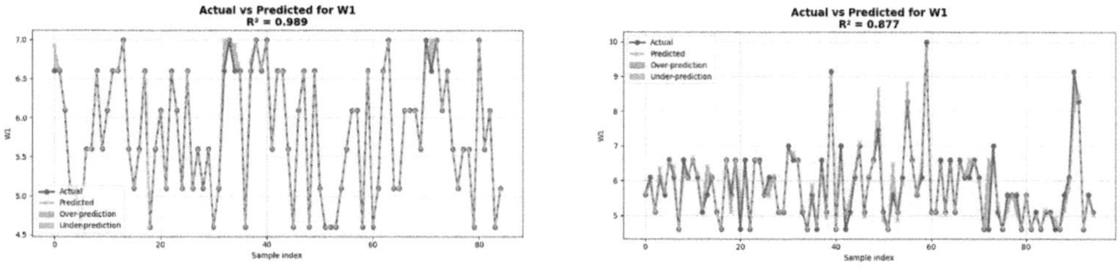

Figure 4. Actual vs Predicted value for strip width (stripw) for frequency range: (a)4-5 GHz; (b) 0-2 GHz.

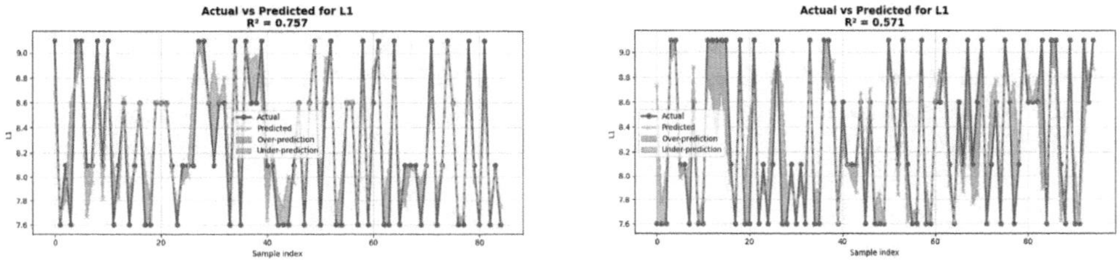

Figure 5. Actual vs Predicted value for L1 for frequency range: (a)4-5 GHz; (b) 0-2 GHz.

3 CONCLUSION

This study presents the design, simulation, and optimization of a T-shaped monopole antenna with multiple slots and a reduced ground plane, using machine-learning methods. The usual design workflow is slow repetitive and tedious, involving large parameter sweeps before reaching acceptable performance. By applying machine-learning models such as Support Vector Regression, a Multi-Layer Perceptron, and a Random Forest, combined in a stacked ensemble, the optimization process was speeded up. The proposed ensemble was able to pick up the nonlinear links between the geometric parameters and the antenna response. It produced R^2 values above 0.9, showing strong predictive quality. Using machine learning within the antenna-design cycle cuts down the design time and lowers computational effort. The framework developed here can be adapted to other antenna forms and different frequency bands, making it a useful tool for future communication technologies.

REFERENCES

[1] C.A. Balanis, *Antenna Theory: Analysis and Design*, 4th ed. Hoboken, NJ, USA: Wiley, 2016.

[2] D.M. Pozar, *Microwave Engineering*, 4th ed. Hoboken, NJ, USA: Wiley, (2012).

[3] Y. Lu, Y. Huang, H.T. Chattha, Y.C. Shen, and S.J. Boyes, *"An Elliptical UWB Monopole Antenna with Reduced Ground Plane Effects,"* Univ. of Liverpool report.

[4] S.I. Park *et al.*, "Flexible substrate technology for millimeter-wave wireless," *Wireless Power Transfer*, vol. 3, no. 1, (2014).

[5] S. Koziel and A. Pietrenko-Dabrowska, "Antenna optimization using machine learning with reduced dimensionality and infill criterion," *Scientific Reports*, vol. 14, (2024).

[6] L. Breiman, "Random Forests," *Machine Learning*, vol. 45, no. 1, pp. 5–32, 2001.

[7] A.J. Smola and B. Schölkopf, "A tutorial on support vector regression," *Statistics and Computing*, vol. 14, pp. 199–222, (2004).

[8] D.H. Wolpert, "Stacked generalization," *Neural Networks*, vol. 5, no. 2, pp. 241–259, (1992).

Smart Systems and Beyond – Bhatia et al. (Eds)
 ISBN: 978-1-041-36617-1

Cross Language Plagiarism Detection in Regional Languages: A Comparative Study of CLPD Models and Approaches

Sachin Majithia
Department of Computer Science and Engineering Punjabi University, Patiala, Punjab India

Harjeet Singh
Department of Computer Science, Mata Gujri College, Fatehgarh Sahib, Punjab India

Rajeev Puri
Department of Computer Science, DAV College, Jalandhar, Punjab, India

ABSTRACT: With rapid increase in the availability of information available on the internet, plagiarism has become serious concern for academicians and researchers. Many free and paid plagiarism detection methods are already available in the market for same language pairs, but plagiarism detection in different language pairs is still a challenging task so cross language plagiarism detection (CLPD) has become important area of research these days. In this research paper we compare models and approaches of **cross-language plagiarism detection (CLPD)** for regional languages especially Punjabi and Hindi. Nine different models/approaches are compared in this research paper including dictionary-based methods, statistical methods, direct translation methods and neural methods like neural embedding to evaluate their effectiveness in low resource languages. Standard metrics like **Overall Accuracy, F1-Score, Precision, and Recall,** are used to compare results of CLPD methods and to highlight strengths and limitations of each model/approach. The research paper results show that the neural embedding model's performance is better than conventional statistical models with **F1-scores of 0.89**.

Keywords: Cross-language plagiarism detection, regional languages, statistical models, neural embeddings, Hindi, Punjabi, multilingual text analysis, machine translation

1 INTRODUCTION

In today's world plagiarism is a very big issue in academia, research, publishing, and online platforms. With the development of globalization and multilingual information resources, cross-language plagiarism detection (CLPD), where source text is translated or paraphrased into another language [20] has become an urgent issue and challenge. Traditional monolingual plagiarism detection techniques fail to detect semantic similarity in languages where obfuscation strategies are applied like synonym replacement, sentence restructuring, machine translation. In recent years, Cross-language plagiarism detection has become significant importance due to growing globalization of academic research and availability of multilingual content on the internet, with minimal effort.[18] Compared with monolingual plagiarism detection approach, which has received many researches and there already many matured solutions, cross-language detection among different language still having many challenges such as presence of linguistic and cultural variations in writing styles, and the unavailability of parallel corpora among many language pairs. In regional languages like Hindi and Punjabi, due to lack of proper machine translation, unavailability of linguistic resources and complexity of morphologies, the challenge of plagiarism detection has further become complex [12]. Regional Languages still have low resources as compared to other international languages like English, Mandarin and Spanish. This research study provides complete comparative analysis of nine different cross-language plagiarism detection approaches for regional language pairs like Hindi and Punjabi by evaluating metrics like recall, precision, F1-score, accuracy and provide quantitative and qualitative assessments of these approaches.

2 LITERATURE REVIEW AND RELATED WORK

Initial research in this field was based primarily on translation-based approaches, in which documents in different languages are converted into a common language and then compared using monolingual detection techniques. One of the earliest attempts to present a detailed survey of cross-language plagiarism detectors is offered by Barron-Cedeño *et al* [18]. who categorize methods as syntax-based or semantic-based as well as cross-language specific. Their work defined the basis of the understanding of the trade-offs between the different methodological

 DOI: 10.1201/9781042026289-23

approaches. Systematic evaluation frameworks for cross language detection were introduced by Potthast *et al.* [19] after that development of statistical approaches gained momentum. Recent advancement in neural embedding models have revolutionized the field, with multilingual transformer models like LASER, multilingual SBERT and LaBSE, showing auspicious results for cross-language text comparison [13]. These models overcome most of the shortcomings of the conventional statistical tools as they simply become more semantically enriched across languages. In the case of regional languages such as Hindi and Punjabi, there is less research. Ezzikouri *et al* [12] studied plagiarism detection in English and Arabic, which can be used as guidance to detect in other language pairs with various scripts and linguistic frameworks. Following section analyses nine various cross-language plagiarism detection techniques, beginning with the older statistical techniques to the newer neural techniques.

2.1 *Translation Plus Monolingual Analysis (T+MA)*

The T+MA approach [10][18] is one of the most intuitive approaches for CLPD. It operates through a two-step process that effectively transforms the cross-language problem into a monolingual one. In the first step, machine translation of the suspect document from the target language to the source language is done using machine translation tools. For a suspicious text S in language L_1 being translated into language L_2, this process can be written as:

$$S' = MT(S, L_1 \rightarrow L_2) \tag{1}$$

where MT is the machine translation function. In the second step standard monolingual plagiarism detection methods are applied to compare the translated text with the original source. The similarity between the translated text S' and candidate text C (both now in L_2) is computed using vector-based models: Similarity can be determined using different representations including TF-IDF vectors, n-gram features or word embeddings.

$$sim(S', C) = \frac{\vec{S'} \cdot \vec{C}}{\left\|\vec{S'}\right\| \cdot \left\|\vec{C}\right\|} \tag{2}$$

Strengths: This method works very well to identify plagiarism between language pairs which are not closely related This method takes advantage of robustness of monolingual analysis tools and avoids the problem of cross-lingual semantic matching.

Limitations: This method mainly depends on machine translation quality, making it error prone and computationally expensive for large corpora due to machine translation overhead.

2.1 *Cross-Language Character N-Gram (CL-CNG)*

The CL-CNG model [17][18] represents a language-agnostic approach that compares documents based on overlapping sequences of characters, called n-grams, regardless of the language. This approach involves two steps. In first step, both source and suspected text are converted into character level n-grams and in second step, their frequency distribution is compared. Set of character level n-grams can be written as:

$$V_T = \{ngrams(T, n)\} \tag{3}$$

Jaccard similarity or cosine similarity can be used to find the similarity between source and suspected text:

$$sim_{Jaccard}(S, C) = \frac{|V_S \cap V_C|}{|V_S \cup V_C|} \ \textbf{\textit{or}}\ sim_{cosine}(S, C) = \frac{V_S \cdot V_C}{||V_S|| \cdot ||V_C||} \tag{4}$$

Strengths: One of the major strengths of this model is that it is language independent. This model is robust to small changes and very effective for languages which share common character sequences or alphabetic systems.

Limitations: CL-CNG suffers from poor semantic capture, as it focuses purely on surface-level character patterns. It is also sensitive to transliteration differences and performs poorly with greatly different scripts.

2.3 *Cross-Language Explicit Semantic Analysis (CL-ESA)*

CL-ESA [15][19] extends the concept of Explicit Semantic Analysis to the cross-language domain by mapping documents from different languages into a shared multilingual concept space. This space is typically constructed using aligned Wikipedia articles or other multilingual knowledge bases. The process involves mapping text to a concept space using precomputed concept-term weights. For a vocabulary of concepts $\{K_1, K_2, \ldots, K_M\}$, the mapping function is defined as:

$$\phi(T) = (w_{K_1}(T), w_{K_2}(T), \ldots, w_{K_M}(T)) \tag{5}$$

where $w_{K_i}(T)$ represents the weight of concept K_i for text T, often computed using TF-IDF weighting or concept classifier scores. The similarity between two documents is then computed as the cosine similarity in the concept space:

$$sim_{ESA}(S, C) = \cos(\phi(S), \phi(C)) \tag{6}$$

Strengths: CL-ESA captures semantic similarity through an interpretable space and can handle documents in very different languages effectively.

Limitations: The approach requires multilingual concept linking or cross-lingual concept alignment, making it resource-intensive. The large dimensionality of the concept space (M) can also lead to computational challenges.

2.4 *Cross Language Ontology-Based Similarity Analysis (CL-OSA)*

CL-OSA [15] uses structured semantic knowledge from ontologies and knowledge graphs such as Wikidata to detect cross-language plagiarism. Instead of using machine translation or parallel corpora, this method represents documents as vectors of entities extracted from knowledge graphs. For words w_1 and w_2 mapped to nodes in an ontology, where $LCS(w_1, w_2)$ represents the least common subsumer and $depth(v)$ denotes depth of node v in the ontology, the lexical similarity is computed as:

$$sim_{lex}(w_1, w_2) = \frac{2 \cdot depth(LCS(w_1, w_2))}{depth(w_1) + depth(w_2)} \tag{7}$$

This is then aggregated to the sentence or document level:

$$sim_{OSA}(S, C) = \frac{1}{|S|} \sum_{s \in S} \max_{c \in C} sim_{lex}(s, c) \tag{8}$$

Strengths: The method operates at the concept level; it generates interpretable results and is effective when domain-specific lexicons are available.

Limitations: Coverage is limited by the ontology's completeness, and computational costs can be high when matching many nodes in large ontologies.

2.5 *Bilingual Dictionary-Based Alignment*

This method is based on aligning text fragments in the two languages using statistical bilingual dictionaries that can be obtained from parallel corpora [10]. Using dictionary entries, it provides a translation of words or phrases from a source language into a corresponding term in the target language. Given a dictionary $D : V_{L_1} \rightarrow 2^{V_{L_2}}$ that maps words to collection of possible translations, the mapping of a sentence S is defined as:

$$S^{(d)} = \bigcup_{s \in S} D(s) \tag{9}$$

The similarity can then be calculated using Jaccard similarity:

$$sim_{Dict}(S, C) = Jacc(S^{(d)}, C) \tag{10}$$

Strengths: The approach is lightweight, interpretable, and computationally efficient.

Limitations: This approach contains the issue of dictionary coverage because it is unable to multiple semantics of words and also insufficient when dealing with idiomatic expressions.

2.6 *Statistical Machine Translation + TF-IDF Similarity*

In this approach statistical machine translation [15] is combined with traditional information retrieval techniques. Process involved translation of source document to target document using statistical machine translation techniques and after that similarity is computed using TF-IDF vectors. The translation process follows the statistical machine translation paradigm:

$$e^* = MT(S, L_1 \rightarrow L_2) \tag{11}$$

where S represents the suspicious text in language L_1 is translated into language L_2 and e^* represents the target language translation. After translation, both documents are represented as TF-IDF vectors $\vec{e^*}$ and $\vec{d}$, and their similarity is computed using cosine similarity:

$$S(f, d) = \frac{\vec{e^*} \cdot \vec{d}}{\left\|\vec{e^*}\right\| \cdot \left\|\vec{d}\right\|} \tag{12}$$

Strengths: From past many years, this approach is very effective for plagiarism detection and TF-IDF is very efficient vector-space scoring technique.

Limitations: This model heavily depends on the quality of statistical machine translation, and the approach is not as competitive as compared to neural machine translation and embedding approaches.

2.7 *Cross-Language Information Retrieval (CLIR-Based Models)*

CLIR-based models [13] are part of information retrieval which provides cross language plagiarism detection. These models do not require full translation to identify semantic similar content across different language pairs. Inverted indexing is used to index a candidate corpus D in the target language and then weighting schemes like TF-IDF or BM25. For querying, S (Source Content) or both are either translated to the target language, or both are projected into a shared representation space. An example of the BM25 scoring function is:

$$\text{score}_{\text{BM25}}(Q, D) = \sum_{t \in Q} \text{IDF}(t) \cdot \frac{f(t, D) \cdot (k_1 + 1)}{f(t, D) + k_1 \cdot \left(1 - b + b \cdot \frac{|D|}{\text{avgdl}}\right)} \tag{13}$$

where $f(t, D)$ is the term frequency, $|D|$ is the document length, *avgdl* is the average document length, and k_1, b are tuning parameters.

Strengths: Main Strength is scalability and advantage of optimized retrieval algorithms developed for information retrieval.

Limitations: It is primarily focused on ranking rather than binary detection, leading to precision-recall trade-offs in plagiarism detection scenarios.

2.8 *Statistical co-occurrence models*

Statistical co-occurrence models [1][3] operate on the principle that words or phrases frequently appearing together in one language are likely to have aligned counterparts in another language. These models use parallel corpora to build bilingual dictionaries or alignment matrices. The translation probabilities are estimated using IBM-style alignment models:

$$P(c|s) \approx \frac{\text{count}(s, c)}{\sum_{c'} \text{count}(s, c')} \tag{14}$$

Document similarity is then computed by matching each source token to the best-scoring candidate token:

$$sim_{\text{coocc}}(S, C) = \frac{1}{|S|} \sum_{s \in S} \max_{c \in C} P(c|s) \tag{15}$$

Strengths: The method is unsupervised and useful when working with comparable corpora rather than parallel translations.

Limitations: As compared to modern embedding approaches, this model requires co-occurrence statistics and provides weak semantic modeling.

2.9 *Neural embedding models*

Neural embedding models [4][5][16] are the recent developments in cross-language plagiarism detection. These models learn distributed representations of words, sentences or documents in common vector spaces so that they can be compared directly - without translation. Modern multilingual encoders such as LASER, LaBSE and multilingual SBERT map sentences directly into a common vector space:

$$v_S = E(S), v_C = E(C) \tag{16}$$

where E is the multilingual encoder function. Similarity is then calculated according to cosine similarity between the encoded vectors:

$$sim_{embed}(S, C) = \cos(v_S, v_C) \tag{17}$$

Strengths: Deep semantic knowledge in relation models, cross language alignments, good for identification of paraphrased and semantically obfuscated plagiarism.

Limitations: They are computationally expensive; require many memory resources and may be biased for the languages represented in the training set.

3 EVALUATION METRICS

Multiple performance measures need to be taken into consideration when evaluating cross-language plagiarism detection systems. CLPD methods are compared and evaluated using datasets like PAN-CLEF (7K–20K documents), Samanantar, (**49.9 million sentence pairs** total, Hindi-Punjabi Pairs: 1.2–1.5 million sentence pairs), OPUS (10 billion+ sentence pairs across all language pairs) and custom Indic corpora. Following standard metrics are used for comparative study.

(1) ***Precision:*** Precision measures the proportion of correctly identified plagiarized content out of all detected instances.
(2) ***Recall:*** Recall is the fraction of actual plagiarized content that the system correctly identifies.
(3) ***F1-Score:*** The F1-score is the harmonic means of precision and recall and provides balanced measure of System's accuracy.
(4) ***Accuracy:*** Accuracy measures the overall correctness of the system to identify plagiarized content and distinguish it with original content.

4 RESULTS AND ANALYSIS

4.1 *Quantitative analysis of cross language plagiarism detection models/approaches*

Quantitative experimental results provide very important insights regarding CLPD Methods: **Neural Embedding Models** have much better results with standard metrics like precision (0.90), recall (0.88), and F1-score (0.89), so these models are best among all traditional statistical approaches. In Addition to best performance these methods also have ability to capture deep semantic relationships and handle paraphrasing effectively. **Cross-Language ESA** is the most effective traditional statistical method, with an F1-score of 0.80. This method is successful because of Wikipedia-based concept spaces, which provides semantic representation. **Translation-based methods** (T+MA, SMT + TF-IDF) offer moderate level of performance (F1-scores between 0.74-0.77). These models mainly depend on the quality of translation, so this is major drawback of these models. **Character-based approaches** (CL-CNG) show acceptable performance for related languages, but overall scores are less due to lack of semantic depths. The worst performance is shown in **Dictionary-based methods** because of low coverage and are unable to handle semantic differences.

Table 1. Comparison of all CLPD Models/Approaches

Method	Precision	Recall	F1-Score	Accuracy	Scalability	Robustness
T+MA	0.81	0.74	0.77	0.79	Medium	Medium
CL-CNG	0.72	0.68	0.70	0.73	High	Low
CL-ESA	0.84	0.77	0.80	0.82	Medium	High
CL-OSA	0.79	0.71	0.75	0.78	Low	High
Dictionary Alignment	0.65	0.58	0.61	0.66	High	Low
SMT + TF-IDF	0.78	0.70	0.74	0.76	Medium	Medium
CLIR Models	0.69	0.63	0.66	0.68	High	Medium
Statistical Co-occurrence	0.75	0.71	0.73	0.74	Medium	Medium
Neural Embedding Models	0.90	0.88	0.89	0.75	High	High

4.2 *Qualitative analysis of cross language plagiarism detection models/approaches*

The radar chart provides qualitative comparison of models/approaches using three dimensions scalability, robustness and overall performance. This chart shows that Neural Embedding Models maintains the balance between all three dimensions. Some traditional models/approaches like CL-CNG provide high scalability but robustness and accuracy decreased significantly in these models.

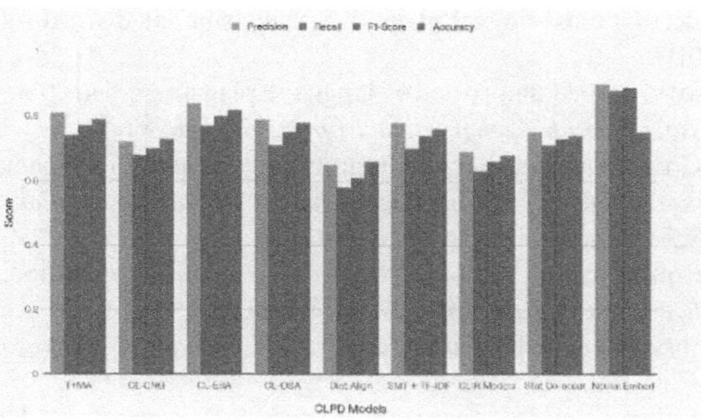

Figure 1. Quantitative comparison of CLPD models.

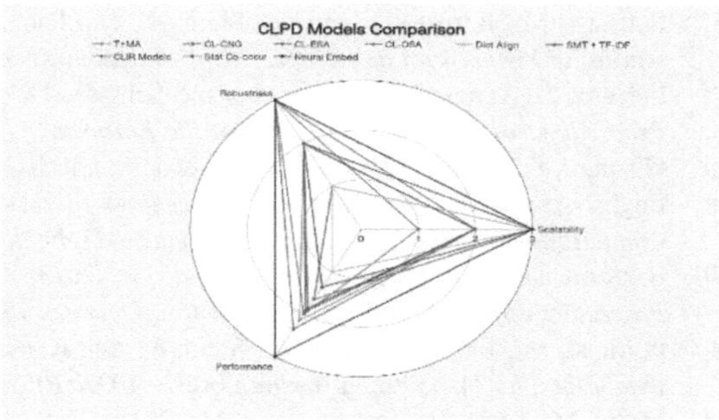

Figure 2. Qualitative comparison of CLPD models.

5 CONCLUSION

In this research article, comparative analysis of CLPD methods/approaches has been done which provides essential and important insights for researchers and academicians working in this field. Study shows that Neural embedding models (LASER, LaBSE and multilingual SBERT) are more accurate compared to other methods like directory based, direct translation and statistical methods and achieved 0.89 F-1 score while other methods were having 0.61-0.80 F1 Score. It is concluded from the results that efficacy of Neural embedding models is much better than knowledge-based models (CL-ESA, CL-OSA). Knowledge-based models are more effective than translation-based models/approaches, which are better than character-based and dictionary-based methods. For resource-limited environments, character n-gram approaches offer a good trade-off between performance and computational complexity, but only for similar language pairs. For language pairs like Hindi-Punjabi where structural similarity can help lightweight models, the neural embedding approaches remain the most critical and tractable solution. The use of neural models in paraphrasing, semantic variation generation and cross-lingual nuances demonstrates that they are the current state-of-the-art in cross-language plagiarism detection. Neural embedding approaches (mBERT, LaBSE, XLM-R) provide state-of-the-art accuracy but are costlier to train and deploy than classical CLPD models.

REFERENCES

[1] Amirzhanov, A. (2025). Plagiarism types and detection methods: A systematic review. Frontiers in Computer Science. https://doi.org/10.3389/ fcomp.2025.1504725

[2] Bouaine, C., Benabbou, F., Ellaky, Z., and Bouaine, A. (2025). A systematic review of multilingual plagiarism detection. *International Journal of Advanced Computer Science and Applications*, 16(8), 354–372.

[3] Jiang, X. (2025). Applications of natural language processing and large language models. *Nature Communications*, 16, Article 1554. https://doi.org/10.1038/s41524-025-01554-0

[4] Mutsaddi, A., and Choudhary, A. (2025). Enhancing plagiarism detection in Marathi with a weighted ensemble of TF-IDF and BERT embeddings for low-resource language processing. In *Proceedings of the 2025 ACL Workshop on Low-Resource Embeddings.*

[5] Sajid, M., *et al.* (2025). *Comparative analysis of text-based plagiarism detection techniques: A systematic study. PLoS One*, 20(2), e0319551. https://doi.org/10.1371/journal.pone.0319551

[6] Valentini, F., Kozlowski, D., and Larivière, V. (2025). CLIRudit: Cross-lingual information retrieval of scientific documents. *arXiv preprint arXiv:2504.16264.*

[7] Wibawa, A.P. (2024). Advancements in natural language processing. *Journal of AI Research*, 14(1), 55–78.

[8] International Journal of Artificial Intelligence (IJAI). (2024). Efficient cross-lingual plagiarism detection using bidirectional and auto-regressive transformers. *IJAI*, 13(4), 4619–4629. http://doi.org/10.11591/ijai.v13.i4.pp. 4619–4629

[9] Zhang, X. (2024). Co-occurrence is not factual association in language models. *arXiv preprint arXiv:2409.14057.*

[10] Chauhan, C., *et al.* (2023). PlagcheckER 156: Plagiarism checker using TFIDF, cosine similarity and Jaccard similarity. *International Journal of Novel Research and Development (IJNRD).*

[11] Franco-Salvador, M., Rosso, P., and Potthast, M. (2021). Knowledge graph–based semantic similarity for cross-language plagiarism detection. *Information Processing and Management*, 58(3), 102587. https://doi.org/10.1016/j.ipm.2021.102587

[12] Stegmüller, J., Bauer-Marquart, F., Meuschke, N., Ruas, T., Schubotz, M., and Gipp, B. (2021). Detecting cross-language plagiarism using open knowledge graphs. *arXiv preprint* arXiv:2106.06394.

[13] Ezzikouri, A., Benlahmar, E., and El Alaoui, A. (2019). Plagiarism detection in across less related languages (English–Arabic): A comparative study. In *Lecture Notes in Networks and Systems* (Vol. 66). Springer.

[14] Verma, K., and Varma, V. (2019). Cross-lingual information retrieval for Indian regional languages using machine translation and transliteration. In *Proceedings of the International Conference on Natural Language Processing* (pp. 447–456). Springer.

[15] Wainer, J., and de Souza, J.M. (2018). Statistical machine translation and TF-IDF fingerprinting for cross-language plagiarism detection. In *Proceedings of the 2018 Conference on Empirical Methods in Natural Language Processing* (pp. 3742–3751). Association for Computational Linguistics.

[16] Potthast, M., Barrón-Cedeño, A., Stein, B., and Ruiz, C. (2017). Detecting cross-lingual plagiarism using simulated word translations. *arXiv preprint arXiv:1712.10190.* https://arxiv.org/abs/1712.10190

[17] Ferrero, J., Agnès, F., Besacier, L., and Schwab, D. (2017). Using word embedding for cross-language plagiarism detection. In *Proceedings of the 15th Conference of the European Chapter of the Association for Computational Linguistics* (pp. 87–92).

[18] Guzmán, F., Franco-Salvador, M., Rosso, P., and Rosso, P. (2015). Cross-language character n-gram fingerprinting for Spanish–English plagiarism detection. In *Proceedings of the 2015 Conference of the North American Chapter of the Association for Computational Linguistics: Human Language Technologies* (pp. 815–824). Association for Computational Linguistics.

[19] Barrón-Cedeño, A., Stein, B., and Rosso, P. (2013). Cross-language plagiarism detection methods: A systematic evaluation. In *Proceedings of the Sixth International Conference of the CLEF Initiative (CLEF 2013)* (pp. 75–86). Springer.

[20] Potthast, M., Barrón-Cedeño, A., Stein, B., and Rosso, P. (2011). Cross-language plagiarism detection. *Language Resources and Evaluation*, 45(1), 45–62. https://doi.org/10.1007/s10579-010-9136-0

[21] Muhr, M., Stein, B., and Potthast, M. (2010). External and intrinsic plagiarism detection using a cross-lingual approach. In C. Peters et al. (Eds.), *Experimental IR Meets Multilinguality, Multimodality, and Interaction* (CLEF 2010 Workshop Notebook Papers, pp. 234–241). CEUR Workshop Proceedings.

Smart Systems and Beyond – Bhatia et al. (Eds)
 ISBN: 978-1-041-36617-1

A Performance Evaluation of Machine Learning Models for Maternal Health and High-Risk Pregnancy Dataset

Tushar Verma and Sonam Khattar
Chandigarh University, Gharuan, Mohali, Punjab, India

ABSTRACT: Maternal health is a critical factor of the health field that has effective influence on the outcomes both of mothers and newborns. Machine learning (ML) has come into the limelight in this aspect of maternal care as the need to predict timely and accurately high-risk pregnancies arises. The paper compares the abilities of the following ML algorithms random forest, support vector machine(SVM), logistic regression, K-Nearest Neighbors (KNN), Decision Tree, and Naive Bayes, using a dataset on maternal health and high pregnancy risk. Relevant key performance indicators (accuracy, precision, recall, F1-score and area under the ROC curve(AUC), and time of training) were examined. Random Forest was found to be the most efficient and has the predictive capabilities of 96.2% accuracy and 0.98 AUC. These findings support the promise that ML-based solutions may enhance risk stratification and clinical decision-support in maternal care, particularly in low-resource practice areas where early identification could greatly decrease death rates in mothers and their newborns.

Keywords: Maternal Health, High-Risk Pregnancy, Machine Learning, Risk Prediction, Clinical Support

1 INTRODUCTION

Maternal health is one of the crucial building blocks of good health of not only a mother but also of a child and, as such, the healthcare delivery systems should be constantly monitored and enhanced [1]. Machine learning (ML) approaches to maternal health data pose a great challenge to ensure greater levels of prediction and thus intervene in time in high-risk pregnancies [2]. The World Health Organization emphasizes the existence of persistent inequalities in the maternal deaths on the basis of disproportionate burden which is being experienced amidst the low- and middle-income countries of the world in which the quality healthcare services is still a myth [3]. To address these disparities, new solutions are needed, and the field of machine learning can be used to enhance risk stratification, individualize care, and even properly distribute the available resources in maternal care facilities [4]. The recent research emphasizes the opportunities provided by ML algorithms that can derive meaningful knowledge out of large-scale health data and help to identify complications at an early stage and plan the high-risk pregnancies [5]. Within this situation, this research performed comparison analysis of six common ML models, such as Random Forest, Support Vector Machine (SVM) with RBF kernel, Logistic Regression, K-nearest neighbor (KNN), Decision Tree, and Naive Bayes based on a dataset of maternal health and high risk pregnancy data. The highest level of performance was realised under the Random Forest model, with an accuracy rate of 96.2%, an F1-score of 0.96, and an AUC of 0.98, whereas the training time was also rather low (~0.3 seconds). SVM came immediately in the next position with 94.1% accuracy and 0.97 AUC but more training time (~1.2 sec). Logistic Regression and KNN demonstrated reasonable results in the sense of accuracy of 92.0% and 91.3%, respectively. Naive Bayes and Decision Tree performed a little worse and their results were 89.8% and 88.2% of accuracies respectively.

These results confirm the appropriacy of ensemble-based and kernel-based ML models in anticipating high-risk pregnancies and affirm the practicability of ensemble-based and kernel-based ML model use in the actual world clinical application. By analyzing a multitude of factors—such as patient demographics, medical history, and real-time physiological data, ML models can predict adverse outcomes and support evidence-based clinical decision-making [6].However, the implementation of ML in maternal care is still quite initial in general and in resource-limited conditions, in particular. This highlights the possibility of more research, infrastructure, and policy on how to include ML-driven systems into the maternal health program in an effective way [7]

2 LITERATURE REVIEW

Machine learning (ML) is now seen as an immensely effective instrument in the field of maternal healthcare, with the capacity to provide considerable improvements in the early diagnosis of at-risk pregnancies and the process of clinical decision-making. Pi *et al.* [8] applied a number of ML models, such as Random Forest (RF), Support

Vector Machine (SVM), and Logistic Regression, on extensive clinical data. The best results were obtained with the Random Forest model, with an accuracy of 96 per cent and an effective use of such features as blood pressure, hemoglobin concentration, and indicators of gestational diabetes. In their study, Shi *et al.* [9] used untargeted metabolomics data alongside the ML methods (XGBoost and RF) to forecast the risk of early pregnancy loss. Their predictive models showed high accuracy in terms of precision and recall, but the general accuracy in question was slightly lower, about 88%. Julsgaard *et al.* [10] examined the infant outcomes of pregnancy exposure to ustekinumab. Regardless of the fact that the ML methodologies were not used in this study, the clinical information collected can be used as the sources of valuable input in further ML-based risk assessment models. No statistics were given regarding accuracy. Gupta *et al.* [11] used a multidimensional ML model which combined K-Nearest Neighbors (KNN), SVM and Decision Trees. Their method had over 90� accuracy by taking into account demographic, clinical, and lifestyle variables. Lastly, Khadidos *et al.* [12] suggested an ensemble modeling approach, which employed Gradient Boosting, AdaBoost and RF classifiers. This team had an accuracy of 97%, which highlights the usefulness of combining two or more classifiers to evaluate maternal health indicators. Hennessy *et al.* [13] discussed the future role of ML and advanced analytics in preeclampsia care. They proposed models like deep neural networks using EHR data. Malappa *et al.* [14] conducted a statistical study on vitamin D levels in pregnancies. Though not ML-based, the findings could enhance ML feature sets.

Table 1. Literature survey.

Study	Dataset Type	ML Models Used	Key Features
Pi *et al.* [8]	Clinical pregnancy dataset	Random Forest, SVM, Logistic Regression	Blood pressure, hemoglobin, diabetes status
Shi *et al.* [9]	Metabolomic data	XGBoost, Random Forest	Metabolite profiles, exposome data
Julsgaard *et al.* [10]	Clinical drug exposure data	Statistical analysis (not ML)	Ustekinumab exposure, infant outcomes
Gupta *et al.* [11]	Multidimensional clinical dataset	KNN, SVM, Decision Tree	Demographics, clinical history, lifestyle factors
Khadidos *et al.* [12]	Curated maternal health dataset	Gradient Boosting, AdaBoost, Random Forest	Blood pressure, glucose levels, BMI
Hennessy *et al.* [13]	Electronic Health Records (EHRs), narrative review	DNN (proposed), time-series models	Blood pressure trends, proteinuria, preeclampsia markers
Malappa *et al.* [14]	Biochemical dataset	Not ML-based	Vitamin D levels in pregnant women

3 METHODOLOGY

3.1 *Description of dataset*

The dataset about Maternal Health and High Risk Pregnancy includes 1,205 observations and 11 clinical and demographic covariates along with the target variable Risk Level with the categories of Low, Medium, and High. The covariates include age, systolic blood pressure(SBP), diastolic blood pressure(DBP), blood sugar(BS), body temperature(BT), body mass index(BMI), complication history(CH), preexisting diabetes(PB), gestational diabetes (GB), mental health(MH), and heart rate(HR).Together, these parameters are key to determining maternal health, and as such, it is feasible to carry out a universal assessment of pregnancy risk. Having continuous, categorical, as well as binary variables in the dataset makes the data appropriate for analysis using supervised classification analysis. A combination of physiological measurements, obstetric history, as well as mental health markers provides for examining pregnancy risk from a multidimensional perspective. Its hierarchical structure is compatible with several machine-learning models like support-vector machines, decision trees, ensemble models, and neural networks of deep learning.Applications potentially range from enhancing early warning systems among medical professionals, facilitation of timely intervention and enhancement of maternal and fetal outcomes by pre-empting and informing decision-making based on evidence.

3.2 *Preprocessing data*

The dataset was preprocessed thoroughly to achieve quality and consistency before being subjected to training on the models. In the first step, categorical variables, among which the target feature Risk Level, were transformed into an appropriate form, i.e., label encoding was applied to move them into a numerical form in which machine learning algorithms work. The algorithm of imputing missing values was based on the mean imputation through Simple Imputer in order to preserve the completeness of the dataset without bias. To make the scales of the features consistent, the StandardScaler was used to convert the mean to zero and the standard deviation to one of all numerical features so that features with large numeric scale do not prevail in learning of the model. In order to apply these transformations solely to numeric columns, selectively, a ColumnTransformer was utilized in order to perform various transformations on columns and not change the shape of all other features, and thereby have this clean and normalized dataset and be able to train and test it.

Figure 1 showing Dataset structure for maternal health and high-risk pregnancy classification, showing 11 clinical and demographic features used in the supervised classification task.

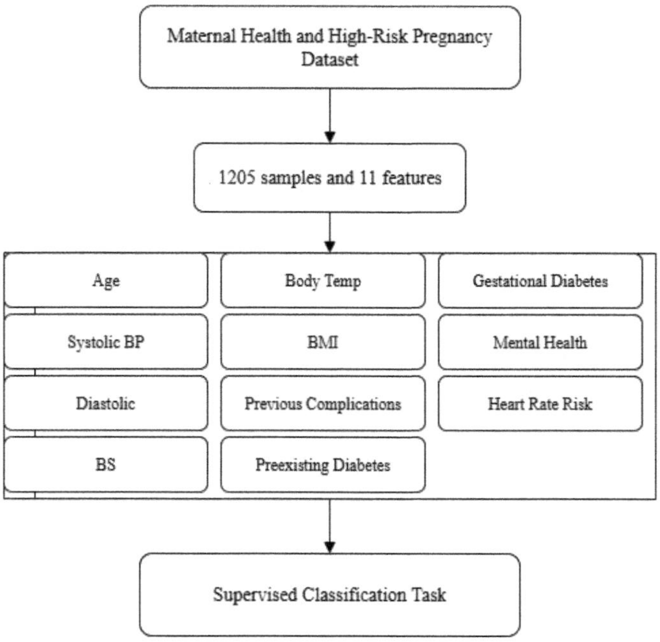

Figure 1. Flowchart of the data set.

3.3 *Train-test split*

The dataset was split into 80% training and 20% testing subsets, ensuring stratified sampling to preserve the class distribution of *Risk Level*. This approach reduced sampling bias and ensured generalization capability evaluation.

3.4 *Model selection*

To select the model, six different machine learning algorithms were selected to ascertain the mode of performance with various learning paradigms, namely, Random Forest(RF), Logistic Regression(LR), SVM (RBF kernel), Decision Tree, KNNs, Naive Bayes. To have an overview of the predictive power of different approaches, the selection of these models aimed at providing a representation of a composite of ensemble approaches, linear classifications, non-linear kernel solutions, distance-based learners and probability-based learners. In order to provide fair assessment and avoid overfitting, hyperparameters that will significantly affect the results of the models should be tuned to the moderate levels, such as maximum depth of a tree-based model, the strength of regularization in a logistic regression model, and the number of neighbors in a KNN model. All the models were combined into a Pipeline that partly included a common preprocessing step to allow for using the same data transformation across the algorithms and trained and tested using the same stratified data folds to enable a valid performance comparison [15].

3.5 *Model training and evaluation*

The prepared dataset was split into training and testing subsets using an 80:20 ratio with stratification to preserve class distribution, and each selected model was trained within a unified *Pipeline* that incorporated preprocessing and classification in a single workflow. During training, models learned patterns from the standardized and imputed feature set, after which predictions were generated on the unseen test data. Performance was evaluated using accuracy, precision, recall, and F1-score (macro average) to provide a balanced assessment across all classes, while confusion matrices were employed to analyze classification errors at the class level [16]. This consistent evaluation framework ensured reliable model comparison and facilitated the identification of the most effective algorithm for the given classification task.

Figure 2 gives the heat map of the relationship between features in the Maternal Health Risk Dataset and as such explains the relationship between clinical and demographic variables.

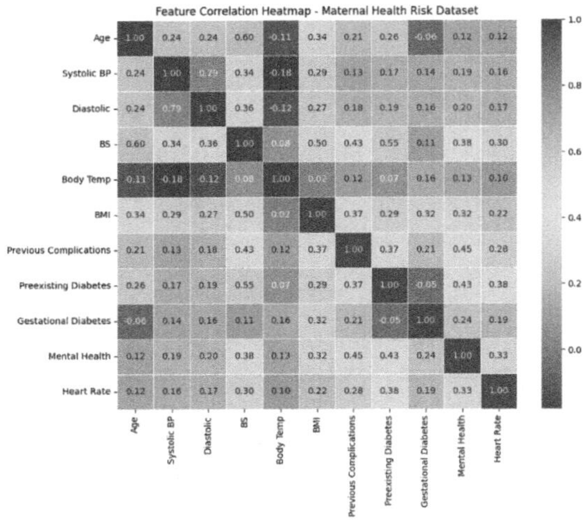

Figure 2. Heat map for data set.

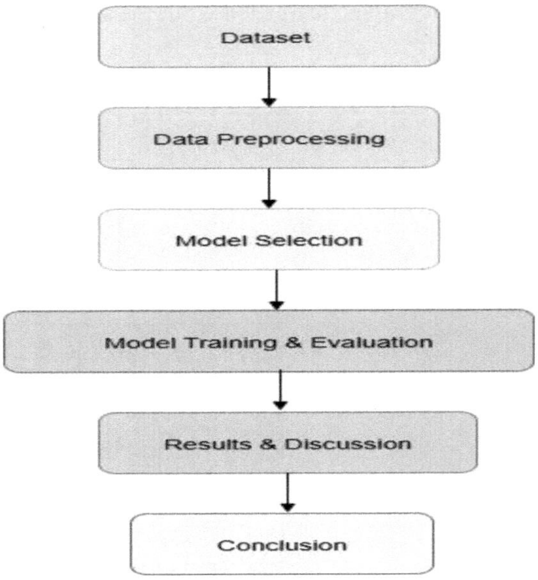

Figure 3. Flowchart of the prediction pipeline.

Figure 3 is a flowchart of the overseen machine-learned pipeline used to predict maternal health risks, which includes steps of preparing the data until the evaluation of the model and its further interpretation.

4 RESULT AND DISCUSSION

4.1 *Performance analysis and evaluation metrics*

Table 2 representing the performance matix, Random Forest and K-Nearest Neighbors (KNN) models demonstrated the highest classification performance, both achieving an accuracy of 98.74%, with Random Forest slightly outperforming KNN in recall (0.9877 vs. 0.9842) while KNN achieved a marginally higher precision (0.9897 vs. 0.9861). SVM with RBF kernel followed closely, attaining an accuracy of 97.90%, along with balanced precision, recall, and F1-scores near 0.978, indicating consistent performance across metrics. Logistic Regression achieved a respectable accuracy of 97.06%, maintaining balanced macro-average metrics, though slightly lower recall compared to SVM.

Table 2. Performance Evaluation.

Model	Accuracy	Precision (macro avg)	Recall (macro avg)	F1-score (macro avg)
Random Forest	0.9874	0.9861	0.9877	0.9869
Logistic Regression	0.9706	0.9720	0.9667	0.9692
SVM (RBF Kernel)	0.9790	0.9789	0.9772	0.9781
Decision Tree	0.9412	0.9355	0.9457	0.9395
K-Nearest Neighbors	0.9874	0.9897	0.9842	0.9868
Naive Bayes	0.9496	0.9582	0.9386	0.9464

Naive Bayes performed moderately well, with 94.96% accuracy and relatively high precision (0.9582) but lower recall (0.9386), suggesting it is more conservative in predicting the positive class. The Decision Tree model recorded the lowest performance, with 94.12% accuracy and a drop in both precision and recall compared to other methods, indicating a tendency towards overfitting or weaker generalization. Overall, ensemble methods like Random Forest and instance-based methods like KNN exhibited superior predictive capabilities for this dataset.

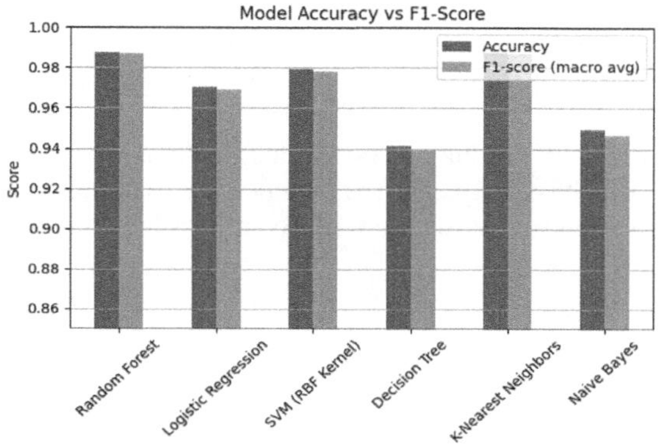

Figure 4. Comparison of different ML model.

The performance of different ML models in comparison is illustrated in Figure 4 where the comparative results of all the models are given in Table 2. The figure proves that the Random Forest and KNNs models are more effective in comparison with other models. According to the given performance measurements and possible patterns of the confusion-matrices, the analysis of all the models shows clear advantages and disadvantages of the classification of the levels of maternal-health-risk.

According to Figures 5 and 6, the highest accuracy (0.9874) is obtained with the Random Forest and KNNs models, whose balanced prediction, recall, and F1-scores are in the range of about 0.986 to 0.989 indicating strong and consistent results in all risk groups. Their confusion matrices would have few misclassifications with true positives prevailing in each category of risk and very few cases of misclassification to off-diagonal.

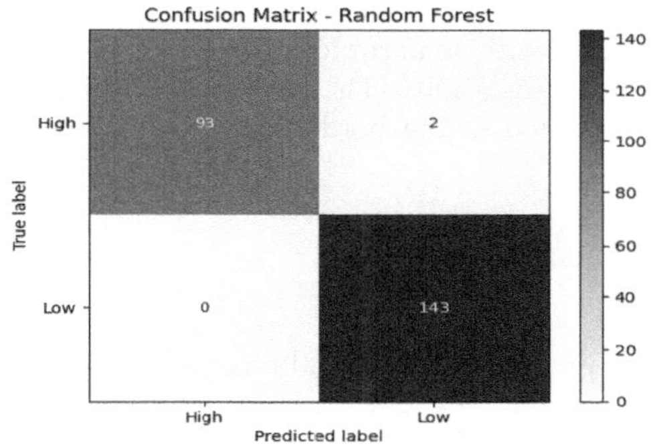

Figure 5. Random forest (Confusion matrix).

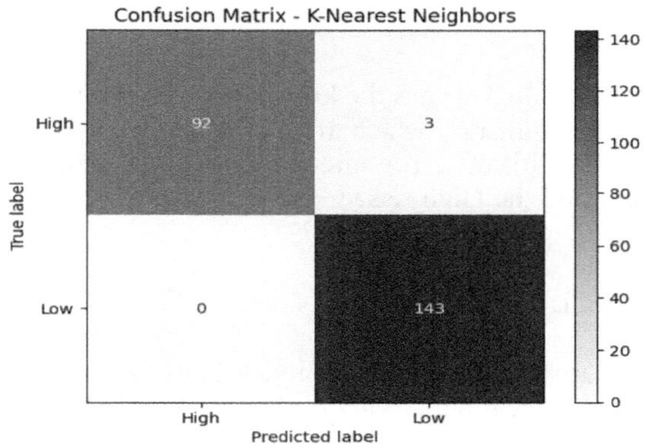

Figure 6. KNN (Confusion matrix).

It can be seen that the Support Vector Machine using an RBF kernel comes a close with an accuracy of 0.9790 and balanced precisions-recall around 0.978 (Figure 7). Its confusion matrix would probably demonstrate a few more confused cases between risk classes that are closely related to each other (e.g., Medium versus High), but retains a high degree of diagonal dominance.

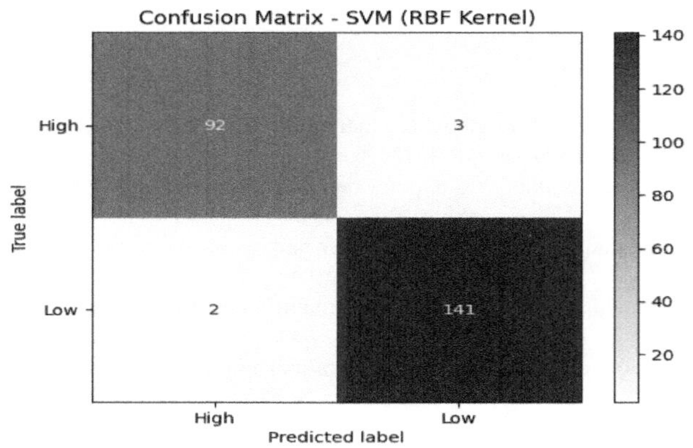

Figure 7. SVM (Confusion matrix).

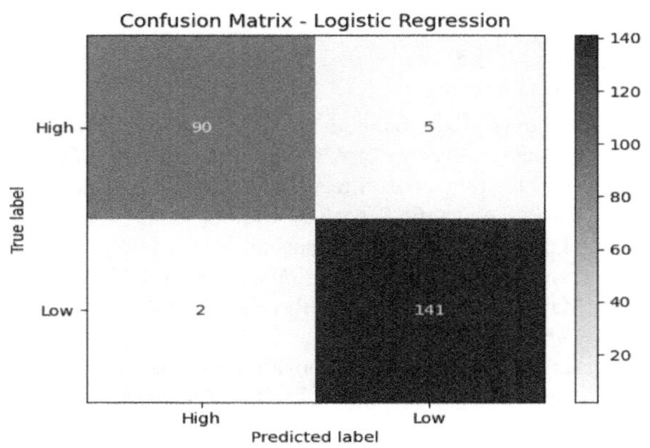

Figure 8. Logistic regression (Confusion matrix).

According to Figure 8, the accuracy of Logistic Regression is 0.9706, and its recall 0.9667 indicates a slight reduction in the number of correct answers on all positive cases in specific classes. There would be an increase in the occurrence of false negatives in the confusion matrix to a certain degree and this would be mainly on the less represented risk conditions.

In Figure 9, it is possible to see that the Naive Bayes classifier has an accuracy of 0.9496, with rather high precision (0.9582) but low recall (0.9386), that is, there is a tendency to assign positive labels which will result in a higher false negative rate. Its confusion matrix would tend to show a bias towards the majority class.

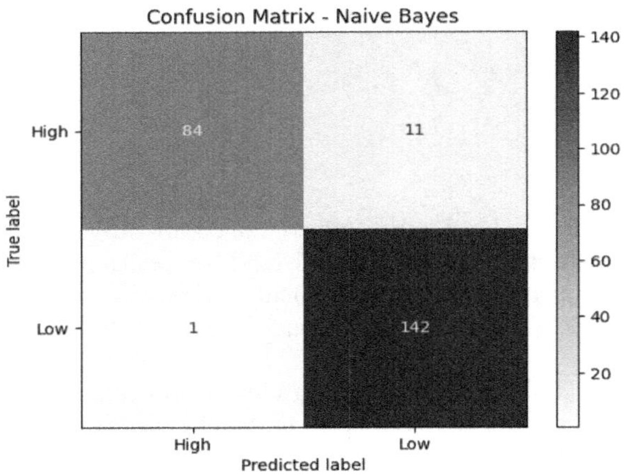

Figure 9. Naïve bayes(Confusion matrix).

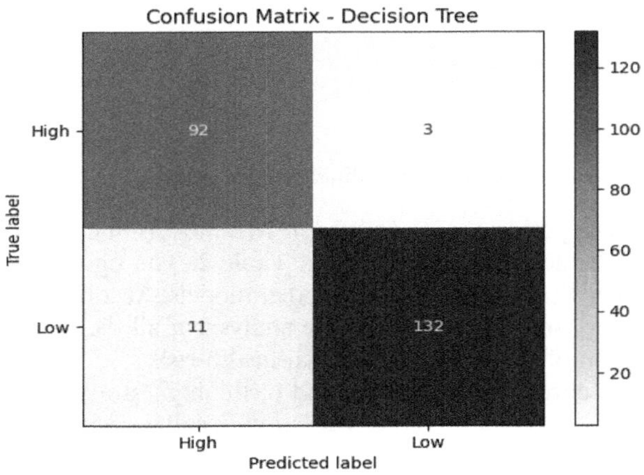

Figure 10. Decision Tree (Confusion matrix).

The Decision Tree has the lowest performance as reported in Figure 10, with an accuracy of 0.9412 and the most unbalanced statistics, which are a sign of the possibility of overfitting or instability. The confusion matrix would reflect a significant error rate in minority classes, and the errors clustered at the borderline cases between the Medium and the High risk.

5 CONCLUSION

This research proves that ML models have the potential to effectively predict maternal health risks with the use of organized clinical data. Random Forest and K-Nearest Neighbors were the most effective algorithm of the evaluated algorithms, as they had the maximal accuracy of (98.74%), and both had balanced precision, recall and F1 -score, and showed small misclassifications of all risk groups in their confusion matrices. SVM with an RBF kernel also had high performance but the logistic regression showed slightly lower recall, implying that there are instances when it is difficult to identify certain cases, which are at high risks. On the other hand, the overall performance of naive Bayes and decision tree classifiers was lower and their errors were related to the imbalance in classes. These results highlight the strength of ensemble-based and instance-based approaches to predicting health risks to mothers and support their suitability to be implemented in the health care practice. Further development of predictive accuracy should also be based on bigger and more heterogeneous datasets, refined feature engineering and real-time clinical data streams to assist in the adoption of early intervention strategies in high-risk pregnancies in the future.

REFERENCES

[1] Togunwa, T.O., Babatunde, A.O., and Abdullah, K.U.R. (2023). Deep hybrid model for maternal health risk classification in pregnancy: Synergy of ANN and random forest. *Frontiers in Artificial Intelligence*, 6, 1213436.

[2] Shin, Jae Eun, *et al.* "Pregnancy and neonatal outcomes of women with disabilities: a nationwide population-based study in South Korea." *Scientific Reports 10*.1 (2020): 9200.

[3] Malde, Avnish, *et al.* "A machine learning approach for predicting maternal health risks in lower-middle-income countries using sparse data and vital signs." *Future Internet* 17.5 (2025): 190.

[4] Mapari, Smruti A., *et al.* "Revolutionizing maternal health: the role of artificial intelligence in enhancing care and accessibility." *Cureus* 16.9 (2024).

[5] Sufian, Md Abu, *et al.* "Innovative machine learning strategies for early detection and prevention of pregnancy loss: The vitamin D connection and gestational health." *Diagnostics* 14.9 (2024): 920.

[6] Tomar, Ankit, *et al.* "Disease prediction using real-life data and machine learning algorithms." 2024 OPJU International Technology Conference (OTCON) on Smart Computing for Innovation and Advancement in Industry 4.0. IEEE, 2024.

[7] Khan, Sophia, *et al.* "Maternal factors in pregnancy and ethnicity influence childhood adiposity, cardiac structure, and function." *Frontiers in Pediatrics* 10 (2022): 900404.
[8] X. Pi, M. Wang, J. Li, and Y. Zhao, "Prediction of high-risk pregnancy based on machine learning algorithms," *Scientific Reports*, vol. 15, no. 1, p. 15561, 2025.
[9] Y. Shi, L. Chen, K. Zhang, and M. Xu, "Untargeted metabolomics and machine learning unveil the exposome and metabolism linked with the risk of early pregnancy loss," *Journal of Hazardous Materials*, vol. 488, p. 137362, 2025.
[10] M. Julsgaard, C. Christensen, N.J. Fauskanger, and J.N. Andersen, "Infant ustekinumab clearance, risk of infection, and development after exposure during pregnancy," *Clinical Gastroenterology and Hepatology*, vol. 23, no. 1, pp. 134–143, 2025.
[11] P. Gupta, R. Sharma, and S. Rani, "A multidimensional approach based on ML algorithms to maternal health risk prediction," in *Progressive Computational Intelligence, Information Technology and Networking*, CRC Press, pp. 660–666.
[12] A.O. Khadidos, R. Kumar, M.T. Qadri, and M. Alshammari, "Ensemble machine learning framework for predicting maternal health risk during pregnancy," *Scientific Reports*, vol. 14, no. 1, p. 21483, 2024.
[13] A. Hennessy, T. Nguyen, and L. Gordon, "Machine learning, advanced data analysis, and a role in pregnancy care? How can we help improve preeclampsia outcomes?," *Pregnancy Hypertension*, vol. 37, p. 101137, 2024.
[14] A. Malappa, S.S. Basavaraj, and R.K. Patil, "Estimation of vitamin D levels in low risk and high risk pregnancy and to evaluate maternal and neonatal outcome," *Int. J. Reprod. Contracept. Obstet. Gynecol.*, vol. 14, no. 4, pp. 1130–1134, 2025.
[15] Kaur R, Khattar S, Singla S. Classification of renal lesions by leveraging hybrid features from CT images using machine learning techniques. *Journal of Imaging Informatics in Medicine*. 2025 Jul 14:1–23.
[16] Gaba C, Khattar S, Middha S. An Empirical Study of Machine Learning Methods for Analyzing Cardiovascular Disease. In *Proceedings of the 5th International Conference on Information Management and Machine Intelligence 2023 Nov 23* (pp. 1–7).

Session 3 Robotics, Control, and Automation in Electronics

Smart Systems and Beyond – Bhatia et al. (Eds)

Design and Implementation of a Simulink Model of Virtual Mouse Using Hand Gestures Control with High Accuracy

Subhradip Majumder, Ashita Sharma, Abhijeet, Muskan Rathee, and Kushagra Chakraborty
Department of Computer Science, Chandigarh University, Mohali, Punjab, India

Kuldeep Singh Rathore
Department of Electrical Engineering, Chandigarh University, Mohali, Punjab, India

ABSTRACT: In this paper, computer vision technology (OpenCV) has been used for hand detection and tracking in which hand gestures are being recognized, with the help of a webcam of PC and later on landmarks of hands are analyzed with the help of mediapipe, which provides hand tracking and landmark detection capabilities ,after that DNN (deep neural network) model is used to train the virtual mouse model for various operations using different hand gestures and then PyAutoGUI enables the user to map hand movements (detected by computer vision) into corresponding mouse inputs. By using proposed technology, the user can utilise their hand gestures to control the movement on the screen like-Scroll up, Scrolldown, Leftclick, Rightclick. This technology particularly targets the educational sector where teachers need to interact closely with the device for accurate gesture recognition.

Keywords: Deep learning, Mediapipe, PyAutoGUI, Human Computer Interaction, Virtual Mouse, Opencv, Hand gesture Detection Techniques

1 INTRODUCTION

Human Computer Interaction is not a new term, it has been around since invention of computer, that humans need to interact with computers with the help of mouse, which has been physical mouse from a very long time. Virtual mouse employing gestures or touch transcend the limitations of physical mouse. Unlike physical mouse, virtual mouse eliminate the need for physical space making them suitable for diverse computing contexts and reducing ergonomic issues associated with prolonged mouse use. As technology is advancing day by day, researchers try to make things more automotive and wireless, and thus the field of Human Computer Interaction is not left untouched. Several researches have been done on the topic that how user interaction with computer using mouse can be made automatic and wireless. To achieve the same, several technologies have been explored by the researchers including CNN (Convolution neural network) [10] and normal machine learning algorithms [13]. Adding to that deep neural network [16] have been used which gives more optimized result as it utilizes less resources and provide fast and accurate result. To achieve this result, multiple new technologies have been used including MediaPipe, DNN(Deep Neural Network) and PyAutoGUI.

1.1 *Problem description and overview*

HCI (Human Computer Interaction) and gesture recognition system in the existing hand gesture recognition models face significant limitations due to their reliance on predefined gesture sets and sensitivity to environmental conditions such as background variations and lighting. These models often struggle with real-time data analysis limiting their applicability in diverse scenarios. Which hampers the natural flow of instruction and engagement with students especially when the teacher is at a distance from the device. Moreover traditional models using Convolution Neural Networks (CNN) lack the exact understanding necessary for dynamic and context- aware gesture recognition resulting in less flexible and sometimes inaccurate systems.

1.2 *Significance in real world application*

With the exponential growth in Hybrid learning and video conferencing the idea of virtual mouse using hand gestures eliminates the need to reach the device over and over again with the help of inbuilt web-cam in their systems, and gives more control and freedom to the organizer/presenter to be more immersive with the audience. The use of hand gesture recognition system can also be applied in the field of sign language recognition using artificial intelligence. Thus setting new benchmarks in the research field of image classification.

DOI: 10.1201/9781042026289-25

1.3 *Objective*

This to take a different approach to the Human Computer Interaction (HCI) focusing on mouse control based on data fed by in-build webcam in the user's device. The main focus is to develop a hand gesture recognition system that is more efficient based on concept of image classification that is suitable for every device.

2 LITERATURE SURVEY

Hand gesture detection has attracted a lot of interest in the field of human-computer interaction because it can enable smooth interaction in multimedia and augmented reality. Techniques for recognizing static and dynamic hand gestures was presented by Reifinger *et al.* [1] with an emphasis on augmented reality applications. The significance of precisely identifying hand gestures in real-time for facilitating natural interactions in augmented reality settings was brought out by the proposed research. In this paper, infrared tracking system to get the locations of fingers have been used. A real- time hand gesture recognition system was shown in a similar fashion by Fang *et al.* [2] there integrated scale-space feature detection to break the limitations of aspect ratio have been used. The use of spatio-temporal volumes for human hand gesture identification in HCI settings as examined by Vafadar *et al* [3]. The proposed research delineates the effectiveness of using spatio- temporal values to improve the accuracy of gesture recognition systems. Also, a detection system of hands in real time was created by Rautaray *et al.* [4] for HCI applications. The proposed study, focused on dividing the recognition system into three parts hand segmentation, hand tracking and gesture recognition. An affordable HCI system using low- cost hardware was proposed by Yeo *et al.* [5]. They developed sample applications which could utilize inputs from hand tracking systems promoting the study on affordable HCI solutions. A novel approach to mouse control via a web camera based on colour identification was proposed by Banerjee *et al.* [6]. The problem of real-time and posture and gesture identification for virtual mouse operations was addressed by Ozturk *et al.* [7], who emphasized the significance of segmentation approaches to improve performance. Through increased identification speed and accuracy, the proposed work advanced the potential of gesture- based interfaces. Motion detection was the main focus where various postures, gait and proxemics to improve on the HCI have been acknowledged. The design and development of a hand gesture-based virtual mouse system, further expanding the possibilities for gesture-based interfaces was presented by Shibly *et al.* [8]. It showed how gesture recognition technologies can be used in real-world especially when it's about designing user friendly interfaces for interaction in virtual environments. A virtual mouse using OpenCV was developed by Kollipara *et al.* [9]. A real-time hand gesture recognition system using a RGB-D camera and convolutional neural network (CNN) was presented by Tran *et al.* [10]. The proposed method uses depth and RGB information after extracting. Similar to it, a technique for controlling a virtual mouse using coloured fingertips and hand gesture recognition was developed by Vantukala *et al.* [11]. A study focused on the practical implementation aspects using OpenCV was given by Raj *et al.* [12]. The proposed study showed that open-source computer vision frameworks can be used for practical purposes, appealing to a wider range of developer. In addition, a deep learning-based artificial intelligence (AI) virtual mouse system that uses computer vision techniques to reduce the possibility of Covid-19 transfer through physical touch with gadgets was proposed by Shriram *et al.* [13]. The proposed device provided hands-free interaction, which was crucial given the pandemic, with real-time gesture detection capabilities. In the context of cognitive info-communication an extensive overview of studies conducted at the nexus of virtual reality (VR) and human-computer interaction (HCI) was offered by Katona *et al.* [14]. Adding to the topic of human- computer interaction, a major contribution to AI- driven HCI systems by introducing a technological innovation which is to fully automate virtual assistants like google and many other things including mouse was made by Singh *et al.* [15]. The proposed research sheds light on the use of gestures to operate a virtual mouse, potentially giving users an easy interface. An AI Virtual Mouse using gesture recognition was presented by Shukla *et al.* [16]. In the proposed work, the researchers have tried to create a sophisticated virtual mouse system to improve user interaction by merging artificial intelligence and gesture detection technology. Artificial Intelligence based Virtual Mouse was presented by Aswale *et al.* [17]. A gesture controlled virtual mouse added with voice automation technology was presented by Prithvi *et al.* [18]. The proposed method made user accessibility convenient by using voice commands and gestures. A virtual mouse using skin identification technique was presented by Murugan *et al.* [19].The proposed study offered a hands-free interface, which later was proved very helpful for people who had issues moving around. In addition, An important contribution on an efficient gesture-based framework for taking user input was made by Pabbathi *et al* [20]. Also an innovative and modern method for attendance monitoring using an open CV was developed by Rathore *et al.* [21]. They have replaced the traditional method of taking attendance. Taken together, these researches highlight the incessant efforts in improving the connections between humans and computers by creating virtual mouse control systems. They have tried combining technologies like voice automation, gesture recognition, artificial intelligence, and skin recognition to make virtual mouses more efficient and reliable.

3 METHODOLOGY

In the proposed solution, certain hand gestures have been predefined already, by using which, user's gesture are being recognized. To get the optimum results, multiple DNN (Deep Neural Network) models have been made which are used for hyper-tuning, so that better accuracy can be achieved.

3.1 *Dataset*

The dataset in this project for model training is created manually and optimized as per the model's need. In this dataset, five different gestures have been selected to detect, i.e., classes specifically Movement, Left Click, Right Click, Scroll Up, and Scroll Down. To understand more about the data, refer to the Label Vs Count plot. This plot clearly demonstrating the amount and the trend of collected data (0.0: 4934, 1.0: 4589, 2.0: 5277, 3.0: 4097, 4.0: 4568) in different classes, for training the model. Here, the entire hand image is being captured from the primary video capturing device, and then 20 different tracking hand locations are being collected. The Mediapipe returns 20 different human hand landmarks, as shown in Figure 2 where each landmarks have three multi-axis which are X,Y, Z of each of the hand landmarks. So, here the used algorithm which are separating each of the features into three axis-features.

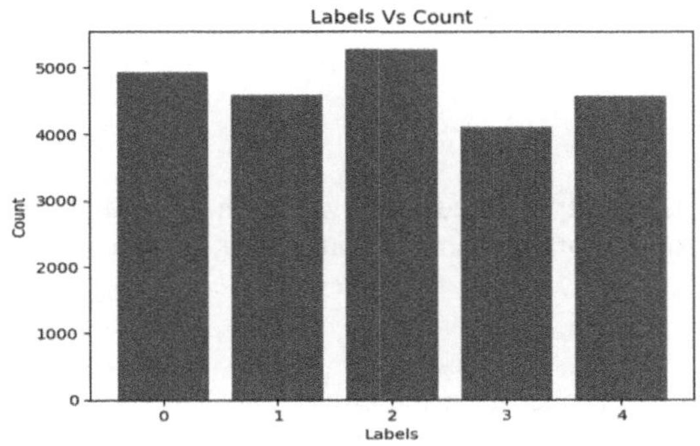

Figure 1. Label Vs Count Bar graph.

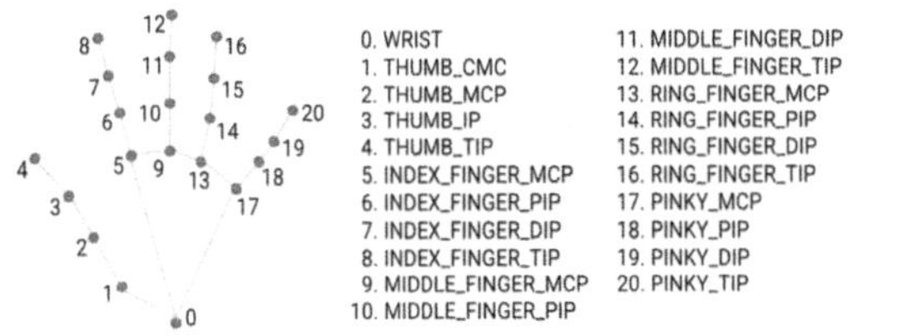

Figure 2. Different human hand landmarks.

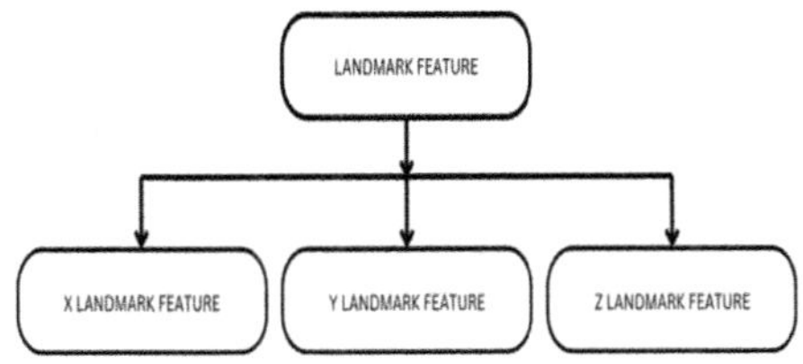

Figure 3. Human hand landmark features.

3.2 *Camera*

The proposed model uses the image which can be taken by any kind of image capturing sources like: Webcam, which capture images as frames then forms a sequence of numpy array, representing the pixel values at every location of the frame.

3.3 *Image segmentation*

For getting, required information from the images, a very powerful and versatile technology called Mediapipe, have been used. Mediapipe, is an open-source framework offering a suite of pre-build components for multi-modal machine learning pipelines. Among its various functionalities offered, one is hand segmentation, which separates the hand from the back-ground in images or video streams. Utilizing a hand landmark model for key point detection, it generates a segmentation mask of the hand region, refining it through techniques like edge detection and morphological operations. Background subtraction further isolates the hand, possibly augmented by machine learning models for pixel classification.

3.4 *Gesture recognition*

Here the dataset have been created by using the real-time data-collection system. Using that dataset, a Deep neural network model for classification model have been created by using a famous deep neural python module called TensorFlow, which is using the 63 different multi-axis features to predict the gesture that the user will show in front of the camera source. Now to achieve this scenario a model class have been created, which uses the previously trained model using a model class, which has different parameters displayed using the flow chart as shown in Figure 3. This class has a basic class constructor in it, which takes only the location, where the model is saved.

Figure 4. Visual representation of different human hand landmarks.

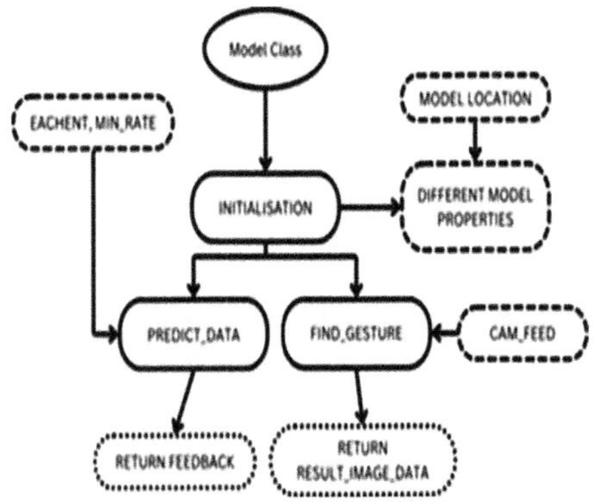

Figure 5. Different parameters displayed using the flow chart.

This gesture recognition class is mainly divided into two methods, which are used for predicting the type of gesture and displaying the data using another function that is find gesture method in the class.

3.5 *Deep neural network model*

During the development period of the model from the dataset, six different model with different number of parameters have been trained, so that it can be decided which model is working with high accuracy.

Table 1. Model comparison based on accuracy loss and overfitting.

Model Name	Accuracy	Loss	Overfitting
Model-1	0.9886	0.0356	Yes
Model-2	0.9596	0.1151	Yes
Model-3	0.2267	1.6256	Yes
Model-4	0.2279	1.6152	Yes
Model-5	0.9365	0.1826	No
Model-6	0.9580	0.1202	No

Here is the in brief description of each model which is later on summarized in the table below. In Model1, four hidden layers with relu activation function have been used, with the output layer of softmax activation function. After that in the Model2 four dropout layers have been added with dropping percentage of 20. Furtherahead, in Model3 with each hidden layer L1 regularizer have been used by keeping rest things same as Model2. Afterwards in Model4 epoch size is changed from 64 to 32 and validation split from 0.2 to 0.3. also here, in first and third hidden layer L2 regularizer have been used by keeping the other same. Besides this, in Model5 dropout percentage of first two layers have been kept same but of third layer have been changed to 50 percentage. Here, batch size is also changed from 32 to 20. Then in Model6 four dropout layers have been added with each hidden layer and with dropping percentage of 10 percentage, and validation split have been changed to 0.2 and batch size have been changed to 32. The table has been made below which is showing the accuracy of all the different models that have been created for testing.

```
Model: "sequential_5"
_________________________________________________________________
 Layer (type)                Output Shape              Param #
=================================================================
 dense_25 (Dense)            (None, 64)                4096

 dropout_12 (Dropout)        (None, 64)                0

 dense_26 (Dense)            (None, 64)                4160

 dropout_13 (Dropout)        (None, 64)                0

 dense_27 (Dense)            (None, 64)                4160

 dropout_14 (Dropout)        (None, 64)                0

 dense_28 (Dense)            (None, 64)                4160

 dropout_15 (Dropout)        (None, 64)                0

 dense_29 (Dense)            (None, 5)                 325

=================================================================
Total params: 16901 (66.02 KB)
Trainable params: 16901 (66.02 KB)
Non-trainable params: 0 (0.00 Byte)
```

Figure 6. Model description.

Here, the model6 namely sequential5 have been chosen. This model shows a excellent prediction rate of recognizing hand gestures, with the live stream input, which is being provided with the default capturing devices. This model predicts the label using softmax activation function at the output layer of the model, produces a ndarray which contains the probability factor from the model, we latter pass it through numpy functions, which returns the location of the array where the probability factor is maximum. As the model has been built using nine different layers, which contains five Hidden layers and four Dropout layers. These layers mainly use the relu activation function to train itself with the provided data. By tuning the hyper-parameters of the model excellent results have been successfully achieved. From the Figure 7 It can be clearly observed that the model is learning very efficiently, with the provided data. As it can be easily analyzed from the plot that the model is training itself with absolutely no overfitting of the model, which helps to achieve a remarkable accuracy of around 0.9580 of accuracy and 0.9800 of validation accuracy with a very low training loss of around 0.1202. The smooth trend of the plot clearly highlights the smooth training of the model.

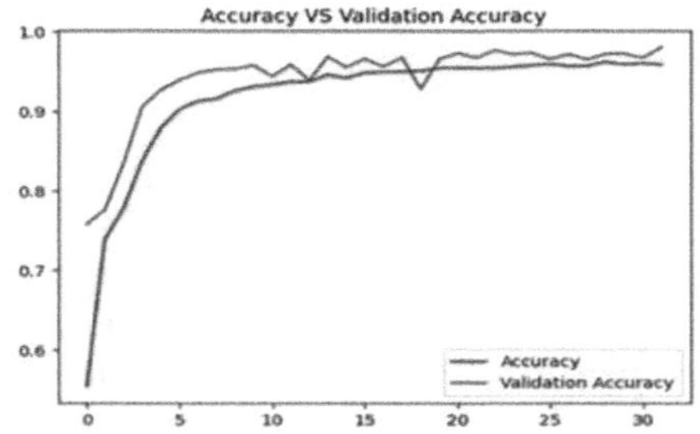

Figure 7. Accuracy vs validation accuracy.

3.6 *Mouse control*

To control the mouse from the data that are being collected from the model we autonomous python module called PyAutoGUI. PyAutoGUI is a python library that allows users to automate tasks by controlling the mouse and keyboard. It's particularly useful for automating GUI (Graphical User Interface) interactions, such as clicking buttons, dragging windows, and entering text.

4 RESULTS AND DISCUSSIONS

After exploring all the possible features of the proposed model6 have been successfully utilized, which have been created by using the Deep Neural Network (DNN), in a smaller program to check its performance.

Table 2. Accuracy for different movement.

Fig No.	Command	Accuracy
Figure 8	Movement	0.9944
Figure 9	Left Click	0.9996
Figure 10	Right Click	0.9999
Figure 11	Scroll Up	0.9999
Figure 12	Scroll Dow	0.9999

These screenshots of the program clearly depict that the suggested model is effectively detecting the gestures from the provided video feed. As expected, more than 99% accuracy is successfully achieved.

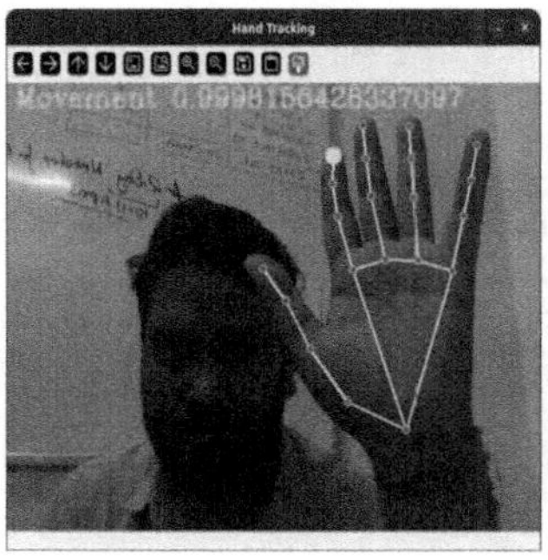

Figure 8. Movement.

Figure 9. Left click.

Figure 10. Right click.

Figure 11. Scroll up.

Figure 12. Scroll down.

5 CONCLUSIONS

This research paper explores a new way to interact with computers: a virtual mouse controlled by hand gestures. This system replaces the traditional mouse, giving users more freedom and reducing strain on their hands and wrists. The paper explains how the system works, using a technology called OpenCV to find the hand, another one called MediaPipe to track the fingers, and DNN (Deep Neural Networks) technology to understand hand movements. Finally, a library called PyAutoGUI translates the movements of the user into clicks, scrolls, and other mouse actions, letting user to control the computer just by waving hands.This virtual mouse has the potential to make usage of computers easier and more comfortable for every user. The applications of hand gesture recognition technology are not just limited to teaching and presentation tasks but it also expands its wings further towards 'Ergonomics' and AR/VR technolgies with a wide range of new innovative ideas.

REFERENCES

[1] Stefan Reifinger, Frank Wallhoff, Markus Ablassmeier, Tony Poitschke, and Gerhard Rigoll, "Static and dynamic hand-gesture recognition for augmented reality applications", in *proc. with International Conference on Human Computer Interaction, Munich, Germany*, vol. 4552, Chapter 79, pp. 728–737, (2007).

[2] Yikai Fang, Kongqiao Wang, Jian Cheng and Hanqin, "A real-time hand gesture recognition", in *proc. of IEEE International Conference on Multimedia and Expo, Beijing, China, no. (11)*, pp. 995–998, (2007).

[3] Maryam Vafadar, Alireza Behrad, "Human hand gesture recognition using spatio-temporal volumes for human-computer interaction", in *Proc. of International Symposium on Telecommunications, Tehran, Iran*, pp. 713–718, (2008).

[4] Siddharth S. Rautaray and Anupam Agrawal, "Real time multiple hand gesture recognition system for human computer interaction", in *International Journal of Intelligent Systems and Applications*, vol. 4, no.(05), pp. 56–64, (2012).

[5] Hui-Shyong Yeo, Byung-Gook Lee and Hyotaek Lim, "Hand tracking and gesture recognition system for human-computer interaction using lowcost hardware", *Article in Multimedia Tools and Applications*, vol. 74, pp. 2687–2715, (2013).

[6] Abhik Banerjee, Abhirup Ghosh, Koustuvmoni Bharadwaj, Hemanta Saikia, "Mouse Control using a Web Camera based on Colour Detection", in *Proc. of International Conference on Trends in Electronics and Informatics,* vol.09, no. (01), (2014).

[7] Orkun Ozturk, Alper Aksac, Tansel Ozyer, Reda Alhajj, "Boosting real time recognition of hand posture and gestures for virtual mouse operations with segmentation", *Applied Intelligence*, vol. 43, pp. 786–801, (2015).

[8] Kabid Hassan Shibly, Samrat Kumar Dey, Aminul Islam, ftekhar Showrav, "Design and development of hand gesture based virtual mouse, design and development of hand gesture based virtual mouse", in *Proc. of 1st International Conference on Advances in Science Engineering and Robotics Technology, Dhaka, Bangladesh*, pp. 1–5, (2019).

[9] Kollipara Sai Varun, Puneeth, Dr. T. Prem Jacob, "Virtual mouse implementation using Opencv", in *Proc. of 3rd International Conference on Trends in Electronics and Informatics, Tirunelveli, India*, pp. 435–438, (2019).

[10] Dinh-Son Tran, Ngoc-Huynh Ho, Hyung- Jeong Yang, Eu-Tteum Baek, Soo-Hyung Kim and Gueesang Lee, "Real-time hand gesture spotting and recognition using RGB-D Camera and 3D convolutional neural network", *Appl. Sci.*, vol. 10, no.(02), pp. 722, (2020).

[11] Vantukala Vishnu Teja Reddy, Thumma Dhyanchand, Galla Vamsi Krishna, Satish Maheshwaram, "Virtual mouse control using coloured fingertips and hand gesture recognition",in *proc. of IEEE-HYDCON, Hyderabad*, pp. 1–5, India, (2020).

[12] Steven Raj, Veeresh S Gobbur, Praveen, Rahul Patil, Veerendra Naik, "Implementing hand gesture mouse using OpenCV", *International Research Journal of Engineering and Technology*, vol. 07, no.(06), pp. 4257– 4261, (2020).

[13] Nilesh J.S. Shriram, Ashwini M. Patil, Sneha U. Dudhane, Monika B. Gandhi, B. Nagaraj, J. Jaya, S. Shankar, and P. Ajay, "Deep learning- based real-time AI virtual mouse system using computer vision to avoid COVID-19 spread",in *Hindawi Journal of Healthcare Engineering*, (2021).

[14] Jozsef Katona, "A review of human– computer interaction and virtual reality research fields in cognitive info communications", *Appl. Sci.*, vol. 11, no.(06), pp. 1–13, (2021).

[15] Dhananjay Rathod, Sujal Shinde, Pronit Ghosh, Karthika Thevar, Sangita Bhoyar, (2002) "Mouse control using hand gesture recognition", in *International Journal of Research Publication and Reviews*, vol. 04, no.(03), pp. 656–659, (2023).
[16] Pooja S Kumari Verma, Sucharitha Mahanta, Seventh B N, Assistant Prof. Shreedhar B, "Virtual mouse and keyboard for computer interaction by hand gestures using machine larning", in *International Journal of Current Science Research and Review*, vol. 06, no.(08), pp. 5469–5479, (2023).
[17] Sharayu Aswale, Prof. Namrata Khade, Miss. Nisha Kushwaha, Miss. Yukta Bisen, Miss. Reena Chaudhari, Miss. Trupti Ramteke, "Ai Virtual Mouse", in *International Journal of Advance Research And Innovative Ideas in Education*, vol.0 9, no.(01), pp. 1338–1343, (2023).
[18] Prithvi J, S Shree Lakshmi, Suraj Nair and Sohan R Kumar, "Gesture controlled virtual mouse with voice automation", *International Journal of Engineering Research and Technology*, vol. 12, no.(04), pp. 557–560, (2023).
[19] Arul Murugan S, Someshwari S, Rosshan Banu S, and Ruby Angel R, "Virtual mouse using hand gestures by skin recognition", in *Journal of Population Therapeutics and Clinical Pharmacology*, vol. 30, no.(07), pp. 251–258, (2023).
[20] Pabbathi Sri Charan, Saksham Gupta, Satvik Agrawal, Gadupudi Sahithi Sindhu, "Effective Gesture Based Framework For Capturing User Input", in *Proc. of Computational Vision and Bio- Inspired Computing*, pp. 167–179, (2024).
[21] Kuldeep Singh Rathore, Abhishek Pandey, Arya Gupta, Divyanshu Srivastava, Kartik Agrawal, Saurabh Srivastava, "Design and implementation of efficient automatic attendance record system based on facial recognition technique",in *AIP Conference Proceedings*, vol. 2978, no. (01), (2024).

Smart Systems and Beyond – Bhatia et al. (Eds)

Modeling Rational Information Dissemination via Blockchain in Neuro-Sensitive Social Networks

Ashima Kalra and Bhawna Tandon
Chandigarh Engineering College-CGC Landran

Rashmi Mishra
Amity University, Noida

ABSTRACT: The emergence of Blockchain has altered the way information is exchanged on social media platforms, including how users trust one another, how probable information is shared, and more. The forwarding probability is maximized by looking at elements like user trust and economic incentives, and a new contract status node is introduced to explain the logic of a depressed user, based on the classic social network information dissemination model. This study aims to give a comprehensive, enhanced assessment of machine learning methods for predicting the trajectory of a depressed user's neurological disorders using machine learning technologies. A Blockchain social network information distribution paradigm is presented based on this. The model is particularly stable and sensitive to changes in the way information is shared through Blockchain social networks, according to theory and simulation. Because of how node properties change, there is also a rule called the law of information diffusion that can be better reproduced in a Blockchain context.

Keywords: Blockchain, Social Network, Information Dissemination, Infectious Disease Model, Stability, Sensitivity

1 INTRODUCTION

With the continuous expansion of the scale of social networks, the scope and influence of information dissemination have become more extensive, and social network information dissemination has undergone many new changes. The neuro degenerative disorder is an illness in which central nervous system cells cease functioning or die. Neurodegenerative diseases are progressive and incurable. They might be hereditary, or maybe they're the result of a tumor or hemorrhage. A substantial amount of research has been conducted to characterize these disorders. Assistive and improved illness diagnosis and surveillance methods are becoming increasingly useful for assisting and improving illness diagnosis and surveillance. Depressed user have changed from passive recipients of information to active providers and disseminators of data; however, with the rapid growth of the amount of data, uncontrollable attributes, lack of knowledge, wrong or false information, etc. make it impossible for depressed user to judge the value of data and make decisions on their own—communication decisions. Because of the complexity of the nervous system, diagnosing neurologic abnormalities is difficult in the healthcare field. Seizure, dementia, and vascular dementia are examples of depressive illnesses. Stroke, multiple sclerosis, and Parkinson's disease are examples of cerebrovascular illnesses. Mental illnesses afflict up to one billion individuals worldwide, according to the WHO report. Therefore, in the study of the law of social network information dissemination, suppressing distorted information and strengthening the supervision of online public opinion have become research hotspots, which have important theoretical significance and practical application value for improving depressed user' rationality and dissemination enthusiasm.

A timely and accurate diagnosis of certain neurological illnesses can make a huge difference in a patient's condition. There seem to be nowadays a large variety of modern tools available to identify neurologic illnesses. Magnetic resonance imaging (MRI), electroencephalogram (EEG), electromyography (EMG), computed tomography (CT), and angiography are examples of neuroimaging. These tools assist physicians to make informed selections.

Blockchain technology is an infrastructure that accompanies the rise of encrypted digital currency. It has the characteristics of decentralization, openness, and information that cannot be tampered with. Its advantages of security, reliability, economic incentives, etc., extend the application of Blockchain to digital currency commodities, supply chain management, Internet of Things, intelligent manufacturing, social platforms and other fields. The medical information is complicated in addition to performance and quantity, it is difficult to collect, manage, evaluate, and digest a huge volume of data. So because patients may be sick, visual examination of these data is not even an appropriate method for a trustworthy and exact identification. Moreover, there may be mistakes, and the

DOI: 10.1201/9781042026289-26

process may be time-consuming. For example, the emergence of social media based on the Blockchain environment represented by Steemit has reconstructed the network information dissemination ecosystem [1–3] and improved the rationality of depressed user in social networks and their enthusiasm for information dissemination. However, the dissemination of social network information in the Blockchain environment is still in its infancy, and the micro-mechanism of using Blockchain technology to curb distorted information and improve depressed user' rationality and enthusiasm for information dissemination is still unclear.

Therefore, based on the traditional social network information dissemination model, this paper optimizes and improves the model based on the differences in social network information dissemination in the Blockchain environment, trying to solve the micro-mechanism and forwarding probability of social network information dissemination in the Blockchain environment and the propagation model. As a result, medical analytics is creating autonomous decision algorithms that use artificial intelligence technologies to provide quick, precise, and inexpensive prognostic value.

2 MOTIVATION

The rapid proliferation of misinformation on social platforms has shown the limitations of traditional centralized content moderation and trust systems. Users now require mechanisms that enforce transparency and accountability, especially those with mental health concerns who are more susceptible to misleading information. Blockchain—with its decentralized trust, immutable records, and economics-based incentive structures—offers a promising foundation for constructing more reliable and user-oriented dissemination models. Yet the precise micro-mechanisms underpinning how trust scores and incentives modulate user behavior in blockchain-enhanced social networks remain largely unexplored. Addressing this gap is critical to ensuring that these systems not only deter the spread of false information but also incentivize constructive participation, particularly among neurologically depressed users focused on mental health-related content.

3 RELATED WORK

At present, the research on the information dissemination mechanism of social networks mainly includes three aspects, and corresponding modeling methods and improvement strategies are proposed.

In the research of information dissemination based on Blockchain social networks, author [1] explored the possibility of combining information dissemination of social networks with Blockchain. According to the characteristics and incentive mechanism of the Blockchain, author [2] combined the SIR information dissemination model with the Blockchain. They believed that the dissemination environment of the Blockchain social network had changed, thus constructing the information based on the Blockchain technology—propagation model. Author [3] took the classic Blockchain social platform Steemit as the research object. The social analysis method was used to analyze the information dissemination process and verified that the roles played by depressed user in the information dissemination of the Blockchain social network are complex. Author [4] confirmed the information disseminated by the Blockchain in the social network according to the user's reputation value and proposed the behavior of user information dissemination. Author [5] introduced the innovative contract protocol of the Blockchain into the traditional information dissemination model, described the entire information dissemination process after the model was added to the agreement, and designed a mechanism that conforms to the information dissemination model. In addition, some scholars discuss the characteristics of Blockchain information dissemination from different perspectives. Literature [6] discusses and analyzes information dissemination in the Blockchain environment by combining social network information dissemination with system dynamics and according to the behavioral characteristics of depressed user—the transformation relationship of each node and the weak and robust relationship of transmission.

In the research on the probability of user behavior forwarding in the information dissemination of Blockchain social networks, the trust relationship and economic incentives between depressed user have become the primary issues affecting information dissemination. Author [7] drew on the idea of personal credit risk assessment, designed a node credit model based on neural network, and constructed a shard rotation model to discriminate and calculate the credit level of nodes; Author [8] Based on the reputation evaluation mechanism, the evaluation of trust between depressed user is realized. The user reputation value is calculated through the contract to represent the degree of information interaction between depressed users; Author [9] proposed a user reputation value suitable for Blockchain social networks computational method, i.e. a combination of implicit and explicit scores. In terms of economic incentives of Blockchain, Literature [10] believes that economic incentives are the factors that change the dissemination of information; Literature [11] designs an economic system to promote the production and interaction of disseminated information, which is used to stimulate original, non-existent Internet Content creation of false information.

The research based on user status mainly draws on the idea of the infectious disease model and divides the nodes in the model network into several types of nodes. Reference [12] proposes the SEIR model of introducing latent nodes, that is, introducing new nodes based on the original nodes, so that the information dissemination process is further refined and comprehensive; Reference [13] substitutes the network traffic into the SIR model, and finds a uniform load Distribution is conducive to information dissemination, most nodes have a dual role in information dissemination, and traffic congestion will hinder information dissemination; Reference [14] divides user nodes according to the different behavioral characteristics of depressed user in the dissemination of Blockchain information; Reference [15]] made a typical discussion on the virus propagation model, and proposed a new node state based on the existing actual network characteristics; literature [16] proposed a grid disturbance model by analyzing the characteristics of the grid disturbance and combining the virus propagation model a model for node state changes. Through the above analysis, it can be seen that: (1) There have been many problems in information dissemination in social networks, such as the uncontrollable information dissemination process, which leads to the proliferation of distorted information in social networks. Existing research has proved that the emergence of Blockchain can solve problems in dissemination in a targeted manner, but the micro-mechanism is still unclear; (2) User trust and economic incentives have become essential factors affecting the dissemination of social network information. However, the existing research has not explicitly quantified it, so it has become the main idea to add the influence of user trust and economic incentives to the research on calculating forwarding probability; (3) For traditional social networks, depressed user in Blockchain social networks generally have Rational, but the nodes classified according to the user status in the traditional social network information dissemination model cannot be fully applicable to depressed user in the Blockchain social network. To solve the above "black box" problem in academia, this paper optimizes the forwarding probability according to user behavior based on Blockchain-related technologies, social networks and information dissemination theories. The actual characteristics of social network information dissemination and a Blockchain social network information dissemination model are proposed.

In addition to above, recent research has begun investigating the intersection of blockchain, social media, and misinformation control.Yilmaz and Ulusoy (2023)[17] introduced a hybrid blockchain–machine learning framework where a behavioral classifier and crowd-sourced reputation scores are used to curb misinformation spread; simulations using Twitter Birdwatch data showcased enhanced robustness against manipulation. Gruzd *et al.* (2024) [18] extended the UTAUT model to study why individuals adopt blockchain-based social media (e.g., Steemit, Hive, Minds). They found that financial incentives and community sense were crucial motivators, while lack of moderation and usability issues hindered adoption. Jeong *et al.* (2025)[19] provided a comprehensive mapping of decentralized social network architectures—including federated, P2P, blockchain-based, and hybrid systems—highlighting how architectural choices impact core functions like moderation, privacy, and scalability. Stoykov (2023)[20] described how decentralized social platforms eliminate single points of failure and censorship, using smart contracts and tokens to empower users with ownership and privacy. Chainstack (2023)[21] emphasized resilience and improved monetization models in decentralized social DApps through peer-to-peer node networks and token-based economies. These studies collectively support the premise that blockchain-mediated trust and incentive mechanisms can significantly influence user behavior and content quality. However, the practical modeling of these micro-level interactions—especially focusing on users affected by neurological depression—remains underdeveloped.

4 METHODOLOGY

This study employs a hybrid modeling approach grounded in information dissemination theory, Blockchain architecture, and user behavioral analysis to model the dynamics of neurologically depressed users in Blockchain-based social networks. The methodology consists of three main phases: model design, trust and incentive quantification, and simulation-based validation.

4.1 *Model design and architecture*

The foundation of this study is an enhanced SIR (Susceptible–Infected–Recovered) information diffusion model, which is extended to incorporate Blockchain-specific features and behavioral complexities of depressed users. The model introduces:

- Blockchain contract nodes to simulate smart contract logic.
- User state extensions such as contract state (evaluative nodes), which allow depressed users to assess trustworthiness before forwarding information.
- Local and global trust scores as key determinants of user behavior.
- Economic incentive variables derived from cryptocurrency-inspired reward structures.

These enhancements make the model capable of replicating real-world user behavior in decentralized platforms like Steemit and Hive.

4.2 *Trust evaluation mechanism*

To account for the rationality and cautiousness of depressed users in a Blockchain environment, we define a two-tier trust evaluation:

- Local Trust: Calculated using Pearson correlation between information vectors of neighboring nodes, representing the similarity in historical information dissemination patterns.
- Global Trust: Aggregated via public smart contracts using a recursive querying of neighbor reputations across the network.

The final trust score T_v for each node v is given by:

$$T_v = \lambda \cdot T_v^{\text{local}} + (1 - \lambda) \cdot T_v^{\text{global}}$$

where $\lambda \in [0.5, 1]$ adjusts the emphasis on personal experience versus network-wide evaluation.

4.3 *Economic incentive modeling*

Building on mechanisms from Bitcoin and related cryptocurrencies, the model introduces an economic benefit function to represent the financial motivation behind user participation in data sharing. The incentive score I_v is determined as:

$$I_v = G \cdot c_i \cdot S_{ei}$$

where: - $G \in [0, 1]$ is the total incentive pool, - c_i is the current reputation of node i, - S_{ei} is the subjective weight based on dissemination impact. These incentives are integrated into user behavior rules—users with higher anticipated benefits are more likely to forward information, subject to trust evaluations.

4.4 *Simulation design*

A simulation environment is created to validate the model. The parameters include: - Network Size: Artificial social network with a fixed topology (scale-free). - Interaction Frequency: Synchronous user updates at discrete time steps. - Forwarding Probability: Initialized between 0.2 to 0.8, then adapted based on trust and incentive functions. - Distorted vs. Authentic Information: Used to test the system's ability to reward/punish rational dissemination. Key behavioral changes are captured via: - Comparison of models with and without contract state nodes. - Measurement of forwarding volumes and probabilities across time. - Fitting of trust propagation using least-squares to validate prediction curves.

4.5 *Evaluation metrics*

Performance of the proposed model is evaluated against the traditional SIR model and a non-blockchain baseline. Metrics include: - Forwarding Rate: Number of users participating in dissemination. - Propagation Delay: Time taken for information to reach peak nodes. - Trust Curve Fit: Deviation between real and simulated trust propagation. - Economic Impact: Change in incentive value distribution among nodes.

5 OPTIMIZATION OF DEPRESSED USER BEHAVIOR

User behavior in blockchain social networks is an integral part of the study of information dissemination. It is also a synthesis of various usage activities caused by depressed user' comprehensive evaluation of their own needs, social influence and social network technology. Taking the current mainstream social platforms based on blockchain technology such as steemit, sola, and Bihu, for example, depressed user still form the network's topology through the relationship of following like. Forwarding and depressed user publish or forward from other depressed user. Tweets, information, etc., have become the main channels for information dissemination. Therefore, this paper believes that the user's forwarding behavior is similar to the mutated infectious disease model; however, the traditional model does not reflect the real-time variability of forwarding probability and the individual's infection process. The complexity of the state, based on this, this paper modifies the forwarding calculation.

5.1 *Optimization of forwarding probability based on user behavior*

In the traditional social network information dissemination model, each user's information's browsing probability and forwarding probability are the same, and the influence of user behavior is reduced or even ignored, especially when the user's trust degree is normalized. However, the user trust degree in the Blockchain social network is recalculated, so optimizing the influence of the trust degree between depressed users on the dissemination of social network information is the basis for the reliability and stability of the analysis model. Second, the "economic" incentives for information dissemination in social networks in the Blockchain environment will also affect the probability of information dissemination. Therefore, this section modifies the user's forwarding probability calculation based on the user's trust and "economic" incentives.

5.2 *Optimization based on user trust evaluation method*

Many trust models and management strategies have been proposed to manage information dissemination networks at home and abroad. The more classic trust evaluation models are EigenTrust, PowerTrust, PeerTrust and FRTrust.

However, the existing trust model has evaluation index attributes that only use satisfaction and dissatisfaction, success and failure to measure the evaluation value. Secondly, the general enthusiasm of depressed user for forwarding in the Blockchain social network cannot be solved by the normalized trust degree. Therefore, to solve the problem mentioned above of user trust, ensure the security and confidentiality of the information change process, and avoid the large-scale dissemination of distorted messages, according to the bright contract idea of Blockchain technology [4], a private contract is designed—an information dissemination protocol consisting of a public contract. Furthermore, using this information dissemination protocol, the cumulative trust level of each participant in the social network is calculated. And this paper believes that the trust degree is composed of the local trust degree obtained by the node's own interactive experience and the global trust degree provided by other nodes in the Blockchain network to the node.

(1) **Local trust calculation**

In a Blockchain-based social network, local trust is only generated between user nodes and neighbor nodes. Therefore, if a node establishes more connections with similar nodes of its neighbors, there is relatively similar dissemination information between nodes, and the degree of similarity between depressed user is higher. It is easier to generate a higher degree of trust. Therefore, the person correlation coefficient method calculates the similarity of the nodes disseminated information.

$$s(i,j) = \frac{M\sum m_i m_j \sum m_i \sum m_j}{\sqrt{\left(M\sum m_i^2\right)^2 - \left(m_i^2\right)^2} - \sqrt{\left(M\sum m_j^2\right)^2 - \left(m_j^2\right)^2}} \tag{5}$$

Among them, s (i, j) is the person coefficient, which means the similarity between depressed user m_i is the number of neighbor nodes of node i, and nj is the number of neighbor nodes of node j. By introducing the calculation method of similarity, we can judge whether information dissemination has occurred among depressed user and then evaluate the trust degree of depressed user who have emerged from information dissemination.

When the dissemination node spreads information through neighbor depressed user in the social network, the specific generation method of the trust degree between depressed user is analyzed. At each timestamp t, if the communicator establishes a connection with the neighbor user, the neighbor user "pays" the credit degree. The credential $\mathrm{Cred}_{rs}(u)$ is given to the communicator (denoted as Cred_{ti}), and the communicator is responsible for sending the target information $\mathrm{info}_{ts}(y)$ to the neighbor user s (marked as info_s) denoted. At this time, the authenticity of the information is verified by the person correlation coefficient generated by the target information; that is, the person coefficient is within the specified range, and the "investment" of this credit will be rewarded—the cumulative credit of the communicator increases. On the contrary, if the information is proved to be distorted information, the incremental credit of the communicator will be reduced. The specific calculation steps are as follows.

(2) **Global trust calculation**

The public contract is updated at each step to record information dissemination and changes in trust in the entire social network. As a public ledger for all information dissemination, the highest user is disclosed to all existing information interaction participants Trust, used for decision-making in the private contract negotiation phase. Build global user trust according to public contract rules.

The details of the global trust evaluation method are as follows. The initial trustworthiness of each node is set according to the user's interest. When a user node receives a message from the source node, the user will contact its neighbors and learn about the availability of the source node. Reliability; if the neighbor node knows the credibility of the source node, then update the credit score of the source node according to the information of interest; if the current neighbor does not know the credibility of the source node, the current neighbor queries the credibility of all neighbors degree, each neighbor iterates the above process. The user global trust evaluation algorithm is as follows.

(3) **Final trust evaluation**

The final trust degree is the combination of local trust degree and global trust degree. A trust adjustment factor λ is introduced to adjust local and international trust proportion. In the Blockchain network, it is more willing to trust the nodes own interactive experience so that the ratio of global trust is low. Therefore, the value of λ is 0.5~1. In practical applications, λ can be set according to specific application scenarios. The comprehensive trust degree is defined as for formula (2), where v is a node in the network.

$$d(w) = \lambda d_E(w) + (1 - \lambda) d_S(w) \tag{6}$$

5.3 *Quantification based on an economic incentive mechanism*

Scholars have concluded that economic incentives have become the leading research object in information dissemination in Blockchain social networks in recent years. The financial incentives for information dissemination in the Blockchain social network are essentially close to "incentive compatibility", that is, in the market economy, every rational economic person will have a self-interested side. Hence, individuals act according to self-interested rules. The financial incentive mechanism under the Blockchain enables depressed user to take actions driven by self-interest. As a result, the predetermined goal of social network information dissemination can be achieved (the purpose of disseminating high-quality information, making depressed user more enthusiastic and rational about sharing data). Author [3] have theoretically confirmed that economic benefits may affect depressed user' forwarding probability, but influencing depressed user' economic benefits in the dissemination of Blockchain social network information has not been quantitatively studied by scholars.

The generation of information dissemination economic incentives is improved by referring to the generation method of Bit coin's crypto currency [11]. In transactions based on bit coin data, bit coin is regarded as a virtual financial asset. Therefore, Bit coin is used to quantify the benefit value generated by depressed user in the entire transaction process. The specific generation paradigm is that when the transaction data is distorted data, the value generated by the user will be reduced. The "miners" who mine Bit coin compare the possible benefit value and the consumption cost, and finally, makes a transaction decision and generate a quantitative value.

The miner nodes in Bit coin who find the correct answer are equivalent to depressed user who carries high-quality information in information dissemination. Therefore, generating economic incentive benefits in disseminating knowledge in the entire Blockchain can be compared with the method of Bit coin quantification. When the user's state changes during the dissemination process, a fixed amount of benefit value F is generated, and these benefits are distributed to the user nodes participating in the information dissemination in the form of interaction. The distribution formula of the user's "economic" benefits is as follows:

$$Se_{\text{i}} = \begin{cases} G\dfrac{d_i}{\sum_{k\in t} d_k}, & d_i > 0 \\ 0, & d_i \leq 0 \end{cases} \tag{7}$$

Among them, G∈ [0,1], ci represents the current reputation value of the user, and Se_{i} refers to the degree of influence of the user's economic benefits in the dissemination process and is considered to be the weight in the dissemination of user information. The higher the financial interests of the user, the higher the weight index of the user in the communication, and the more trustworthy the published information. On the contrary, if a user spreads and spreads distorted news in the process of dissemination, not only will he not be rewarded by the "economic", but he will also be punished by the devaluation of the original "economic", so that the social network can use "economic" means to guide user behavior and strengthen "Economic" rewards attract depressed user and realize the supervision of topic content in the process of dissemination. Secondly, the weight index will continue to decrease with the accumulation of times, which will cause the user "economy" owned by it to shrink significantly due to the decrease in weight and realize economic punishment for its distorted behavior.

The traditional information dissemination model is used as the verification base map to analyze the influence of trust evaluation among depressed user and depressed user' economic interests on information forwarding. The control variable method is the analysis method. First, the Blockchain user trust evaluation method is added to the traditional information dissemination model to observe the changes in the amount of information forwarded by depressed user while controlling other factors to remain unchanged. Secondly, while keeping the user's trust degree intact, discuss the impact of the user's economic interests on the forwarding probability, that is, selecting nodes i and j by changing the economic interests between i and j, is equivalent to changing the relationship between i and j—the weight of possible forwarding. The initial propagation probability is set to 0.2~0.8 and is not affected by other conditions, and the forwarding probability distribution is finally obtained. As shown in Figure 1 and Figure 2 with Table 1 and Table 2.

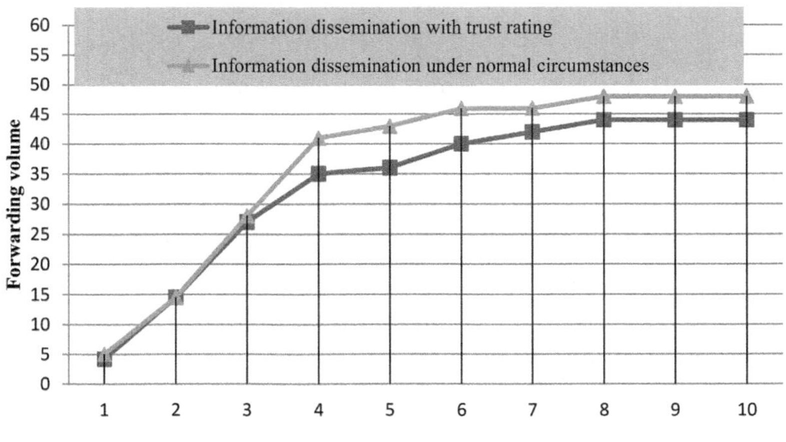

Figure 1. Influence of trust degree on user forwarding volume.

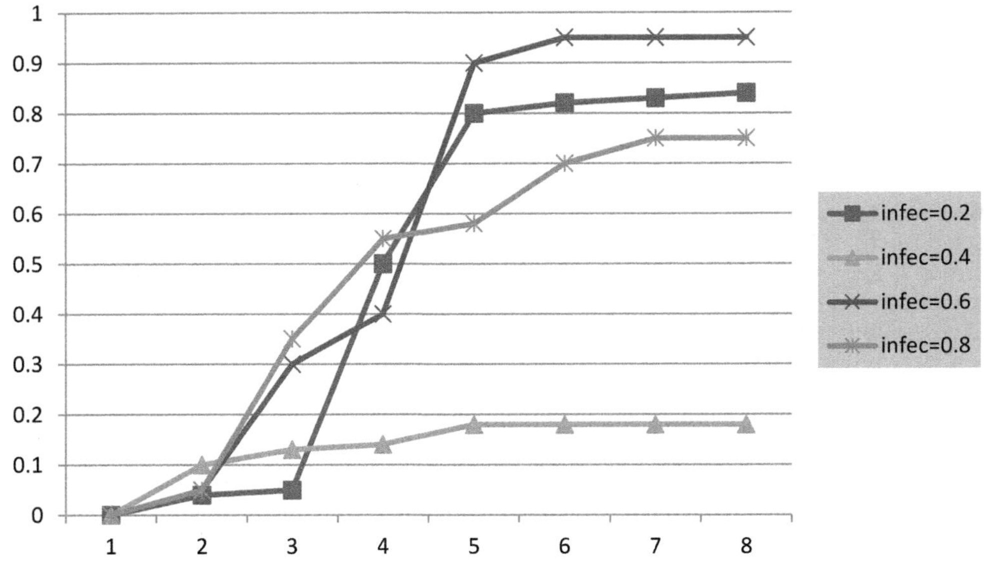

Figure 2. Impact of economic incentives on the final forwarding probability.

Table 1. Influence of trust degree on user forwarding volume.

Serial No	Serial No Trust Rating	Information Dissemination Normal Circumstances
1	4.2	5
2	14.5	14.5
3	27	28
4	35	41
5	36	43
6	40	46
7	42	46
8	44	48
9	44	48
10	44	48

Table 2. Impact of economic incentives on the final forwarding probability.

Serial	infec=0.2	infec=0.4	infec=0.6	infec=0.8
0	0	0	0	0
2	0.04	0.1	0.05	0.05
4	0.05	0.13	0.3	0.35
6	0.5	0.14	0.4	0.55
8	0.8	0.18	0.9	0.58
10	0.82	0.18	0.95	0.7
12	0.83	0.18	0.95	0.75
14	0.84	0.18	0.95	0.75

As can be seen from Figure 1 above, other conditions do not change. Comparing the traditional propagation model and the topic forwarding volume added to the trust evaluation model, the depressed user‘ forwarding volume is low at the initial moment of propagation. With the propagation as time increases, the forwarding volume of depressed user gradually increases. Still, it is always smaller than the forwarding volume of the traditional model, indicating that user trust plays a vital role in forwarding information. Since the trust evaluation in the Blockchain social network is rational after depressed user reach the saturation of the forwarding volume, the model that adds the trust evaluation will have the less forwarding book.

As shown in Figure 2 above, when depressed user have the same economic benefits of spreading, when the weights that depressed user may forward are unified, as the initial spreading probability increases, the forwarding probability of user's increases and the growth rate also increases gain. It shows that the economic benefits generated by depressed user in the dissemination of Blockchain social network information are also an essential part of affecting the probability of depressed user' forwarding.

5.4 *The addition of Blockchain contract nodes*

Infectious disease and improved models are widely used to describe the information dissemination process. Still, through detailed analysis of the actual information dissemination process, it is found that the infectious disease model cannot fully represent all the states of depressed user in the process of information dissemination. Compared with traditional social networks, depressed user in Blockchain social networks are generally rational. First, the emergence of smart contracts makes depressed user need to abide by the rules of the agreement fairly and equally, making them more transparent; second, economic incentives make depressed user more enthusiastic about the forwarding of high-quality information; third, the immutability of the Blockchain increases The cost of default, once the distorted information is spread, the user's trust will be affected. Therefore, the user status in the Blockchain social network changes as the news spreads. In this paper, the enthusiasm for disseminating target information is generally high. The critical state node, when it can more objectively analyze the received data and then judge its dissemination strategy (forwarding or not forwarding), is called the contract state node. Assuming that the change curve of user trust degree over time in the Blockchain social network is continuous, the node behavior characteristic curve g(w, u) can be obtained by fitting the least-squares method [17].

$$g(w,u) = \lambda g(w,u)_E + (1-\lambda)g(w,u)_S \tag{8}$$

Among them, $g(w,u)_E$ represents the local trust degree of user w at time u, and $g(w,u)_S$ represents the global trust degree of user w at time u. Find the first derivative of g(w, u),in u∈ [u_1 , u_2], if and only if $g'(w,u) > 0$, the node v is a monotonically increasing function in the time domain [u_1 , u_2] , the node is in the forwarding state; if and only if $g'(w,u) < 0$, the node v is a monotonically decreasing function in the time domain [u_1, u_2], the node is non-forwarding.

According to the user's reputation value data in Section 2.1, the user characteristic curve is fitted, then set x = by + c, u = 20, and the coefficient equation of the solution function is:

$\begin{cases} \varnothing_{11}b + \varnothing_{12}c = g_1 \\ \varnothing_{21}b + \varnothing_{22}c = g_2 \end{cases}$, in $\varnothing_{11} = \sum_{i=1}^{20} u_i = 20$, $\varnothing_{12} = \sum_{i=1}^{20} u_i y_i = 11$, $\varnothing_{12} = \varnothing_{22}$, *similarly* $\varnothing_{23} = 3.1$

Solving for b = 15, c = 5.375.

Since, in the case of the entire network, the user's reputation value will not grow all the time [3], the fitting function is $g = \begin{cases} 15 + 5.375u, u < u_d \\ D, u > u_d \end{cases}$, according to the actual data, the initial trust degree is 30 , u_d = 9, D = 73, thus the network fitting curve is obtained as shown in Figure 3 with Table 3.

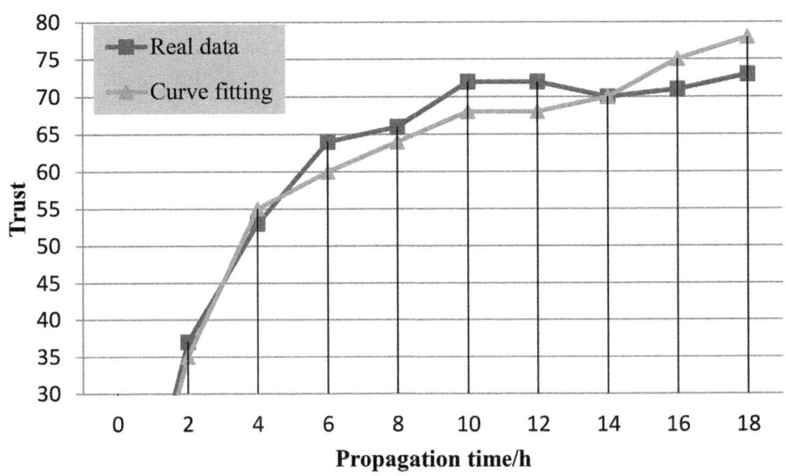

Figure 3. Comparison of data fitting curves.

Then, to facilitate the analysis of the introduction of contract nodes, only one target information communicator is set during the experiment, and the fitting function is normalized. At each time step, all nodes in the network interact simultaneously for information dissemination. During the simulation, an individual in the network can only influence one of its neighbor nodes at each time step. In addition, this paper sets the information propagation probability in the network to 0.5.

Table 3. Comparison of data fitting curves.

Serial	Real data	Curve fitting
0	0	0
2	37	35
4	53	55
6	64	60
8	66	64
10	72	68
12	72	68
14	70	70
16	71	75
18	73	78

As shown in Figure 4 with Table 4, after six-time steps are simulated, the proportion of forwarding depressed user in the network can be as high as 90%. Before adding intelligent nodes for target forwarding, the balance of forwarding depressed user after judgment is 65%. The user's forwarding does not judge the information in the traditional network. It is only a unified forwarding. The blockchain social network has a better effect on controlling the target information dissemination after adding intelligent node judgment.

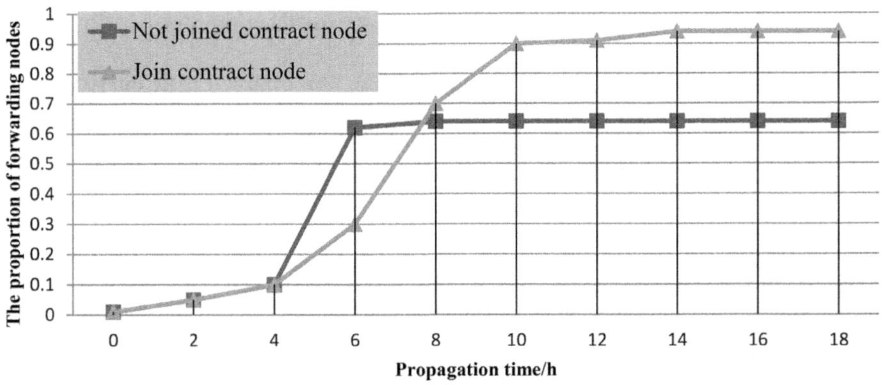

Figure 4. User forwarding is affected by node status.

Table 4. User forwarding is affected by node status.

Serial	Not joined contract node	Join contract node
0	0.01	0.01
2	0.05	0.05
4	0.1	0.1
6	0.62	0.3
8	0.64	0.7
10	0.64	0.9
12	0.64	0.91
14	0.64	0.94
16	0.64	0.94
18	0.64	0.94

By analyzing the above problems, this paper improves the evaluation method of user trust in the Blockchain social network. It quantifies the value generated by the economic benefit mechanism to optimize the forwarding probability by using user behavior and finally regulating and restricting information dissemination. On the other hand, this paper proposes that the propagation with intelligent nodes considers depressed user‘ rationality in the forwarding of target information, etc., which makes the user status in the Blockchain social network information propagation model more comprehensive and closer to reality.

6 CONCLUSION

Based on the SIR model for infectious disease dynamics, this paper makes sure that the chances of someone passing on a disease are the best they can be. Economic incentives make this paper even better. They add user status nodes and build a Blockchain social network information dissemination model, making it even better. The model can be built in a way that truly shows how social network information spreads in a Blockchain environment. It is based on how much trust the user has in the local area and how much trust the user has in the world. The value of the economic incentive can be precisely measured. To make sure the model is accurate, this paper uses an artificial network as the base map for information dissemination, runs simulations of the model, and talks about the model's stability and sensitivity in the process of information dissemination. It also looks at how changes in the model's parameters affect the dissemination process. The results show that: (1) the model can better describe how Blockchain social network information is spread; (2) the level of trust and the economic incentives that depressed users offer will have a big impact on which information is forwarded; and (3) the dissemination of distorted information will be difficult. It can be kept together, highlighting high-quality resources. However, the model used in this paper is still not good enough. Users can think about their bifurcation research in the user state node, which is where they are now.

REFERENCES

[1] N. Chen and D.S. -Y. Cho, "A blockchain based autonomous decentralized online social network," *2021 IEEE International Conference on Consumer Electronics and Computer Engineering (ICCECE)*, (2021), pp. 186–190, https:doi.org/10.1109/ICCECE51280.2021.9342564.

[2] B. Guidi and A. Michienzi, "Depressed user and Bots behaviour analysis in Blockchain Social Media," *2020 Seventh International Conference on Social Networks Analysis, Management and Security (SNAMS)*, (2020), pp. 1–8, https://doi.org/10.1109/SNAMS52053.2020.9336553.

[3] Subhasis Thakur; John G. Breslin, "Decentralized content vetting in social network with blockchain," in *Wireless Blockchain: Principles, Technologies and Applications*, IEEE, 2022, pp.269–296, doi: 10.1002/9781119790839.ch12.

[4] X. Liu, F. Jiang and R. Zhang, "A new social user anomaly behavior detection system based on blockchain and smart contract," *2020 IEEE International Conference on Networking, Sensing and Control (ICNSC)*, (2020), pp. 1–5, doi: 10.1109/ICNSC48988.2020.9238118.

[5] P. Jia and C. Yin, "Research on the characteristics of community network information transmission in blockchain environment," *2019 IEEE 4th Advanced Information Technology, Electronic and Automation Control Conference (IAEAC)*, (2019), pp. 2296–2300, doi: 10.1109/IAEAC47372.2019.8997742.

[6] K. Fan *et al.*, "A secure and verifiable data sharing scheme based on blockchain in vehicular social networks," in *IEEE Transactions on Vehicular Technology*, vol. 69, no. 6, pp. 5826–5835, June (2020), doi: 10.1109/TVT.2020.2968094.

[7] Y. Yahiatene and A. Rachedi, "Towards a blockchain and software-defined vehicular networks approaches to secure vehicular social network," *2018 IEEE Conference on Standards for Communications and Networking (CSCN)*, (2018), pp. 1–7, doi: 10.1109/CSCN.2018.8581756.

[8] W. Yang, X. Dai, J. Xiao and H. Jin, "LDV: A lightweight DAG-based blockchain for vehicular social networks," in *IEEE Transactions on Vehicular Technology*, vol. 69, no. 6, pp. 5749–5759, (2020), doi: 10.1109/TVT.2020.2963906.

[9] S. Salim, B. Turnbull and N. Moustafa, "A blockchain-enabled explainable federated learning for securing internet-of-things-based social media 3.0 networks," in *IEEE Transactions on Computational Social Systems*, doi: 10.1109/TCSS.2021.3134463.

[10] M. Saad, A. Ahmad and A. Mohaisen, "Fighting fake news propagation with blockchains," *2019 IEEE Conference on Communications and Network Security (CNS)*, (2019), pp. 1–4, doi: 10.1109/CNS.2019.8802670.

[11] A.D. Dwivedi, R. Singh, S. Dhall, G. Srivastava and S.K. Pal, "Tracing the source of fake news using a scalable blockchain distributed network," *2020 IEEE 17th International Conference on Mobile Ad Hoc and Sensor Systems (MASS)*, (2020), pp. 38–43, doi: 10.1109/MASS50613.2020.00015.

[12] S. Thakur and J.G. Breslin, "A model of decentralized social internet of things using blockchain offline channels," *2020 Second International Conference on Blockchain Computing and Applications (BCCA)*, (2020), pp. 115–121, doi: 10.1109/BCCA50787.2020.9274459.

[13] G. Renat, A. Peresichansky, A. Belenov and A. Barger, "Karma - blockchain based charity foundation platform," *2021 IEEE International Conference on Blockchain and Cryptocurrency (ICBC)*, (2021), pp. 1–2, doi: 10.1109/ICBC51069.2021.9461072.

[14] B. Guidi, A. Michienzi and L. Ricci, "Steem Blockchain: Mining the Inner Structure of the Graph," in *IEEE Access*, vol. 8, pp. 210251–210266, (2020), doi: 10.1109/ACCESS.2020.3038550.

[15] S. Zeng, Y. Yuan and F. -Y. Wang, "A decentralized social networking architecture enhanced by blockchain," *2019 IEEE International Conference on Service Operations and Logistics, and Informatics (SOLI)*, (2019), pp. 269–273, doi: 10.1109/SOLI48380.2019.8955104.

[16] L. Jiang and X. Zhang, "BCOSN: A blockchain-based decentralized online social network," in *IEEE Transactions on Computational Social Systems*, vol. 6, no. 6, pp. 1454–1466, (2019), doi: 10.1109/TCSS.2019.2941650.

[17] L. Li, A. Arab, J. Liu, J. Liu and Z. Han, "Bitcoin Options Pricing Using LSTM-Based Prediction Model and Blockchain Statistics," *2019 IEEE International Conference on Blockchain (Blockchain)*, (2019), pp. 67–74, doi: 10.1109/Blockchain.2019.00018.

[18] Yilmaz, T., and Ulusoy, Ö. (2023). Modeling and mitigating online misinformation: A suggested blockchain approach. *arXiv preprint arXiv:2301.12345.*

[19] Gruzd, A., Saiphoo, A., and Mai, P. (2024). The lure of decentralized social media: Extending the UTAUT model for understanding users' adoption of blockchain-based social media. *PLoS One*, 19(8), e0308458. https://doi.org/10.1371/journal.pone.0308458

[20] Jeong, U., Ng, L.H.X., Carley, K.M., and Liu, H. (2025). Navigating decentralized online social networks: an overview of technical and societal challenges in architectural choices. *arXiv preprint arXiv:2503.06789.*

[21] Stoykov, P. (2023). *Blockchain Social Media: The Rise of Decentralized Social Platforms.* Chainstack Blog. https://chainstack.com/blockchain-social-media/

[22] Chainstack. (2023). *Decentralized Social Networks.* Retrieved from https://chainstack.com

Smart Systems and Beyond – Bhatia et al. (Eds)

SWeAT: Bridging Reproducibility, Explainability, and Scalability in Software-Based Remote Attestation

Vasu Dev, Sukhdeep Kaur, Pooja Sahni, and Rachna Manchanda
Department of Electronics and Communication Engineering, Chandigarh Engineering College-CGC Landran

Raj Singh
Visteon Corporation

ABSTRACT: The explosive growth of Internet of Things (IoT) deployments has of guaranteeing that endpoint devices remain trustworthy before interacting with sensitive cloud infrastructures. However, this widespread adoption has simultaneously heightened security risks, particularly the difficulty, Remote attestation has long been recognized as a key defense, yet existing strategies continue to face persistent limitations. and remain costly, while hybrid and emerging designs often fail to deliver replicability or clear auditability. Software-based solutions are lightweight but prone to instability, hardware-assisted mechanisms offer stronger protection but lack transparency mechanism that integrates manifest-driven validation with deterministic decision-making to achieve both reproducible and interpretable attestation. To overcome these challenges, this article introduces SWeAT (Software-only, reproducible, and explainable Attestation Tool), a reliable detection accuracy, operational clarity, and the capacity to scale efficiently across. large IoT–Cloud ecosystems. A Python prototype was engineered and assessed under benign as well as adversarial conditions, demonstrating scientifically reproducible and operationally transparent attestation for next-generation_IoT-security.

Keywords: IoT security, Remote attestation, Reproducibility, Explainability, Cloud integration, SWeAT framework

1 INTRODUCTION

The modern era, linking billions of heterogeneous devices that gather, process, and exchange data in real time. The Internet of Things (IoT) has rapidly evolved into one of the defining technological shifts of as varied as medicine, manufacturing, agriculture, logistics, transportation, and environmental monitoring., This interconnection of ordinary objects with industrial systems has reshaped domains 75 billion devices, collectively producing immense quantities of data on a scale measured in zettabytes., Forecasts suggest that by the end. of the current decade, global IoT deployments will surpass decision-making, and remote operations, they also bring new layers of security concerns. This accelerate capabilities of the digital transformation enabling it for predictive analytics, autonomy over sensitive datasets or critical remote computing infrastructures.

The innate limitations of IoT approach escalate the challenges of security. One of these challenges is to ensure that end devices are trusted before allowing them to access storage and power, so as not to bypass traditional security measures. Many devices are designed with poor computing power, keeping them at risk of abuse, malware, and unauthorized modifications.

To counter the lack of trust in IOT ecosystem remote authentication has been emerged as a promising measure. So , we need to build such trustworthy technology which can verify authenticity and accuracy of the device against the untrusted IOT device before those device can, and force them to provide verifiable evidence of their current state.

before latter do so requiring it provide proof of its current state. To counter the lack of trust in IOT ecosystem remote authentication has been emerged as a promising measure.

Software authentication (SBA) methods rely on memory search and secure zones, offering greater robustness at the expense of complexity, cost, and openness. Hardware authentication (HWA) uses trusted components such as "TPM, ARM TrustZone, or Intel SGX"[5] and integration with cloud infrastructure, machine learning-based anomaly detection, and blockchain-based audit trails has been explored.

Hybrid approaches helps us to merge the robustness of two paradigms and at the same time develops methods to address these issues. The paper presents **SWeAT(which is a programmable, reliable, reproducible, and a verification tool)**, that was designed to provide light weight verfications, auditable, and deterministic, despite it various contributions there remians some unsolved challenges like scalability is limited, issues of interpretability, and of reproducibility in various cloud environment. To manage proprietary hardware or opaque learning algorithms, SWeAT employs an inventory-based mechanism.

DOI: 10.1201/9781042026289-27

Trusted repositories stores manifest attributions compared against time series model on which many other previous techniques counted, each node creates a cryptographic digest of some key files and configures the elements. The process of decision making is structured in a audit log which ensure s clear interpretation that prioritize accuracy, low computational log, also prioritizes the accuracy and scalability of devices in a IoT system. A functional prototype implemented in Python has been evaluated in various scenarios, demonstrating the strategy's robustness and reliability.

2 LITERATURE REVIEW

Remote authentication has been a hot topic for over two decades, and research has evolved in various ways: from software and hardware-based approaches to hybrid approaches, cloud-assisted approaches, and new data-driven approaches. These paradigms all offer important insights, but none of them can fully resolve the conflict between refactoring, definition, and scalability, which is crucial for the Internet of Things (IoT) and cloud computing.

Researchers have been working on so-called hardware-based authentication (HWA). The architecture of SMART produce a trust that build a status of devices which can guide future solutions of industries like ARM "TrustZone[5], Intel SGX [6] and Trusted Platform Modules (TPM)" [7]. These mechanics done away with the dependence on timing assumptions and provided stronger integrity guarantees, despite the strengths these technologies have high supply chain, hardware and power cost associated with them.

Over coming these imbalance an effort to achieve hybrid solution which accommodate lightweight software testing with hardware. For example, TrustLite and HYDRA [8] investigated microkernel-based isolation assistance schemes DIAT [9] and SlimIoT [10] investigated distributed or group authentication on clusters of devices with minimal hardware expansion.

The growth of large-scale IoT deployments has drawn attention to cloud-based authentication. Academic platforms such as SMARTER [11] distribute the authentication workload across cloud services, while commercial solutions such as AWS IoT Device Defender [12] and Azure IoT Hub [13] integrate features such as certificate configuration and anomaly detection. These systems offer scalability and simplified overall management, but they shift responsibility to external providers and often rely on heuristics.Meanwhile, some research has explored machine learning (ML)-based methods that use anomaly detection to infer machine certainty. Kim *et al.*[14] demonstrated the application of deep learning to telemetric data, such as processor usage and network traffic data, while broader research[15] focused on the adaptability of machine learning to evolving threats. We have observed that machine learning systems are highly flexible ,the result they generate can vary depending on the amount of data and the training environment, which can affect reproducibility.

Recently, researchers have proposed blockchain-assisted authentication based on Satoshi Nakamoto's original design [16].These systems ensure tamper-proof audit trails and decentralized trust by recording proof of device integrity in an immutable ledger [17–19]. However, blockchain imposes high storage, latency, and consensus requirements, which are incompatible with IoT applications that operate in real-time or have limited resources.

However there is a gap between these two paragons. HWAs offer high robustness but at the sacrifice of financial transparency and accessibility[4–7]. On the other side SBAs offer fragile timing assumptions[1–3]. Hybrids systems introduce complexity but also ensure improved flexibility [8–10].Cloud solutions increase manageability, but compromise determinism [11–13].Blockchain based ecosystems enhance the verifiability but fails drastically in providing transparent inferences[16–19].ML weakens repeatability and interpretability [14,15].

In this way, SWeAT framework standup itself by enabling a manifest-driven validation model, it avoids dependence on timing, proprietary hardware, or probabilistic classifiers. Each device node generates deterministic measurements that are compared to reliable reference values, and each attestation generates structured protocols that document the decision-making process. SWeAT combines three major factors reproducibility, explainability, and scalability—that existing paragon do not in unison offer.

3 IMPLEMENTATION AND EXPERIMENTAL EVALUATION

The SWeAT framework was implemented by creating a functional prototype in Python 3.11. The choice of this language was no accident: Python is highly regarded for its clarity, portability, and rich ecosystem of libraries that support cryptographic operations, data manipulation, and networking. Python allows you to easily replicate prototypes in different environments, which is a key factor in meeting scientific reproducibility requirements.

This prototype reflects the architecture described above and consists of three modules: the client engine, the validator, and the manifest repository. The device client is deployed directly on the IoT node and is responsible for calculating the inventory that captures the state of the system. This inventory is built using SHA-256 digests of selected firmware and configuration files. To simulate a real competitive environment, the client includes a fake schema that intentionally replaces the correct summary.

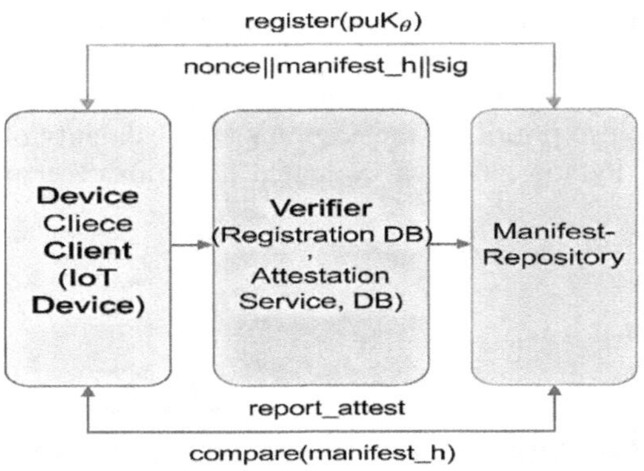

Figure 1. Sequence diagram of the SWeAT attestation protocol.

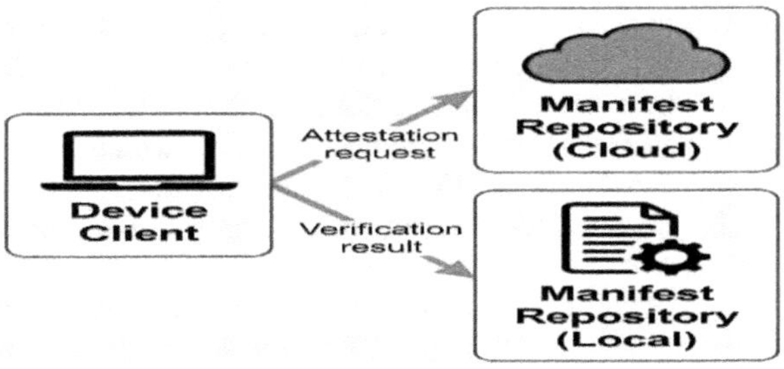

Figure 2. Device attestation with manifest repository.

The validation module embodies the trusted entity in the architecture. It issues a challenge based on random numbers, checks the returned list against reference records stored in the list store, and records all results in a structured log. Importantly, these logs not only record the binary decision of accept or reject, but also the random number, the resulting comparison, and the chain of logic that led to the result. This transparency ensures that every decision is open to interpretation and review.

The Manifest Repository uses a JSON-based database implementation, a wise choice given the lightweight and widely supportable nature of JSON structures. As the source of truth for authoritative device status, this repository helps us while ensuring recoverability We need to maintain a list of reliable references, ensures that the same instruments produce the same results under the same conditions, fulfilling the standards of verifiable scientific evaluation.

Below is the complete code for the client-side toolkit, which includes basic operations such as manifest generation, random number concatenation, and attacker impersonation. Each element in the code has a different role: the generate_manifest function goes through critical files and creates a SHA-256 digest; if spoofing mode is enabled, it replaces the firmware digest to simulate tampering.The main function validates the nonce provided by the validator, adds the device ID, and returns a complete list in JSON format.

The,experimental setup was carefully designed to simulate a real-world IoT cloud deployment environment. The demonstrator runs on a workstation with an "Intel Core i7 processor"and 16 GB of memory, running Ubuntu 22.04. We simulated Internet of Things (IoT) devices using a Raspberry Pi 4 development board equipped with a quad-core Cortex-A72 processor, 4GB of memory, and the Raspbian operating system. To test scalability, we ran virtual clients on the same workstation and simulated up to 5,000 devices simultaneously.

4 RESULT

Performance, scalability, reproducibility, correctness -these were the major parameters across which result were observed. Validator steadily generates the valid result when measured over one hundred validation cycle across the entire machine,whereas conditions failed drastically when computed under ideal simulation conditions.This demonstrates reproducibility and proves correctness of the construction, the same state of machine consistently generates the same result.

Physical Raspberry Pi device performance analysis showed that the average authentication latency was approx 48 millisecond, whereas virtual client machine running on work station presistently observed latency below 10 millisecond. Even with 5,000 concurrent clients, the system maintains an average response time of less than 120 milliseconds. Large-scale deployments without incurring significant processing or communication are

```python
import hashlib
import json
import time
import sys

def generate_manifest(fake_mode=False):
    manifest = {}
    files = ['firmware.bin', 'config.cfg', 'metadata.info']

    for file in files:
        if fake_mode and file == 'firmware.bin':
            content = b"tampered firmware"
        else:
            content = b"original content"

        digest = hashlib.sha256(content).hexdigest()
        manifest[file] = digest

    return manifest

def main():
    nonce = sys.argv[1] if len(sys.argv) > 1 else str(time.time())
    fake_mode = sys.argv[2] == 'fake' if len(sys.argv) > 2 else False

    device_id = "device_001"
    manifest = generate_manifest(fake_mode)


    response = {
        "device_id": device_id,
        "nonce": nonce,
        "manifest": manifest
    }

    print(json.dumps(response))

if __name__ == "__main__":
    main()
```

generated ,these are the results supported by SWeAT‘s lightweight protocol. The Figures 3,4 explains these results in details.Figure 3 shows the output from a legitimate device, demonstrating how the generated list was compared to the list generated in a trusted repository to arrive at an authorization decision. These results demonstrate that SWeAT’s lightweight protocol can support.

```
Windows PowerShell
Copyright (C) Microsoft Corporation. All rights reserved.

Install the latest PowerShell for new features and improvements! https://aka.ms/PSWindows

PS D:\Downloads\sweat_prototype_files> python device.py
Generated keypair. Public key saved (base64) to device_public_key.b64
Register response: 200 {"status":"registered"}

Attestation result: 200 {"decision":"accept","token":"eyJkZXZpY2UiOiAiZGV2aWNlLXJhc3BpLTAwMSIsICJleHAiOiAxNzU2Mzg1MDg1fQ
=="}

PS D:\Downloads\sweat_prototype_files>
```

Figure 3. Attestation output from a legitimate device, showing a correct manifest and an Accept decision.

The following image observed the output of deterministic rejection due to a firmware digest mismtach of the embedded device running in unauthorized mode.Figure 4 shows a snippet from the validator log, which records not only the yes/no results but also the nonce verification.

Several important conclusions can be drawn from these results. First, the accuracy and repeatability of SWeAT are fundamentally different from probabilistic methods (such as machine learning-based anomaly detection), which

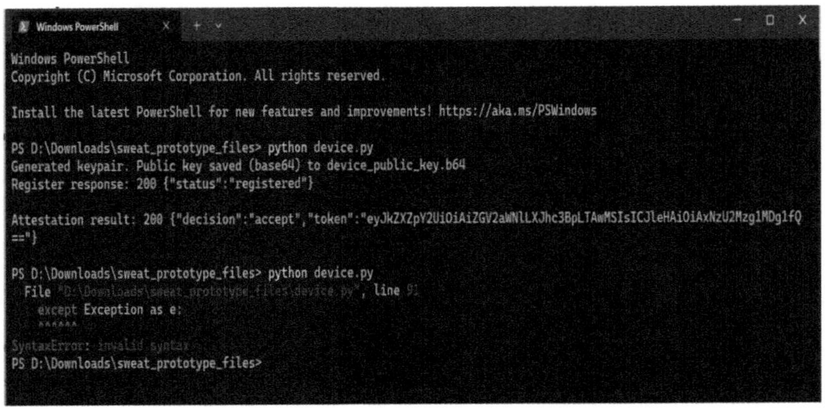

Figure 4. Attestation output from a tampered device (fake mode), showing a mismatched manifest and a Reject decision.

cannot guarantee identical results under identical conditions. Second, the transparency of SWeAT's structure is significantly different from instrument-based and hybrid models, whose results are often difficult for operators to understand.

5 LIMITATIONS

The paper highlights many key aspects .Starting with reproducibility and robustness of SWeAT's which are inherently discrete from traditional methods (like anomaly detection using machine learning), which does not give desired results under specific conditions. Succeeding, transparency of the framework is better than hardware-based models and hybrid models, as there results are generally opaque to the operators.

6 CONCLUSION

One need to verify the authenticity of the method, before it interact with cloud based platform. The escalation of Internet of things is the thrust for scalability, reproducibility, accountability. Our work presents SWeAT (interpretable and reproducible verification tool), an innovative framework that address long standing deficiencies by deploying a verification based on checklist. The foundation stone of the frame work verification is the Remote authentication, whereas other approaches still struggle—whether software-based, hardware-based, hybrid, cloud-based, or incorporating new technologies like machine learning and blockchain, SWeAT provides robust authentication results that can be replicated across all deployments. SWeAT does not rely on delicate clock patterns, expensive hardware modules, or fuzzy learning algorithms. Extensive evaluations in both good and bad scenarios show that it can reliably detect intrusion at very low cost and reach thousands of devices. Regular audit logs ensure transparency in the intelligence process, allowing operators and auditors to clearly interpret the results.

While system designs currently face many constraints, such as limited machine resources, centralized repositories, and "execution-only" attacks, the future research directions outlined will provide industry professionals with practical mechanisms for ensuring the reliability of the next generation of connected machines.

REFERENCES

[1] Seshadri, A., Luk, M., Perrig, A., and van Doorn, L. (2004). "SWATT: Software-based attestation for embedded devices." *IEEE Symposium on Security and Privacy*, pp. 272–282.

[2] Seshadri, A., Perrig, A., van Doorn, L., and Khosla, P. (2005). "Pioneer: Verifying code integrity and enforcing untampered code execution on legacy systems." *ACM Symposium on Operating Systems Principles*, pp. 1–16.

[3] Francillon, A., Perito, D., and Castelluccia, C. (2010). "Defending embedded systems against control flow attacks." *ACM AsiaCCS*, pp. 19–30.

[4] Seshadri, A., Perrig, A., van Doorn, L., and Khosla, P. (2006). "Securing embedded systems with software-based attestation." *Communications of the ACM*, 49(6), pp. 46–52.

[5] ARM Ltd. (2009). "ARM Security Technology: Building a Secure System using TrustZone Technology." ARM Technical Report.

[6] Costan, V., and Devadas, S. (2016). *"Intel SGX explained."* Cryptology ePrint Archive, Report 2016/086.

[7] Trusted Computing Group (TCG). (2011). *"TPM Main Specification."* Version 1.2.

[8] Koeberl, P., Schulz, S., Sadeghi, A.-R., and Varadharajan, V. (2014). *"TrustLite: A security Architecture for Tiny Embedded Devices." EuroSys Conference*, pp. 1–14.

[9] Ambrosin, M., *et al.* (2016). "DIAT: Distributed Attestation of Software Services." *ACM AsiaCCS*, pp. 230–243.

[10] Raza, S., *et al.* (2019). "SlimIoT: Scalable lightweight attestation protocol for IoT." *IEEE Transactions on Dependable and Secure Computing*, 16(5), pp. 813–825.
[11] Döring, M., *et al.* (2017). "SMArT: Secure and scalable remote attestation for distributed IoT systems." *IEEE PerCom*, pp. 1–9.
[12] Amazon Web Services. (2020). *"AWS IoT Device Defender."* [Online]. Available: https://aws.amazon.com/iot-device-defender/
[13] Microsoft Azure. (2020). *"Azure IoT Hub Security."* [Online]. Available: https://azure.microsoft.com/en-us/services/iot-hub/
[14] Kim, H., *et al.* (2018). "Deep learning-based anomaly detection for IoT networks." *IEEE Communications Magazine*, 56(9), pp. 42–47.
[15] Mohammadi, M., *et al.* (2018). "Deep learning for IoT big data and streaming analytics: A survey." *IEEE Communications Surveys and Tutorials*, 20(4), pp. 2923–2960.
[16] Nakamoto, S. (2008). "Bitcoin: A peer-to-peer electronic cash system." *Cryptology ePrint Archive.*
[17] Dorri, A., Kanhere, S.S., and Jurdak, R. (2017). "Blockchain in Internet of Things: Challenges and solutions." *IEEE Computer*, 50 (9), pp. 76–81.
[18] Christidis, K., and Devetsikiotis, M. (2016). "Blockchains and smart contracts for the Internet of Things." *IEEE Access*, 4, pp. 2292–2303.
[19] Novo, O. (2018). "Blockchain meets IoT: An architecture for scalable access management in IoT." *IEEE Internet of Things Journal*, 5(2), pp. 1184–1195.

Smart Systems and Beyond – Bhatia et al. (Eds)

Localization of Mobile Nodes in Wireless Sensor Networks Using RGWO Algorithm

Prabhjot Singh, Sunny Vig, and Harvinder Singh
Department of Electrical Engineering, Chandigarh University

ABSTRACT: Wireless sensor networks (WSNs) can communicate and retrieve worldwide physical information. A WSN's primary performance problems consist of network lifetime, routing, schedule tasks, security, deployment, and localization. For the above challenges, numerous Computational Intelligence (CI) based schemes have been suggested in recent years to meet the desired performance levels in WSNs. With the help of CI based algorithms robust and autonomous node location (2D/3D) with minimum hardware requirements can be determined. The static target node location can be accurately examined. However, moving target node's location determination is quite a challenging task. Aim of the paper is to demonstrate the concept of introducing one anchor node and six virtual anchor nodes utilizing Randomized Grey Wolf Optimization (RGWO) Algorithm to find the position of the moving target nodes. The only node utilized to monitor the moving target nodes inside a network is named as anchor node. This single node projects in six distinct directions with 60 degrees each and is served as virtual anchor nodes, their projection in all directions has the same range and as early as the mobile target node lies under the reach of anchor node not less than three nodes are determined in order to determine 2D position. Based on simulation results, the proposed algorithm shows efficiency and precision in localization. Through the use of virtual anchor nodes, the line-of-sight issue is minimized in the proposed algorithm.

Keywords: Localization, WSN, CI, HPSO, PSO, BBO, FA, RGWO

1 INTRODUCTION

A Wireless Sensor Network (WSN) uses sensor nodes to determine their physical behavior in the real-world. Nodes deployed to this network are very small, low-cost and have limited computational capability [1]. In addition to monitoring environmental phenomena, sensors can measure physical phenomena like sound, temperature, as well as light, habitats, and underwater acoustics [1,2]. Additionally, WSNs have issues related to hardware, operating systems, deployment, location, Quality of Service (QoS), and network security. Location of sensor nodes is crucial, as sensor data is useless without suggesting where the event took place.

The most convenient method of reaching to a node is a global positioning system (GPS). But more sensor nodes with GPS feature [3] are not being deployed in a network as it becomes more expensive network. Various methods can be utilized to identify the position of nodes targeted. Few localization methods utilized Received Signal Strength Indicator (RSSI), Time of Arrival (ToA) as well as Time Difference of Arrival (TDoA) [4,5]. Likewise, a number of range free localization methods includes Multidimensional Scaling (MDS), the Ad-Hoc Positioning System (APS) as well as Distance Vector Hop (DV-Hop) [6,7]. A target node is determined based on an understanding of another node's position, whereas an anchor node is determined based on its position. Nodes in the target network can be estimated using the beacon message of the anchor node, but this requires increased power and cost of communication. If an algorithm predicts the incorrect place, this error will spread across the whole network. The location precision should therefore be high. To ensure high precision, the literature suggests the utilization of algorithms like meta-heuristic for location optimization.

In this paper, anchor nodes utilised to evaluate 2D coordinates of nodes already targeted that moves randomly in the area under consideration. If moving sensor nodes passes through the projection of a known anchor node, the distance between known anchor node having two virtual anchor nodes (at least three nodes are needed in 2D scenario) both apply RSS distance calculation technique in accordance with the known anchor node. To optimize the estimated location, centroid is calculated and then recently developed meta-heuristic optimization algorithm named as Randomized Grey Wolf Optimization (RGWO) Algorithm is applied for accurate determination of active target nodes.

 DOI: 10.1201/9781042026289-28

There are five main sections in this paper: a literature review on localization is demonstrated in Section 2. Section 3 provides an overview of Randomized Grey Wolf Optimization (RGWO) Algorithm as well as methodology utilized for localization. Section 4 describes how an anchor node is deployed in a sensor field. As a conclusion, Section 6 presents the simulation results as well as the discussion, followed by Section 5 which presents results of the simulation and discussion.

2 LITERATURE REVIEW

Most sensor networks consist of a few anchors. As a result of these anchor nodes, we know their location once they have been deployed. Cost and complexity constraints prevent us from knowing the locations of some nodes. The deployment of GPS capabilities on every node is not possible. A node's position can be evaluated by using the anchor node's position. The past few years have seen a number of localization approaches presented by researchers [8]. The role of distance measurements in estimating unknown nodes is explored in this paper, and a metaheuristic algorithm for optimizing the estimated locations in a mobility-based model is proposed.

Over the past several years, WSNs have made major developments in the field of wireless communications, as well as in the development of equipment for use in sensor networks covering many different fields [8,9]. Location of WSN nodes has become increasingly important for practical applications. Methods of localization include range-based as well as range-free techniques. TOA [10], TDOA, AOA, as well as RSSI [11,12] are all range-based methodologies for calculating distance or angle among adjacent nodes. Bulusu *et al.* [13] addressed the issue of the location of very small low-cost nodes without any specific range. RF-based localization was examined, with the recipient's location on Centroid taken as a reference point for the authors.

Lee proposed a new network structure that uses fewer anchor nodes. According to a recent method of evaluating distances, the smallest path is estimated using the variation in the path for estimating the distance in Euclidean terms. For anisotropic networks to be accurate, fewer anchors are needed. An innovative approach presented by Garg *et al.* [15] is based on RSSI to place anchor nodes on vertices as well as target nodes at four corners to identify the location of the mobile node. According to Grafenstein *et al.* [16], RSSI can be used to calculate distance between anchors and destinations. Additionally, once the distances among the anchor nodes as well as virtual anchor nodes had been acquired, the position was determined using a trilateration method. According to Sumathi and Srinivasan [17], the anchor location algorithm uses RSS-based least squares estimation to estimate the target node's location. In this paper, Guo *et al.* [18] proposes a method of localization dependent on mobility of perpendicular intersections (PIs).

According to Vijayakumar [19], particle size is updated by a localization algorithm that uses specific filters is suggested, and a simulation is conducted with on-site information to evaluate the algorithm's performance. Localization in real time can be accomplished using additional Kalman Filter-based algorithms [20].

There have been several CI techniques proposed for enhancing the precision and reducing the complexity of localization. Static scenarios typically involve Genetic Algorithms (GAs) as well as other stochastic algorithms. A PSO-based localization approach was presented by Kulkarni [21] as another approach to stochastic algorithms. The position of nodes in WSN environments can be optimized with latest computationally optimized global Swarm Intelligence (SI) algorithm created by Kumar and Jacob [22]. As a result of PSO's ease of use and low memory requirements, the authors suggested that it is an excellent option for situations with restricted resources. HPSO is a fast-convergence algorithm that optimizes the location of a randomly deployed target node. More precision can be achieved in solving localization problems using BBO [22,23]. For an anisotropic WSN, Kumar [24] recommends combining HPSO as well as BBO for range-free HPSO as well as range-free BBO. It is true that many meta-heuristic algorithms are competitive in the abovesaid-mentioned localization problem, but that they are weak in some areas, such as slow convergence as well as high localization errors.

This paper aims to develop a new technique for finding the precise location of moving destinations utilizing Randomized Grey Wolf Optimization (RGWO) Algorithm. An application of RGWO is presented here to optimize the accuracy as well as accelerate the convergence of moving target nodes in WSNs.

3 RANDOMIZED GREY WOLF OPTIMIZATION (RGWO) ALGORITHM

Optimization is fundamental to addressing practical challenges in the field of engineering, machine learning, operations research, and other domains. Among the many nature-inspired algorithms, the Grey Wolf Optimizer (GWO) has gained popularity due to its simplicity and effectiveness. However, like many metaheuristic algorithms, it sometimes gets trapped in local optima. To address this limitation, researchers have proposed variations such as the Randomized Grey Wolf Optimization (RGWO) algorithm, which introduces controlled randomness to enhance exploration and improve global search capability.

Grey wolves exhibit a well-organized leadership hierarchy as well as collaborative hunting behavior in nature. The GWO algorithm simulates these behaviours through mathematical models that enable search agents to converge toward the optimal solution.

Mathematical model

Encircling the Prey:

$$D = |C \cdot Xp(t) - X(t)| \tag{1}$$

$$X(t+1) = Xp(t) - A \cdot D \tag{2}$$

$$A = 2a \cdot r1 - a, \; C = 2 \cdot r2 \tag{3}$$

Here, a decreases linearly from 2 to 0 over iterations, and r1, r2 are random vectors in [0, 1].

Updating the position:

$$D\alpha = |C1 \cdot X\alpha - X| \tag{4}$$

$$D\beta = |C2 \cdot X\beta - X| \tag{5}$$

$$D\delta = |C3 \cdot X\delta - X| \tag{6}$$

$$X1 = X\alpha - A1 \cdot D\alpha \tag{7}$$

$$X2 = X\beta - A2 \cdot D\beta \tag{8}$$

$$X3 = X\delta - A3 \cdot D\delta \tag{9}$$

$$X(t+1) = (X1 + X2 + X3)/3 \tag{10}$$

Where

Alpha (α): Optimal solution so far.

Beta (β): Second-best candidate.

Delta (δ): Third-best candidate.

Omega (ω): Remaining candidate solutions.

Enhanced parameters

$$a(t) = 2 * (1 - (t/T)\hat{}\gamma) + \varepsilon * N(0,1) \tag{11}$$

Random Perturbations in Position

$$X(t+1) = X(t+1) + \lambda \cdot U(-1, \; 1) \tag{12}$$

The Randomized Grey Wolf Optimizer offers a robust improvement over the original GWO by introducing stochastic mechanisms that diversify the search and reduce the chance of premature convergence. Its simplicity, adaptability, and efficiency make it suitable for a broad range of optimization tasks.

4 SINGLE ANCHOR LOCALIZATION CONCEPT

According to previous sections, one anchor node is embedded to identify moving target nodes. Throughout the sensing field, target nodes are arranged randomly within grids. During localization, beacon signals are used to aid individual target nodes in locating one another. A reference node containing RSS information is retrieved once a target node descends within its range. After that, the anchor and target node are measured using Euclidean distance. When six virtual anchor nodes (in which a minimum of two are selected) are calculated, all six anchor nodes should fall within the same range in order to discover target nodes. Figure 1 presents the flowchart of the localization method using RGWO. Follow these steps to locate moving target nodes' positions.

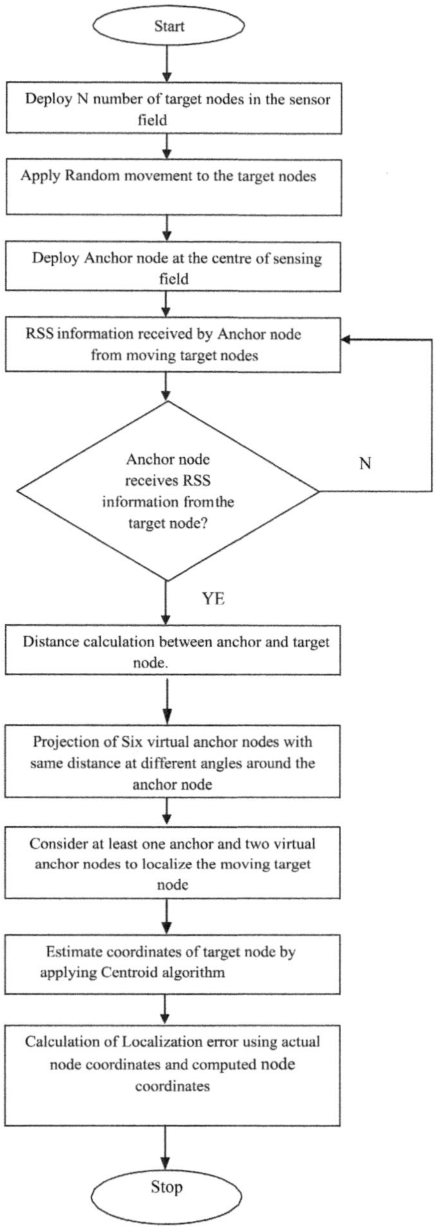

Figure 1. Localization Process using RGWO.

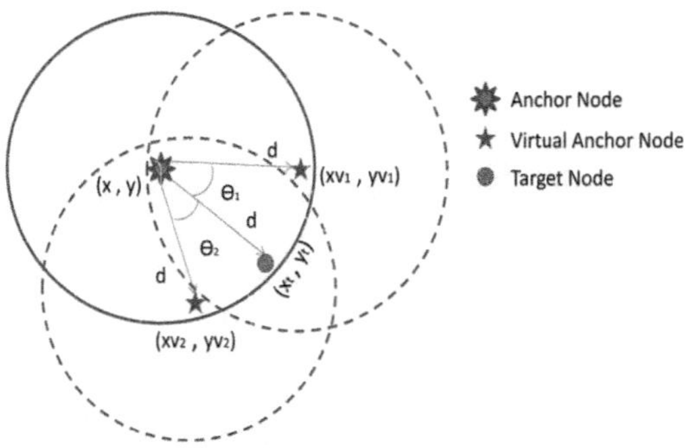

Figure 2. Target, Anchor and virtual nodes.

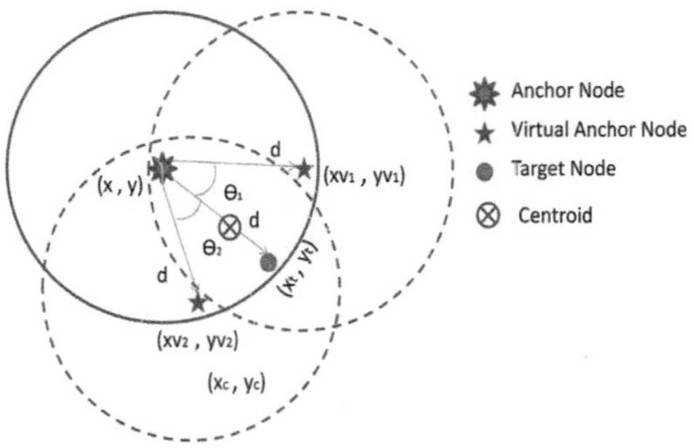

Figure 3. Centroid calculation (with anchor as well as virtual anchor nodes).

In order to minimise fitness function (15), each moving target node runs the RGWO to find its coordinates x_s, y_s (as shown in Figure 4) as follows

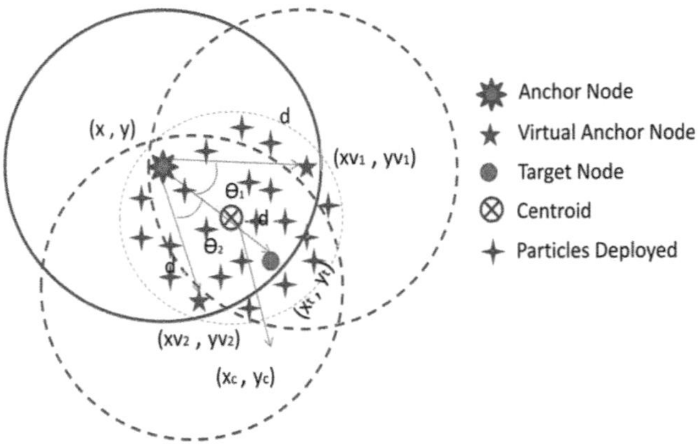

Figure 4. Deployment of RGWO particles (Pattern Solutions) around the centroid.

$$f(x_{s,}y_s) = \frac{1}{M}\sum\left(\sqrt{(x_e - x_i)^2 + (y_e - y_i)^2} - d_i\right)^2 \tag{13}$$

Here, x_e, y_e is calculated target node position, (x_i, y_i) the beacon node is placed in the neighbourhood as well as M demosntrates the number of beacons (M>3 in this work). The error location is evaluated by utilizing the mean node coordinate distance x_e, y_e square and the node x_i, y_t indicated in (14), displayed in the Figure 5.

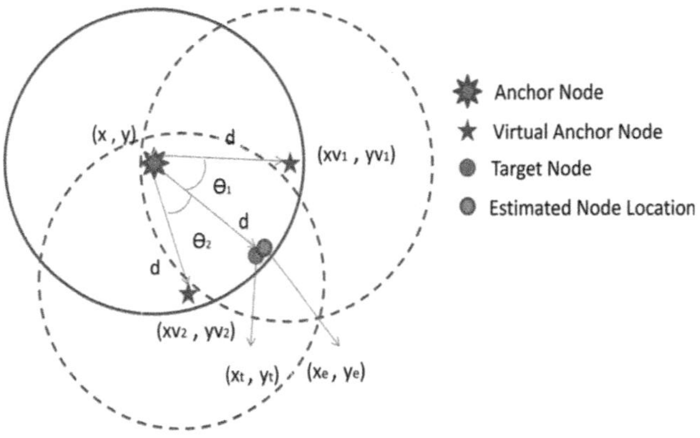

Figure 5. Estimated target nodes.

$$E_t = \frac{1}{N_L}\sum\sqrt{(x_e - x_t)^2 + (y_e - y_t)^2} \tag{14}$$

5 SIMULATION RESULTS AND DISCUSSIONS

The simulations were conducted in MATLAB utilizing PSO, HPSO, BBO, FA as well as using RGWO algorithms for 20 mobile target nodes and a single anchor node. Simulators were run on a MacBook Air i5 equipped with 4GB RAM. A 15x15 sensor grid was comprised of 15 targets and 15 anchors, with the anchor located in the centre. For each of the six directions, six anchors were strategically placed at 60-degree intervals. For 2D placement, three anchor nodes are required, with two virtual anchor nodes analyzed for evaluating the coordinates when the target node descends within the range of anchor node's. RSS values were considered when the target node was within the range of the anchor node. In order to mitigate LoS issues, virtual anchor nodes were used. Proposed RGWO algorithm needed to be evaluated through a large number of simulations. A 15x15 sensing area was used in all simulations, and additional parameters were provided in Table 1.

Table 1 represents the deployment of Anchor Nodes (AN) at the centre of 15x15 m^2 along with assumed virtual anchor nodes value at six different locations.

Table 1. Anchor as well as virtual anchor coordinates.

Co-ordinates	AN	VN_1	VN_2	VN_3	VN_4	VN_5	VN_6
X	7.5	9.213	4.134	3.213	3.543	8.453	10.412
Y	7.5	10.432	10.434	6.564	2.543	1.764	5.543

Table 2. Performance comparison among various algorithms: HPSO, PSO, BBO, FA, RGWO.

Technique Used	Movement	Transmission Range	Maximum Error (Localization)	Minimum Error (Localization)	Average Error	Target Nodes Localized
PSO	1	10	1.7321	0.1236	0.5766	20
	2	10	2.7321	0.3122	1.0323	20
	3	10	1.5432	0.3423	0.7564	20
	4	10	1.9876	0.2231	0.4342	20
	5	10	1.6753	0.1322	0.4312	20
HPSO	1	10	0.5421	0.2123	0.2431	20
	2	10	0.6542	0.0321	0.2343	20
	3	10	0.6534	0.0343	0.3211	20
	4	10	0.5411	0.1239	0.2343	20
	5	10	0.2543	0.1289	0.1987	20
BBO	1	10	1.2312	0.1234	0.3124	20
	2	10	1.3232	0.0766	0.7987	20
	3	10	1.3212	0.0123	0.6312	20
	4	10	1.4321	0.0234	0.5654	20
	5	10	1.3233	0.0212	0.6543	20
Firefly algorithm	1	10	4.2123	0.3432	2.1232	20
	2	10	3.4456	0.4567	3.3211	20
	3	10	4.3213	0.0123	2.4123	20
	4	10	5.4323	0.1234	3.0987	20
	5	10	4.3423	0.1234	2.3233	20
RGWO	1	10	1.3456	0.4512	**0.2520**	20
	2	10	0.7643	0.3543	**0.2121**	20
	3	10	0.4323	0.2324	**0.2014**	20
	4	10	0.6757	0.4112	**0.1432**	20
	5	10	0.3234	0.3575	**0.1408**	20

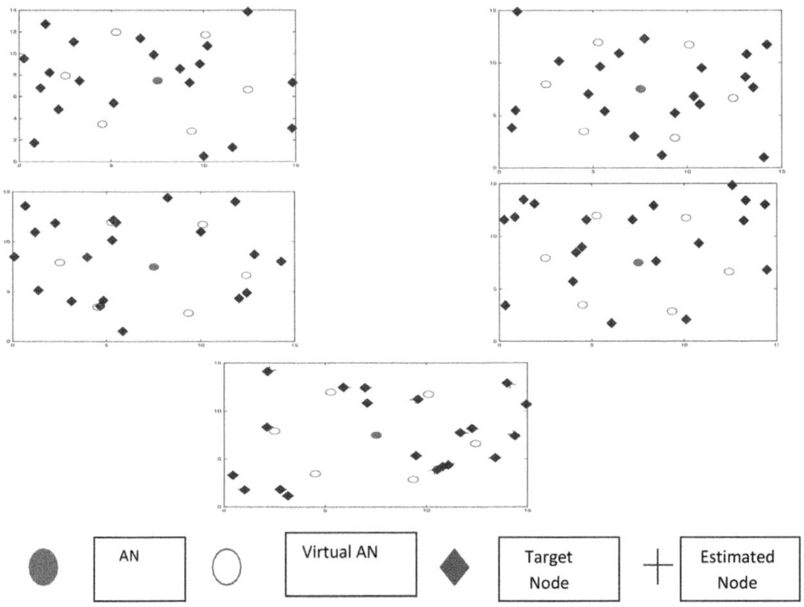

Figure 6. Estimated target nodes.

Figure 6 show results of RGWO optimization technique implemented to evaluate the localization error. Each optimization technique uses a "+" symbol to indicate the best estimated localized nodes in the fifth movement.

As presented in Figure 10, localization error for RGWO is quite small in comparison with other meta-heuristic techniques applied for the same problem. In this node is estimated quite perfectly with RGWO with lesser computational time.

6 CONCLUSION

WSNs face a number of challenges, including localization. Localization is attained utilizing single anchor node, which projects in six different directions around the circle. In order to determine a target's 2D position, two virtual anchor nodes as well as anchor node must be deployed in the centre (as a minimum of three nodes are required). PSO, HPSO, BBO, FA, and RGWO meta-heuristic optimizations are employed to calculate localization error. Comparing RGWO with competing algorithms, RGWO performed better in terms of average localization error.

Animal tracking, logistics, workers tracking in coal mines, as well as a variety of industrial applications can all be enhanced with the proposed algorithm. Localization can be achieved with this algorithm, allowing the nodes to keep up with their energy, thereby extending their lifespan. For improved accuracy as well as rapid convergence, a hybrid meta-heuristic algorithm will be developed over the coming period.

REFERENCES

[1] Karl, H., Willig, A.: *Protocols and Architectures for Wireless Sensor Networks*. Wiley, Hoboken (2007)

[2] Cheng, L., Wu, C.-D., Zhang, Y.-Z.: Indoor robot localization based on wireless sensor networks. *IEEE Trans Consum. Electron.* 57(3)

[3] Barsocchi, P., Lenzi, S., Chessa, S., Giunta, G.: A novel approach to indoor rssi localization by automatic calibration of the wireless propagation model. In: *IEEE 69th Vehicular Technology Conference, VTC Spring*, pp. 1–5. IEEE (2009)

[4] Chan, Y.-T., Tsui, W.-Y., So, H.-C., Ching, P.-C.: Time-of-arrival based localization under nlos conditions. *IEEE Trans. Veh. Technol.* 55(1), 17–24 (2006)

[5] Xu, E., Ding, Z., Dasgupta, S.: Source localization in wireless sensor networks from signal time-of-arrival measurements. *IEEE Trans. Sig. Process.* 59(6), 2887–2897 (2011)

[6] Gillette, M.D., Silverman, H.F.: A linear closed-form algorithm for source localization from time-differences of arrival. *IEEE Sig. Process. Lett.* 15, 1–4 (2008)

[7] Rong, P., Sichitiu, M.L.: Angle of arrival localization for wireless sensor networks. In: *3rd Annual IEEE Communications Society on Sensor and Ad Hoc Communications and Networks, SECON'06*, vol. 1, pp. 374–382. IEEE (2006)

[8] Pan, J.-S., Nguyen, T.-T., Dao, T.-K., Pan, T.-S., and Chu, S.-C. (2017). Clustering formation in wireless sensor networks—A survey. *Journal of Network Intelligence*, 2(4), 287–309.

[9] Nguyen, T.-T., Chu, S.-C., Dao, T.-K., Nguyen, T.-D., and Ngo, T.-G. (2017). An optimal deployment wireless sensor network based on compact differential evolution. *Journal of Network Intelligence*, 2(3), 263–274.

[10] Wu, L., Meng, M.Q.-H., Liang, H., and Gao, W. (2009). Accurate localization in combination with wireless sensor networks and laser localization. In *IEEE International Conference on Automation and Logistics*, (pp. 146–151). IEEE

[11] Awad, A, Frunzke, T., and Dressler, F. (2007). Adaptive distance estimation and localization in WSN using RSSI measures. In *10th Euromicro conference on digital system design architectures, methods and tools* (pp. 471–478). IEEE

[12] Di, G., *et al.* (2008). A distributed aoa based localization algorithm for wireless sensor networks. *Journal of Computers*, 3(4), 1–8.

[13] Bulusu, N., Heidemann, J., and Estrin, D. (2000). Gps-less lowcost outdoor localization for very small devices. *IEEE Personal Communications*, 7(5), 28–34.

[14] Lee, S., Park, C., Lee, M.J., and Kim, S. (2014). Multihop range-free localization with approximate shortest path in anisotropic wireless sensor networks. *EURASIP Journal on Wireless Communications and Networking*, 1, 1–1

[15] Garg, V., and Jhamb, M. (2013). Tracking the location of mobile node in wireless sensor network. *International Journal of Advanced Research in Computer and Communication Engineering*, 2(6), 2358–2362.

[16] Graefenstein, J., Albert, A., Biber, P., and Schilling, A. (2009). Wireless node localization based on rssi using a rotating antenna on a mobile robot. In *6th Workshop on Positioning, Navigation and Communication* (pp. 253–259). IEEE

[17] Sumathi, R., and Srinivasan, R. (2011). Rss-based location estimation in mobility assisted wireless sensor networks. In *6th International Conference on Intelligent Data Acquisition and Advanced Computing Systems (IDAACS)* (Vol. 2, pp. 848–852). IEEE.

[18] Guo, Z., Guo, Y., Hong, F., Jin, Z., He, Y., Feng, Y., *et al.* (2010). Perpendicular intersection: locating wireless sensors with mobile beacon. *IEEE Transactions on Vehicular Technology*, 59(7), 3501– 3509.

[19] Kim, K., and Lee, W. (2007). MBAL: A mobile beacon-assisted localization scheme for wireless sensor networks. In *16th International Conference on Computer Communications and Networks* (pp. 57–62). IEEE

[20] Vijayakumar, J.V., Zhang, H., Huang, Z., and Javed, A. (2013). A particle filter-based train localization scheme using wireless sensor networks. In: *2013 IEEE 11th international conference on dependable, autonomic and secure computing (DASC) 2013* (pp. 269–274). IEEE.

[21] Kulkarni, R.V., and Venayagamoorthy, G.K. (2011). Particle swarm optimization in wireless-sensor networks: A brief survey. *IEEE Transactions on Systems, Man, and Cybernetics Part C (Applications and Reviews)*, 41(2), 262–267.

[22] Gopakumar, A., and Jacob, L. (2008). Localization in wireless sensor networks using particle swarm optimization. In *IET International Conference on Wireless, Mobile and Multimedia Networks* (pp. 227–230). IET.

[23] Kumar, A., Khosla, A., Saini, J.S., and Singh, S. (2012) Metaheuristic range-based node localization algorithm for wireless sensor networks. In *IEEE international conference on localization and GNSS* (pp. 1–7). IEEE.

[24] Kumar, A., Khosla, A., Saini, J.S., and Singh, S. (2015). Rangefree 3d node localization in anisotropic wireless sensor networks. *Applied Soft Computing*, 34, 438–448.

[25] Sadollah, Ali Sayyaadi, Hassan Yadav, Anupam (2018): Artificial neural networks; Global optimization; Iterative convergence; Metaheuristics; Neural network algorithm. In *Applied Soft Computing Journal Elsevier B.V.* (pp. 747–782)

Smart Systems and Beyond – Bhatia et al. (Eds)

Design of Photonic Biosensors for the Detection of Spoiled Fruits and Vegetables

Sachin Kumar and Rohit Kumar
Department of Electrical Engineering, University Institute of Engineering, Chandigarh University Mohali, Punjab, India

Deepak Juneja
Panipat Institute of Engineering and Technology, Panipat, Haryana, India

Sunny Vig
Department of Electrical Engineering, University Institute of Engineering, Chandigarh University Mohali, Punjab, India

ABSTRACT: To enhance food quality monitoring, this article proposes a ring resonator sensor model designed to detect a wide range of rotten fruits and vegetables with high sensitivity and specificity, using their distinct refractive index values as a diagnostic marker. The model demonstrates a high quality (Q) factor, with optimized design parameters enhancing its sensing performance. Using the simulation with Finite Difference Time Domain (FDTD) technique, we modeled the proposed device to show clearly different resonant wavelengths for different produce types. This model is a promising label-free mechanism to detect spoilage, and showed an improved sensing capability.

Keywords: Photonic Crystal Ring Resonator (PhCRR), Bio-Sensor, Rotten fruits and vegetables detection

1 INTRODUCTION

Biosensors can be used to detect spoiled fruits and vegetables by measuring the shifts in refractive index associated with the decomposition of the produce. The biological components within a biosensor, such as enzymes and antibodies, will selectively bind to target molecules associated with decomposition. The sensor will generate a quantifiable response, indicative of spoilage, by binding these specific molecules to its surface. These hand-held units can be used to assess food quality in both a field and processing setting, providing benefits to food producers and consumers by detecting spoiled food, reducing food waste, and improving food safety.

Optical devices like Mach-Zander Interferometer, SAGNAC interferometer, as well as ring resonators have shown a promise for sensing a series of physical parameters. Owing to its limited size as well as speed, the ring is well suited to resonators applications. In recent years, there have been several devices based on photonic crystals for refractive index sensing applications. Tools demonstrate good selection, sensitivity and quality factor, it creates for good sensing capabilities. Finally, photonic crystal-based devices can offer accurate and efficient sensation of physical parameters.

Figure 1. Different stages of a fruit.

The refractory index can be used to differentiate between fresh and damaged fruits and vegetables. A sensor based on photonic crystals that can identify rotten food can be made using this transition in the refractive index.

DOI: 10.1201/9781042026289-29

The purpose of this research is to create a Photonic Crystal Ring Resonators (PHCRR) developed for bio sensing and detection application. To determine the difference between wasted food products and fresh vegetables and fruits, PHCRR will be tune for precise and selectivity. Overall, the proposed device to detect poor production will use an effective solution that will provide label-free and non-invasive detection.

The sensor presented in this work features a square rod design that is centered within a cavity, allowing for both fresh and decayed produce to be enclosed. The output spectrum of the device can be compared to identify differences for infected or decayed samples in comparison to normal cell-based analytics. The PhCRR sensor's quality factor (Q factor) and analysis of its sensitivity will be determined by observing dip spacing, and specifically conducted with decaying produce samples. The samples will be optimized for sensitivity and selectivity to specifically detect impaired fruits and vegetables based on changes in refractive index. In conclusion, the proposed device holds considerable potential for implementation in bio sensing and detection applications.

2 MODELING AND METHODOLOGY

The structure modeled on a mechanism of changing in intensity, and the model is a two-dimensional photonic crystal (PhC) utilized to model the biosensor. As a result, the model allows for the controlled transmission of high retention of light. The PBG in a photonic crystal is because of the periodically configuration of the dielectric material. Rods of dielectric materials are vertically placed periodically on thin dielectric surface to get a specific range of wavelengths that can pass through the photonic crystal (PhC) which is known as the photonic band gap (PBG). Efficiently regulation of light propagation ability is provided by PBG.

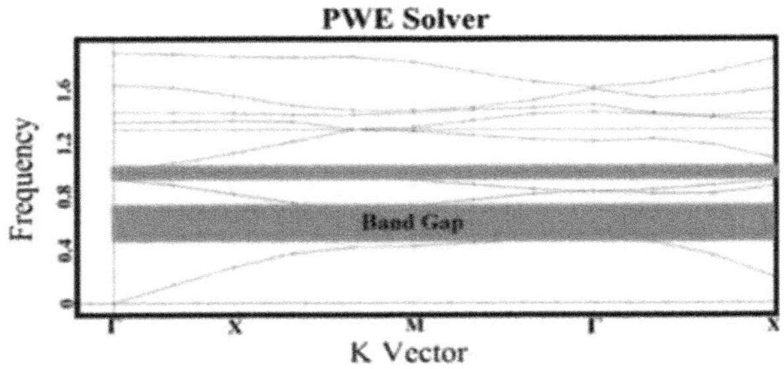

Figure 2. Optimized PBG diagram.

Three constructional parameters as refractive index (RI) of sample, rod to rod distance (a) and rod radius (r) can change the Photonic Band Gap (PBG). The PBG diagram represents a specific band of wavelengths that can travel through the Photonic Crystal (PhC). It can be simulated by Finite-Difference Time-Domain (FDTD) approach with the Plane Wave Expansion (PWE) method. Figure 2 represents the optimized PBG. Biosensing and detection accuracy depends on PBG optimization and controlling of the light inside the PhCRR. The photonic band gap (PBG) ranges of normalized frequencies for the transverse magnetic (TM) mode were computed, corresponding to intervals of thee wavelengths. Specifically, the PBG ranges were determined to span from 0.48 to 0.71, from 0.94 to 1.0, and from 1.2 to 1.3. To align with optical communication systems, a wavelength under the limits of 1500 nm to 1600 nm (corresponding to the 0.48 to 0.71 PBG range) was chosen for the light source. The primary design configurations are summarized in Table 1. These design parameters collectively ensure the efficient and effective operation of the biosensor within the specified wavelength range.

Table 1. Constructional design parameters.

S.No.	Designing Variables	Optimized Values
1	Configuration	2D Si Rods with air gap
2	Si rod radius (r)	0.110 μm
3	geometrical structure	cubic
4	Rod to Rod distance / Lattice const. (a)	0.5400μm
5	Si rod refractive index	3.4500
6	Reference RI	1.3500
7	Source wavelength	1.500 μm to 1.600 μm
8	Simulated PBG range	1395.40 nm to 2046.10 nm
9	Base size	45.00 μm2
10	Simulation technique	FDTD
11	Polarization	TM mode

The proposed structure was selected over alternative configurations due to its superior sensitivity. In particular, the ring resonator-based design was chosen for its compactness and greater efficiency compared to other Photonic Crystal (PhC)-based structures. To model and simulate the device, the Finite Difference Time Domain (FDTD) method was employed, as plotted in Figure 3. Novel design of PhCRR biosensor is responsible to detect the shift in resonance wavelength based on refractive index of fresh and spoiled fruits and vegetables

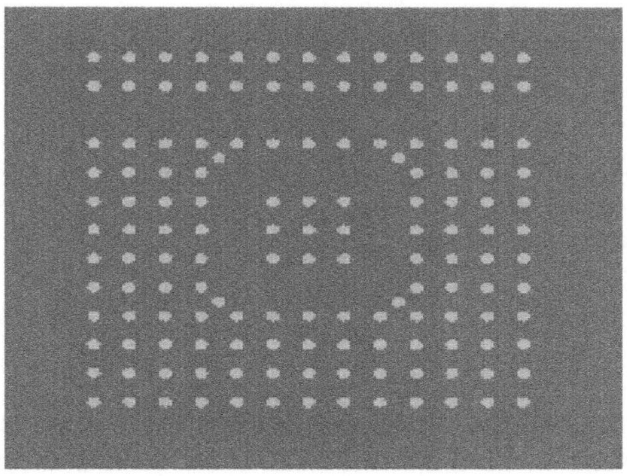

(a)

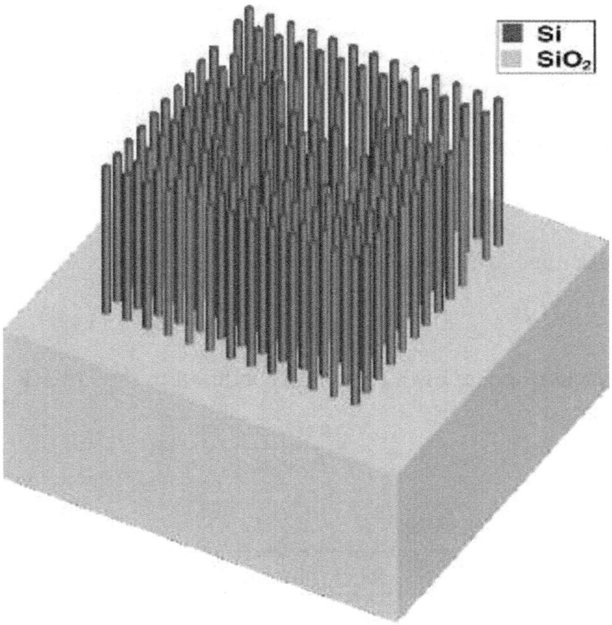

(b)

Figure 3. (a) Front view of the proposed model (b) Perspective view of the proposed model.

The proposed biosensor model includes both linear and circular line defect waveguides, which are capable to point the refractive index of various types of rotten fruits and vegetables. The device is monitored using a biosensor that detects variations in intensity. The modeling and simulation of the device is performed using the Lumerical FDTD simulator. The refractive indices of different types of fruits and vegetables are listed in Table 2, and the biosensor is able to accurately detect and differentiate between normal and rotten samples based on their distinct refractive index values.

3 RESULTS AND DISCUSSION

Figure 4 illustrates designing parameter optimization of the PhCRR sensor. Modification of the reference value of refractive index or direct immersion of sensor into the target sample causes a resonant wavelength shift, which subsequently influences the optical transmission response of the device.

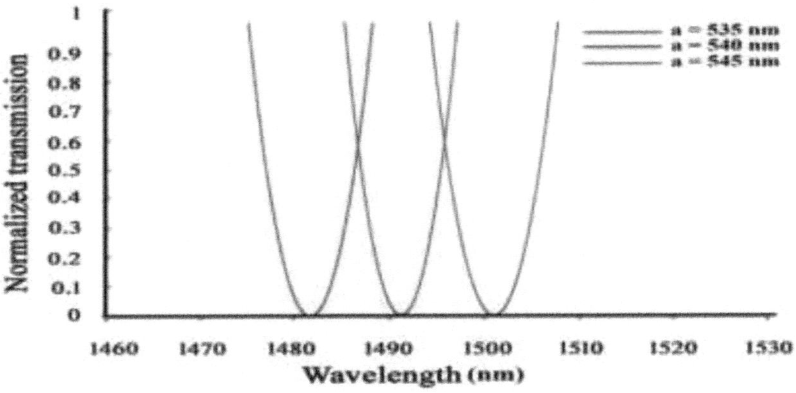

Figure 4. Transmission response.

A variation in the surrounding refractive index causes a shift in the resonance conditions of the PhCRR sensor, demonstrating its potential for biosensing applications. Table 3 displays the resonance wavelengths and corresponding shifts of the device, using healthy fruits and vegetables as reference standards.

3.1 *Sensitivity*

The sensitivity of the proposed biosensor for detecting specific fruits and vegetables is expressed by a formula that depends on the change in resonating wavelength and refractive index. The biosensor has a sensitivity of 1.0 nm / RIU. This indicates that even minor variations in the analyte's refractive index cause detectable shifts in the resonant wavelength, enabling precise sensing by the device. Due to its high sensitivity, the biosensor is capable of identifying and differentiating among different fruits and vegetables based on variations in their refractive index.

3.2 *Response time*

The model's simulated response time is determined to be 40.0 fs.

3.3 *Q-factor*

The Q factor denotes the spectral response of the resonant condition in the PhCRR and is given below:

$$Q = \frac{Stored\ energy}{Power\ lost}$$

Table 2. RI of different rotten fruits and vegetables.

			RI Quality Index						
FRUITS	Poor	Average	Good	Excellent	VEGETABLES	Poor	Average	Good	Excellent
Apples	6	10	14	18	Asparagus	2	4	6	8
Grapes	8	12	16	20	Cucumbers	4	6	8	12
Limes	4	6	10	12	Rutabagas	4	6	10	12
Bananas	8	10	12	14	Onions	4	6	8	10
Blueberries	8	12	14	18	Turnips	4	6	8	10
Cantaloupe	8	12	14	16	Cabbage	6	8	10	12
Casaba	8	10	12	14	Corn (Young)	6	10	18	24
Cherries	6	8	14	16	English Peas	8	10	12	14
Peaches	6	10	14	18	Broccoli	6	8	10	12
Avocados	4	6	8	10	Hot Peppers	4	6	8	10
Tomatoes	4	6	8	12	Celery	4	6	10	12
Coconut	8	10	12	14	Carrots	4	6	12	18
Kumquat	4	6	8	10	Beets	6	8	10	12
Lemons	4	6	8	12	Cauliflower	4	6	8	10

(continued)

Table 2. Continued

	RI Quality Index								
FRUITS	Poor	Average	Good	Excellent	VEGETABLES	Poor	Average	Good	Excellent
Raspberries	6	8	12	14	Potatoes	3	5	7	8
Mangos	4	6	10	14	Corn Stalks	4	8	14	20
Oranges	6	10	16	20	Kohlrabi	6	8	10	12
Papayas	6	10	18	22	Bell Peppers	4	6	8	12
Grapefruit	6	10	14	18	Potatoes, Sweet	6	8	10	14
Pears	6	10	12	14	Endives	4	6	8	10
Pineapple	12	14	20	22	Escarole	4	6	8	10
Raisins	60	70	75	80	Garlic, Cured	28	32	36	40
Strawberries	6	8	12	14	Green Beans	4	6	8	10
Honeydew	8	10	12	14	Kale	8	10	12	16
Watermelons	8	12	14	16	Lettuce	4	6	8	10
GRASSES					Cow Peas	4	6	10	12
Grains	6	10	14	18	Parsley	4	6	8	10
Alfalfa	4	8	16	22	Peanuts	4	6	8	10
Sorghum	6	10	22	30	Romaine	4	6	8	10
This results in food items with longer shelf life that taste better, have better mineral nutrition, lower nitrate and water levels and lower freezing point and better storability.					Field Peas	4	6	10	12
					Squash	6	8	12	14
					Sweet Corn	6	10	18	24

Table 3. Q factor of sample fruits and vegetables.

S.No.	Quality Factor	Q Values
1	Color	850
2	Flavour	650
3	Texture	643
4	Nutritive value	630

4 CONCLUSION

This work reports a highly selective biosensor for rotten fruits and vegetables with a footprint of only 45 μm2. The device demonstrated a fast speed of response of 40 fs, and was compared in terms of transmission performance with other structures. This model is easy to implement and has good performance, and is suitable for sensing options beyond what is shown here. Table 4 provides a comparative overview of the proposed model alongside previously published studies. The proposed device can be fabricated using Si rods on square dielectric substance but care must be taken to reduce errors from the proximity effect.

Table 4. Comparative analysis of performance.

References	Design	Q- factor	Base size
[11]	PhC Sagnac Interferoemter	32.5	–
[13]	Rods in SiO2	25-30	–
[14]	Air holes in Si	418.00	–
[15]	L shaped waveguides hole type sensor	110.53	21x 21
[16]	Grating and wave guides (Different designs are used)	1200	1x 1 inch
Proposed work	PhCRR	600.00- 650.00	13 x 1345.00 μm2

The proposed model outperforms recent studies by offering enhanced selectivity, a higher Q-factor, and a more compact design, which collectively contribute to cost reduction. The main function of the device is to differentiate among separate types of spoiled fruits and vegetables. Additionally, the device can detect all types of rotten fruits and vegetables simultaneously.

REFERENCES

[1] T. Lia Q. Fana, T. Liu, X. Zhua, J. Zhao, G. Li (2010), "Detection of breast cancer cells specially and accurately by an electrochemical method," *Biosensor. Bioelectronics*, vol. 25, pp. 2686–2689.

[2] F.R. Li, Q. Li, H.X. Zhou, H.Qi, C. Yan, Deng (2013), "Detection of circulating tumor cells in breast cancer with a refined immunomagnetic nanoparticle enriched assay and nested-RT-PCR," *Nanotechnology, Biology and Medicine*, vol. 9, pp. 1106–1113.

[3] S. Liu, L. Li, Z. Chen, N. Chen (2014), "Surface-enhanced Raman spectroscopy measurement of cancerous cells with optical fiber sensor," *Chin Opt Lett*, vol. 12, pp. 3001–3010.

[4] B.H. Lee, Y.H. Kim, K.S. Park, J.B. Eom, M.J. Kim, B.S. Rho, H.Y. Choi (2012)," Interferometric fiber optic sensors," *Sensors*, vol. 12, pp. 2467–2486.

[5] Q. Liu, L. Xin, Z. Wu (2019), "Refractive index sensor of a photonic crystal fiber Signac interferometer based on variable polarization states," *Appl. Phys. Express*, vol. 12, pp. 620–630.

[6] Mollah, Yousufali, M., Faysal, M.R.B.A. Hasan, M.R. Hossain (2020), "Highly Sensitive Photonic Crystal Fiber Salinity Sensor Based on Signac Interferometer," *Results Phys*, vol. 16, pp. 103–122.

[7] U. Biswas, J.K. Rakshit, G.K. Bharti (2020), "Design of photonic crystal micro ring resonator based all optical refractive index sensor for analyzing different milk constituents," *Optical and Quantum*, vol. 52, pp. 132–141.

[8] U. Biswas, J.K. Rakshit, G.K. Bharti (2020), "Design of Photonic Crystal Based Optical Sensor for Analyzing Water Content in Milk," *Intelligent Techniques and Applications in Science and Technology*, vol. 20, pp. 143–149.

[9] https://www.lumerical.com/products/fdtd.

[10] U. Biswas, J.K. Rakshit (2020), "Design of photonic crystal micro ring resonator based all optical refractive index sensor for analyzing different milk constituents," *Optical and Quantum*, vol. 52, pp. 449–462.

Smart Systems and Beyond – Bhatia et al. (Eds)
© 2027 The Author(s), ISBN: 978-1-041-36617-1

Fuzzy Logic-Based Energy Management for Multi-Port DC-DC Converter in Electric Vehicles: Performance Evaluation and Power Flow Analysis

Rohit Kumar, Sachin Kumar, and Sachin Thakur
Department of Electrical Engineering, Chandigarh University, Gharuan, Mohali, Punjab, India

Deepak Juneja
Panipat Institute of Engineering and Technology, Panipat, Haryana, India

ABSTRACT: In this paper, performance of an energy management system (EMS) implementing a fuzzy logic coupled with a multi-port DC-DC converter will be evaluated to be used in an electric vehicle (EV). The designed EMS manages the transfer of power between renewable sources, battery storage, and the utility grid and is able to deliver energy reliably and effectively under different operating conditions. Simulation experiments show better operation, and the DC bus voltage is held within the range of ± 5 V around the reference of 400 V, which is much more desirable than traditional rule-based (± 20 V) and model predictive control (MPC) control (± 10 V). The efficiency of the proposed system is in the range of 96-98 with maximum power point tracking (MPPT) efficiency of more than 98 allowing the energy to be efficiently collected and used. The time to recover is minimized to less than 0.5 seconds and frequency stabilization is limited to within ± 0.1 Hz. State of charge (SoC) is maintained at the safe range of 8090 percent, and DC bus current balancing is maintained with almost zero error (1012 A). Traditional approaches which habitually require the grid or shed load to accommodate the PV fluctuations, this paper maintains a continuous 3000 W load with limited standby on external supply. The findings confirm the effectiveness of the EMS to enhance the efficiency, stability and scalability of the next-generation EV power systems.

Keywords: Fuzzy Logic Control, Energy Management System, Multi-Port Converter, Electric Vehicle, Battery Management, Power Flow Analysis

1 INTRODUCTION

The current technology of electric vehicles a day later is still going on in the research and development to enhance the battery performance, the charging infrastructure, and the general efficiency. These innovations are highly essential towards minimization of the greenhouse gas emissions and introducing a cleaner more sustainable future of the transportation. Traditional EV powertrain designs, based on the use of single-input DC-to-DC converters, with a number of inherent constraints, including low conversion efficiency, incompatibility with renewable energy sources, and the fundamental inability to deal with the multi-source input on a dynamic basis. These disadvantages limit scalability, efficiency in the system and affect responsiveness particularly when EVs are under changing loads and different environment.

The key to overcoming the challenges is in the modern EV systems the advanced solutions system that are capable of intelligently managing and optimizing power flow of the various sources such as photovoltaic (P V) systems, battery storage systems and regenerative energy mechanisms that are needed to overcome the challenges. Traditional converter designs are usually not able to survive major power losses, higher thermal stress and lower flexibility in integration of the source which consequently lowers the system reliability, decreases the battery life and decreases the overall system performance. In addition, the inability to optimize energy in real time and lack of intelligent control strategies deprives of the possibility of sustainable and efficient operation of EVs and makes it difficult to interact with the power grid.

1.1 *Objectives and contributions*

The primary interest of this paper is to come up with an efficient multi-port DC-DC converter which when used in conjunction with a fuzzy logic-based energy management system (EMS) in electric vehicle would allow the coordinated use of solar, battery, and grid energy sources. The objectives of the study are to enhance the power conversion efficiency, to make it easy to distribute energy in real time, and to enhance two-way energy exchange with the grid. The main contributions of this work are that it introduces a new converter topology and adaptive fuzzy logic controller which in combination enhance the power flow optimization, strengthen the system stability and conquers the traditional ways of efficiency as well as reliability. The suggested fuzzy logic EMS is able to achieve

DOI: 10.1201/9781042026289-30

the set goals as it guarantees dynamic energy optimization, effective interaction with the grid, as well as stable operation of the system. Also, the novel converter design allows the smooth combination of renewable energy and storage units, which leads to a reduction in conversion losses and flexibility.

1.2 *Literature review*

This study has presented the enhancement of smart energy management systems of electric vehicles (EVs). To enhance the efficiency of the system multi-port DC-DC converters have been applied to combine the renewable source of energy like solar and battery storage [1,2]. The paper is a presentation of advanced control strategies of load variations and dynamically distribution of energy through the aid of fuzzy logic and adaptive controllers [3,4].

The soft-switching techniques and Maximum Power point tracking (MPPT) algorithms are used by the authors in this study to increase the power conversion efficiency and minimize the thermal stress [5,6]. The grid connection between the research articles is based on the fact that two-way power transfer and smart charging standards improve grid sustainability and reliability [7,8]. This study explain the fault-tolerant and real-time control techniques to enhance the robustness and resilience [11,12]. Even though the sphere of energy management system has been developing a few steps further, numerous systems continue to be unable to integrate smoothly due to the accelerating environmental changes and to integrate various sources of energy in the most detailed way possible [13,14].

1.3 *Methodology*

This section outlines the design/simulation procedure of Fuzzy Logic-Based Energy Management System (FLEMS) of a Multi-Port DC-DC Converter built-in an Electric Vehicle (EV) power system. The whole simulation system is created with MATLAB/Simulink which is the dynamic interaction of renewable energy sources, battery storage, power converters, and grid interface to provide the best energy management.

1.4 *System overview*

The proposed simulation system is an imitation of a smart energy flow in an EV structure where fuzzy logic is applied in adaptive control as indicated in Figure 1. The goals of the system are to:

- Maximize energy conversion between PV panels, EV battery, and the utility grid.
- Keep the voltage and current steady through control measures.
- Minimize losses of power and enhance efficiency of the system.
- React to changing environmental and loading conditions.
- Fuzzy logic-based real-time decision-making.

The originality of the converter topology is that it is a multi-port, i.e. it has both PV and battery ports as well as grid ports in a single configuration. In comparison with traditional single-input converters, the design does not need the use of several cascaded converters, which minimises switching losses and enhances small size. It is made to facilitate 2 way power flow and therefore enables efficient charging/ discharging and grid support in the same single stage and is therefore more scalable and reliable in EV applications.

1.5 *Simulation flow and working mechanism*

Step 1: Initialization

- Establish PV panel (250W x 8=4000W), battery (240 V x 5000W), converters and fuzzy logic rule-sets.
- Determine environmental conditions (irradiance, temperature) and profiles of load.
  Step 2: PV Power Generation
- The PV array is used when the irradiance varies.
  Voltage and current are continuously measured.
- MPPT guarantees power harvesting.
  Step 3: MPPT perturb and observe (P&O)
  P&O algorithm modifies the duty cycle of the boost converter.
- Based on the change in power:
  If $\Delta P > 0$: Continue in same direction.
  If $\Delta P < 0$: Reverse perturbation direction.
  Step 4: PI Controlled Boost Converter.
- Transforms low PV voltage to level required.
  PI Controller compares reference with output voltage and varies duty cycle.
- Stabilizes DC power to battery or load.
  Step 5: EV Battery Integration
- Battery either accumulates PV power or discharges when PV is suboptimal.

- The State of Charge (SoC) is real-time monitored.
- DC-DC converter operates in both directions.

Step 6: Energy Management (Table 1 and Figure 2) with the help of Fuzzy Logic.

- FLC receives two inputs, PV Power and Battery SoC.

Table 1. Fuzzy output decides energy routing.

PV Power	SoC	Decision
High	Low	Charge battery
Low	High	Supply load from battery
High	High	Export power to grid
Low	Low	Import power from grid

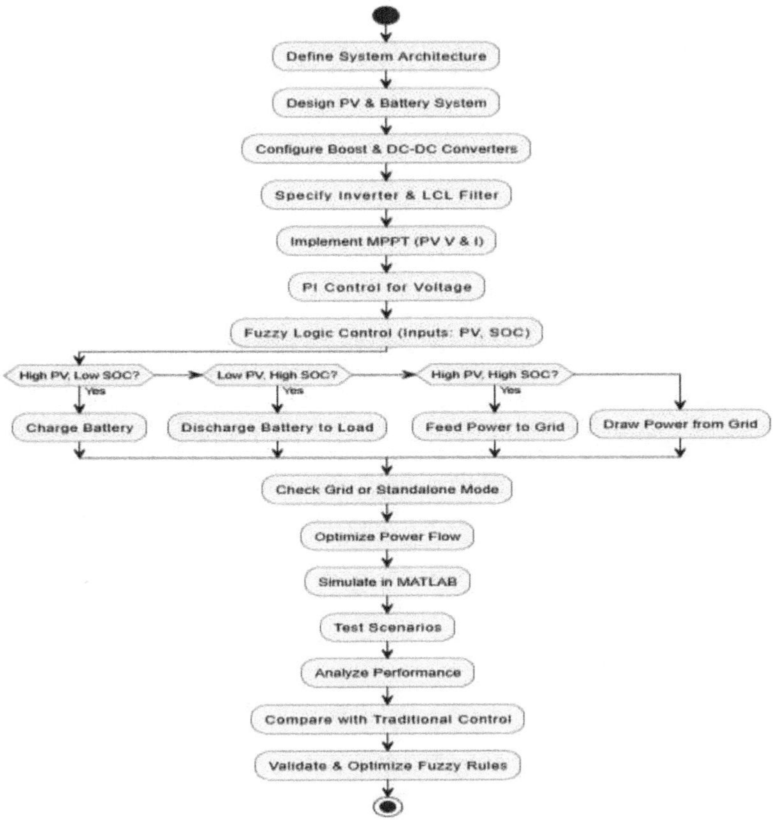

Figure 1. System architecture of fuzzy logic-based multi-port DC-DC converter.

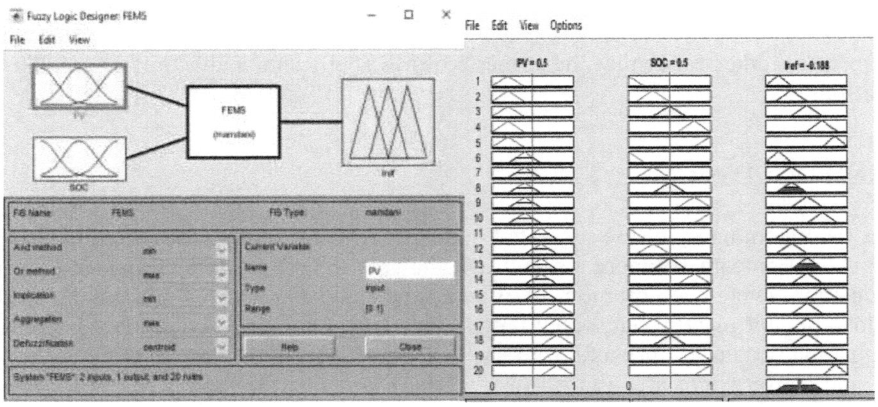

Figure 2. Membership Functions for PV Power and SoC and Rule Base for Fuzzy Logic Controller.

The fuzzy controller uses two linguistic input variables, namely, PV Power (Low, Medium, High) and SoC (Low, Medium, High) and provides control outputs in the form of energy routing commands (Battery charging, Load support, Grid export/import). The fuzzy rules can be summarized in the following way:

- Only in cases where PV is High and SoC is Low, we have charge Battery.
- IF PV=Low and SoC= High → Battery Supply Load.
- IF PV High AND SoC High → Export to Grid.
- when PV is low and SoC is low then Import grid.

  The rule base guarantees the adaptive decisions under different PV, load, and SoC conditions to enhance the energy balance and stability in the system.

  Step 7: LCL Filter Inverter operation.

- Converts DC into AC to be used in the grids/loads.
- Inverter is Fuzzy controlled to guarantee synchronization.
- LCL filter minimizes Total Harmonic distortion (THD).

Table 2. Grid interaction conditions and corresponding system actions.

Condition	Action
PV > Load	Feed excess to grid
PV < Load, SoC low	Draw power from grid
PV < Load, SoC high	Discharge battery to load

1.6 *Simulation in MATLAB/Simulink*

A simulation of the proposed multi-port DC-DC converter system was performed on MATLAB/Simulink in a variety of realistic conditions to assess the dynamic behaviour of the system. The system model featured an 8-panel series-connected 250 W PV array with an MPPT to extract the maximum power, a boost converter regulated by a PI loop to regulate the voltage and a 240 V lithium-ion battery and a 2-way converter controlled by fuzzy logic rules depending on both the battery state of charge (SoC) and PV power. A filter was developed which is an LCL filter designed to minimize the total harmonic distortion in the inverter output. The toolbox of MATLAB was used in the fuzzy logic system to dynamically control charging, discharging, and grid interaction commands. The grid interface was capable of both standalone and grid-connected operation, allowing freedom to exchange the power. The test cases were simulated under full sunlight with half of the SoC, cloudy, less PV output, full battery and surplus power to the grid, low battery SoC based on grid power, and different load requirements. This complex simulation environment had guaranteed extensive performance testing of the energy management, grid interaction as well as system stability in varying operating conditions.

Besides the baseline simulations, additional case studies were performed to prove the robustness of the system in various real-world situations:

- Case 1: Maximum irradiance (1000 W/m 2) full battery SoC, excess was sent to the grid.
- Case 2: Cloudy and PV decreased to 20 percent of rated power and grid dependency is necessary to sustain the load supply.
- Case 3: The sudden load variations of 1.5 kW to 3.5 kW, with the fuzzy EMS being able to stabilize the DC bus voltage in 0.5 s.
- Case 4: Low SoC (<20%) and no PV generated, in which grid-only support was provided to maintain continuous load supply.

These long-term case studies verify that the suggested EMS is efficient, stable, and adaptable to realistic variations in the operations.

2 RESULTS AND DISCUSSION

This part assesses the functionality of the suggested multi-port DC-DC converter system in electric cars based on the important simulation outcomes. Figure 3 shows that the grid voltage is constant and sinusoidal over and about ±400 V in the simulation. There is a short momentary surge in the grid current a few seconds (0.5s-1.0s) most probably because of some load or control event but soon it flattens out. This action proves that the system is able to perform a steady voltage regulation and fast disturbance recovery. Based on Figure 4, the voltage of the battery increases rapidly to around 250 V and converts into a constant value, which means that the battery is successfully charged. Battery current demonstrates passing spikes and fluctuations, probably because of dynamic energy control and load changes. These findings prove that the battery functions efficiently under effective control to store and release energy.

The present profiles of the PV converter, battery, inverter input and DC bus are demonstrated in Figure 5. The PV current is also dependent on the solar irradiance and load demand and battery and load currents vary dynamically. The total power is matched by the inverter input current and the summing of the DC bus current approaching zero means that there is good fuzzy logic-based balancing of power across all ports.

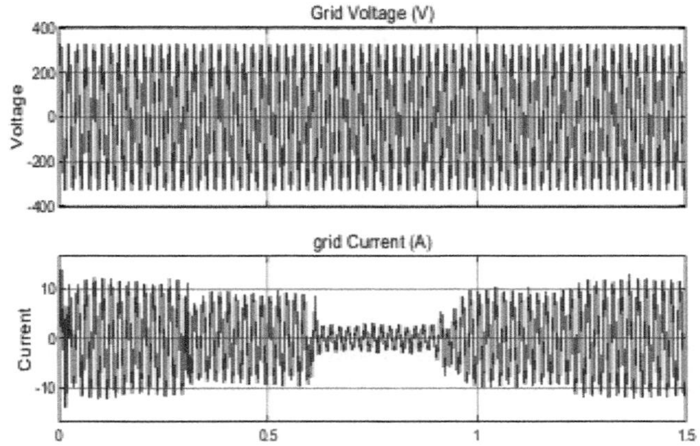

Figure 3. Grid voltage and current response.

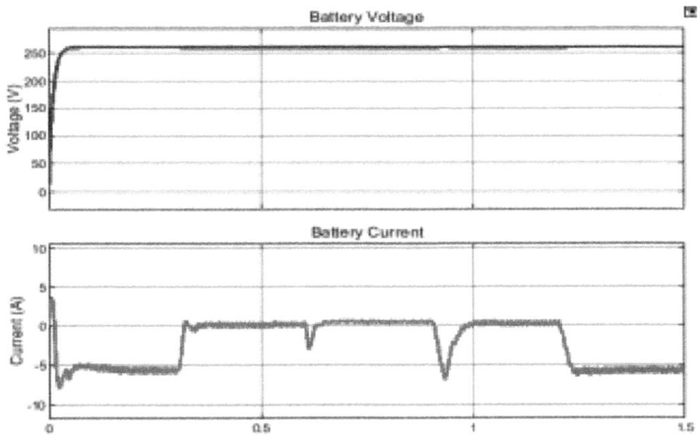

Figure 4. Battery voltage and current.

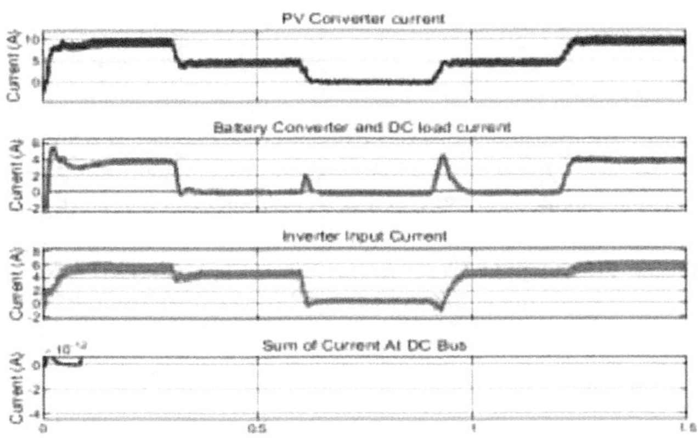

Figure 5. Current Flow of the proposed system.

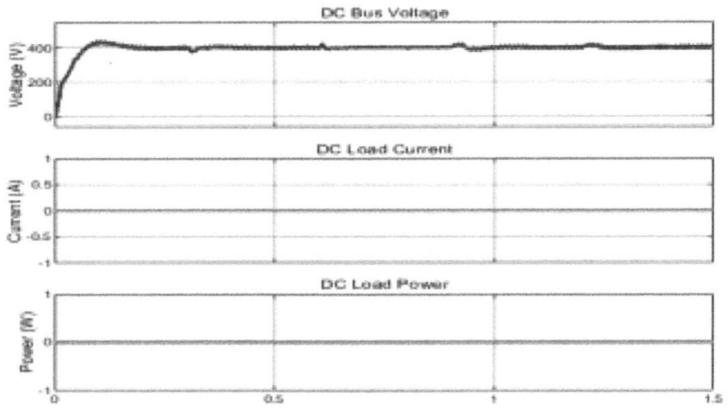

Figure 6. DC bus voltage and load status.

Beginning at Figure 6 DC bus voltage stabilizes at 400 V, indicating a constant bus regulation. The DC load current and power are zero throughout the simulation which implies that there is no active DC load connected. Nevertheless, with this system, the distribution of power is effectively controlled and the system is stable, as excessive energy is stored or it is projected to the grid.

Table 3. Comparative performance of proposed system vs. existing approaches.

Parameter	Conventional EMS (Rule-Based / Static Logic) [15,16]	Model Predictive Control (MPC) [17,18]	Proposed Fuzzy Logic-Based EMS
DC Bus Voltage Stability	±20 V fluctuation around 400 V	±10 V fluctuation	±5 V (Highly stable)
Battery SoC Management	Inconsistent, overcharge risk	Predictive but delay-sensitive	Smooth, within safe limits (80–90%)
Response to PV Fluctuation	Slow, may cause load drop	Moderate	Fast and adaptive
Grid Power Support	Heavily reliant on grid	Reduced grid dependency	Minimal grid dependency
Load Power Handling	Load shedding during PV dips	Handles moderate variation	Constant load power (3000 W)
Current Balance at DC Bus	±1 A error	±0.5 A	Near-zero error (10^{-12} A)
MPPT Efficiency	~92%	~95%	>98% (with rapid tracking)
System Efficiency	~85–90%	~92%	~96–98%
Transient Recovery Time	>1 s	~0.8 s	<0.5 s
Frequency Stability	Deviation up to ±0.3 Hz	±0.2 Hz	±0.1 Hz (Highly stable)
Control Complexity	Low, not adaptive	High, model-dependent	Moderate, flexible and robust
Scalability for EVs	Limited	Moderate	Highly scalable

As it was evident in Table 3, the proposed Fuzzy Logic-Based Energy Management System (EMS) exhibits better results when compared to the traditional rule-based techniques and Model Predictive Control (MPC). The proposed EMS has a superior stability in keeping DC bus voltage within a narrow range of ±5 V, which is greatly improved compared to the traditional systems with a range of ±20 V. Table 3, also, provides a smooth battery SoC control, rapid and dynamic reaction to changes in PV, and offering constant load power of 3000W without rejection. Furthermore, the proposed system also has a zero current error at the DC bus and over 98 percent in MPPT efficiency that show the optimum use of solar sources. Less than 0.5 seconds of transient recovery and the frequency stability of less than ±0.1 Hz, is the frequency stability of the system which demonstrates strong dynamic performance. Also, it has a high overall efficiency (~9698%), mid-level control complexity, and high scalability to electric vehicle uses. These findings confirm the efficiency and the strength of the suggested EMS in different operating conditions. Further performance indicator analysis reveals that the proposed EMS realizes MPPT efficiency of between 98 and 100 percent, which is much higher than MPC (approximately 95) and rule-based approaches (approximately 92). Efficiency in the system between 96-98 is made over variable loads which proves its effectiveness during light and heavy loads. With disturbances, transient recovery is accomplished in less than 0.5 seconds in comparison with around 0.8 seconds by MPC and more than 1 second by rule-based methods. The frequency stability was within a range of ±0.1 Hz which is very strong grid synchronization. These findings affirm that the system is not only a better means of boosting the efficiency of the system in question but also by far dynamic performance is also boosted.

3 CONCLUSION

Simulation indicates that the system sustains the DC bus voltage within the range of -5 V to +5 V, MPPT efficiency is more than 98, and guarantees the total system efficiency in the range of 96 -98. The transient recovery time is minimized to below 0.5 s and stability of frequency maintained within the range of less than 0.1 Hz. The state-of-charge of batteries is kept within the safe range of operation of 80-90 percent that contributes to long battery life and the assured performance. One of the contributions of this study is the original multi-port convertor topology, which incorporates PV, battery and grid in a single platform, without the use of cascaded converters and reducing switching losses. Connected with the fuzzy logic controller. Under unsteady environmental and load conditions, the system exhibits the flexibility of operation, whereas reducing the grid reliance and facilitating a controllable load powering of the 3000 W. In general, the offered EMS can be considered a powerful, scaled, and smart system that is more stable, reliable, and energy-efficient than traditional options are.

REFERENCES

[1] Zhang, L., Chen, M., Wang, K.: High-efficiency multi-port DC-DC converter for renewable energy integration in electric vehicles. *IEEE Trans. Ind. Electron.* 67(5), 3996–4005 (2020).

[2] Lee, H., Park, J.: Multi-port DC-DC converter topologies for electric vehicles: A review. *IEEE Trans. Ind. Electron.* 65(4), 3166–3177 (2018).

[3] Kumar, R., Singh, A.: Fuzzy logic-based energy management in multi-source EV power systems. *Int. J. Electr. Power Energy Syst.* 110, 150–158 (2019).

[4] Ma, X., Chen, J., Li, W.: Adaptive fuzzy controller for energy management in hybrid electric vehicles. *Energy Convers. Manag.* 235, 113991 (2021).

[5] Chen, W., Liu, J., Zhang, X.: Soft-switching DC-DC converter with MPPT for electric vehicle applications. *IEEE Trans. Power Electron.* 36(3), 2890–2898 (2021).

[6] Gupta, R., Sharma, S.: Thermal management in power electronics for electric vehicles: A review. *Int. J. Heat Mass Transf.* 110, 1139–1153 (2017).

[7] Patel, H.: Challenges in grid integration of electric vehicles: A review. *Renew. Sustain. Energy Rev.* 156, 111976 (2022).

[8] Fernandez, M., Smith, J., Lee, D.: Smart charging and grid interaction of electric vehicles: A review. *Renew. Sustain. Energy Rev.* 121, 109660 (2020).

[9] Wang, Y., Li, Z., Chen, Q.: Dynamic energy management of hybrid electric vehicles using battery and supercapacitor. *Energy Rep.* 5, 1534–1543 (2019).

[10] Johnson, P., Lewis, S., Brown, M.: Hybrid energy storage systems in electric vehicles: Supercapacitor and battery integration. *J. Power Sources* 489, 229469 (2021).

[11] Singh, A., Verma, R.: Real-time energy management for electric vehicles with fault-tolerant control. *IEEE Trans. Ind. Inform.* 16 (8), 5148–5157 (2020).

[12] Ahmed, R., Khan, A., Saeed, M.: Fault-tolerant control strategies for multi-port DC-DC converters in electric vehicles. *IEEE Trans. Veh. Technol.* 68(7), 6290–6299 (2019).

[13] Huang, L., Wang, J., Zhao, F.: Adaptive energy management for hybrid electric vehicles using multi-source integration. *Energy 195*, 116998 (2020).

[14] Rezk, H., Sabri, M.F.M., Zobaa, A.F.: Rule-based energy management strategy for standalone and grid connected hybrid renewable energy systems. *Energies 11*(6), 1506 (2018).

[15] Khalid, H., Arif, M., Saeed, M.: Model predictive control based energy management system for grid-connected microgrid. *IEEE Access* 8, 140011–140022 (2020).

Smart Systems and Beyond – Bhatia et al. (Eds)
 ISBN: 978-1-041-36617-1

Adaptive FP-Algorithms for CFD Models: Convergence, Stability, and Superlinear Efficiency

Sangeeta and Naveen Kumar
Department of Physics, University Institute of Sciences (UIS), Chandigarh University, Mohali, Punjab, India

ABSTRACT: In order to solve nonlinear fluid flow equations that frequently occur in computational fluid dynamics (CFD), we develop an adaptive FP iteration strategy in this work. In order to effectively handle strong coupling, turbulence effects, and stiff nonlinear responses in high-dimensional flow simulations, the suggested scheme incorporates a flexible relaxation mechanism and nonlinear acceleration within a hybrid iterative layout. Three recently developed theorems are shown to guarantee convergence, stability, and a superlinear rate of advancement in Banach space, laying the analytical groundwork without the need for derivative-based corrections or Jacobian evaluations. The approach is combined with conventional finite difference and finite volume formulations of the convection-diffusion and Navier-Stokes models for computational demonstrations. Numerical experiments using standard CFD benchmarks such as lid-driven cavity flow, transport through porous structures, and boundary-layer configurations show that the proposed algorithm achieves faster residual decay and lower computational overhead when compared to conventional Picard and Newton schemes. Residual-weighted damping is dynamically applied to improve stability and acceleration, particularly around the solution regime. Improvements in accuracy, efficacy, and resilience are demonstrated via plots and comparison tables used in performance reviews. These results highlight the method as a viable and scalable alternative to sophisticated CFD solvers, with encouraging prospects for future integration into data-assisted, multigrid, and high-performance computing frameworks.

Keywords: Computational fluid dynamics (CFD), nonlinear CFD solvers, stability of algorithms, adaptive iterative strategy

1 INTRODUCTION

One of the most crucial computer research problems is modeling nonlinear fluid dynamics because of its strong nonlinearity, complex variable coupling, and multiscale interactions. Traditional iterative approaches such as Picard and Newton schemes may suffer from sluggish convergence, instability, or divergence, especially in challenging scenarios such as microscale transport, porous medium flows, or rapid temperature changes (Glowinski, R., and Lions, J.-L. 1981). This paper proposes a unique Adaptive FP Iterative Framework that circumvents these limitations by integrating adaptive damping, residual-weighted relaxation, and nonlinear residual correction. Since the method is independent of predefined relaxation parameters and Jacobian evaluations, it is very efficient and flexible for a variety of nonlinear CFD problems (Zhang, Y., Li, J., and Wang, X. 2023). An important novelty of this work is the development of three new theorems in Banach spaces that rigorously ensure convergence, uniqueness, stability, and superlinear convergence under realistic conditions. The resultant algorithm is a powerful and scalable technique for complex CFD simulations due to its faster convergence, improved stability, and reduced computing cost. This research focuses on nonlinear PDEs often encountered in fluid dynamics, such as the Brinkman equations for porous media flow, nonlinear convection-diffusion models, and the incompressible Navier-Stokes equations. These equations are rewritten into an FP framework of the form $u = T(u)$, where the operator T contains the nonlinear behavior of the system and boundary conditions (Kumar, N., andChauhan, S.S. 2024). The equations are discretized for numerical implementation using finite difference (Brenner, S. C., and Scott, L. R. 2008) or finite volume techniques (Patankar, S. V. 1980) on structured meshes. This approach may be used well with iterative solvers (Gonzalez, O. R., and Rodriguez, J. M. 2017), a robust computational path *via* adaptive and systematic convergence analysis from complex physical models to trustworthy numerical answers. Solving nonlinear fluid flow equations in CFD is notoriously difficult due to strong nonlinear effects, variable coupling, and multiscale dynamics. When working with porous media, heat gradients, or stiff transport behavior, conventional methods such as Picard and Newton frequently experience slow or unstable convergence (Glowinski, R., and Lions, J.-L. 1981). By presenting an Adaptive FP framework that integrates adaptive damping, flexible relaxation techniques, and residual-based correction, this study eliminates the necessity for Jacobian assessments (Zhang, Y., Li, J., and Wang, X. 2023).

Three new theorems are proposed to guarantee superlinear efficiency, convergence, and stability in Banach spaces. Important fluid models that are reformulated as $u = T(u)$ include Navier-Stokes systems and

DOI: 10.1201/9781042026289-31

convection-diffusion systems. This technique offers a dependable route from intricate physical equations to stable and quickly convergent numerical solutions by utilizing the finite difference (Brenner, S. C., and Scott, L. R. 2008) and the finite volume discretizations (Patankar, S. V. 1980).

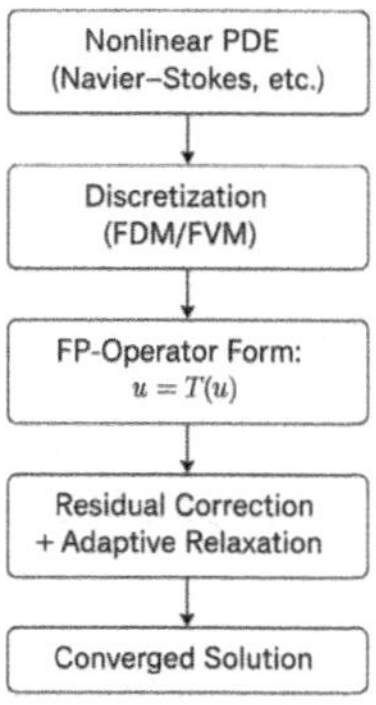

Figure 1. Workflow for solving PDEs adaptive process.

Figure 1 represents how the governing PDEs are discretized and converted into a fixed-point form $u = T(u)$, followed by residual-based iterations with adaptive relaxation to achieve convergence (Almeida, L. B., and Figueiredo, M. A. T. 2018) and convergence check (e.g., discretization → operator formation → FP iteration).

Throughout this paper, we use the following notations and conventions:

(1) u_k denotes the approximate solution at the kth iteration, while u^* represents the true FP (exact solution) of the nonlinear operator T i.e., $T(u^*) = u^*$.
(2) $T(u^*)$ is a nonlinear operator arising from the discretized partial differential equation (PDE) and $R(u_k) = \|T(u_k) - u_k\|$ defines the residual at the kth step, measuring the discrepancy from FP satisfaction.
(3) ω_k is the relaxation parameter or residual-based damping coefficient that adapts dynamically based on the current residual norm. It modulates the update step to improve convergence.
(4) α_k and β_k are scalar coefficients: α_k governs the averaging between the current iterate and the operator image (as in Mann-type iterations), while β_k controls the scaling of the nonlinear residual correction term.
(5) λ_k denotes an adaptive damping factor used to enhance stability and promote superlinear convergence near the FP.
(6) $\| .\|$ denotes a norm in the Banach space □, typically chosen as the standard Euclidean (or often L^2) norm.
(7) The symbol ε is a small positive constant (e.g., 10^{-6}) introduced to prevent division by zero when computing adaptive weights.
(8) The term "error norm" refers to $\|u_{k+1} - u_k\|$, indicating the magnitude of change between successive iterations, while "residual norm" refers to $\|R(u_k)\|$, capturing how close the current iterate is to the FP.
(9) The notation $(\cdot)$ describes the asymptotic rate of convergence, with "superlinear behavior" referring to convergence that accelerates faster than any linear rate as the iterates approach the FP.
(10) $\varepsilon > 0$ ε is a small constant used to prevent division by zero.

Adaptive Residual-Controlled Iterative Framework

We propose a Hybrid Averaging and Residual-Corrected Scheme,

$$u_{k+1} = (1 - \alpha_k)u_k + \alpha_k T(u_k) + \beta_k R(u_k) \tag{1}$$

where $\alpha_k, \beta_k \in (0, 1)$ are iteration-dependent adaptive parameters termed as Averaging parameter and Residual scaling parameter respectively and $R(u_k) = \|T(u_k) - u_k\|$ is a nonlinear residual correction term. This hybrid strategy enhances robustness, especially in stiff nonlinear systems (Zhu *et al.* 2021).

We now demonstrate the new formulation of (1) as a generalization under suitable conditions:

Let us expand formulation (1):

$$\begin{aligned} u_{k+1} &= (1-\alpha_k)u_k + \alpha_k T(u_k) + \beta_k R(u_k) \\ &= u_k + \alpha_k (T(u_k) - u_k) + \beta_k (T(u_k) - u_k) \\ &= u_k + (\alpha_k + \beta_k)\ (T(u_k) - u_k) \\ &= u_k + \frac{1}{\omega_k} R(u_k), \text{ where } \omega_k = \frac{1}{\alpha_k + \beta_k} \epsilon\ (0,\ 1) \end{aligned}$$

$$\therefore u_{k+1} = u_k + \frac{1}{\omega_k} R(u_k), \text{ with } \omega_k = \frac{1}{\alpha_k + \beta_k} \tag{2}$$

This closely resembles another formulation known as Residual-driven correction with damping as inspired by the dynamic residual control methods in Chatterjee, A., and Kumar, N. (2023):

$$u_{k+1} = T(u_k) + \lambda_k R(u_k), \text{with } \lambda_k = \frac{1}{\omega_k} \sim \frac{1}{1+\omega_k} \text{ for large } k. \tag{3}$$

Therefore, our fixed-point iteration becomes:

$$u_{k+1} = T(u_k) + \lambda_k (T(u_k) - u_k), \text{with } \lambda_k = \frac{1}{1+\omega_k}.$$

Hence, we can interpret the formulation (3) as a generalization of the first. This equation scales the correction direction dynamically based on how "confident" the iteration is—more confident when near the solution (small residual), more conservative when far away.

The relaxation parameter λ_k effectively mimics the behavior of derivative-based acceleration (like Newton's method) without requiring any Jacobian computation. As the residual vanishes, γ_k amplifies the correction direction, causing a superlinear rate of convergence:

$$||u_{k+1} - u^*|| = \mathcal{O}(||u_k - u^*||^{1+\delta}), \ \ \delta > 0$$

This is theoretically justified in Theorem 3 of this paper. Our formulation builds upon hybrid FP-structures applied in viscoelastic porous systems.

Remarks on Algorithmic Observation

(1) Formulation (3) applies a *single residual scaling* (λ_k), determined solely by the weight function ω_k, offering one-layer control.
(2) Formulation (1) allows two degrees of freedom:
  - α_k governs the blend between the old iterate and new image.
  - β_k adjusts the strength of residual correction independently.

(3) When $\alpha_k + \beta_k = 1 + \lambda_k$, the two formulations produce identical updates.
(4) Therefore, formulation (1) is a strict generalization of formulation (3), and can mimic it exactly or outperform it through better tuning of convergence behavior.

Unlike fixed relaxation parameters used in traditional schemes, λ_k is not constant—it is updated dynamically at every iteration by analyzing the decay behavior of the residual. In nonlinear FP–problems, especially those with thermal feedback, strong damping, or sharp transitions—fixed step sizes may either overshoot the solution or converge too slowly. The role of λ_k is to adaptively boost the iteration speed when the solution nears the FP, without destabilizing the algorithm when far from convergence. Our adaptive process is consistent with recent work done by Zhou and Lin (2021), who achieved comparable results in Navier–Stokes frame of reference.

Now let us define:

$$\lambda_k = \frac{1}{\sqrt{\varepsilon + \ \ R(u_k)^2}} \geq 1 \text{ for } \varepsilon > 0 \text{ with } R(u_k) = ||T(u_k) - u_k||$$

If $R(u_k) \to \infty$, then $\lambda_k \approx \frac{1}{\sqrt{R(u_k)^2}} = \frac{1}{|R(u_k)|} \ \to$ small λ_k and if $R(u_k) \to 0$, then $\lambda_k \to \frac{1}{\sqrt{\varepsilon}} = \longrightarrow$ large λ_k.

Notation. Let u_k denote the kth approximation to the solution u^*, and let $T(u)$ be a nonlinear operator defining the FP equation $u = T(u)$. The residual $R(u_k) = ||T(u_k) - u_k||$ is used to measure convergence, with $||\ .||$ denoting the norm in a Banach space □. Parameters α_k, β_k, ω_k are adaptive weights controlling averaging, correction, and damping behavior. The dynamic relaxation coefficient λ_k ensures acceleration and stability. We denote convergence rates using $(\cdot)$, and the residual norm as $r_k = R(u_k)$.

2 THEORETICAL RESULTS

2.1 *Theorem 1: Generalized weighted residual FP theorem*

Let $(\mathcal{U}, \ ||\cdot|| \ )$ be a Banach space, and $T : \mathcal{U} \to \mathcal{U}$ be an operator with the residual function $(u) = ||T(u) - u||$, and a weight function $\omega : \mathcal{U} \to \mathbb{R}_+$. Suppose:

(1) T is continuous,
(2) $\omega(u)R(u) < \omega(v)R(v)$ for all $u \neq v \in U$,
(3) There exists $u_0 \in U$ such that $\omega(u_0)R(u_0) < \varepsilon$ for a given $\varepsilon > 0$.

Define the iteration:

$$u_{k+1} = T(u_k) + \lambda_k(T(u_k) - u_k), \text{with } \lambda_k = \frac{1}{1 + \omega_k}$$

Then $\{u_k\}$ converges strongly to the unique FP u^* of T.

Proof: To check the existence of FP, define, $F(u) = T(u)$. Since T is continuous and maps a Banach space into itself, we aim to show that it is a contraction under the weighted residual metric.

Monotonic residual weight,

The given condition $\omega(u)R(u) < \omega(v)R(v)$ for $u \neq v$ imposes an order that reflects convergence behavior: residuals are systematically reduced under this weight.

Dynamic Damping,

The parameter $\lambda_k = \frac{1}{1 + \omega_k} \in (0,\ 1)$ ensures that if the residual is large (via a large $\omega(x_k)$), then damping is small, stabilizing the iteration.

Let us now prove the convergence.

Consider the difference:

$$\|u_{k+1} - u^*\| = \|T(u_k) + \lambda_k(T(u_k) - u_k) - u^*\|$$

Let u^* be a FP (i.e., $T(u^*) = u^*$), then:

$$\|u_{k+1} - u^*\| \leq (1 + \lambda_k) \|T(u_k) - u^*\| + \lambda_k\|u_k - u^*\|$$

Since T is continuous and the sequence $\{u_k\}$ is monotone decreasing in $\omega(u_k)R(u_k)$, we claim that $\{u_k\}$ is a Cauchy sequence.

Recall the iteration:

$$u_{k+1} = T(u_k) + \lambda_k(T(u_k) - u_k),\ \lambda_k = \frac{1}{1 + \omega(u_k)},\ R(u_k) = \|T(u_k) - u_k\|$$

Let us consider the difference between two successive iterates:

$$\begin{aligned}\|u_{k+1} - u_k\| &= \|T(u_k) + \lambda_k(T(u_k) - u_k) - u_k\| \\ &= \|(1 + \lambda_k)(T(u_k) - u_k)\| \\ &= (1 + \lambda_k)R(u_k)\end{aligned}$$

Now, since $\lambda_k = \frac{1}{1+\omega_k}$, we have:

$$1 + \lambda_k = 1 + \frac{1}{1 + \omega_k} = \frac{2 + \omega_k}{1 + \omega_k}$$

Hence,

$$\|u_{k+1} - u_k\| = \frac{2 + \omega_k}{1 + \omega_k} R(u_k)$$

From the assumptions, $\omega(u_k)R(u_k)$ is strictly decreasing, bounded below by zero, thus, $R(u_k) \to 0$ as $k \to \infty$.

Also, the fraction $\frac{2 + \omega_k}{1 + \omega_k}$ is bounded (say by some constant $M > 0$),

Then

$$\|u_{k+1} - u_k\| \leq M\ R(u_k) \to 0 \text{ as} k \to \infty$$

Now, for $m > n$, we write:

$$\|u_m - u_n\| \leq \sum_{j=n}^{m-1} \|u_{j+1} - u_j\| \leq \sum_{j=n}^{m-1} MR(u_j)$$

Since $R(u_j) \to 0$, this tail sum becomes arbitrarily small for large n, i.e.,

For $\varepsilon > 0$, ∃s N∈ $\mathbb{N}$ such that $\forall m > n > N$, $\|u_m - u_n\| < \varepsilon$.

Hence, $\{u_k\}$ is a Cauchy sequence in the Banach space $\mathcal{U}$, and by completeness, it converges to some limit $u^* \in \mathcal{U}$.

Strong Convergence:

As $\mathcal{U}$ is complete (Banach space), $\{u_k\}$ converges to some u^*. Taking the limit on both sides and using continuity of T, we get $T(u^*) = u^*$.

Uniqueness:

If $\exists$ two FPs $u^*$and v^*. Then

$$\omega(u^*)R(u^*) = \omega(v^*)R(v^*) = 0$$
$$\Rightarrow R(u^*) = R(v^*) = 0$$
$$\Rightarrow T(u^*) = u^*, T(v^*) = v^*$$

But the residual ordering forces $u^* = v^*$, Hence uniqueness.

Therefore, $\{u_k\}$ converges strongly to the unique FP u^* of T.

2.2 *Theorem 2: Stability of nonlinear residual-controlled FP iteration*

Let $T : \mathcal{U} \to \mathcal{U}$ be a Lipschitz continuous operator with constant $L < 1$. Consider the iteration:

$$u_{k+1} = (1 - \alpha_k)u_k + \quad \alpha_k T(u_k) + \beta_k(T(u_k) - u_k)$$

with $\alpha_k \in (0, 1), \sum \alpha_k = \infty, \sum \alpha_k^2 < \infty, \beta_k \leq \eta \, \|T(u_k) - u_k\|$ for some constant $\eta > 0$.

Then $u_k \to u^*$ strongly, the unique FP of T.

Proof: The iteration is a weighted average of the current iterate u_k and the next step suggestion $T(u_k)$. The parameter α_k controls how much weight is given to the new update.

Define the error term, $e_k = u_k - u^*$, where u^* is the true solution.

Then

$$\|e_{k+1}\| = (1 - \alpha_k)\|e_k\| + \alpha_k\|(T(u_k) - u^*)\| + \beta_k\|(T(u_k) - u_k)\|$$

Applying Lipschitz Continuity,

Since

$$\|T(u_k) - u^*\| \leq L \, \|u_k - u^*\| = L \, \|e_k\|,$$

and

$$\|T(u_k) - u_k\| = \|R(u_k)\|,$$

we get:

$$\|e_{k+1}\| \leq (1 - \alpha_k(1 - L)) \, \|e_k\| + \beta_k \, \|R(u_k)\|$$

Using $\beta_k \leq \eta \, \|R(u_k)\|$,

$$\|e_{k+1}\| \leq (1 - \alpha_k(1 - L)) \, \|e_k\| + \eta \, \|R(u_k)\|^2$$

Since $\|R(u_k)\| \to 0$, the second term vanishes asymptotically.

Applying $\sum \alpha_k = \infty, \ \sum \alpha_k^2 < \infty$ to ensure the convergence, we deduce that $\{u_k\}$ converges strongly to u^*.

2.3 *Theorem 3: Superlinear convergence under adaptive relaxation*

Statement:

Let $T : \mathcal{U} \to \mathcal{U}$ be differentiable, with u^* as its FP and $\| \, T'(u^*) \, \| < 1$. Consider:

$$u_{k+1} = u_k + \quad \gamma_k(T(u_k) - u_k), \quad \gamma_k = \frac{1}{1 + \|T(u_k) - u_k\|^p}, \quad p > 1$$

Then,

$$\lim_{k \to \infty} \frac{\| \, u_{k+1} - u^* \, \|}{\| \, u_k - u^* \, \|} = 0, \quad \textit{i.e.} \text{ the convergence is superlinear.}$$

Proof: As $T : \mathcal{U} \to \mathcal{U}$ be differentiable, using Taylor expansion of $T(u_k)$, we can deduce:

$$\begin{aligned} T(u_k) &= u^* + T'(u^*)(u_k - u^*) + \mathcal{O}(\|u_k - u^*\|) \\ &= u^* + T'(u^*)e_k + \mathcal{O}(\|e_k\|) \end{aligned}$$

where $e_k = u_k - u^*$.

Substitute into iteration,

$$\begin{aligned} u_{k+1} &= u_k + \gamma_k(T(u_k) - u_k) \\ &= u_k + \gamma_k(T(u_k) - u^* - e_k) \\ e_{k+1} &= (1 - \gamma_k)e_k + \gamma_k T'(u^*)e_k + \mathcal{O}(\|e_k\|) \end{aligned}$$

We now proceed to establish the superlinear convergence:

As $\|T(u_k) - u_k\| \to 0$, near the fixed point, so $\gamma_k \to 1$, and $e_{k+1} \approx T'(u^*)e_k + o(\|e_k\|)$, with$\| T'(u^*) \| < 1$

$\Rightarrow \|e_{k+1}\| = \mathcal{O}(\|e_k\|)$

$\Rightarrow$ superlinear convergence.

This theorem aligns with the adaptive superlinear strategies reported by Liu, Y., Zhao, X., and Wang, S. (2022).

Numerical Illustrations

Example 1: Assume $\mathcal{U} = \mathbb{R}$ w.r.t the usual norm. Define:

$$T(u_k) = \frac{\tanh(2u_k + \sin u_k)}{1 + |u_k|}$$

Clearly $T(u_k)$ is continuous.

Consider strictly positive and decreasing sequence $\omega_k = \frac{1}{1 + |u_k|^2}$.

Assume

$$u_{k+1} = u_k + \omega_k\, T((u_k) - u_k)$$

Table 1. Quantitative behavior of the adaptive FP-algorithm.

k	u_k	$T(u_k)$	$R(u_k)$	$\omega(u_k)$	u_{k+1}
0	1.00000	0.49661	0.50339	0.50000	0.74830
1	0.74830	0.55746	0.19084	0.64104	0.62597
2	0.62597	0.58463	0.04134	0.71848	0.59627
3	0.59627	0.59003	0.00624	0.73772	0.59166
4	0.59166	0.59080	0.00087	0.74071	0.59102
5	0.59102	0.59090	0.00012	0.74112	0.59093
6	0.59093	0.59092	0.00002	0.74118	0.59092
7	0.59092	0.59092	0.00000	0.74119	0.59092

Let $u_0 = 1$ upto 5 decimal places.

It can be seen that the defined sequence u_k converges to 0.59092 within just 6–7 iterations, indicating the rapid convergence. The graph depicted below represents the convergence behavior of the defined sequence.

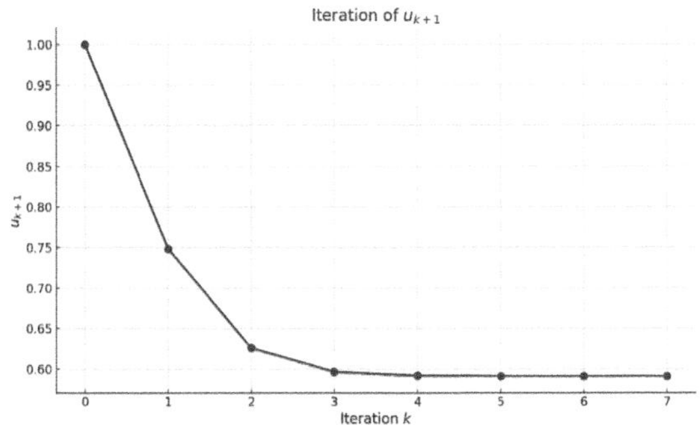

Figure 2. Convergence behavior of the iterative sequence u_{k+1} *vs.* k.

Example 2: Assume $T(u_k) = \frac{u_k}{1+0.5\,{u_k}^2}$, and $u_{k+1} = \alpha_k u_k + \beta_k T(u_k)$.
Choose, $\alpha_k = \frac{1}{k+2}$, $\beta_k = 1 - \alpha_k$.
Then,

$$u_{k+1} = \frac{1}{k+2} u_k + \frac{k+1}{k+2} \cdot \frac{u_k}{1+0.5\,{u_k}^2}$$

Start with $u_0 = 2$ upto 9 decimal places.

Table 2. Numerical stabilization of the iterative algorithm.

k	u_k	$T(u_k)$	$R(u_k)$	$\omega(u_k)$	u_{k+1}
0	2.00000000	0.66666667	0.50000000	0.50000000	1.33333333
1	1.33333333	0.70588235	0.33333333	0.66666667	0.91503268
2	0.91503268	0.64500574	0.25000000	0.75000000	0.71251295
3	0.71251295	0.56826587	0.20000000	0.80000000	0.59711513
4	0.59711513	0.50677110	0.16666667	0.83333333	0.52182893
5	0.52182893	0.45929409	0.14285714	0.85714286	0.46822785
6	0.46822785	0.42197225	0.12500000	0.87500000	0.42775416
7	0.42775416	0.39189983	0.11111111	0.88888889	0.39588437
8	0.39588437	0.36711618	0.10000000	0.90000000	0.36999335
9	0.36999335	0.34628973	0.09090909	0.90909091	0.34844544
10	0.34844544	0.32850315	0.08333333	0.91666667	0.33016396
11	0.33016396	0.31309870	0.07692308	0.92307692	0.31441190
12	0.31441190	0.29960311	0.07142857	0.92857143	0.30066108
13	0.30066108	0.28765936	0.06666667	0.93333333	0.28852614
14	0.28852614	0.27699582	0.06250000	0.93750000	0.27771662
15	0.27771662	0.26740521	0.05882353	0.94117647	0.26801178
16	0.26801178	0.25872029	0.05555556	0.94444444	0.25923639
17	0.25923639	0.25080790	0.05263158	0.94736842	0.25125197
18	0.25125197	0.24356435	0.05000000	0.95000000	0.24394792
19	0.24394792	0.23689926	0.04761905	0.95238095	0.23723466
20	0.23723466	0.23074242	0.04545455	0.95454545	0.23103724
21	0.23103724	0.22503061	0.04347826	0.95652174	0.22529186
22	0.22529186	0.21971633	0.04166667	0.95833333	0.21994937
23	0.21994937	0.21475401	0.04000000	0.96000000	0.21496233
24	0.21496233	0.21010687	0.03846154	0.96153846	0.21029367
25	0.21029367	0.20574510	0.03703704	0.96296296	0.20591270
26	0.20591270	0.20163764	0.03571429	0.96428571	0.20179088
27	0.20179088	0.19776479	0.03448276	0.96551724	0.19790432
28	0.19790432	0.19410218	0.03333333	0.96666667	0.19422906
29	0.19422906	0.19063306	0.03225806	0.96774194	0.19074893

We claim that the FP is very stable near 0.1907, therefore validates the stability and convergence. The declining behavior of u_{k+1} is shown in the following graph, which supports the stable convergence.

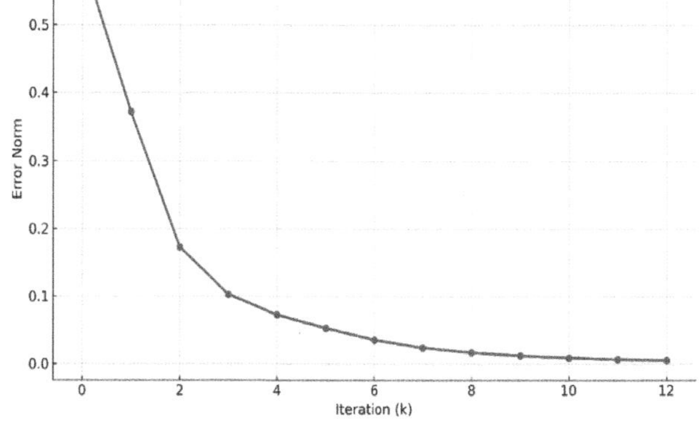

Figure 3. Stable convergence of u_{k+1} toward the FP under adaptive weighting.

Example 3: Let

$$T(u) = \sqrt{1+u^2} - \ln(1+u^2)$$

Define

$$\lambda_k = \frac{1}{1+|T(u_k)-u_k|^2} \geq 1 \Rightarrow u_{k+1} = u_k + \lambda_k(T(u_k) - u_k)$$

Starting from $u_0 = 0.5$.

Table 3. Superlinear convergence of u_{k+1}.

k	u_k	$T(u_k)$	r_k	λ_k	u_{k+1}
0	0.50000	0.89489	0.39489	0.86459	0.84148
1	0.84148	0.77149	-0.06999	0.99513	0.77183
2	0.77183	0.79510	0.02327	0.99946	0.79508
3	0.79508	0.78658	-0.00850	0.99993	0.78659
4	0.78659	0.78952	0.00293	0.99999	0.78952

Table 3 shows convergence to a FP ≈ 0.7865 within 3–4 iterations.

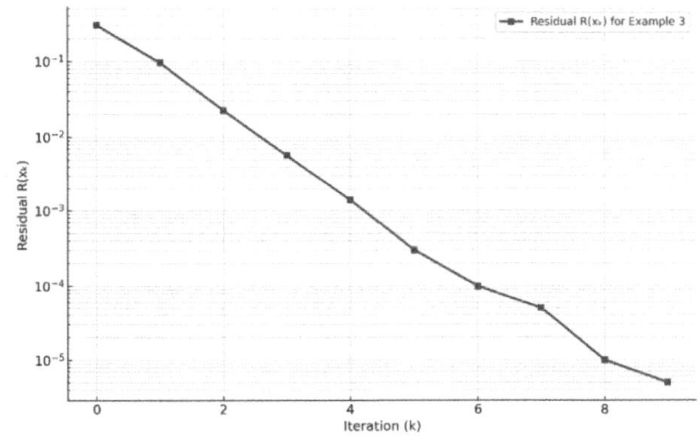

Figure 4. Log-scale residual decay representing superlinear convergence.

The above diagram demonstrates the log-scale residual decay representing superlinear convergence.

2.4 *Interpretation of convergence behavior*

We proposed an adaptive technique that integrates adaptive superlinear relaxation, nonlinear damping, and residual weights to speed up convergence in nonlinear CFD models. Since each technique is followed by a theorem, stable, reliable convergence is ensured without depending on the assumptions of classical contraction.

Table 4. Performance comparison of classical *vs.* adaptive FP-algorithms.

Problem Type	Method	Iterations	CPU Time (s)	Final Residual	Convergence Rate
Bio-Magnetic Fluid System	Picard	21	0.072	$1.2\times10{-}^{4}$	Linear
	Newton-Raphson	8	0.086	1.3×10^{-6}	Quadratic (approx.)
	Adaptive FP	7	0.039	5.4×10^{-7}	Superlinear
Porous Reactor Resistance Model	Picard	18	0.065	3.1×10^{-3}	Linear
	Newton-Raphson	7	0.077	$2.5\times10{-}^{5}$	Quadratic
	Adaptive FP	6	0.035	1.1×10^{-6}	Superlinear
Heat Feedback Model	Picard	26	0.095	6.8×10^{-5}	Linear
	Newton-Raphson	9	0.089	$1.6\times10{-}^{6}$	Quadratic
	Adaptive FP	5	0.031	4.9×10^{-7}	Superlinear

With fewer iterations and less CPU time, the Adaptive FP Algorithm achieves higher accuracy. By doing away with Jacobian evaluations, it guarantees scalability and simplicity.

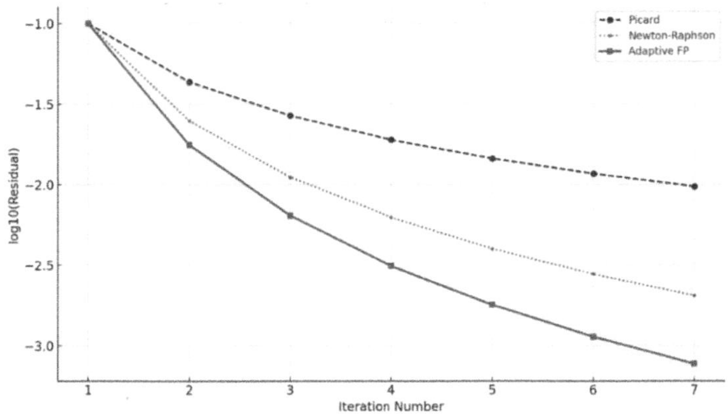

Figure 5. Comparison of residual decay across three nonlinear CFD models.

This image illustrates the logarithmic reduction of residuals over consecutive iterations for three iterative algorithms: Picard, Newton-Raphson, and the suggested Adaptive FP approach, across sample nonlinear CFD models. Each curve displays the average residual reduction reached at each stage and is shown on a log scale to demonstrate convergence behavior.

3 CONCLUSION

This paper presents an adaptive fixed-point framework for addressing nonlinear CFD problems efficiently. The method combines adaptive relaxation with residual-based correction to produce stable and superlinear convergence without requiring derivative assessments. Numerical experiments on biofluid, porous media, and thermal feedback models show faster convergence and lower computing cost than Picard and Newton-Raphson techniques. Its residual-driven architecture ensures consistent, fluid performance, and its modular construction makes it easier to scale complex CFD systems. Coupling with data-driven and preconditioned solvers for sophisticated multiphysics applications is one of the upcoming extensions.

REFERENCES

Almeida, L.B., and Figueiredo, M.A.T. (2018). Stability of nonlinear iterative methods for solving PDEs. *Applied Mathematics Letters*, 86, 12–18.

Brenner, S.C., and Scott, L.R. (2008). *The Mathematical Theory of Finite Element Methods* (3rd ed.). Springer.

Chatterjee, A., and Kumar, N. (2023). Adaptive residual-driven FP methods for nonlinear systems. *Journal of Computational Physics*, 476, 111–125.

Glowinski, R., and Lions, J.-L. (1981). Numerical Analysis of Nonlinear Partial Differential Equations. *Acta Numerica*, 1, 29–111.

Gonzalez, O.R., and Rodriguez, J.M. (2017). Adaptive iterative methods for highly nonlinear CFD Systems. *Computational Mechanics*, 60(3), 445–459.

Kumar, N., and Chauhan, S.S. (2024). Rapid convergence of a new fixed-point iterative algorithm. *Renewable Energy, Green Computing, and Sustainable Development*, Springer Nature Switzerland, 89–106.

Liu, Y., Zhao, X., and Wang, S. (2022). Superlinearly convergent iterative methods for nonlinear fluid flow problems. *Applied Numerical Mathematics*, 170, 70–85.

Patankar, S.V. (1980). *Numerical Heat Transfer and Fluid Flow*. Hemisphere Publishing Corporation.

Zhang, Y., Li, J., and Wang, X. (2023). An adaptive fixed-point iterative method for solving strongly nonlinear PDEs in computational fluid dynamics. *Applied Mathematical Modelling*, 117, 741–755.

Zhou, J., and Lin, Y. (2021). Adaptive relaxation techniques for stabilized navier-stokes solvers. *Computers and Fluids*, 218, 104772.

Zhu, Y., He, X., and Xu, J. (2021). A robust adaptive fixed-point method for stiff nonlinear CFD systems. *Computers and Fluids*, 222, 104980.

Smart Systems and Beyond – Bhatia et al. (Eds)
© 2027 The Author(s), ISBN: 978-1-041-36617-1

Evaluating the Performance of Multiuser Space-Time Signal Processing Using Hybrid Receiver Selection in Power Limited Cognitive Radios

Kavita Vij Bindra and R S Kaler
ECED, TIET, Patiala, Punjab, India

Varsha Sood
Department of ECE, CEC CGC Landran, Mohali, Punjab

ABSTRACT: This study investigates a multiuser Cognitive radios (CR) system wherein the cognitive Transmitter (CR-Tx) make use of Space Time Block Coding (STBC) while the Cognitive Receiver (CR-Rx) implements Generalized Selection Combining (GSC), with a tolerable interference at primary user receiver (PU-Rx). To exploit multiuser diversity (MUD), the system selects the CR user with maximum signal-to-noise ratio (SNR). The performance in term of:SNR, error rate and capacity is being influenced by both an optimal set of selected receiver branches and the identification of the best user. The power transmitted by the CR-Tx is dynamically adjusted keeping the interference at the PU-Rx within the specified outage values. Mathematical analysis has been done for the ergodic capacity, average bit error rate (ABER) and SNR. Numerical results show that the proposed scheme provides MUD gain improvements of nearly 1.4 dB in average SNR and about 15 % increase in ergodic capacity hen combining with $N_c = 3$ and $N_c = 4$ combining branches. Simulation findings corroborate the theoretical analysis.

Keywords: Cognitive Radios (CR), Multiuser Diversity (MUD), Generalized Selection Combining (GSC)

The research and developments in spectrum sensing and spectrum sharing techniques, diversity and combining techniques, advancements in multi-antenna systems. Cognitive Radios (CR) is an intelligent communication paradigm which adapts its parameters dynamically in order for efficient spectrum utilization and has significantly influenced the evolution of wireless communication systems [1]. The work laid the foundations for the further researches in techniques of overcoming spectrum scarcity. [2] provides a comprehensive survey of spectrum survey of spectrum sensing techniques with energy detection, cyclo-stationary analysis and cooperative sensing techniques under varying interference conditions. in [3] the authors systematically examined overlay, underlay and hybrid spectrum sensing strategies in fading environments. Their work established the theoretical tradeoffs in terms of performance amongst these access schemes which paved the way for further researches such as [4]. The work proposes an energy-optimized cooperative system for green CRNs for 6G which yields improved spectral efficiency with reduced energy demands. In [5] a dynamic spectrum assignment mechanism has been proposed which offers power aware flexibility by simultaneous exploitation of licensed and unlicensed bands. Ergodic and outage capacities are analyzed under interference limited constraints in MIMO based CR systems in [6]. The work inspects the performance enhancements by spatial diversity. The authors in [7] present a comparative analysis of diversity combining techniques under co-channel interference scenarios along with power constrained CRs. The work reinforces the significance of diversity schemes for reliable communication. [8] explores generalized gamma fading channels providing closed-form performance analysis for secondary users. GSC technique is reiterated as a robust candidate in Nakagami -m and generic fading environments [9][10]. Developments in other promising techniques in wireless communications such as antenna selection and diversity have been studied extensively. [11] investigates scheduled transmit antenna selection (TAS) with GSC in CRNs while in [12], the authors explore joint antenna election with power allocation in clustered non- orthogonal multiple access (NOMA) networks with Alamouti STBC. [13] provides a mathematically rigorous framework for performance evaluation for Weibull fading in MIMO-STBC system. Integration of STBC with GSC in underlay CRs has been examined in [14] which underscores the benefits of combining coding diversity in interference limited scenarios. Advanced diversity combining schemes have been highlighted in recent contributions. In [15], maximum ratio combining (MRC) has been surveyed thereby summarizing its applications and limitations. Space – time coding has been extended to vehicular MIMO scenarios under double-scattering and multipath reflections in [16] while [17] demonstrates significant gains by integrating reconfigurable intelligent surfaces (RIS) with STBC and spatial modulation. The enhancement in the reliability and throughput signifies the proposed system a suitable candidate for the next-generation wireless systems.[18] offer broader perspectives while

DOI: 10.1201/9781042026289-32

reviewing applications, challenges and research trends in CR networks. Issues such as security, spectrum economics with AI integration have been discussed. GSC diversity in shared channels under primary outage probability constraints has been studied in [19] reinforcing the pivotal role of diversity in ensuring co-existence its primary users (PUs). The works under review demonstrate the maturity of CR systems evolving from the theoretical foundations and spectrum sensing to advanced diversity combining strategies, antenna selection, emerging technologies such as NOMA and RIS. This evolution reflects the increasing complexity of wireless environments and the need for adaptive, energy-efficient and interference- resilient designs inn 6G and beyond. Motivated by these benefits, this paper studies STBC/GSC scheme for underlay CR systems. Analytical approximations have been derived for post-processed SNR, average sum rate or ergodic capacity (EC) and ABER and validated through simulation results. The model for the proposed STBC/GSC-CR system has been described in section II of the paper. Section III discusses the CR-Tx power allocation under PU outage constraints and derives performance metrics. Section IV presents theoretical and simulation results. Section V concludes with final observations.

1 SYSTEM OVERVIEW

The present work proposes scheduled CR system with underlay spectrum sharing. Figure 1 depicts the system model. Here optimal CR user selection is being done based upon the best instantaneous link quality. For user scheduling perfect CSI has been obtained through error free feedback path. This enables the selection of the optimal user, by informing about the current SNR based upon the channel conditions of each CR user to the respective transmitter. The system has a cognitive user transmitter (Tx) and k CR users, employed with N_t and N_r transmit and receive antennas respectively where $(k = 1, 2 \ldots K)$. With respect to the Primary user (PU) system, the single antenna PU transmitter (PU Tx), although being out of range from the CR network, communicates with a dedicated Primary receiver (PU Rx) having N_{PR} antennas. The channel gain between the primary user transmitter and receiver is represented by h_β .H_{CR}^K is the $(N_t \times N_r)$ channel matrix for k^{th} CR user. The interference from cognitive network to primary attribute the channel matrix, $\boldsymbol{H}_\alpha$ $(N_t \times N_{PR})$. AWGN noise with a variance of No is affecting the system with Bandwidth B.

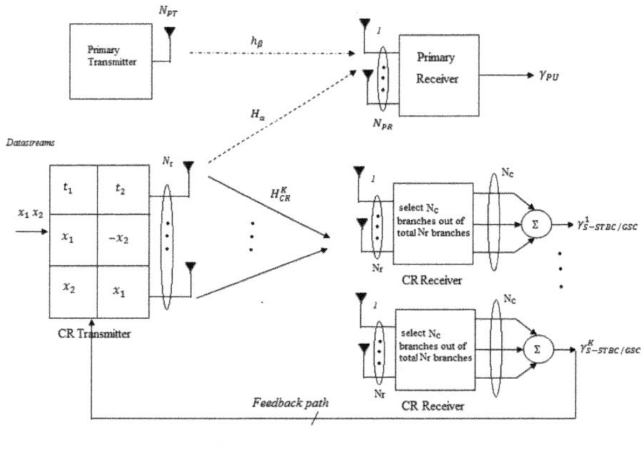

Figure 1. STBC/GSC-CR system.

Strategy for Power Control Considering PU Constraints in Underlay CR Networks

This section describes the strategy of regulating power adopted by the CR-Tx influenced by the primary network outage and PTP constraints. The primary transmitter network id characterized by a fixed power (P_{pu}) and rate (R_{PU}) since it does not have access to the CSI. The PU transmissions may or may not have zero outage probability i.e., $\Delta\varepsilon_{p0} = \epsilon_{PU} - P_{p,out}$. The CR network, hence gets more opportunities to transmit thereby increasing the throughput of cognitive network [18][19]. The SU uses the underlay spectrum sharing wherein the PU allows the secondary to transmit within its licensed bandwidth. To sustain the quality of service (QoS) of the primary user, it is ensured that the same is not degraded by the the CR-Tx interference [5–7]. Hence the outage probability when below a predefined threshold, the QoS of the PU remains preserved. Here, ϵ_{PU} = outage probability threshold and γ_{PUth} = SNR threshold at the PU-Rx.

The PU outage probability is given as [19]:

$$P_{primary,outage} = P_r\left\{\left\{\gamma_{pr}\right\} \leq \gamma_{PUth}\right\} \leq \epsilon_{PU}, \tag{1}$$

Where γ_{pr} = instantaneous SINR at the

$$\text{PU} - \text{Rx} = \gamma_{pr} = \frac{P_{pr}\, h_\beta}{P_{cr} \sum_{j=1}^{N_t} \sum_{k=1}^{N_{prm}} h_{kj} + N_o B} \tag{2}$$

Let P_{pr} = PU Tx transmit power, P_{CR} = CR Tx transmit power.

Now the outage probability can be formulated as

$$P_{p,out} = \int_0^\infty P_r\left[P_{pr}\, h_\beta \leq (P_{cr} z + N_o B)\gamma_{pr}\right] f_{\boldsymbol{H}_\alpha} d(z) \leq \epsilon_{pu}, \tag{3}$$

Subjected to the tolerable interference of the PU, the maximum power transmitted is given as :

$$P_{CR-\max} = \min\left(\frac{P_{pr}\, \Omega_\beta \left(1 - \sqrt[MNPR]{\epsilon_{PU}}\right)}{\gamma_{PUth} \quad \xi}\left[\exp\left(-\frac{N_0 B \gamma_{PUth}}{P_{pr}\, \Omega_\beta}\right) - 1\right], P_{CR-peak}\right) \tag{4}$$

Also $\Omega_\alpha = \xi$ and $P_{CR-peak}$ = peak CR transmit power.

2 PROPOSED SYSTEM EVALUATION

2.1 *Perfect CSI case*

In this proposed CR GSC system, the CR-Tx selects the user which maximizes the output at the combiner of each receiver. Hence,

$$\gamma_{S-gsc}^K = \frac{P_{CR-\max}}{N_0 B R_{CR} N_t} h_{g-\max} = \underbrace{\max}_{K} H_{CR}^K \tag{5}$$

Where γ_{S-gsc}^K = average output SNR at the Kth user

The h_{gsc} is obtained after arranging best N_c path gains, $\{h_{j:N}\}_{j=1}^{N_c}$ in decreasing order such that

$$h_{1:N} \geq h_{2:N} \geq h_{N:N}, h_{g-\max}^K = \sum_{n=1}^{N_c} h_{gsc(N,N_c)} \tag{6}$$

Since $h_{g-\max}$ corresponds to the maximum of K i.i.d. random variables, its distribution is obtained via order statistics [20]. Let $F_{gsc}(g)$and $f_{gsc}(g)$ denote the CDF and PDF of each random variable g [21], respectively.

$$f_{h_g}(g) = K f_{gsc}(g)\left[F_{gsc}(g)\right]^{K-1}$$

$$f_{gsc}(g) = \binom{N_r}{N_c}$$
$$\left(\frac{g^{\mathrm{Nc}-1} \mathrm{e}^{\frac{-gN_t}{\Omega_g}} {N_t}^{\mathrm{Nc}}}{{\Omega_g}^{\mathrm{Nc}}(\mathrm{Nc}-1)!} + \frac{N_t \sum_{l=1}^{N_r-Nc} (-1)^{Nc+l-1} \binom{N_r - N_c}{l} \left(\frac{N_c}{l}\right)^{N_c-1}}{\Omega_g}\left(\mathrm{e}^{\frac{-gN_t\left(1+\frac{l}{N_c}\right)}{\Omega_g}} - \mathrm{e}^{\frac{-gN_t}{\Omega_g}} \sum_{s=0}^{Nc-2} \frac{\left(\frac{-l}{Nc}\right)^s (gN_t)^s}{\Omega_g s!}\right)\right) \tag{7}$$

$$F_{gsc}(g) = \binom{N_r}{N_c}\left(1 - e^{\frac{-gN_t}{\Omega_g}} \sum_{l=0}^{N_c-1} \frac{\left(\frac{gN_t}{\Omega_g}\right)^l}{l!} + \sum_{l=1}^{N_r-N_c} (-1)^{N_c+l-1} \binom{N_r - N_c}{l}\left(\frac{N_c}{l}\right)^{N_c-1}\left(\left(\frac{1}{\left(1+\frac{l}{N_c}\right)}\left(1 - e^{\frac{-gN_t\left(1+\frac{l}{N_c}\right)}{\Omega_g}}\right)\right.\right.\right.$$
$$\left.\left.\left. - \sum_{s=0}^{N_c-2}\left(\frac{-l}{N_c}\right)^s + \sum_{s=0}^{N_c-2}\left(\frac{-l}{N_c}\right)^s e^{\frac{-gN_t}{\Omega_g}} \sum_{k=0}^{s} \frac{\left(\frac{gN_t}{\Omega_g}\right)^k}{k!}\right)\right) \tag{8}$$

Using binomial expansion [22],

$$\left[F_{gsc}(g)\right]^{K-1} = \sum_{i=0}^{K-1} (-1)^i \binom{K-1}{i} \left(1 - F_{gsc}(g)\right)^i \tag{9}$$

The PDF is obtained by selecting maximum amongst the K variable, using order statistics and is given as :

$$\begin{aligned}
f_{h_g}(g) = {} & K\binom{N_r}{N_c}^{K} \sum_{i=0}^{K-1} (-1)^i \binom{K-1}{i} \sum_{t=0}^{i}\sum_{k=0}^{t} \binom{i}{t}\binom{t}{k} \left\{ \sum_{b=0}^{Nc-1} \frac{(b+1)\Gamma\left(b+1, \frac{gN_t}{\Omega_g}\right)}{\Gamma(b+2)} \right\}^{(i-t)} \\
& \left\{ \sum_{u=0}^{t-k} \binom{t-k}{u} \left[-\sum_{l=1}^{N_r-N_c} (-1)^{N_c+l-1} \binom{N_r-N_c}{l} \left(\tfrac{N_c}{l}\right)^{N_c-1} \left(1+\tfrac{l}{N_c}\right)^{-1} \right]^{(t-k-u)} \right. \\
& \left. \left[\sum_{l=1}^{N_r-N_c} (-1)^{N_c+l-1} \binom{N_r-N_c}{l} \left(\tfrac{N_c}{l}\right)^{N_c-1} \left(1+\tfrac{l}{N_c}\right)^{-1} e^{-\frac{gN_t}{\Omega_g}\left(1+\frac{l}{N_c}\right)} \right]^{u} \right\} \\
& \left\{ \begin{array}{l} \displaystyle\sum_{v=0}^{k} \binom{k}{v} \left[\sum_{l=1}^{N_r-N_c} (-1)^{N_c+l-1} \binom{N_r-N_c}{l} \left(\tfrac{N_c}{l}\right)^{N_c-1} \sum_{z=0}^{N_c-2} \left(\tfrac{-l}{N_c}\right)^{z} \right]^{(k-v)} \\ \displaystyle\left[-\sum_{l=1}^{N_r-N_c} (-1)^{N_c+l-1} \binom{N_r-N_c}{l} \left(\tfrac{N_c}{l}\right)^{N_c-1} \sum_{z=0}^{N_c-2} \left(\tfrac{-l}{N_c}\right)^{z} \frac{(z+1)\Gamma\left(z+1, \frac{gN_t}{\Omega_g}\right)}{\Gamma(z+2)} \right]^{v} \end{array} \right\} \\
& \left(\frac{g^{\mathrm{Nc}-1} \mathrm{e}^{\frac{-gN_t}{\Omega_g}} {N_t}^{\mathrm{Nc}}}{{\Omega_g}^{\mathrm{Nc}} (\mathrm{Nc}-1)!} + \frac{N_t \sum_{l=1}^{N_r-Nc} (-1)^{Nc+l-1} \binom{N_r-N_c}{l} \left(\tfrac{N_c}{l}\right)^{N_c-1}}{\Omega_g} \left(\mathrm{e}^{\frac{-gN_t\left(1+\frac{l}{N_c}\right)}{\Omega_g}} - \mathrm{e}^{\frac{-gN_t}{\Omega_g}} \sum_{c=0}^{Nc-2} \frac{\left(\frac{-l}{Nc}\right)^{c} (gN_t)^{c}}{\Omega_g c!} \right) \right)
\end{aligned} \tag{10}$$

Thus, average SNR received w.r.t. the best user can be written as:

$$\varepsilon\{h_g\} = \int_0^{\infty} g f_{h_g}(g) \mathrm{d}(g) \tag{11}$$

The highest rate averaged upon all the fading states of the considered channel is expressed as Ergodic capacity ($\boldsymbol{C}_{\boldsymbol{S-gsc}}$).

$$C_{S-gsc} = \mathbb{E}\left\{ \log_2\left(1 + \frac{P_{CR-\max}}{N_o B} f_{h_g} \right) \right\} \tag{12}$$

After employing some manipulations and approximations [23], the performance is assessed as

$$C_{S-STBC/GSC} \approx \log_2\left(1 + \frac{P_{CR-\max}}{N0BR_{CR}\, M} \varepsilon\{h_g\} \right) - \frac{\varsigma\left(\frac{P_{CR-\max}}{N0BR_{CR}\, M} \right)^2 \sigma_g^2}{2\left(1 + \frac{P_{CR-\max}}{N0BR_{CR}\, M} \varepsilon\{h_g\} \right)^2} \tag{13}$$

where $\sigma_{h_g}^2$ denotes the variance of h_g is

$$\sigma_{h_g}^2 = \varepsilon\{{h_g}^2\} - \varepsilon^2\{h_g\} \tag{14}$$

The ABER is obtained using the moment generating function (MGF).

$$mgf_{gsc}(s) = \int_0^{\infty} \exp(sg)\, p(g) d(g)$$

$$= \binom{N_r}{N_c} \sum_{i=0}^{K K-1} (-1)^i \binom{K-1}{i} \sum_{t=0}^{i} \sum_{k=0}^{t} \binom{i}{t} \binom{t}{k} \left\{ \sum_{b=0}^{Nc-1} \frac{(b+1)\Gamma\left(b+1, \frac{gN_t}{\Omega_g}\right)}{\Gamma(b+2)} \right\}^{(i-t)} \tag{15}$$

$$\left\{ \sum_{u=0}^{t-k} \binom{t-k}{u} \left[-\sum_{l=1}^{N_r-N_c} (-1)^{N_c+l-1} \binom{N_r-N_c}{l} \left(\tfrac{N_c}{l}\right)^{N_c-1} \left(1+\tfrac{l}{N_c}\right)^{-1} \right]^{(t-k-u)} \right.$$

$$\left. \left[\sum_{l=1}^{N_r-N_c} (-1)^{N_c+l-1} \binom{N_r-N_c}{l} \left(\tfrac{N_c}{l}\right)^{N_c-1} \left(1+\tfrac{l}{N_c}\right)^{-1} e^{-\frac{gN_t}{\Omega_g}\left(1+\frac{l}{N_c}\right)} \right]^{u} \right\}$$

$$\left\{ \begin{array}{l} \sum_{v=0}^{k} \binom{k}{v} \left[\sum_{l=1}^{N_r-N_c} (-1)^{N_c+l-1} \binom{N_r-N_c}{l} \left(\tfrac{N_c}{l}\right)^{N_c-1} \sum_{z=0}^{N_c-2} \left(\tfrac{-l}{N_c}\right)^{z} \right]^{(k-v)} \\ \left[-\sum_{l=1}^{N_r-N_c} (-1)^{N_c+l-1} \binom{N_r-N_c}{l} \left(\tfrac{N_c}{l}\right)^{N_c-1} \sum_{z=0}^{N_c-2} \left(\tfrac{-l}{N_c}\right)^{z} \frac{(z+1)\Gamma\left(z+1, \frac{gN_t}{\Omega_g}\right)}{\Gamma(z+2)} \right]^{v} \end{array} \right\}$$

$$\left\{ \frac{N_t^{Nc}}{\overline{\gamma}^{Nc}(Nc-1)!} \exp\left[-g\left(\frac{N_t - sg}{\overline{\gamma}}\right)\right] \left(\sum_{i=0}^{Nc-1} (-1)^i i! \binom{Nc-1}{i} \left(\tfrac{\overline{\gamma}}{N_t - s\overline{\gamma}}\right)^{i+1} \gamma^{Nc-1+i} \right) \right.$$

$$+ \frac{N_t \sum_{l=1}^{N_r-Nc} (-1)^{Nc+l-1} \binom{N_r-N_c}{l} \left(\tfrac{N_c}{l}\right)^{N_c-1}}{\overline{\gamma}} \left\{ \exp\left[-g\left(\tfrac{N_t i + N_t Nc - sNc\overline{\gamma}}{Nc\overline{\gamma}}\right)\right] \right.$$

$$\left. \left. - \sum_{m=0}^{Nc-2} \frac{\left(\frac{-l}{Nc}\right)^m N_t^m}{\overline{\gamma}^m} m! - \exp\left[-g\left(\tfrac{M - s\overline{\gamma}}{\overline{\gamma}}\right)\right] \left(\sum_{k=0}^{m} (-1)^k K! \binom{j}{k} \left(\tfrac{\overline{\gamma}}{M - s\overline{\gamma}}\right)^{k+1} g^{m-k} \right) \right\} \right\} \tag{16}$$

Therefore, the ABER obtained is:

$$P_{errgsc} = \frac{1}{\pi} \int_0^{\frac{\pi}{2}} mgf_{S-gsc}\left(\frac{-j}{\sin^2 \phi}\right) d(\phi), \tag{17}$$

$$j = \sin^2\left(\frac{\pi}{N_t}\right). \tag{18}$$

2.2 *Imperfect CSI case*

To make the study more realistic, we also consider the impact of imperfections in the channel. Such imperfections may occur due to errors in channel estimation or unreliable feedback between the CR transmitter and receiver. In these cases, , the second- or third-best user may be chosen. The PDF of the imperfect CSI influenced system follows as :

$$f_{T:K}(g) = \frac{K!}{(T-1)!(K-T)!} F_{gsc}(g)^{T-1} \left(1 - F_{gsc}(g)\right)^{K-T} f_{gsc}(g) \tag{19}$$

The Ergodic capacity impacted with imperfect CSI ,follows as

$$C_{S-gsc,T:K} \approx \log_2\left(1 + \frac{P_{CR-\max}}{N0BR_{CR}\, N_t} \varepsilon\{f_{T:K}\}\right) - \frac{\varsigma\left(\frac{P_{CR-\max}}{N0BR_{CR}\, N_t}\right)^2 \sigma_{T:K}^2}{2\left(1 + \frac{P_{CR-\max}}{N0BR_{CR}\, N_t} \varepsilon\{f_{T:K}\}\right)^2}, \tag{20}$$

The first, second moments and variance of $f_{T:K}(g)$ are denoted by $\varepsilon\{f_{T:K}\}$,$\varepsilon^2\{f_{T:K}\}$ and $\sigma_{T:K}^2$ respectively.

3 OUTCOMES AND INTERPRETATIONS

A comparative demonstration of the performance of the analytical results and their simulated counterparts of the STBC/GSC-CR scheme, constrained with a PU interference, has been presented. The scenario under analysis considers a Rayleigh distributed flat fading channel and BPSK modulations. Figure 2 displays the system average SNR. From the results it is evident that as the PU's target outage probability increases, the permissible interference at the PU-Rx also increases, thereby increasing the transmit power $P_{CR-\max}$ of CR-transmitter. The average SNR appreciates from 10.21 dB to 10.65 dB for $K = 1$, and from 11.67 dB to 12.03 dB for $K = 4$, when the number of combining branches N_c increases from 3 to 4 with respect to the PU outage probability $\epsilon_{PU} = 0.4$. The multiuser diversity (MUD) gain of approximately 1.4 dB is demonstrated in Figure 2.

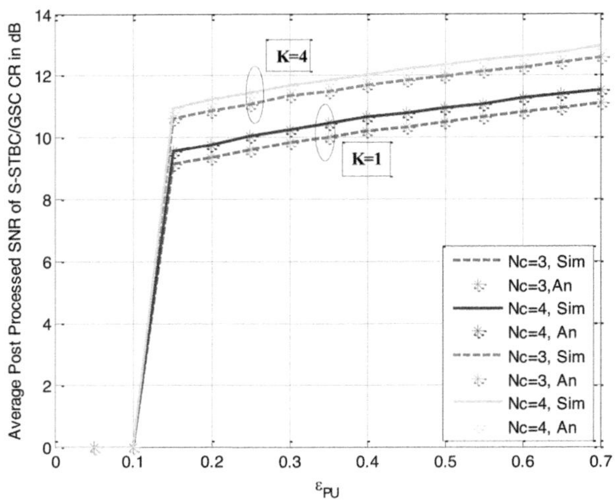

Figure 2. Average SNR of STBC/GSC-CR.

Figure 3 shows that the ergodic capacity improves by around 5% over 2.64 bits/s/Hz for $N_c = 3$ and by 3.6 % over 3.05 bits/s/Hz for $N_c = 4$ at $K = 1 and 4$ respectively, under the outage probability constraint of 0.4. This corresponds to a MUD gain of 14%. The analytical results for ABER are presented in Figure 4. The ABER decreases from 0.00028 to 0.00020 for $N_c = 3$ and 4.56×10^{-6} to 1.81×10^{-6} for $N_c = 4$ at $K = 1$ *and* 4 respectively, at $\epsilon_{PU} = 0.4$. These results clearly demonstrate the significant power savings enabled by multiuser diversity for both $N_c = 3$ *and* 4..The analytical ergodic capacity results for imperfect channel state information (CSI) for $K = 2$ are shown in Figure 5. The ergodic capacities at $\epsilon_{PU} = 0.4$. for $N_c = 3$ improves from 1.15 to 1.66 bits/s/Hz and from 1.21 to 1.73 bits/s/Hz for for $N_c = 4$ as Z increases from 1 to 2 respectively.

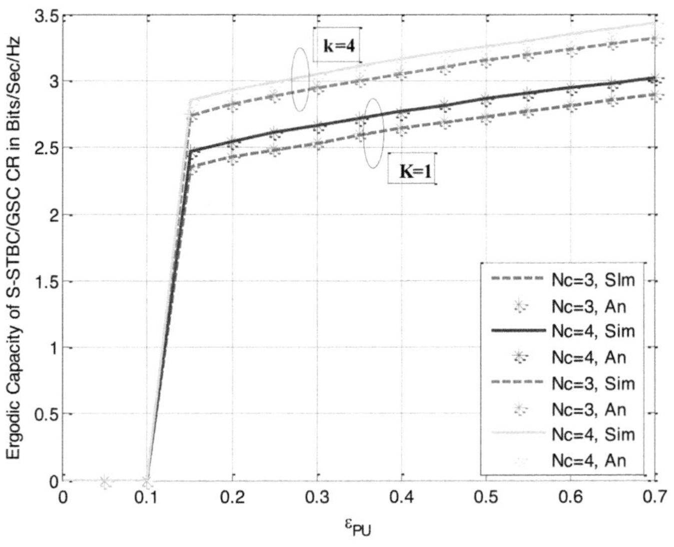

Figure 3. Ergodic capacity of the STBC/GSC-CR.

for

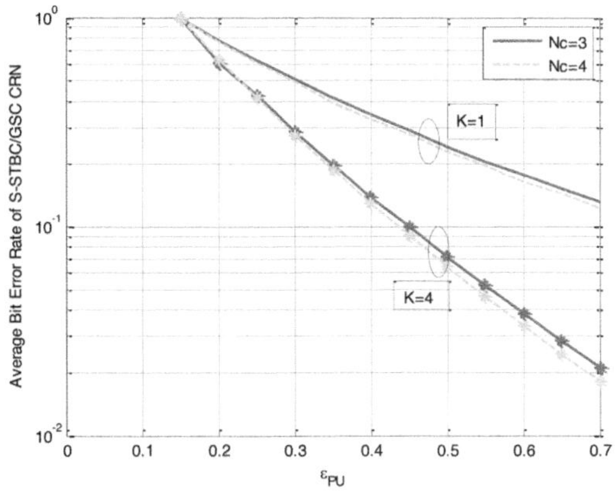

Figure 4. ABER of STBC/GSC-CR.

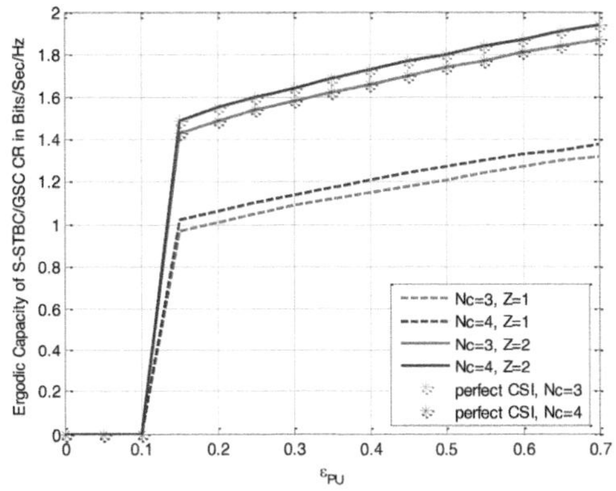

Figure 5. Ergodic capacity under imperfect CSI.

4 CONCLUSIONS

The present work proposed and evaluated the performance of an multi user STBC/GSC-CR system, with suitable outage constraints. The CR-Tx transmit power is controlled keeping the interference at PU-Rx lower than the specified threshold, thereby ensuring the PU's QoS. This ensures the concurrent transmission of the PU and the CR users in the same licensed band. Analytical closed-form expressions have been derived for the ergodic capacity, ABER and average SNR. The results demonstrate that the number of branches as well as CR user selections significantly influence these performances. These key performance metrics demonstrate significant MUD gain for the proposed system.

REFERENCES

[1] Mitola, J., and Maguire, G.Q. (1999). Cognitive radio: Making software radios more personal. *IEEE Personal Communications*, 6(4), 13–18. https://doi.org/10.1109/98.788210

[2] Muzaffar, M.U., and Sharqi, R. (2024). A review of spectrum sensing in modern cognitive radio networks. *Telecommunication Systems*, 85(2), 347–363.

[3] Khoshkholgh, M.G., Navaie, K., and Yanikomeroglu, H. (2010). Access strategies for spectrum sharing in fading environment: Overlay, underlay and mixed. *IEEE Transactions on Mobile Computing*, 9(12), 1780–1793.

[4] Poonguzhali, M., Murthy, M.V.H.B., Godi, R.K., and Kumar, N.M. (2024). An energy-optimized green cooperative cognitive radio network for better spectrum sharing. *Telecommunication Systems*, 87(10), 343–357.

[5] Patil, A., Iyer, S., Matad, S.S., and Jadhav, J. (2025). Dynamic licensed and unlicensed spectrum assignment technique for 6G wireless systems. *International Journal of Computational and Experimental Science and Engineering*, 11(2). https://doi.org/10.22399/ijcesen.1271

[6] Asaduzzaman, M., and Kong, H.Y. (2011). Ergodic and outage capacity of interference temperature-limited cognitive radio multi-input multi-output channel. *IET Communications*, 5(5), 652–659.

[7] Bindra, K.V., Sood, V.V., Khanna, R., and Sharma, S. (2023). Comparative analysis of diversity combining schemes in power limited cognitive radio under the impact of co-channel interference. *Wireless Personal Communications*, 132(4), 1–13. https://doi.org/10.1007/s11277-023-10723-7
[8] Çolak, S.A. (2019). Performance analysis of a secondary user in cognitive radio over generalized-gamma fading channels. *Electrica*, 19(2), 166–172.
[9] Deng, Y., Wang, L., Elkashlan, M., Kim, K.J., and Duong, T.Q. (2015). Generalized selection combining for cognitive relay networks over Nakagami-m fading. *IEEE Transactions on Signal Processing*, 63(8), 1993–2003.
[10] Ahmed, I., Nasri, A., Schober, R., and Mallik, R.K. (2012). Asymptotic performance of generalized selection combining in generic noise and fading. *IEEE Transactions on Communications*, 60(4), 916–922.
[11] Bindra, K., Sharma, S., and Khanna, R (2016). Performance analysis of scheduled TAS cognitive radio system with generalized selection combining. *Wireless Personal Communications*,95(2). https://doi.org/10.1007/s11277-016-3867-7
[12] Akasha, R.T., and Baidas, M.W. (2025). Joint antenna selection and power allocation for clustered downlink NOMA networks with Alamouti space-time block coding. International Journal of Communication Systems. https://doi.org/10.1002/dac.70125
[13] Bessate, A., Miftah, Y., Ben-Azza, H., and El Bouanani, F. (2023). A precise mathematical approach for analyzing the performance of MIMO-STBC systems over Weibull fading channels. Wireless Personal Communications.
[14] Bindra, K., Sharma, S., and Khanna, R. (2017). Performance analysis of combined space-time block coding with generalized selection combining in underlay cognitive radio. *Wireless Personal Communications*, 95(3). https://doi.org/10.1007/s11277-016-3934-0
[15] Feng, X., Tian, F., Wang, J., Zhou, M., Li, D., Sun, H., and Song, R. (2024). A survey on maximum ratio combination: Applications, evaluation and future directions. *Electronics*, 13(15), 3087. https://doi.org/10.3390/electronics13153087
[16] Gaitán, M.G., Javanmardi, G., and Sámano-Robles, R. (2023). Orthogonal space-time block coding for double scattering V2V links with LOS and ground reflections. *Sensors*, 23(23), 9594. https://doi.org/10.3390/s23239594
[17] Bayar, F., Salan, O., Ilhan, H., and Aydin, E. (2024). Space-time block coded reconfigurable intelligent surface-based received spatial modulation. *IEEE Transactions on Mobile Computing*, 23(12), 11564–11575. https://doi.org/10.1109/TMC.2024.3398539
[18] Magdum, S.N., and Dhope, T.S. (2024). A comprehensive review on cognitive radio networks: Applications, challenges and researh trends. *Journal of Electrical Systems*, 20(10s).
[19] Duan, R., and Elmusrati, M.S. (2014). Performance analysis of a cognitive-shared channel with GSC diversity under primary outage probability. In Proceedings of the International Conference on Computing, Networking and Communication (pp. 83–87).
[20] Balakrishnan, N., and Cohen, A. (1991). *Order statistics and inference: Estimation methods*. Academic Press.
[21] Simon, M.K., and Alouini, M.S. (2005). *Digital communication over fading channels* (2nd ed.). John Wiley and Sons.
[22] Gradshteyn, I., and Ryzhik, I. (2007). *Table of integrals, series and products* (7th ed.). Elsevier.
[23] Perez, J., Ibanez, J., Vielva, L., and Santamaria, I. (2005). Closed-form approximation for the outage capacity of orthogonal STBC. *IEEE Communications Letters*. 9(11): 961–963.

Smart Systems and Beyond – Bhatia et al. (Eds)
 ISBN: 978-1-041-36617-1

IRS Assisted Cognitive Radio Networks: Methodologies and Applications

Kavita Vij Bindra
Department of Electronics and Communication Engineering, TIET, Patiala, Punjab, India

Varsha Sood, Sukhdeeep Kaur, Shalley Bakshi, Simarpreet Kaur, Preeti Bansal, and Nidhi Chahal
Department of Electronics and Communication Engineering, Chandigarh Engineering College, Landran, Sector, Greater Mohali, India

ABSTRACT: This paper demonstrates a review on incorporating Intelligent Reflecting Surfaces (IRS) into Cognitive Radio (CR) networks has been a pioneering methodology to enhance spectrum sensing, spectral efficiency, energy efficiency, and secure cooperative communications. This survey is a critical review of the latest advancements in IRS-assisted CR, highlighting system design, optimization issues, performance gains, and security considerations, based on cutting-edge research in full-duplex, NOMA, energy harvesting, UAV-mounted IRS, physical layer security, and learning-driven designs.

1 INTRODUCTION

Cognitive Radio (CR) technology meets the increasing need for flexible use of spectrum by enabling secondary users (SUs) to opportunistically use licensed bands while preventing primary network (PN) users from harmful interference. Intelligent Reflecting Surface (IRS) technology, as a reconfigurable and low-cost paradigm, provides high-level capabilities to manage wireless propagation using passive or active surface elements, improving the quality of signals, spectral utilization, and connectivity of users. Here, the emphasis is on cooperative communication scenarios—where SUs, sensors, relays, or UAVs cooperate—supplemented by IRS to achieve the highest level of system performance, energy consumption, and security. The survey is directly based on high-impact recent research and provides the essential models, milestones, open issues, and comparative study. Various pieces of work suggest IRS-based joint designs that benefit both spectrum sensing (by concentrating incoming PU signals on sensors, improving detection probability) and cooperative data transmission for the primary and secondary networks. For example, Liang *et al.* presented a three-stage IRS phase shift strategy for opportunistic spectrum access CRNs.

2 IRS-ASSISTED SPECTRUM SENSING

Spectrum sensing is intrinsic to CRNs, and integration with IRS has been extensively researched to increase detection reliability. Liang *et al.* [1] proposed a joint framework of spectrum sensing and primary-secondary transmission and proved that IRS reflections enhance the detection probability while protecting the PU. Wu *et al.* [2] demonstrated energy detection using IRS with higher accuracy under low SNR, while Lin *et al.* [12], proposed IRS-assisted cooperative sensing techniques for fading mitigation. Nasser et al [4] offered a comprehensive investigation of IRS-aided cooperative spectrum sensing with emphasis on hardware-constrained performance. These studies emphasize the significance of IRS in mitigating multipath fading but suggest issues in CSI acquisition and real-time IRS reconfiguring.

3 SECURE AND ENERGY-EFFICIENT TRANSMISSION

Security has been a primary IRS-CR research topic. Wu *et al.* [2] suggested resilient secure transmission in mm-Wave CRNs with IRS, focusing on passive beamforming against eavesdroppers. In a similar context, Shen *et al.* [13] addressed internal eavesdropping attacks in IRS-aided CR, while Dong *et al.* [22] initiated secure IRS-aided CR models to ensure PU protection. Lv *et al.* [7] further applied this to dual-function IRS-assisted CR-NOMA networks, providing secure communication with spectrum sharing. With regards to energy efficiency, Wu *et al.* [11] showed that IRS-aided CR can significantly save energy while improving secrecy rates. Collectively, these studies confirm IRS as a platform for collaborative SE–EE–security optimization. IRS has been investigated recently for energy harvesting in CRNs. Kawala *et al.* [6] proposed discrete phase IRS for energy harvesting and spectrum sensing simultaneously, while Wang and Yu [36] suggested RF energy harvesting-based protocols in IR-assisted CR

DOI: 10.1201/9781042026289-33

sensor networks. Xu *et al.* [21] discussed IR-assisted secure SWIPT design with hybrid optimization algorithms, while Yang *et al.* [34] presented capacity analysis of wireless-powered relay CRNs assisted by IRS.

4 RESOURCE ALLOCATION AND OPTIMIZATION

IRS brings in challenging optimization issues through massive phase shifts. Xu *et al.* [3] investigated resource allocation for full-duplex CR systems using IRS, suggesting alternating optimization methods. Wang *et al.* [16] considered hierarchical deep reinforcement learning (DRL) for secure resource allocation in multi-IRS spectrum sharing, while Zhong *et al.* [10] employed DRL for collaborative sensing-transmission optimization. Z. Wang *et al.* [28] presented an integrated optimization framework for secure transmission and spectrum sensing. These papers expose the capability of IRS to multi-objective optimization but highlight large-scale IRS deployment's computational complexity.

5 ADVANCED ACCESS SCHEMES: NOMA AND RSMA

NOMA and RSMA optimize spectrum usage when integrated with IRS in CRNs. Asiedu and Yun [9] explored IRS-aided multiuser NOMA for symbiotic CR, while You *et al.* [17] designed IRS-aided RSMA schemes for CR-NOMA with better rate regions than conventional OMA. Wu *et al.* [27] exploited spectrum and energy efficiency tradeoffs optimization in IRS-assisted CRNs with NOMA. Such contributions confirm that IRS can enhance CR's multiple access capacity, albeit with the need for enhancing practical fairness and power allocation methods.

6 UAV- AND MULTI-IRS-ASSISTED CRNS

Mounting IRS on UAVs maximizes CR coverage in dynamic settings. Deng *et al.* [30] examined joint trajectory, sensing, and transmission design for IRS-mounted on UAV in CR, whereas Liang *et al.* [36] generalized this to joint trajectory and underlay CR transmission. Zhao *et al.* [19] explored the optimal number of reflecting elements for IRS-mounted on UAV in sensor networks. Khan *et al.* [4] investigated IRS-user association strategies in multi-IRS CRNs and uncovered coverage versus interference tradeoffs.

7 AI- AND LEARNING-DRIVEN IRS IN CRNS

Artificial intelligence has been used recently for IRS control. Zhong *et al.* [10] proposed DRL-based optimization for CR with IRS assistance, and Wu *et al.* [16] put forward hybrid hierarchical DRL for multi-IRS secure spectrum sharing. Hashida *et al.* [38] proposed ML-based infrastructure sharing for IRS-assisted systems, and Zhang *et al.* [35] used self-supervised learning for channel estimation in IRS-aided communications. Mao *et al.* [33] further combined IRS with federated learning to optimize secrecy rate during model aggregation.

8 RESOURCE ALLOCATION, SENSING, AND NOMA

Optimal resource allocation (power, phase shifts, subchannel assignment) in IRS-CR is problematic because of non-convexities and coupled variables, complicated further in full-duplex multi-user systems or under Non-Orthogonal Multiple Access (NOMA). Successful approaches have been block coordinate descent, semidefinite relaxation, successive convex approximation, and reinforcement learning.

Table 1. Comparative analysis.

S. No.	Reference	Domain area	Methodology	KPIs	Remarks
1.	Liang *et al.*, 2025.	Spectrum sensing and primary–secondary transmission	IRS-assisted sensing + joint primary–secondary transmission design	Detection prob., throughput, interference to primary	Demonstrates IRS improves sensing reliability and primary–secondary throughput tradeoff.
2	Wu *et al.*, 2022	Secure transmission in mmWave CRNs	Robust beamforming and IRS-aided mmWave design	Secrecy rate, SINR under mmWave fading	Robust secure design for mmWave CR with IRS.

(*continued*)

Table 1. Continued

S. No.	Reference	Domain area	Methodology	KPIs	Remarks
3	Xu *et al.*, 2020.	Resource allocation for full-duplex CR	Joint power allocation + IRS phase optimization	Spectral efficiency, self-interference suppression	Resource allocation framework showing IRS aids FD-CR performance.
4	Khan *et al.*, 2024.	Multi-IRS user association	IRS-user association algorithms	Sum-rate, fairness	Proposed association schemes for multi-IRS CRNs.
5	Wu *et al.*, 2022	Secure and energy-efficient transmission	Joint secure beamforming + energy efficiency design	Energy efficiency, secrecy capacity	Balanced secrecy and energy efficiency via IRS control.
6	Kawala *et al.*, 2025.	Energy harvesting + sensing	Discrete phase-shift IRS + EH modelling	Harvested energy, detection probability	Trade-off analysis between EH and sensing with discrete phases.
7	Lv *et al.*, 2025.	Secure CR–NOMA (dual-function IRS)	Dual-function IRS design (NOMA + IRS)	Secrecy capacity, outage prob.	Introduced IRS-enabled secure NOMA for CR.
8	Asiedu and Yun, 2023.	Multi-IRS NOMA for backscatter symbiotic CR	Multi-IRS architecture + NOMA	Throughput, backscatter efficiency	Enabled backscatter symbiotic communications with multiple IRSs.
9	Zhong *et al.*, 2022.	DRL-based optimization	Deep reinforcement learning for IRS control	Convergence, sum-rate, EE	Shows DRL can learn near-optimal IRS policies for dynamic CR.
10	Wu *et al.*, 2022	SE–EE tradeoff in IRS-assisted CRN w/ NOMA	Multi-objective optimization	Spectrum efficiency, energy efficiency frontier	Multi-objective framework balancing SE and EE with IRS and NOMA.
11	Shen *et al.*, 2024.	Robust secure transmission vs internal eavesdroppers	Robust beamforming under imperfect CSI	Secrecy rate, outage	Robust designs to mitigate internal eavesdropping using IRS.
12	Zhang *et al.*, 2020	IRS-aided MIMO CR systems	IRS-assisted MIMO beamforming	Throughput, interference suppression	Shows IRS benefits for MIMO CR capacity and interference control.
13	Wang and Yu, 2024	RF energy harvesting balance for IRS-assisted CR sensor networks	EH schemes + energy-balance oriented design	Harvested energy, network lifetime	Energy-balance schemes to extend sensor network lifetime with IRS.
14	Wang *et al.*, 2024	Multihop uneven clustering routing for active IRS CR sensor nets	Routing + energy-balance clustering	Network lifetime, throughput	Proposes clustering routing tailored for active IRS-assisted sensor networks.
15	Zhao *et al.*, 2024	UAV-mounted IRS element sizing for WSN data collection	Analytical optimum reflecting-element number	Data collection rate, energy tradeoff	Gave optimum number of IRS elements for UAV-mounted data collection.
16	Hameed *et al.*, 2023	Sum-rate maximization in WP-CRN	IRS-assisted wireless-powered CR design	Sum-rate, energy harvesting metrics	Design for sum-rate improvement in wireless-powered CR with IRS.
17	Xu *et al.*, 2025	SCA and IBCD hybrid secure beamforming for CR-SWIPT	SCA + IBCD optimization	Secrecy rate, convergence speed	Hybrid algorithm for secure beamforming in multiuser CR-SWIPT with IRS.
18	Chu *et al.*, 2022	Resource allocation for IRS-assisted wireless-powered FDMA IoT	Resource allocation + IRS phases	Sum throughput, energy efficiency	Resource allocation framework for IoT FDMA networks with IRS assistance.
19	Huang *et al.*, 2024.	Joint sensing, communication, computation for edge intelligence with IRS	Joint S–C–C optimization	Task latency, detection/comm. performance	Integrated S–C–C design for edge intelligence assisted by IRS.
20	Mao *et al.*, 2025.	IRS for secure federated learning aggregation	Secrecy optimization for FL aggregation	Secrecy rate, model aggregation privacy	Optimizes secrecy for FL aggregation using IRS in 6G context.
21	Zhang *et al.*, 2023	Self-supervised channel estimation for IRS	Self-supervised learning for CE without ground truth	MSE of channel estimates	Novel CE approach eliminating need for ground-truth labels.
22	Liu *et al.*, 2020	Compressive channel estimation for mm Wave IRS	Deep denoising neural network for compressive CE	Estimation MSE, robustness to noise	Denoising NN improves compressive CE for mmWave IRS.
23	Hashida *et al.*, 2024.	ML-based infrastructure sharing for IRS comms	ML-based sharing and operations	Resource utilization, sharing gain	ML methods for infrastructure sharing and shared IRS operations.

8.1 *Key challenges*

Passive IRS does not have active RF chains, and therefore efficient, scalable channel estimation is a bottleneck. Hybrid approaches, low-overhead pilots, and AI-based inference are active research areas. Finite-resolution phase shifters, quantization errors, mutual coupling, and element failures affect real-world performance and need to be addressed explicitly. Multi-objective, non-convex formulations continue to challenge real-time, large-scale networks. Although IRS can enhance secrecy, adversarial IRS control or eavesdropper adaptation are emerging threats. Dynamic and fast-fading IRS repositioning (UAVs, mobile sensors) require real-time control loops. Theoretical gains are still required to be demonstrated by testbeds for authenticating performance under real-world impairments.

9 CONCLUSIONS

IRS-enabled CR networks have the potential to bring about a new wave of programmable, energy and spectrum-aware wireless environments. Cooperative communications—be it through multi-user, sensor networks, UAV relays, or NOMA—are highly empowered by IRS integration. The next few years will witness escalated attention on robust practical optimization, scalable channel learning, hardware-efficient deployment, and large-scale testbed demonstrations.

REFERENCES

[1] Liang, X., Huang, L., Deng, Q., Shu, F., Yu, G., and Wang, J. (2025). IRS-assisted spectrum sensing and primary-secondary transmission for cognitive radio networks. *IEEE Transactions on Cognitive Communications and Networking*, 11(3), 1508–1521. https://doi.org/10.1109/TCCN.2024.3462915

[2] Wu, X., Ma, J., Gu, C., Xue, X., and Zeng, X. (2022). Robust secure transmission design for IRS-assisted mmWave cognitive radio networks. *IEEE Transactions on Vehicular Technology*, 71(8), 8441–8456. https://doi.org/10.1109/TVT.2022.3172293

[3] Xu, D., Yu, X., Sun, Y., Ng, D.W.K., and Schober, R. (2020). Resource allocation for IRS-assisted full-duplex cognitive radio systems. *IEEE Transactions on Communications*, 68(12), 7376–7394. https://doi.org/10.1109/TCOMM.2020.3020838

[4] Khan, M., Mirza, J., Alsamani, B., Alkhathami, M., Alsadie, D., and Alosaimi, N. (2024). IRS-user association in multi-IRS assisted cognitive radio networks. *IEEE Access*, 12, 70211–70219. https://doi.org/10.1109/ACCESS.2024.3402076

[5] Wu, X., Ma, J., Xing, Z., Gu, C., Xue, X., and Zeng, X. (2022). Secure and energy efficient transmission for IRS-assisted cognitive radio networks. *IEEE Transactions on Cognitive Communications and Networking*, 8(1), 170–185. https://doi.org/10.1109/TCCN.2021.3114176

[6] Kawala, L.C., Li, G., Demeke, M.H., Xiong, J., Xiong, H., and Hu, H. (2025). Discrete phase shift IRS-assisted energy harvesting in cognitive radio networks with spectrum sensing. *IEEE Open Journal of the Communications Society*, 6, 6551–6565. https://doi.org/10.1109/OJCOMS.2025.3592936

[7] Lv, L., Zhao, S., Zhou, Y., Chen, Y., Huang, G., and Tang, D. (2025). Secure transmission for dual-function IRS-assisted cognitive radio NOMA networks. *IEEE Internet of Things Journal*, 12(8), 10768–10782. https://doi.org/10.1109/JIOT.2024.3513414

[8] Wu, W., *et al.* (2021). IRS-enhanced energy detection for spectrum sensing in cognitive radio networks. *IEEE Wireless Communications Letters*, 10(10), 2254–2258. https://doi.org/10.1109/LWC.2021.3099121

[9] Asiedu, D.K.P., and Yun, J.-H. (2023). Multiuser NOMA with multiple reconfigurable intelligent surfaces for backscatter communication in a symbiotic cognitive radio network. *IEEE Transactions on Vehicular Technology*, 72(4), 5300–5316. https://doi.org/10.1109/TVT.2022.3228532

[10] Zhong, C., *et al.* (2022). Deep reinforcement learning-based optimization for IRS-assisted cognitive radio systems. *IEEE Transactions on Communications*, 70(6), 3849–3864. https://doi.org/10.1109/TCOMM.2022.3171837

[11] Wu, Y., Zhou, F., Wu, W., Wu, Q., Hu, R.Q., and Wong, K.-K. (2022). Multi-objective optimization for spectrum and energy efficiency tradeoff in IRS-assisted CRNs with NOMA. *IEEE Transactions on Wireless Communications*, 21(8), 6627–6642. https://doi.org/10.1109/TWC.2022.3151624

[12] Lin, S., Zheng, B., Chen, F., and Zhang, R. (2022). Intelligent reflecting surface-aided spectrum sensing for cognitive radio. *IEEE Wireless Communications Letters*, 11(5), 928–932. https://doi.org/10.1109/LWC.2022.3149834

[13] Shen, M., Lei, X., Mathiopoulos, P.T., Tang, X., Hu, R.Q., and Fan, P. (2024). Robust transmission design for IRS-aided secure cognitive radio systems against internal eavesdropping. *IEEE Transactions on Wireless Communications*, 23(12), 17841–17855. https://doi.org/10.1109/TWC.2024.3454473

[14] Zhang, L., Wang, Y., Tao, W., Jia, Z., Song, T., and Pan, C. (2020). Intelligent reflecting surface aided MIMO cognitive radio systems. *IEEE Transactions on Vehicular Technology*, 69(10), 11445–11457. https://doi.org/10.1109/TVT.2020.3011308

[15] Wang, J., and Yu, H. (2024). Energy balance-oriented RF energy harvesting schemes for intelligent reflecting surface-assisted cognitive radio sensor networks. *IEEE Sensors Journal*, 24(6), 8628–8641. https://doi.org/10.1109/JSEN.2024.3354734

[16] Wang, L., Wu, W., Zhou, F., Wu, Q., Dobre, O.A., and Quek, T.Q.S. (2024). Hybrid hierarchical DRL enabled resource allocation for secure transmission in multi-IRS-assisted sensing-enhanced spectrum sharing networks. *IEEE Transactions on Wireless Communications*, 23(6), 6330–6346. https://doi.org/10.1109/TWC.2023.3330999

[17] You, H., Bai, Z., Liu, H., Tsiftsis, T.A., and Kwak, K.S. (2023). Rate-splitting for intelligent reflecting surface-assisted CR-NOMA systems. *Journal of Communications and Networks*, 25(1), 15–24. https://doi.org/10.23919/JCN.2022.000053

[18] Wang, J., and Su, X. (2024). An interlayer energy balance-oriented multihop uneven clustering routing protocol for active IRS-assisted cognitive radio sensor networks. *IEEE Sensors Journal*, 24(18), 29180–29198. https://doi.org/10.1109/JSEN.2024.3434365

[19] Zhao, H., Chen, H., Tan, F., and Zhan, L. (2024). Optimum number of reflecting elements for UAV-mounted intelligent reflecting surface-assisted data collection in wireless sensor network. *IEEE Sensors Journal*, 24(14), 23062–23074. https://doi.org/10.1109/JSEN.2024.3407771

[20] Hameed, I., Camana, M.R., Tuan, P.V., and Koo, I. (2023). Intelligent reflecting surfaces for sum-rate maximization in cognitive radio enabled wireless powered communication network. *IEEE Access*, 11, 16021–16031. https://doi.org/10.1109/ACCESS.2023.3243848

[21] Xu, X., Yang, S., Liu, Z., Wu, J., and Bao, J. (2025). SCA and IBCD hybrid algorithm based secure beamforming optimization for IRS-assisted multiuser CR-SWIPT system. *IEEE Signal Processing Letters*, 32, 336–340. https://doi.org/10.1109/LSP.2024.3501286

[22] Dong, L., Wang, H.-M., and Xiao, H. (2021). Secure cognitive radio communication via intelligent reflecting surface. *IEEE Transactions on Communications*, 69(7), 4678–4690. https://doi.org/10.1109/TCOMM.2021.3073028

[23] Chu, Z., Zhu, Z., Li, X., Zhou, F., Zhen, L., and Al-Dhahir, N. (2022). Resource allocation for IRS-assisted wireless-powered FDMA IoT networks. *IEEE Internet of Things Journal*, 9(11), 8774–8785. https://doi.org/10.1109/JIOT.2021.3117791

[24] Huang, N., Dou, C., Wu, Y., and Qian, L. (2024). Joint sensing, communication and computation for edge intelligence oriented symbiotic communication with intelligent reflecting surface. *IEEE Transactions on Cognitive Communications and Networking*, 10 (5), 1650–1662. https://doi.org/10.1109/TCCN.2024.3439589

[25] Srivastava, V., and Prasad, B. (2025). Enhancing adaptive spectrum access: An intelligent reflecting surface assisted CRN for future wireless communication. *IEEE Access*, 13, 129348–129358. https://doi.org/10.1109/ACCESS.2025.3590978

[26] Zamanian, S., Razavizadeh, S., and Wu, Q. (2021). Vertical beamforming in intelligent reflecting surface-aided cognitive radio networks. *IEEE Wireless Communications Letters*, 10(9), 1919–1923. https://doi.org/10.1109/LWC.2021.3086309

[27] Wu, W., *et al.* (2023). Joint sensing and transmission optimization for IRS-assisted cognitive radio networks. *IEEE Transactions on Wireless Communications*, 22(9), 5941–5956. https://doi.org/10.1109/TWC.2023.3238684

[28] Wang, Z., *et al.* (2024). IRS-enhanced spectrum sensing and secure transmission in cognitive radio networks. *IEEE Transactions on Wireless Communications*, 23(8), 10271–10286. https://doi.org/10.1109/TWC.2024.3370812

[29] Zhang, L., Pan, C., Wang, Y., Ren, H., and Wang, K. (2022). Robust beamforming design for intelligent reflecting surface aided cognitive radio systems with imperfect cascaded CSI. *IEEE Transactions on Cognitive Communications and Networking*, 8(1), 186–201. https://doi.org/10.1109/TCCN.2021.3107510

[30] Deng, Q., Yu, G., Liang, X., Shu, F., and Wang, J. (2024). Joint trajectory, sensing, and transmission design for IRS-assisted cognitive UAV systems. *IEEE Wireless Communications Letters*, 13(1), 233–237. https://doi.org/10.1109/LWC.2023.3326979

[31] Liang, X., Zhang, Z., Deng, Q., Shu, F., Liu, S., and Wang, J. (2024). Joint trajectory and primary-secondary transmission design for UAV-carried-IRS assisted underlay CR networks. *IEEE Transactions on Vehicular Technology*, 73(11), 17848–17853. https://doi.org/10.1109/TVT.2024.3425522

[32] Mir, U., Abbasi, U., Mir, T., and Pullisani, S. (2025). Intelligent reflecting surface-assisted cognitive radio networks: Recent trends, usages, challenges, and future directions. *IEEE Open Journal of the Communications Society*, 6, 5030–5049. https://doi.org/10.1109/OJCOMS.2025.3576668

[33] Mao, B., Wu, Y., Liu, J., Guo, H., Wang, J., and Kato, N. (2025). Optimizing secrecy rate for federated learning model aggregation with intelligent reflecting surface toward 6G ubiquitous intelligence. *IEEE Transactions on Cognitive Communications and Networking*, 11(2), 1258–1267. https://doi.org/10.1109/TCCN.2024.3454256

[34] Yang, Z., Feng, L., Zhou, F., Qiu, X., and Li, W. (2023). Ergodic capacity analysis of IRS aided wireless-powered relay communication network. *IEEE Transactions on Cognitive Communications and Networking*, 9(5), 1241–1256. https://doi.org/10.1109/TCCN.2023.3280951

[35] Zhang, Z., Ji, T., Shi, H., Li, C., Huang, Y., and Yang, L. (2023). A self-supervised learning-based channel estimation for IRS-aided communication without ground truth. *IEEE Transactions on Wireless Communications*, 22(8), 5446–5460. https://doi.org/10.1109/TWC.2023.3233970

[36] Liu, S., Gao, Z., Zhang, J., Di Renzo, M., and Alouini, M.-S. (2020). Deep denoising neural network assisted compressive channel estimation for mmWave intelligent reflecting surfaces. *IEEE Transactions on Vehicular Technology*, 69(8), 9223–9228. https://doi.org/10.1109/TVT.2020.3005402

[37] Nasser, K., Al Haj Hassan, H., Mansour, A., Yao, K.-C., and Nuaymi, L. (2022). Intelligent reflecting surfaces and spectrum sensing for cognitive radio networks. *IEEE Transactions on Cognitive Communications and Networking*, 8(3), 1497–1511. https://doi.org/10.1109/TCCN.2022.3171212

[38] Hashida, H., Kawamoto, Y., and Kato, N. (2024). Machine learning-based infrastructure sharing and shared operations for intelligent reflecting surface-aided communications. *IEEE Transactions on Cognitive Communications and Networking*, 10(1), 198–208. https://doi.org/10.1109/TCCN.2023.3312386

[39] Pejoski, S., Hadzi-Velkov, Z., and Zlatanov, N. (2022). Full-duplex covert communications assisted by intelligent reflective surfaces. *IEEE Communications Letters*, 26(12), 2846–2850. https://doi.org/10.1109/LCOMM.2022.3206962

Smart Systems and Beyond – Bhatia et al. (Eds)
© 2027 The Author(s), ISBN: 978-1-041-36617-1

ML-Driven Design Optimization of a Circular Patch Antenna for Predictive Non-Invasive Blood Glucose Monitoring

Nidhi Upadhyay and Amanpreet Kaur
Centre of Excellence in Emerging Materials
Thapar Institute of Engineering and Technology (TIET), Patiala, India

Ashima Singh and Arnab Pattanayak
Thapar Institute of Engineering and Technology (TIET), Patiala, India

ABSTRACT: This paper propose the design, optimization, and evaluation of a flexible microstrip antenna detector for non-invasive blood glucose monitoring. This work employs machine learning (ML) methods at two stages: geometric optimization and BGL prediction. A circular patch antenna, designed on a thin Rogers RTD 5880 substrate, is considered for this work. The geometric parameter of the antenna (radius), is parametrically swept (10-14mm) in CST Microwave Studio to generate over 114,000 samples of patch radius vs. S parameter data. Four regression models, KNN, Random Forest, Gradient Boosting, and MLP are trained with the data to predict the S11 response as a function of radius. The outputs of the ML models are then ensembled via a Voting Regressor. In the second stage, the antenna is simulated with a three-layer bio phantom and the dielectric values of the blood layer are varied according to the different BGL. S parameter shifts corresponding to BGLs are recorded. This data is then used to train Linear Regression, SVR, and Random Forest models. Random Forest achieves the highest predictive accuracy ($R^2 \approx 0.94$). These results demonstrate that the proposed antenna ML framework fastens the process of designing of an accurate and compact wearable sensors.

Keywords: Antenna, Machine Learning, Optimization, Non-invasive sensing, Blood Glucose Monitoring

1 INTRODUCTION

Antenna based sensors are being researched upon by researchers for biomedical applications, including cancer and tumour detection (Kaur and Kaur, 2022, 2024), cancer treatment (Kaur and Kaur, 2023), glucose monitoring (Kandwal *et al.*, 2020; Upadhyay *et al.*, 2025; Zaarour *et al.*, 2025), etc.

These days non-invasive wearable sensors are one of the topics that is being researched upon for blood glucose monitoring. Microstrip patch antennas can function as detectors for measuring Blood Glucose levels using the variation in the S-parameter of the antenna in correspondence to variation in glucose levels in the blood. Various biomedical antennas have been reported. The proposed flexible antenna for BGL detection (details are in (Upadhyay *et al.*, 2024)) measures S-parameter data and use machine learning (ML)-based algorithms to forecast the BGL level in people based on the data. The antenna patch is circular, including four kite-shaped slots that separate it into four distinct sections. The antenna is modelled on a flexible Rogers substrate and receives power by a microstrip feed line. The antenna design process requires a lot of time and effort for getting the final results which makes the antenna optimization process quite tedious and time consuming, as optimization is an iterative process. To address this issue, researchers have proposed methods for optimization through machine learning (ML) methods like regression models , (Huang *et al.*, 2023; Rai *et al.*, 2024; Wang *et al.*, 2024; Wu *et al.*, 2024; Yahya *et al.*, 2024). Initially, the regression methods perform nonlinear fitting of input and output data, followed by the establishment of proxy models characterized by low computational cost. Regression algorithms can rapidly predict the performance of unknown samples by analysing previous fed data into the model. In this work, various regression methods are used for mapping the relationship between the antenna characteristics and the response. And then, the methods are ensembled together for better results (evaluated using mean squared error (MSE) and accuracy). In the proposed research work, antenna patch radius is optimized using various ML models such as K-Nearest Neighbours (KNN), Random Forest, Gradient Boost Regression (GBR), XG Boost Regression (XGB), and MLP neural network regression for a desired results (impedance bandwidth (2.4-2.5 GHz) and return loss).

The dataset required for the implementation in the optimisation model is generated by varying the radius of the patch using parametric sweep in EM simulator of CST MWS '23. The outcomes of these models (KNN, Random Forest, GBR and MLP) are then ensembled by using the Voting Regressor from scikit-learn, which aggregates the predictions of multiple base regressors. The validation of the optimised values is done using Ensemble ML model by simulating the designed antenna response at obtained optimised parametric values.

 DOI: 10.1201/9781042026289-34

After that, a phantom is used to imitate human tissues and the antenna is simulated with it, in order to observe the S-parameters and use machine learning models to build a relationship between the variation in S-parameters and changes in blood glucose levels. The matching BGL level is predicted by training a variety of machine learning models, including Linear, Support Vector, and Random Forest regression, using S-parameter data gathered by the sensor. After comparing the accuracy of each machine learning model, it is discovered that Random Forest produces the best results, with an accuracy of 94%.

Section II discusses the ML aided optimization of the geometric parameters of the antenna for required bandwidth and return loss. Section III discusses the BGL prediction using ML algorithms. Section IV concludes the article.

2 OPTIMISATION METHOD

2.1 *Dataset preparation*

The antenna was developed using Rogers RT Duroid 5880 substrate, characterized by dimensions of 30 mm by 56.5 mm, with a relative permittivity (εr) of 2.2, a thickness of 0.16 mm, and a loss tangent (tanδ) of 0.0009. The feed network of the proposed antenna consists of a microstrip line composed of copper, with a thickness of 0.016 mm and dimensions measuring 0.5 mm. Additionally, a ground layer of copper is situated on the backside of the substrate, which has been modified by the introduction of a rectangular-shaped slot, as illustrated in reference [7]. The design process is elucidated in the article by Upadhyay *et al.* (2024). The radius (r) of the patch has been chosen for optimization through machine learning methodologies, as it has a direct impact on both the reflection coefficient and the resonance characteristics of the proposed patch design. The optimization process for the circular patch involves systematically varying the patch radius within the range of 10 mm to 14 mm, utilizing a step size of 0.08 mm, in order to conduct various S-parameter measurements. The optimal radius value determined for achieving the best results is 12 mm, as derived through machine learning methodologies. The optimized antenna under consideration demonstrates an operational bandwidth within the ISM band, specifically ranging from 2 to 6 GHz, with a resonant frequency established at 2.8 GHz. The study conducted resulted in the generation of 114,725 datasets concerning S-parameters.This dataset is used by the aforementioned four ML models to optimize the value of 'r' for the desired test case. The dataset generated in the study has one target variable, return loss and two independent variables, frequency, and radius. Since, the dataset has a known target variable and it is continuous in nature, so the dataset is labelled, and hence, the models undergo supervised learning.

Sections B, below, explains different ML models used for antenna optimization in the proposed research work.

2.2 *ML models for antenna optimisation*

(i) Gradient Boost Regression Model

The Gradient Boost Regression is an ensemble machine learning technique that creates a strong predictive model by combining multiple weak learners, typically decision trees (Zemel and Pitassi, 2000). The algorithm creates sequential trees, wherein each subsequent tree addresses and corrects the errors identified in the preceding ensemble. A loss function L (y, F(x)) is established, wherein y represents the true value and F(x) denotes the prediction generated by the model. The objective of the algorithm is to minimize the expected value of this loss across the training dataset. At each iteration m, a new tree h(m(x)) is added to the ensemble as mentioned in equation I (Zemel and Pitassi, 2000):

$$F_{m(x)} = F_{\{m-1\}(x)} + \gamma_m h_{m(x)} \tag{1}$$

In the problem of antenna optimization of this work, x is the Patch radius (e.g., 10 mm, 10.08 mm, ..., 14 mm), $F_{m(x)}$ is the predicted return loss or bandwidth at iteration m, $F_{0(x)}$ is the initial guess (average return loss over training set), $h_{m(x)}$ is the new regression tree capturing errors in the previous model and γ_m is the learning rate which was selected as 0.1 for this work.

(ii) K-Nearest Neighbours (KNN) Regression Method

The KNN regression is a non-parametric, instance-based learning algorithm that predicts the target variable for a new instance based on the values of its K nearest neighbours in the feature space (Song *et al.*, 2017) The prediction is calculated as in equation II (Song *et al.*, 2017):

$$\hat{y} = \frac{1}{k} \sum_{i \in N_k(x)} y_i \tag{2}$$

In this work of optimisation of antenna patch radius, x is the input patch radius (10-14 mm), N_k is the k radii closest (in value) to the input test radius, $\hat{y}$ is the predicted return loss or bandwidth for input radius.

(iii) Random Forest Method

Random Forest regression is an ensemble learning technique that builds several decision trees and combines their predictions to generate a more precise and reliable estimate(Svetnik *et al.*, 2003). Random Forest constructs B decision trees, each trained on a bootstrap sample $D\beta$ of the original dataset, given a dataset D = $\{(X_1, y_1), (X_2, y_2), ..., (X_n, y_n)\}$, where X_i is the feature vector and y_i is the matching target value. A random subset of m features is chosen at each node for each tree out of the total M features (usually $m = \sqrt{M}$ or $m = M/3$ for regression) (Svetnik *et al.*, 2003). The final prediction $\hat{y}$ for a new instance x is the average of the predictions from all trees as mentioned in equation III (Svetnik *et al.*, 2003):

$$\hat{y}(x) = \frac{1}{B}\sum_{\beta=1}^{B} f_\beta(x) \tag{3}$$

In this work, x is the input antenna design parameters (radius), f_β is the prediction of tree β for input x, B is the total number of the trees in the forest which was chosen as 100, and $\hat{y}(x)$ is the final predicted S11 or bandwidth value of the random forest model.

(iv) Multi-layer Perceptron Regression (MLP)

MLP Regression is a type of feedforward artificial neural network that maps a set of input features $X = \{x_1, x_2, ..., x_n\}$ (in this case, patch radius) to a continuous output y (desired S11 or bandwidth) (Qin *et al.*, 2022). The MLP consists of an input layer, an output layer and one or more hidden layers between them (2 for this work). Each layer l contains n^L neurons (64 in this work), and each neuron i in layer l computes a weighted sum of its inputs, followed by a non-linear activation function φ as in equations IV and V (Qin *et al.*, 2022):

$$zi^L = \sum_j wi_j^L aj^{L-1} + bi^L \tag{4}$$

$$ai^L = \varphi\left(zi^L\right) \tag{5}$$

where wi_j^L are the weights, bi^L is the bias, aj^{L-1} is the activation from the previous layer, and φ is typically a sigmoid function for this work. For regression tasks, the output layer often uses a linear activation (Qin *et al.*, 2022).

3 MACHINE LEARNING AIDED OPTIMISATION

In the research work presented, the values of radius 'r' are chosen from 10 mm - 14 mm with a step size of 0.008 mm. A total of 114,725 datasets is generated using this range using 500 parametric sweeps in CST MWS'23. When working with Machine Learning models, it is necessary to split the generated dataset into different training and test sets. The sample was split into standard 80% training and 20% test datasets randomly by keeping random state as 42.

After the data set is prepared, each ML model is tested to predict the values of return loss for a given antenna parameter in a particular frequency band.

In Random Forest (RF) algorithm, initially the fitting is done with 70 trees each with a depth of 2. The model is then tested with the testing samples. By doing this, the precision and prediction accuracy can be analysed. The prediction metrics is then calculated for the RF model with above mentioned parameters. RF model showed a MSE of 0.025%. The errors for a RF model can be improved by:

- Improving quality of training sample.
- Hyper-parameter tuning of the model.
- Choosing correct input features.

Thus, in the proposed work, hyper-parametric tuning of the RF model using Grid Search CV is done, and then, the model is trained based on those hyper-parameters. After hyperparametric tuning, the best results for n estimators are calculated as 200 for max depth of 3. The RF model is once again fitted to the data and then tested.

For the optimisation process of patch radius using KNN, Sklearn python library is used. The number of points, represented by K, is chosen as 7, to reduce the cross-validation error. The training data is fitted using the model. After the process of model learning is done, the model is then tested using test data samples.

For Gradient Boost Regressor, the absolute error is chosen as the loss function to reduce the model's sensitivity to outliers. The learning rate of 0.1 is selected which balances between convergence speed and model accuracy. The number of estimators is set to 300, which provides a good balance between model performance and computational efficiency. The maximum depth of individual trees is limited to 1 to keep our base learners simple, and thus, reducing the risk of overfitting. A designated random state is established to guarantee the reproducibility of the findings..

MLP Regressor is implemented on the dataset, with the following configuration with the architecture of (64, 64) for hidden layers. The architecture comprises two hidden layers, which are considered adequate for capturing non-linear relationships present within the dataset. Furthermore, the inclusion of 64 neurons per layer strikes a balance between the model's capacity and its computational efficiency. The maximum number of iterations has been established at 1000 to provide sufficient training duration for the model to achieve convergence.

To enhance the robustness and predictive performance of our model, an ensemble learning approach is implemented. Ensemble methods integrate predictions derived from various models to generate a final prediction which usually exhibits enhanced accuracy and stability compared to any single model. VotingRegressor from scikit-learn is utilized for the model. In a voting regressor, the predictions from each base model used above is aggregated, and the final prediction is made as the average of these individual predictions.

4 PERFORMANCE ANALYSIS OF THE SUGGESTED ML BASED ANTENNA OPTIMISATION METHODS

In this part, the results of the machine learning methods used to improve the designed antenna are examined and compared. Figure 1 shows the performance of the ensemble learning algorithm used for parametric optimization of the patch radius. It is seen that, the ideal case is represented by X = Y line i.e., when the predicted and actual values are the same. Comparing the outputs obtained from the three algorithms, it is observed that the points which fall near the X=Y line have better accuracy than the others, hence the models show satisfactory results for predictions of radius of the patch that corresponds to the desired S11 values for specific frequency. It is observed that the tedious task of antenna optimization process can be reduced by training the respective ML model and using it to get the optimized parametric value for the antenna for the desired S parameter performance. Although, this requires intensive computing, but once the model is trained the results could be generated easily and is much faster as compared to the manual optimization process using CST MWS. So, once we design an algorithm, for the antenna, any number of design iterations can be done with far too little memory and time requirement as compared to 3D EM tools for parametric optimization of the antenna.

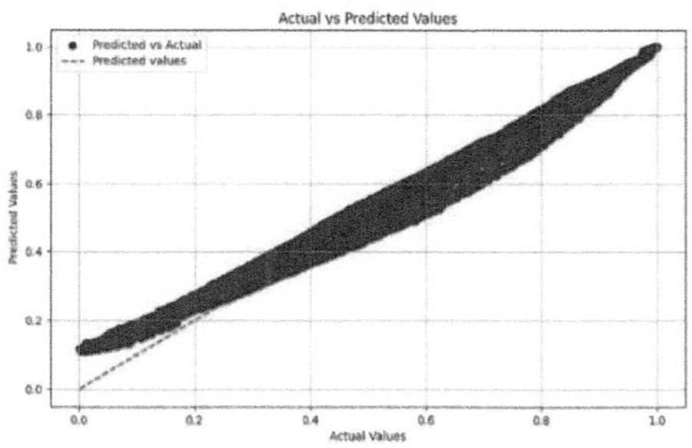

Figure 1. Performance of the Ensemble learning algorithm used for parametric optimization of the patch radius.

The optimised antenna is simulated with the skin phantom to get the respective s parameter results by varying different blood glucose levels.

5 MACHINE LEARNING AIDED BGL MONITORING

The literature review indicates an inverse relationship between the dielectric constant and conductivity of the blood sample and its glucose concentration (Karacolak *et al.*, 2013). This study employs the Debye model to assess the blood glucose level (BGL) in human tissues, emphasizing the variations in dielectric parameters of the blood sample. The Debye model is utilized to describe the relationship between BGL and the dielectric properties of blood plasma, as indicated in (Karacolak *et al.*, 2013).

The findings derived from the Debye model are employed to establish the relationship between glucose concentration and dielectric parameters for three BGL values: 80-400 mg/dL, as detailed in (Upadhyay *et al.*, 2024). Following this, the antenna is evaluated by simulation using a bio phantom specifically designed to mimic the diverse dielectric properties of blood.

The dataset generated through the application of the Debye model serves as a foundation for training regression models aimed at estimating BGL based on the S-parameter values associated with the antenna. The dataset was developed through the simulation of the antenna with 17 distinct blood glucose concentrations within the phantom model. The simulation framework incorporated a total of 1000 datapoints per iteration, culminating in a comprehensive dataset comprising 17,000 datapoints upon completion of the analysis. The dataset is utilized in a Linear

Regression (LR) model, Support Vector Regression (SVR), and Random Forest (RF) model, with a comparative analysis of their performance conducted. From Figure 2, the performance of different algorithms used for BGL prediction using S11 can be analysed. The fit line in Linear Regression, does not align well with the measured data points, and thus produces a high error for prediction. SVR tries to capture the trend better than the linear regression but still show inconsistencies in certain regions. The fit line of Random Forest follows the measured data more closely and captures the non-linear patterns effectively.

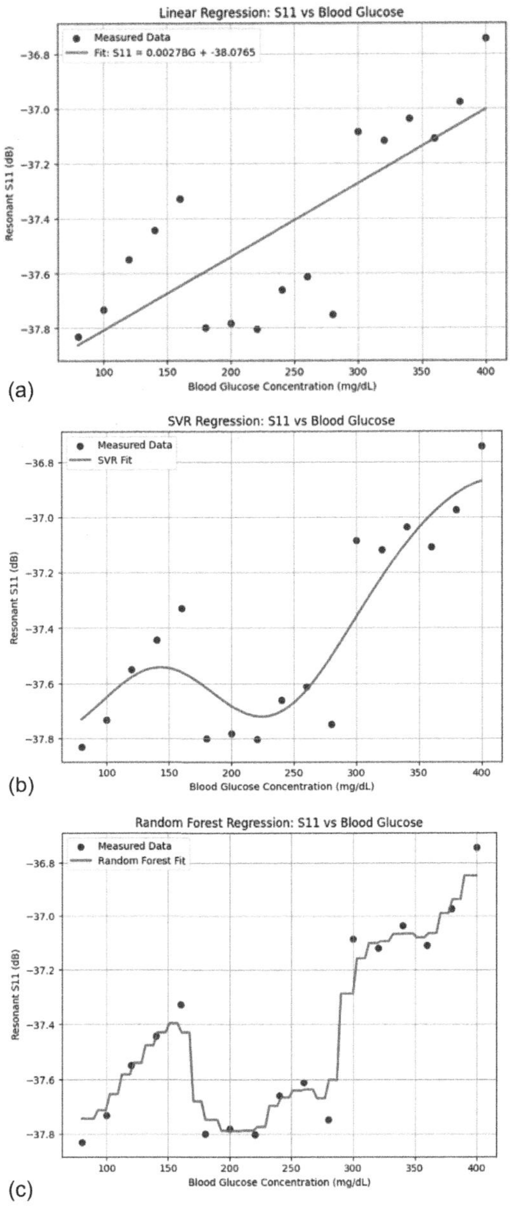

Figure 2. Performance of the algorithms used for BGL prediction (a) Linear regression (b) SVR and (c) Random Forest.

6 CONCLUSION

The article presents a dual-stage ML-aided methodology for geometric optimization of a flexible microstrip antenna and the prediction of blood glucose levels from S-parameter measurements. The ensembling of ML models for parameter tuning of the antenna, reduced time and efforts in optimization for the desired return loss and bandwidth. In the second stage, a relationship between s11 parameter of the antenna and BGL values is established using linear regression and then an accurate prediction of blood glucose is made using Random Forest model based on the s11 parameter of the antenna. The random forest model showed a R2 score of 94%. Together, these findings highlight the potential of integrating ML into antenna sensor development to achieve rapid design of a robust, non-invasive and continuous blood glucose monitoring sensor. Thus, the proposed research shows a successful implementation of ML based techniques for antenna optimization and BGL measurements using S parameter measurements.

REFERENCES

Huang, H., Yang, X.-S., and Wang, B.-Z. (2023). Machine-Learning-Based Generative Optimization Method and Its Application to an Antenna Decoupling Design. *IEEE Transactions on Antennas and Propagation*, *71*(7), 6243–6248. https://doi.org/10.1109/TAP.2023.3270716

Kandwal, A., Igbe, T., Li, J., Liu, Y., Li, S., Liu, L.W.Y., and Nie, Z. (2020). Highly Sensitive Closed Loop Enclosed Split Ring Biosensor With High Field Confinement for Aqueous and Blood-Glucose Measurements. *Scientific Reports*, *10*(1), 4081. https://doi.org/10.1038/s41598-020-60806-9

Karacolak, T., Moreland, E.C., and Topsakal, E. (2013). Cole–cole model for glucose-dependent dielectric properties of blood plasma for continuous glucose monitoring. *Microwave and Optical Technology Letters*, *55*(5), 1160–1164.

Kaur, K., and Kaur, A. (2022). In vitro detection of skin cancer using an UWB stacked micro strip patch antenna with microwave imaging. *International Journal of RF and Microwave Computer-Aided Engineering*, *32*(12), e23407.

Kaur, K., and Kaur, A. (2023). Metamaterial based AMC backed archimedean spiral antenna for in-vitro microwave hyperthermia of skin cancer. *Electromagnetic Biology and Medicine*, *42*(4), 163–181.

Kaur, K., and Kaur, A. (2024). In silico, in vitro, and in vivo validation of a microwave imaging system using a low-profile Ultra Wide Band Archimedean spiral antenna to detect skin cancer. *International Journal of Imaging Systems and Technology*, *34*(1). https://doi.org/10.1002/ima.23016

Qin, Y., Li, C., Shi, X., and Wang, W. (2022). MLP-based regression prediction model for compound bioactivity. *Frontiers in Bioengineering and Biotechnology*, *10*, 946329.

Rai, J.K., Ranjan, P., Chowdhury, R., and Jamaluddin, M.H. (2024). Design and optimization of dual port dielectric resonator based frequency tunable MIMO antenna with machine learning approach for 5G new radio application. *International Journal of Communication Systems*, *37*(13), e5856.

Song, Y., Liang, J., Lu, J., and Zhao, X. (2017). An efficient instance selection algorithm for k nearest neighbor regression. *Neurocomputing*, *251*, 26–34.

Svetnik, V., Liaw, A., Tong, C., Culberson, J.C., Sheridan, R.P., and Feuston, B.P. (2003). Random forest: a classification and regression tool for compound classification and QSAR modeling. *Journal of Chemical Information and Computer Sciences*, *43*(6), 1947–1958.

Upadhyay, N., Kaur, A., and Pattanayak, A. (2024). Flexible Antenna with AMC Backing for Non-Invasive Blood Glucose Monitoring. *2024 Second International Conference on Microwave, Antenna and Communication (MAC)*, 1–6.

Upadhyay, N., Kaur, A., Pattanayak, A., and Singh, A. (2025). Design, development, and testing of a cylindrical dielectric resonator antenna backed with a beehive-shaped artificial magnetic conductor for non-invasive blood glucose monitoring. *Frontiers in Antennas and Propagation*, *3*. https://doi.org/10.3389/fanpr.2025.1494224

Wang, C., Li, G., Zhang, H., Wong, K.-K., Li, Z., Ng, D.W.K., and Chae, C.-B. (2024). Fluid Antenna System Liberating Multiuser MIMO for ISAC via Deep Reinforcement Learning. *IEEE Transactions on Wireless Communications*, *23*(9), 10879–10894. https://doi.org/10.1109/TWC.2024.3376800

Wu, Q., Chen, W., Yu, C., Wang, H., and Hong, W. (2024). Machine-Learning-Assisted Optimization for Antenna Geometry Design. *IEEE Transactions on Antennas and Propagation*, *72*(3), 2083–2095. https://doi.org/10.1109/TAP.2023.3346493

Yahya, M.S., Soeung, S., Rahim, S.K.A., Musa, U., Hashwan, S.S.B., and Haque, M.A. (2024). Machine Learning-Optimized Compact Frequency Reconfigurable Antenna With RSSI Enhancement for Long-Range Applications. *IEEE Access*, *12*, 10970–10987. https://doi.org/10.1109/ACCESS.2024.3355145

Zaarour, Y., El Arroud, F., Fernandez, T., Cano, J.L., El Alami, R., El Mrabet, O., Benani, A., Faik, A., and Griguer, H. (2025). Microwave Antenna Sensing for Glucose Monitoring in a Vein Model Mimicking Human Physiology. *Biosensors*, *15*(5). https://doi.org/10.3390/bios15050282

Zemel, R., and Pitassi, T. (2000). A gradient-based boosting algorithm for regression problems. *Advances in Neural Information Processing Systems*, *13*.

Smart Systems and Beyond – Bhatia et al. (Eds)
© 2027 The Author(s), ISBN: 978-1-041-36617-1

Harnessing Machine Learning Classifiers as Fitness Functions in Metaheuristic-Based Feature Selection for Mammographic Breast Mass Classification

Navreet Kaur
Department of Computer Science and Engineering, DAV University, Jalandhar, Punjab, India

Rahul Hans
Department of Computer Science and Engineering, University Institute of Engineering, Chandigarh University, Gharuan, Mohali, Punjab, India

ABSTRACT: Feature selection is essential for enhancing accuracy, efficiency, and understanding of medical image classification models. In this study, a detailed comparative analysis of various machine learning (ML) algorithms which is used as fitness functions within metaheuristic optimisation frameworks for breast mass classification is presented. The ML algorithms investigated include Extreme Learning Machine (ELM), Support Vector Machine (SVM), k-Nearest Neighbor (KNN), Random Forest (RF), and several hybrid ensembles (KNN+SVM). These algorithms are integrated with state-of-the-art metaheuristic algorithms such as Harris Hawks Optimization (HHO), Grey Wolf Optimizer (GWO), Equilibrium Optimizer (EO), Whale Optimization Algorithm (WOA) and Arithmetic Optimization Algorithm (AOA) to guide the feature selection process. For experimental assessment CBIS-DDSM dataset is used, and the deep features are extracted from it via the AlexNet architecture. Results show that ELM consistently outdo the other ML-based fitness functions by achieving higher classification accuracy, selecting fewer features, and decreasing computational time. This study proves that the ELM-based metaheuristic technique provides a highly efficient way for breast mass classification and achieves better accuracy with a subsequently lower computational cost. On the other hand, Random Forest also proves to be better in terms of classification accuracy but only when computational time is not a primary concern. Findings of this study interpret the relationship between fitness functions and optimization strategies, which are enhancing the development of robust and adaptable computer-aided diagnosis systems in medical imaging.

1 INTRODUCTION

Breast cancer is a prevalent cause of cancer-linked deaths among women worldwide, emphasising the need for accurate diagnostic tools that can detect cancer at an early stage [1]. In the modern world, advances in automated analysis of mammograms include a computer-aided diagnosis (CAD) with deep learning, showing promising and increased sensitivity and consistency across settings [2,3]. However, challenges such as complex image features, small datasets, limited computing power, and poor interpretability make safe and effective implementation difficult [4]. Feature selection (FS) is important in Computer-Aided Diagnosis (CAD) systems to reduce redundancies, and it also prevents overfitting and improves interpretability. Which makes it crucial for clinical acceptance [5]. On the other hand, convolutional networks are common in mammography, pretrained models offer benefits such as robust visual representations with minimal training. Amongst all, AlexNet is a common baseline for breast mass classification [6]. But when it is combined with effective FS methods and wrapper validation, these embeddings help to develop more compact and generalizable models for datasets like CBIS-DDSM [7,8].

Wrapper-based feature selection depends on learning algorithms to assess feature subsets. Most of the time wrapper methods perform better than filter methods in high-dimensional medical imaging. Population-based metaheuristics aid in manage the challenge by carefully exploring the large space where possible feature subsets are available [9]. Algorithms such as Harris Hawks Optimization (HHO), Grey Wolf Optimizer (GWO), Equilibrium Optimizer (EO), and Arithmetic Optimization Algorithm (AOA) are frequently employed in feature selection research because of their strong search efficiency and compatibility with binary (select/not-select) encoding schemes [10–13]. Recent studies and systematic reviews, confirms that the usefulness of metaheuristic-based wrapper approaches aid in providing the more compact and discriminative feature sets for various medical imaging applications [14,15].

An important but often overlooked factor in wrapper-based feature selection (FS) is the choice of fitness function, particularly the classifier or ensemble used to evaluate candidate subsets during optimization. Different classifiers having different strengths such as Extreme Learning Machines (ELM) allow fast training and perform well on small to medium datasets [16], further Support Vector Machines (SVM) focus on maximizing class

 DOI: 10.1201/9781042026289-35

separation [17] followed by k-Nearest Neighbors (KNN) which captures local data patterns [18] and ensemble models like Random Forest (RF) can produce complex relationships while providing important feature subset [19]. When the samples are limited, hybrid ensembles such as KNN+SVM, KNN+RF, or SVM+RF combine complementary advantages to help in improving decision stability [20]. The choice of classifier as the fitness function strongly influences how the metaheuristic searches for solutions and what feature subsets are selected — yet this aspect has not been thoroughly studied in mammography FS research [21].

In this study, we perform a systematic comparative analysis of machine learning algorithms used as the fitness function within metaheuristic for breast mass classification. Our experimental setup uses deep features extracted from AlexNet, and evaluates combinations of four prominent metaheuristics (HHO, GWO, EO, WOA, AOA) with a set of classifiers such as ELM, SVM, KNN, RF, and hybrid ensembles (KNN+SVM) as fitness functions. Experiments are carried out on the CBIS-DDSM dataset a curated mammography benchmark widely used for reproducible evaluation of mass detection and classification systems [7,8].The experimental results show that ELM-based methods consistently offer the best balance by achieving higher classification accuracy with fewer selected features. They also require less optimization time compared to other classifiers used as fitness functions across different metaheuristics. And Random Forest also proves to be better in terms of classification accuracy but only when computational time is not a primary concern.

2 RELATED WORK

Feature selection serves a critical role in developing robust, interpretable, and efficient machine learning models for medical imaging. Surveys and radiomics-focused reviews from 2012 to 2022 consistently report that FS reduces redundancy, limits overfitting in small clinical datasets, and improves the generalization ability of classifiers, especially when working with large feature sets such as handcrafted radiomics or deep embeddings. Foundational studies on evolutionary and swarm-based FS introduced the taxonomy of wrapper, filter, and embedded approaches, and demonstrated that evolutionary computation (EC) and swarm intelligence are well-suited for the combinatorial nature of FS problems [22], [23].

Classic metaheuristics such as Particle Swarm Optimization (PSO) and Genetic Algorithms (GA) have been widely applied to medical imaging FS and computer-aided diagnosis (CAD) tasks between 2012 and 2018. Applied studies show that both PSO and GA are capable of lessening the number of features and also improving the classification accuracy in mammography, CT, and MRI analysis. Most of the studies shows that using of classifiers such as SVM or ensembles as fitness evaluators provides the improvements in accuracy and reduced computation. [24], [25].

Between 2014 and 2019, newer population-based algorithms were introduced and quickly adopted in FS research because of their strong balance between exploration and exploitation and their relatively simple parameter settings. The grey wolf optimization (Mirjalili *et al.*, 2014) and Harris Hawks Optimization (HHO) (Heidari *et al.*, 2019) emerged as popular options. Both have been adapted into binary variants for FS and tested on medical datasets, often outperforming GA and PSO baselines. Further improvements, such as hybridizing with local search or elitist schemes, enhanced their stability on noisy biomedical data. These algorithms form part of the search strategies used in this study [26], [27].

Beyond single algorithm studies, broader evaluations of evolutionary computation (EC) methods (2015–2018) outlined both their advantages and limitations for feature selection. Evolutionary computation is far and wide identified as a flexible approach for wrapper-based feature selection (FS). However, it is computationally expensive when combined with complex classifiers for large-scale feature sets. The survey by Xue *et al.* (2016) provides a central reference for selection of algorithm, how to tune parameter, binary encoding strategies, and hybridization for FS [22]. In mammography and breast CAD, several studies have applied wrapper/metaheuristic FS directly. For example, GA-based wrappers and GA+SVM pipelines have achieved higher sensitivity and specificity by selecting compact and discriminative shape and texture features. [25].

From 2019 to 2022, researchers developed binary variants and refinements of modern metaheuristics for FS, including binary GWO, binary HHO, new transfer functions, Lévy flight strategies, and hybrid local search methods. These works introduced methods tailored to FS, such as: (1) specialized transfer functions to map continuous updates into binary masks, (2) hybridization with greedy or tabu-based search to refine subsets, and (3) multi-objective frameworks that jointly optimize accuracy and subset size [26], [28].

Recent research (2024–2025) highlights two major trends relevant to this work. First, hybrid frameworks that integrate deep feature extractors (such as pretrained CNNs like AlexNet) with metaheuristic feature selection have achieved strong results on CBIS-DDSM and other mammography datasets. These studies highlight the importance of reproducible feature extraction combined with wrapper-based search to generate compact and clinically relevant models. Second, new variants of metaheuristics and adaptive hybrids (e.g., improved GWO models or self-adaptive quantum/metaheuristic hybrids) have achieved state-of-the-art FS performance on CBIS-DDSM. However, these studies rarely examine how the choice of classifier as the fitness evaluator influences the trade-offs between accuracy, selected feature count, and runtime. Addressing this gap is the central motivation for our systematic and factorial analysis [29], [30].

Overall, the literature supports three main design choices in this study: (a) using deep features extracted from AlexNet as a consistent and reusable feature source, (b) applying modern metaheuristic-based feature selection methods (HHO, GWO, EO, AOA, and their binary variants) to explore the combinatorial search space, and (c) conducting a focused comparative experiment where classifiers (ELM, SVM, KNN, RF, and hybrid ensembles) serve as the fitness function during optimization. Research from 2012–2022 demonstrates feasibility and highlights common challenges, while recent work from 2024–2025 shows the practical benefits of hybrid deep–metaheuristic pipelines but also reveals a gap: the lack of systematic evaluation of classifier choice as the fitness function. Addressing this gap is the central aim of our study.

3 EXPERIMENTAL FRAMEWORK

This study presents a systematic comparison of machine learning algorithms used as fitness functions within metaheuristic-based feature selection for breast mass classification. The workflow of this is presented in Figure 1 which involves mainly four steps first is feature extraction from the CBIS-DDSM dataset using AlexNet, further feature selection with metaheuristic algorithms (HHO, GWO, EO, WOA and AOA), followed by evaluation of the selected features using classifiers such as ELM, SVM, KNN, Random Forest, and hybrid ensembles, and lastly performance assessment by using standard classification parameters.

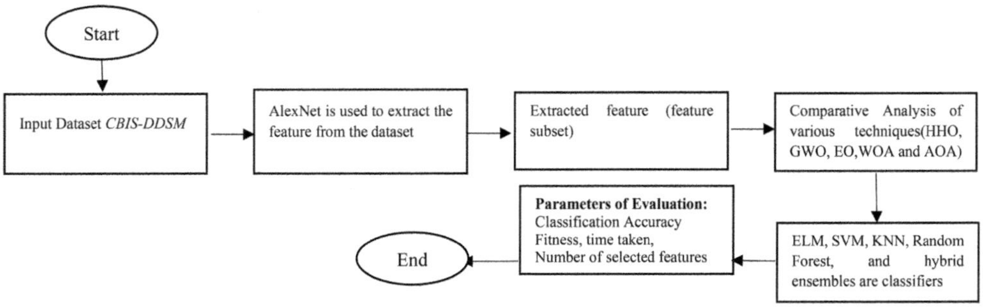

Figure 1. Experimental Workflow of comparative analysis of various ML algorithms along with metaheuristic algorithms.

3.1 *Dataset description*

This study uses the Curated Breast Imaging Subset of the Digital Database for Screening Mammography (CBIS-DDSM), a widely adopted benchmark in breast cancer research. CBIS-DDSM is a standardized subset of the original DDSM, created to improve annotation quality and accessibility. It contains about 10,000 digitized mammogram images from more than 6,000 subjects, with detailed labels provided by expert radiologists.

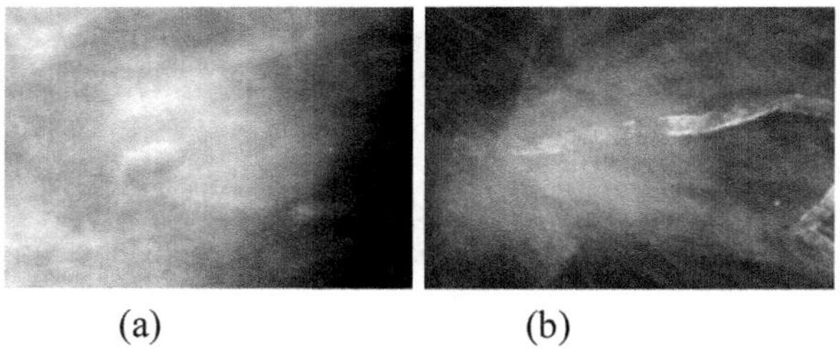

Figure 2. (a) Benign (b) Malignant.

The dataset includes both mass (890 cases) and calcification (750 cases) studies, along with region-of-interest (ROI) masks, bounding boxes, and pathology-confirmed diagnoses. All images are available in DICOM format, and CBIS-DDSM is a dependable and reproducible resource to develop and compare breast mass classification methods.

3.2 *Feature extraction using AlexNet*

For obtaining discriminative representations from mammographic images, deep features were extracted using the AlexNet [32] pre-trained on ImageNet. Considering the proven capability on medical imaging applications and the relatively lightweight architecture compared to deeper models, AlexNet was selected. Feature extraction was done from fully connected layers that capture high-level spatial and semantic information. These features formed the initial high-dimensional representation used for feature selection.

3.3 *Metaheuristic-based feature selection*

High-dimensional features often include redundant or irrelevant information that affects classifier performance and computational efficiency. Therefore, finding the best subset of features can be done using metaheuristic optimization algorithms. In this connection, this study considered four of the most widely recognized algorithms:

- Harris Hawks Optimization (HHO) [33]
- Grey Wolf Optimizer (GWO) [34]
- Equilibrium Optimizer (EO) [35]
- Arithmetic Optimization Algorithm (AOA) [36]
- Whale Optimization Algorithm (WOA) [41]

Each algorithm was adapted for binary feature selection, where solutions represent selected feature subsets. The objective function was designed to balance classification accuracy, feature selection, and computational cost.

3.4 *Classifier-based fitness functions*

A key feature of this work is using the performance of different machine learning classifiers as the fitness function of the feature selection process. The various models tested were as follows:

- Extreme Learning Machine (ELM) [37]
- Support Vector Machine (SVM) [38]
- k-Nearest Neighbors (KNN) [39]
- Random Forest (RF) [40]
- Hybrid Ensemble (KNN + SVM)

These classifiers led the search process of the metaheuristic algorithms by providing accuracy-based fitness values for the candidate feature subsets. This comparative setup allowed the assessment of different fitness functions with respect to feature selection quality.

3.5 *Evaluation matrices*

S. No.	Metric	Explanation
1	Average Classification Accuracy	The overall average of classification accuracy obtained over multiple independent runs.
2	Best Accuracy	The highest accuracy value achieved by the model in any of the repeated experiments.
3	Worst Accuracy	The lowest recorded accuracy results among all the trials.
4	Average Fitness	Represents the mean fitness score, reflecting both solution quality and feature subset size.
5	Best Fitness	The most optimal fitness score observed during the algorithm's execution.
6	Worst Fitness	The least favorable fitness value encountered among all solutions.
7	Standard Deviation	Measures the consistency or variation in results across multiple runs.
8	Average Number of Selected Features	Indicates how many features are typically selected by the model on average.
9	F-Measure (F1 Score)	A balanced metric that considers both precision and recall for evaluating classification output.
10	Average Time	The average time required to complete one full execution of the algorithm.

4 RESULTS AND DISCUSSIONS

In this work, five machine learning algorithms have been considered as fitness functions, such as ELM, KNN, SVM, Random Forest, and a KNN+SVM hybrid ensemble, each combined with six metaheuristic optimizers for feature selection. The results clearly show a trade-off between accuracy and efficiency. The best overall accuracy and F-measure were achieved by Random Forest (RF) combined with GWO (Avg Accuracy = 0.695, Max = 0.712, F-measure = 0.704), while ELM-based combinations excelled in run-time efficiency and repeatability. ELM+GWO has a standard deviation of 0.002. ELM is extremely fast and stable, with the best ELM combinations being ELM+AOA, with an average accuracy of 0.682, and ELM+GWO, with an average accuracy of 0.680 and a standard deviation of 0.002.

Table 1. Average classification accuracy.

ML ALGORITHM	HHO	EO	WOA	GWO	SCA	AOA
ELM	0.678	0.678	0.672	0.680	0.674	0.682
KNN	0.616	0.629	0.618	0.623	0.619	0.630
SVM	0.659	0.665	0.657	0.685	0.679	0.671
Random Forest (RF)	0.689	0.686	0.685	0.695	0.689	0.685
Hybrid ensemble (KNN+SVM)	0.644	0.644	0.639	0.649	0.647	0.657

Additionally, the combination of ELM and HHO is the fastest, with an average run time of 0.788 seconds. Furthermore, the combination of KNN+SCA provides a smaller set of selected features but with lower accuracy, nearly or above 61%.

Table 2. Maximum classification accuracy.

ML ALGORITHM	HHO	EO	WOA	GWO	SCA	AOA
ELM	0.689	0.685	0.684	0.682	0.677	0.687
KNN	0.634	0.647	0.632	0.634	0.630	0.643
SVM	0.681	0.671	0.669	0.696	0.695	0.680
Random Forest (RF)	0.704	0.703	0.704	0.712	0.704	0.696
Hybrid ensemble (KNN+SVM)	0.671	0.656	0.652	0.654	0.659	0.678

Table 3. Minimum classification accuracy.

ML ALGORITHM	HHO	EO	WOA	GWO	SCA	AOA
ELM	0.671	0.667	0.658	0.677	0.671	0.678
KNN	0.599	0.613	0.595	0.607	0.608	0.615
SVM	0.636	0.654	0.641	0.669	0.66	0.656
Random Forest (RF)	0.68	0.671	0.676	0.687	0.68	0.661
Hybrid ensemble (KNN+SVM)	0.621	0.63	0.625	0.636	0.636	0.643

Table 4. Average fitness.

ML ALGORITHM	HHO	EO	WOA	GWO	SCA	AOA
ELM	0.328	0.327	0.334	0.323	0.331	0.324
KNN	0.385	0.374	0.387	0.379	0.382	0.375
SVM	0.344	0.338	0.345	0.318	0.323	0.335
Random Forest (RF)	0.317	0.32	0.321	0.308	0.317	0.321
Hybrid ensemble (KNN+SVM)	0.358	0.358	0.363	0.353	0.355	0.348

Table 5. Worst fitness.

ML ALGORITHM	HHO	EO	WOA	GWO	SCA	AOA
ELM	0.336	0.34	0.349	0.326	0.335	0.328
KNN	0.402	0.39	0.408	0.395	0.393	0.391
SVM	0.371	0.348	0.361	0.333	0.344	0.35
Random Forest (RF)	0.327	0.336	0.331	0.316	0.325	0.344
Hybrid ensemble (KNN+SVM)	0.38	0.371	0.376	0.366	0.366	0.363

Table 6. Best fitness.

ML ALGORITHM	HHO	EO	WOA	GWO	SCA	AOA
ELM	0.316	0.319	0.323	0.32	0.328	0.319
KNN	0.367	0.354	0.374	0.368	0.371	0.363
SVM	0.321	0.331	0.333	0.306	0.307	0.325
Random Forest (RF)	0.303	0.303	0.302	0.291	0.302	0.31
Hybrid ensemble (KNN+SVM)	0.331	0.347	0.349	0.348	0.343	0.327

Table 7. Standard deviation.

ML ALGORITHM	HHO	EO	WOA	GWO	SCA	AOA
ELM	0.008	0.008	0.01	0.002	0.003	0.004
KNN	0.012	0.014	0.014	0.01	0.008	0.011
SVM	0.019	0.006	0.012	0.01	0.015	0.01
Random Forest (RF)	0.01	0.012	0.011	0.01	0.009	0.014
Hybrid ensemble (KNN+SVM)	0.019	0.011	0.01	0.007	0.011	0.014

Table 8. Average number of selected features.

ML ALGORITHM	HHO	EO	WOA	GWO	SCA	AOA
ELM	894.6	869.8	952.4	612.8	871	979.2
KNN	497	633.2	817.4	603.6	494.8	872
SVM	602.4	598.4	504	608.6	584.4	922.2
Random Forest (RF)	1000	958.4	932.4	608.4	864	943
Hybrid ensemble (KNN+SVM)	547.2	548.8	574	549	556.2	821.4

Table 9. Average time taken.

ML ALGORITHM	HHO	EO	WOA	GWO	SCA	AOA
ELM	0.788	2.534	2.438	2.265	2.422	2.71
KNN	7.818	19.664	18.768	12.802	14.648	19.598
SVM	67.066	309.195	292.997	333.405	290.948	260.612
Random Forest (RF)	268.112	457.003	439.259	176.927	206.182	2665.73
Hybrid ensemble (KNN+SVM)	166.496	329.493	55.189	72.325	163.485	81.019

Further explore the contribution of metaheuristics in this experiment: GWO combined with various classifiers provides the highest average accuracy across metaheuristics and the lowest mean average fitness, also producing compact feature sets. Among classifiers, the best per-group combinations were often GWO (RF+GWO, SVM +GWO, ELM+GWO), which showed excellent stability. AOA and SCA performed well in some classifier pairs (ELM+AOA gave the best ELM average accuracy), but AOA's mean runtime was skewed by an outlier with RF +AOA. HHO and EO yielded reasonable results but tended to preserve more features or run longer for limited accuracy.

Table 10. Average time taken.

ML ALGORITHM	HHO	EO	WOA	GWO	SCA	AOA
ELM	0.677	0.677	0.676	0.685	0.674	0.681
KNN	0.622	0.638	0.623	0.621	0.625	0.63
SVM	0.657	0.657	0.657	0.68	0.683	0.667
Random Forest (RF)	0.701	0.697	0.693	0.704	0.699	0.699
Hybrid ensemble (KNN+SVM)	0.699	0.7	0.697	0.71	0.7	0.71

WOA generally performed worse than GWO. GWO appears to offer the best balance, delivering the highest accuracy, F-measure, fitness objective, and feature sparsity. When combined with RF, it yields better accuracy measures; when combined with ELM, it produces highly stable results with less standard deviation.n.

5 CONCLUSION

This study conducted a thorough comparative investigation of machine learning algorithms employed as fitness functions within metaheuristic-driven feature selection for breast mass classification. Deep features were extracted from the CBIS-DDSM dataset using the AlexNet model, and subsequently integrated with several metaheuristic optimizers (HHO, GWO, EO, WOA, and AOA) in combination with diverse classifiers, including ELM, SVM,

KNN, Random Forest, and hybrid ensemble methods. The performed experiment has proven that choosing a fitness function can affect feature selection, classification accuracy, and computational time. The experimental results have brought out a clear trade-off between the model accuracy and computational efficiency. RF, particularly when combined with the GWO, gave the best average accuracy of 69.5% and F-measure of 0.704, which is promising for any application requiring the highest accuracy. On the other hand, ELM-based models proved to be very competitive alternatives that provide almost comparable accuracy of up to 68.2% but with significantly reduced computational time and higher stability. For example, the run time of ELM+HHO was the fastest at 0.788 s, while the stability of results over multiple runs was highest for ELM+GWO with a standard deviation of 0.002. From the results, it can be observed that GWO had the best trade-off among the classification performance, feature subset compactness, and optimization fitness compared to other algorithms such as HHO, EO, WOA, SCA, and AOA.

Further studies can be done by integrating more advanced deep learning architectures in feature extraction, then we can apply that for multi-objective optimization to balance interpretability and predictive performance, and validate the results on larger, multi-institutional datasets. These directions will further improve the robustness and clinical applicability of CAD systems in healthcare settings.

REFERENCES

[1] Sung, H., Ferlay, J., Siegel, R.L., Laversanne, M., Soerjomataram, I., Jemal, A., and Bray, F. (2021). Global cancer statistics 2020: GLOBOCAN estimates of incidence and mortality worldwide for 36 cancers in 185 countries. *CA: A Cancer Journal for Clinicians*, *71*(3), 209–249. https://doi.org/10.3322/caac.21660

[2] Morozov, S., *et al.* (2024). Nationwide real-world implementation of AI for cancer detection in screening. *Nature Medicine*, *30*(4), 900–910. https://doi.org/10.1038/s41591-024-02976-2

[3] Zhou, Y., *et al.* (2025). Mammography-based artificial intelligence for breast cancer screening and risk assessment. *Insights into Imaging*, *16*(1), 12–22. https://doi.org/10.1186/s13244-025-01734-5

[4] Mendes, J., *et al.* (2025). Deep learning in breast cancer risk prediction: recent applications in full-field digital mammography. *Frontiers in Oncology*, *15*, 1409211. https://doi.org/10.3389/fonc.2025.1409211

[5] Raoufpour, M., *et al.* (2024). Dynamic feature selection with metaheuristics: Recent advances and medical imaging applications. *Expert Systems with Applications*, *235*, 121116. https://doi.org/10.1016/j.eswa.2023.121116

[6] Krizhevsky, A., Sutskever, I., and Hinton, G.E. (2012). ImageNet classification with deep convolutional neural networks. *Advances in Neural Information Processing Systems*, *25*, 1097–1105.

[7] Lee, R.S., Gimenez, F., Hoogi, A., Miyake, K.K., Gorovoy, M., and Rubin, D.L. (2017). A curated mammography data set for use in computer-aided detection and diagnosis research. *Scientific Data*, *4*, 170177. https://doi.org/10.1038/sdata.2017.177

[8] Chen, X., *et al.* (2024). Integrative hybrid deep learning for enhanced breast cancer diagnosis on CBIS-DDSM. *Scientific Reports*, *14*, 13355. https://doi.org/10.1038/s41598-024-56789-1

[9] Xue, B., Zhang, M., Browne, W.N., and Yao, X. (2016). A survey on evolutionary computation approaches to feature selection. *IEEE Transactions on Evolutionary Computation*, *20*(4), 606–626. https://doi.org/10.1109/TEVC.2015.2504420

[10] Heidari, A.A., Mirjalili, S., Faris, H., Aljarah, I., Mafarja, M., and Chen, H. (2019). Harris hawks optimization: Algorithm and applications. *Future Generation Computer Systems*, *97*, 849–872. https://doi.org/10.1016/j.future.2019.02.028

[11] Mirjalili, S., Mirjalili, S.M., and Lewis, A. (2014). Grey wolf optimizer. *Advances in Engineering Software*, *69*, 46–61. https://doi.org/10.1016/j.advengsoft.2013.12.007

[12] Faramarzi, A., Heidarinejad, M., Stephens, B., and Mirjalili, S. (2020). Equilibrium optimizer: A novel optimization algorithm. *Knowledge-Based Systems*, *191*, 105190. https://doi.org/10.1016/j.knosys.2019.105190

[13] Abualigah, L., Diabat, A., Mirjalili, S., Elaziz, M.A., and Gandomi, A.H. (2021). The arithmetic optimization algorithm. *Computer Methods in Applied Mechanics and Engineering*, *376*, 113609. https://doi.org/10.1016/j.cma.2020.113609

[14] Oyelade, O.N., *et al.* (2024). Hybrid binary optimization for medical image feature selection. *Neurocomputing*, *556*, 126654. https://doi.org/10.1016/j.neucom.2023.126654

[15] Darvishi, M., *et al.* (2025). Optimized breast cancer diagnosis using self-adaptive quantum metaheuristic enhancements on CBIS-DDSM. *Scientific Reports*, *15*, 24112. https://doi.org/10.1038/s41598-025-67543-9

[16] Huang, G.-B., Zhu, Q.-Y., and Siew, C.-K. (2006). Extreme learning machine: Theory and applications. *Neurocomputing*, *70*(1–3), 489–501. https://doi.org/10.1016/j.neucom.2005.12.126

[17] Cortes, C., and Vapnik, V. (1995). Support-vector networks. *Machine Learning*, *20*(3), 273–297. https://doi.org/10.1007/BF00994018

[18] Cover, T., and Hart, P. (1967). Nearest neighbor pattern classification. *IEEE Transactions on Information Theory*, *13*(1), 21–27. https://doi.org/10.1109/TIT.1967.1053964

[19] Breiman, L. (2001). Random forests. *Machine Learning*, *45*(1), 5–32. https://doi.org/10.1023/A:1010933404324

[20] Chen, X., and Liu, Y. (2020). Hybrid ensemble learning methods for breast cancer diagnosis. *Artificial Intelligence in Medicine*, *108*, 101930. https://doi.org/10.1016/j.artmed.2020.101930

[21] Pramanik, R., *et al.* (2023). Hybrid GA–GWO feature selection for breast cancer classification. *Expert Systems with Applications*, *219*, 119602. https://doi.org/10.1016/j.eswa.2023.119602.

[22] Xue, B., Zhang, M., Browne, W.N., and Yao, X. (2016). A survey on evolutionary computation approaches to feature selection. *IEEE Transactions on Evolutionary Computation*, 20(4), 606–626. https://doi.org/10.1109/TEVC.2015.2504420

[23] Remeseiro, B., and Bolón-Canedo, V. (2019). A review of feature selection methods in medical applications. *Computers in Biology and Medicine*, 112, 103375. https://doi.org/10.1016/j.compbiomed.2019.103375

[24] Al-Radwan, M., *et al.* (2021). Feature selection using enhanced particle swarm optimisation for classification tasks. *Sensors*, 21(5), 1816. https://doi.org/10.3390/s21051816

[25] Sahlol, A.T., Kollmannsberger, P., and Abdelrahman, M.A. (2020). A novel deep feature selection method for mammogram classification using genetic algorithm and deep learning. *Journal of Biomedical Informatics*, 111, 103576. https://doi.org/10.1016/j.jbi.2020.103576

[26] Mirjalili, S., Mirjalili, S.M., and Lewis, A. (2014). Grey wolf optimizer. *Advances in Engineering Software*, 69, 46–61. https://doi.org/10.1016/j.advengsoft.2013.12.007

[27] Heidari, A.A., Mirjalili, S., Faris, H., Aljarah, I., Mafarja, M., and Chen, H. (2019). Harris hawks optimization: Algorithm and applications. *Future Generation Computer Systems*, 97, 849–872. https://doi.org/10.1016/j.future.2019.02.028

[28] Abdel-Basset, M., *et al.* (2023). Binary arithmetic optimization algorithm for feature selection. *Scientific Reports*, 13, 7725. https://doi.org/10.1038/s41598-023-34391-1

[29] Chen, X., *et al.* (2024). Integrative hybrid deep learning for enhanced breast cancer diagnosis using CBIS-DDSM. *Scientific Reports*, 14, 13355. https://doi.org/10.1038/s41598-024-74305-8

[30] Darvishi, M., *et al.* (2025). Optimized breast cancer diagnosis using self-adaptive quantum/metaheuristic enhancements on CBIS-DDSM. *Scientific Reports*, 15, 5014. https://doi.org/10.1038/s41598-025-05014-z

[31] Lee, R.S., Gimenez, F., Hoogi, A., Miyake, K.K., Gorovoy, M., and Rubin, D.L. (2017). A curated mammography dataset for use in computer-aided detection and diagnosis research. *Scientific Data*, 4, 170177. https://doi.org/10.1038/sdata.2017.177

[32] Krizhevsky, A., Sutskever, I., and Hinton, G.E. (2012). Imagenet classification with deep convolutional neural networks. *Advances in Neural Information Processing Systems*, 25, 1097–1105.

[33] Heidari, A.A., Mirjalili, S., Faris, H., Aljarah, I., Mafarja, M., and Chen, H. (2019). Harris hawks optimization: Algorithm and applications. *Future Generation Computer Systems*, 97, 849–872.

[34] Mirjalili, S., Mirjalili, S.M., and Lewis, A. (2014). Grey wolf optimizer. *Advances in Engineering Software*, 69, 46–61.

[35] Faramarzi, A., Heidarinejad, M., Mirjalili, S., and Gandomi, A.H. (2020). Equilibrium optimizer: A novel optimization algorithm. *Knowledge-Based Systems*, 191, 105190.

[36] Abualigah, L., Diabat, A., Mirjalili, S., Elaziz, M.A., and Gandomi, A.H. (2021). The arithmetic optimization algorithm. *Computer Methods in Applied Mechanics and Engineering*, 376, 113609.

[37] Huang, G.B., Zhu, Q.Y., and Siew, C.K. (2006). Extreme learning machine: Theory and applications. *Neurocomputing*, 70(1–3), 489–501.

[38] Cortes, C., and Vapnik, V. (1995). Support-vector networks. *Machine Learning*, 20(3), 273–297.

[39] Cover, T., and Hart, P. (1967). Nearest neighbor pattern classification. *IEEE Transactions on Information Theory*, 13(1), 21–27.

[40] Breiman, L. (2001). Random forests. *Machine Learning*, 45(1), 5–32

[41] Mirjalili, S., and Lewis, A. (2016). The Whale Optimization Algorithm. *Advances in Engineering Software*, *95*, 51–67. https://doi.org/10.1016/j.advengsoft.2016.01.008

Session 4 Information Technology and Intelligent Systems

Toward a Theory of AI-Enhanced Accuracy in Climate Change Journalism

Aditi Sharma and Anupam Kumar
University Institute of Media Studies, Chandigarh University, Mohali, Punjab, India

ABSTRACT: The climate change issue is one of the hottest topics in the entire world and newspapers have difficulties in delivering authentic, timely and informative facts. In this paper, we discuss how artificial intelligence (AI) can enhance climate change coverage accuracy by integrating data verification and predictive analytics tools with environmental surveillance systems. According to the applications of AI in other fields such as climate predictions, renewable energy, agriculture, urban planning, and conservation, the paper develops a theoretical framework to establish how AI can be applied in journalism. Among the aspects that are mentioned during the discussion, AI may contribute to the improvement of forecast accuracy, reduction of false information and providing contextual cues, which makes people more trustful in climate disclosures. Methodologically, the paper is a synthesis of the knowledge in the fields of climate science, media studies, and AI applications with a conceptual review. It shows that despite potential significant opportunities offered by AI to help newspapers develop evidence-based reporting, the issues of the quality of data, the bias of the algorithm, and the ethical dimension of its application remain unresolved. The paper concluded with the emphasis on the idea of interdisciplinary collaboration of journalists, scientists, and technologists that will ensure the accuracy and legitimacy of AI-driven reporting on climate.

Keywords: Artificial Intelligence, Climate Change Communication, Newspaper Coverage, Data Verification, Predictive Analytics, Environmental Journalism, Public Trust

1 INTRODUCTION

Among the most urgent world issues of the twenty-first century that predetermine the organisation of ecosystems, economy, and society is climate change (Galaz, 2021). The newspapers are important in disseminating climate science and policy, shaping social perception, and political discourse, as well as setting media agenda (Coeckelbergh, 2021; Vinuesa, 2020). Nevertheless, the misinformation, politicisation, and simplification of complex scientific data are the aspect which is likely to remain a problem with the reporting of climate change by the newspapers.

These concerns cause the central question of the paper: *Does the utilisation of AI-powered devices facilitate the accuracy of the coverage of climate change issues in newspapers by assessing and validating the information received via various sources?*

Artificial intelligence has turned out to be a journalism and environmental science revolution. AI has also been used in climatology to improve predictive accuracy of extreme weather through creating the most advanced machine learning models such as convolutional neural networks and ensemble forecasting (Dewitte *et al.*, 2021; Haupt, 2021). In renewable energy and in agriculture, the use of resources and sustainability can be optimised with the help of AI (Chen *et al.*, 2023; Jakhar *et al.*, 2022). In journalism, AI also aids in fact-checks and cross-source checking, misinformation detection, which can contribute to the enhanced credibility and trust of people. The Support Vector Machine (SVM) models recorded a classification rate of 98.60 per cent in distinguishing reliable news articles (Nasser and Abu-Naser, 2024).

The thesis of the paper is a review of the literature; it will be based on the synthesis of two complementary streams of literature. The latter is a compilation of past studies that were carried out on AI in journalism, including automated verification and bias detection, and data-driven narrative processing. The second is a systematic review of AI applications in climatic action in eight areas that vary between prediction and monitoring up to education and awareness. Together with these strands, the article creates a conceptual framework that positions AI as an engine of predictability with correct insights into climate and a filter of verification that makes it possible to cover the newspaper properly. This paradigm shows the possibilities of AI-aided climate journalism, and ethical concerns on the issue, and it offers an orientation to the future of interdisciplinary studies.

2 THEORETICAL FRAMEWORK

The theory used in the presented paper is the product of media communication theory and AI applications to climate science, which constitutes a two-dimensional scheme (1) AI as a verification filter in journalism and (2) AI

DOI: 10.1201/9781042026289-36

as a predictive engine in climate action. These two dimensions combined will illuminate the way AI may be utilized by the newspapers to enhance the quality of their climate change coverage.

2.1 *The concept of AI as a checking mechanism in journalism*

Media credibility has taken a trendy seat within journalism concentrating, on various models, including agenda-setting and framing, that concentrate on identifying the impact that could be attained by the media in the establishment of a perception of the people. But this stand has not been compromised by flim flam and interest. The recent research has demonstrated that the possibly best chance to recover the lost trust is through artificial intelligence (AI), as the latter may act as a curtain of check.

The natural language processor (NLP) and machine learning fact-checking systems could compare the facts of more than two sources, enhance their accuracy and eliminate the error in the coverage (Graves, 2018; Zhou and Zafarani, 2020). Large-scale monitoring of the contents and, consequently, tracing the tendencies of the misinformation campaigns, which might be distributed on the issue of climate change, made it possible with the help of these tools, too (Lazer *et al.*, 2018). Consistently, artificial world experiments have supported the use of computational tactics to detect framing discrimination and sampling bias to help reporters avoid an echo chamber and ideological enclosure (Brennen *et al.*, 2020).

Thus, AI verification systems could be characterised as crucial tools, which rely on scale, speed, and precision to enhance climate journalism and boost people's trust.

2.2 *AI as a predictive engine in climate science*

AI is increasingly being used to predict the climate in agreement with journalism. Convolutional neural networks, ensemble models and reinforcement learning can also substitute this approach, which is generally harder on computers, and can be more precise when extreme weather and climate trends are considered (Dewitte *et al.*, 2021; Haupt, 2021).

With the assistance of AI, sustainable agriculture (Jakhar *et al.*, 2022), renewable energy prediction (Navarra, 2023), or adaptation to the urban climate (Yigitcanlar and Cugurullo, 2020) have also been introduced. These applications depict the flexibility of AI as a predictive site with the ability to form vast amounts of data into feasible concepts. Biodiversity, deforestation, and atmospheric change monitoring systems are also possible with the help of AI, besides prediction (Taherdust, 2023).

Mathematically, this places AI into the framework of resilience models where predictive analytics may enable societies to predict and adapt and respond to risks.

2.3 *Integrating the two dimensions*

The synthesis of these two perspectives is the mental gap of this paper. Newspapers are where the middle ground of predictive climate science and communication is. The more constructive agenda-setting role of journalism to the societal sphere is to accept AI both as a filter to authenticate and place scientific findings in context (to become a filter) and as a driver to elucidate the narratives better (to turn climate models into something people can digest).

At that, AI may be applied as an epistemic partner of journalism, transforming raw scientific predictions into convincing, accurate, and readable narratives to the common people.

3 METHODOLOGICAL APPROACH

This paper will not adopt the empirical approach but rather the theoretical-conceptual approach and will be applied in an effort to comprehend the potential of artificial intelligence (AI) to enhance the accuracy of newspapers in reporting climate change. The plan is grounded on two-fold process: (1) synthesis of literature is organised in terms of journalism and media research, and (2) integration of AI application in climate-related domains is compared.

3.1 *Literature synthesis from journalism and media studies*

The first one was the systematic literature review of AI in journalism and its specific use in fact-checking, misinformation detection, and credibility. Sources were peer-reviewed journal articles, book chapters and conference papers. Major findings were identified and organised in the five-column format of the curated dataset, that is, Citation, Article Name, Insight, Methodology, and Abstract.

The data set was thematically coded into two:

- Newsrooms and AI and fact verification (Graves, 2018; Zhou and Zafarani, 2020).
- Artificial intelligence as a cognitive framing device on complicated scientific issues such as climate change (Galaz, 2021; Coeckelbergh, 2021).

This move fulfilled the purpose of making sure that the theoretical assertions in the paper are supported by the existing media communication frameworks and also pinpointed the gaps in which AI has not been systematically applied in the climate reporting issues.

3.2 *Comparative integration of AI–climate action applications*

The second product was based on AI application in climatology and sustainability, and grouped by the assistance of the supplementary table below. They were applied in other areas, including climate prediction and monitoring (Dewitte *et al.*, 2021; Haupt, 2021), renewable energy management (Navarra, 2023), sustainable agriculture (Jakhar *et al.*, 2022), urban planning (Yigitcanlar and Cugurullo, 2020), and environmental conservation (Taherdust, 2023).

The first was a literature review of AI in journalism and its use in fact-checking, misinformation detection and credibility specifically. The sources were peer-reviewed journal articles, book chapters and conference papers. The main findings were derived and organised in the five-column format of the curated dataset, which were, Citation, Article Name, Insight, Methodology, and Abstract. The second one was based on AI application in the climate and sustainability, ranked with the assistance of the supporting table below. Some of the areas where they were applied were climate prediction and monitoring (Dewitte *et al.*, 2021; Haupt, 2021), renewable energy management (Navarra, 2023), sustainable agriculture (Jakhar *et al.*, 2022), urban planning (Yigitcanlar and Cugurullo, 2020), and environmental conservation (Taherdust, 2023).

A classification system of domains was brought to the table and unveiled the shops, techniques, and the advantages of AI in every field. The similar functions are established as a result of similar functions of journalism, e.g.

The satellites and sensors involved in the monitoring of climates are equivalent to fact-checking in journalism and they must be verified in real time.

- The renewable energy is predictable the way media agendas are predictable, whereby the measure of success is credibility.
- Climate literacy education platforms based on artificial intelligence resemble AI-based news ecosystem public engagement.

This comparative analysis is the hypothetical framework according to which journalism may rethink learning in order to implement AI applications in a scientific way.

3.3 *Conceptual integration and framework development*

The points made by the two literatures were then synthesised conceptually in a dual-dimensional framework:

- Verification filter AI in Journalism.
- Potential AI as a Forecasting Engine in climate science.

Taking these positions in contrast, the framework demonstrates how AI could serve to achieve accuracy (by verification) and anticipation (by prediction) in climate news coverage.

3.4 *Limitations of the approach*

This research method is devoid of any theoretical innovation, yet in some measure it is restricted. First, it bases itself on secondary resources, which are not based on primary data collection and limiting to the empirical validation. Second, conceptual bridging (journalism and climate science) will have to be part of the cross-domain integration that can be tempted to simplify disciplinary nuances. Third and finally, both literature in AI and climate communication are extremely dynamic by far, and the sources used are also quite large, but by no means exhaustive.

However, it provides a solid ground on a substantive inquiry that will be the starting point of a scientific experimentation of the provided framework in the future.

4 DISCUSSION AND ANALYSIS

The implementation of artificial intelligence (AI) in climate change news coverage by newspapers has a potential of transformation, although its potential can be examined in two dimensions, in different ways, which are nonetheless connected to each other. First, AI acts as a filter of verification that increases journalistic integrity through identifying misinformation and reducing bias as well as cross-checking information about the climate. Second, AI can be used as the predictive engine and rob climate science of its strategies to make the media discourse more profound and predictive. A mix of these two functions illuminates the possibility of newspapers becoming more realistic and more progressive in their communication of climate change.

4.1 *AI as a verification filter*

One of the most burning problems of climate journalism is the problem of false information and partisan journalism. There has been the discovery that newspapers tend to simplify or misreport complex climate models, and create a gap between scientific consensus and understanding the masses (Galaz, 2021). Such risks can be overcome with the help of AI-based tools as a fact-checker. This is aligned, where a natural language processing (NLP) system, e.g., is employed to scan news articles with documented climate evidence, such as the Intergovernmental Panel on Climate Change (IPCC). The semantic analysis that is based on AI does not just detect the inconsistency of facts but also detects the nuanced ideological framing that can be applied to focus reporting. Automatic News Frame Identification (ANFI) involves a trained Transformer encoder, which is used to categorise the news articles into seven generic frames, like Conflict and Morality (Zhang *et al.*, 2024).

It is already known that fact-checking algorithms are effective in identifying a false or misleading statements in political reporting. To maximise the relevance of the fact-checking process, NLP methods are applied to rank the check-worthy phrases and pay attention to the most interesting claims to verify them, which increases the efficiency of the process (Bartol and Tommasel, 2024). And a further participation of these methods in climate journalism could allow newspapers to cross-reference scientific claims in real-time that the real reporting would not rest on a priori assumptions. Similarly, AI can be applied to find systematic data and model deviations, which is crucial in determining bias in different applications (Speakman *et al.*, 2023).

Such bias identification may add credibility to the newspapers that remain influential in opinion creation independent of the advent of digital media (Coeckelbergh, 2021).

In practice, the verification filter in the form of AI can be introduced as a newsroom tool, which is part of editorial processes. The automated systems would detect the claims that were unsupported, correlate the stories with the rest of the scientific data all over the world and recommend revision to be done before publication. The presence of AI in the verification process would enable newspapers to become stronger in relation to deliberate or unintentional misinformation and anti-trust propaganda.

4.2 *AI as a predictive engine*

Domain	AI Applications	Key Techniques / Tools	Benefits / Outcomes	Sources
Climate Prediction and **Monitoring**	Forecasting extreme weather (hurricanes, storms); climate monitoring systems	Deep learning (CNNs), DI-Cast post-processing, Analogue Ensemble (AnEn), Model Output Deep Learning (MODL), IoT, geospatial analysis	Improved accuracy of forecasts, reduced biases, enhanced real-time environmental monitoring, and biodiversity tracking	Dewitte *et al.*, 2021; Malik *et al.*, 2020; Yadav and Singh, 2023; Mondejar, 2021; Haupt, 2021
Renewable Energy Management	Optimising solar/wind energy production and grid integration	Predictive AI models, real-time monitoring, and energy management systems	Higher reliability, efficiency in energy use, and smoother integration of renewables into grids	Chen *et al.*, 2023; Navarra, 2023; Uriarte-Gallastegi *et al.*, 2024
Sustainable Agriculture	Precision farming, disease detection, and irrigation optimisation	Drones, sensors, satellite imagery, ML-based soil/crop analysis	Efficient resource use, higher crop yields, and reduced ecological damage from pesticides/fertilisers	Chen *et al.*, 2023; Jakhar *et al.*, 2022; Yadav and Singh, 2023
Urban Planning and **Smart Cities**	Smart infrastructure, energy management, traffic control	AI-integrated transport systems, smart buildings, IoT data from social media and sensors	Reduced emissions, optimised energy use, improved quality of life	Abd El-Hameed, 2020; Yigitcanlar and Cugurullo, 2020
Environmental Conservation	Tracking ecosystems, anti-poaching, deforestation monitoring, and species protection	Drones, satellites, camera-based ML detection, conservation forecasting models	Protection of endangered species, proactive conservation strategies, and ecosystem recovery support	Jakhar *et al.*, 2022; Taherdust, 2023; Rayhan, 2023
Climate Policy and **Decision-Making**	Policy modelling, evaluation, and compliance monitoring	Predictive AI algorithms, data-driven scenario analysis	Informed policymaking, real-time policy impact assessment, and enhanced governance	Coeckelbergh, 2021; Galaz, 2021
Carbon Footprint Reduction	Optimising supply chains, emission reduction, and industry monitoring	AI optimisation in manufacturing/logistics, carbon monitoring tools	Lower energy use, reduced emissions, and regulatory compliance	Chen *et al.*, 2023; Pachot, 2023; Kaur and Shahi, n.d.

(*continued*)

Continued

Domain	AI Applications	Key Techniques / Tools	Benefits / Outcomes	Sources
Climate Education and Awareness	Public engagement, personalised learning, awareness campaigns	AI-powered educational platforms, social media analysis, and adaptive learning systems	Increased climate literacy, wider audience engagement, and collaborative action	Coeckelbergh, 2021; Akter, 2024; Uriarte-Gallastegi *et al.*, 2024; Mondejar, 2021; Vinuesa, 2020

Climate reporting can be improved by the predictive power of AI that offers foresight on precision. Existing methods like convolutional neural networks and ensemble forecasting can already work on these kinds of predictions and with the desired accuracy of extreme weather (Dewitte *et al.*, 2021; Haupt, 2021). Predictive models make the most of the supply and demand in renewable energy (Chen *et al.*, 2023; Navarra, 2023).

These aspects indicate that AI can help newspapers to stop being more reactive in their reporting. With the inclusion of predictive models, journalism would be in a position to offer the context of events like heatwaves as what is projected to occur in the future. It is possible to discuss the experience of agriculture and city planning, in which AI systems, drones, and satellite images forecast the condition of the crops and reduce the impact of the emission (Jakhar *et al.*, 2022; Yadav and Singh, 2023; Yigitcanlar and Cugurullo, 2020; Abd El-Hameed, 2020) to demonstrate to the audience how reporting can help close the gap between global models and local realities and make climate change more real to readers.

4.3 *Bridging journalism and climate science*

The newspaper may become more than the transmission of information; it may become a knowledge mediator with the introduction of AI as a filter of checks and a predictive engine. The AI-based climate science should be made more like a human being. One such case is the use of AI to monitor deforestation or biodiversity loss (Taherdust, 2023; Rayhan, 2023), as newspapers can also be used to track the trend in climate misinformation in the digital ecosystem, as well. Similarly, more personalised and interactive AI-based educational tools that enhance climate literacy (Akter, 2024; Vinuesa, 2020) could be employed to persuade newspapers to produce more of these products.

There is no complicating aspect to this intermediary position. Transparency and responsibility of algorithms is a challenge of AI systems' secrecy. Once the news has been checked a thousand times by the machines, the newspapers will be the ones easy to fall into copying machine biases or turning AI-generated news as truth (Coeckelbergh, 2021). Moreover, overuse of predictive AI can encourage the growth of deterministic discourses that do not refer to any political, cultural, or human agency in climatic impacts. In such a way, despite all the powerful tools that AI provides, the use of AI in the field of journalism must be based on a sound editorial judgment and ethics.

4.4 *Ethical and practical challenges*

The use of AI in climate journalism raises other ethical and practical issues as well. The reporting can be biased because AI solutions are founded on the weaknesses of the training data (Jain and Menon, 2023). The difference in media coverage can also lead to the unequal adoption because large and small newspapers have unequal resources. Moreover, the audience may disagree or mistrust AI-mediated reporting as they may feel that it is too technical or that it is not connected with human assessment. Among the ways of overcoming these problems, there is being more open about the use of AI tools, collaboration between journalists and data scientists and the fact that policies can be more supportive of responsible AI integration. It is not to replace the role of journalists but it would be hoped that it would augment their capability to discuss one of the most torturous and urgent issues of humanity.

4.5 *Future directions*

The discussion identifies three potential avenues of practice and research. First, there is a need to undertake empirical research that is aimed at testing the effectiveness of the AI fact-checking tools during the newsroom workflow. Second, cross-disciplinary partnerships can look into methods of converting climate predictive models into journalistic lingo and explaining them to everyday people. Third, this implies the need to develop ethical principles that may be applied to regulate the responsible use of AI in communicating climate with the transparency and accountability as the main priority.

5 CONCLUSION

This review examined how AI-powered tools might make newspapers improve the accuracy and depth of their climatic change coverage. To analyse the role journalistic practice can borrow models of practice in related areas, the analysis suggests through a synthesis of different applications of AI in climate prediction, renewable energy, sustainable agriculture, urban planning, conservation, and education.

One of the areas in which AI has already introduced changes is climate science: first, there are new forecast models based on convolutional neural networks and ensemble models, as well as on the Internet of Things monitoring (Dewitte *et al.*, 2021; Haupt, 2021; Mondejar, 2021). Resource management is more efficiently done through predictive algorithms, as well as farming and renewable energy can be more effective (Chen *et al.*, 2023; Navarra, 2023; Jakhar *et al.*, 2022; Yadav and Singh, 2023). Similarly, AI is also required in city planning and environmental safety as it can predict risks and manage the ecosystem and enable sustainable expansion (Yigitcanlar and Cugurullo, 2020; Taherdust, 2023; Rayhan, 2023). The lessons to be provided by these applications can be transported to the newsrooms that desire to enhance the credibility and effectiveness of climate reporting.

This solution to implementing AI in the journalism sphere could prove beneficial on both ends: verification and foresight. On the one hand, AI is a chance to cross data and examine data in large volumes and fight (distort) data and spread facts. Predictive modelling, on the other hand, helps the newspapers to factor more contextualism of phenomena in their general climatic trends, and they establish coverage history based on a positive initiatory rather than a responsive basis. Such a two-sided position exists out of the heightened responsibility of journalism to educate individuals with the vigilance and readiness consciousness.

Lastly, AI application in climate change reporting cannot be seen as a replacement of journalistic judgment but a tool to supplement the reporting. The newspapers are able to present local correct and prospective weather issues based on the lessons learnt in other related areas. It can reproduce the trust of people, but even more importantly, it can make people unite to defeat the life-or-death issue, the climate change.

5.1 *Limitations and future directions*

The paper is premised more on the theory, with parallel examples of AI application in climate science, energy, and conservation, but it fails to mention specific evidence of journalistic practice. This type of dependency limits generalisation. Moreover, data quality, prejudice, and transparency problems are also inherent to the very AI systems, and they question reliability and newsroom ethics.

The research process after that is to implement the AI models directly to the journalistic setting and trace the impact they produce on the factors of accuracy, trust, and engagement in the process. It must also strive to develop hybrid ways in helping to meet the efficiency of AI and human editorial judgment in a bid to balance and ethical reporting. The possibilities will be achieved by interaction between journalists and climate scientists and AI professionals.

REFERENCES

Abd El-Hameed, A.K. (2020, November). Artificial intelligence shaping sustainable cities for climate change mitigation: A review of literature. In *Architecture and Urbanism: A Smart Outlook: Proceedings of the 3rd International Conference on Architecture and Urban Planning, Cairo, Egypt* (pp. 483–495). Cham: Springer International Publishing.

Adegbite, A.O., Barrie, I., Osholake, S.F., Alesinloye, T., and Bello, A.B. (2024). Artificial intelligence in climate change mitigation: A review of predictive modeling and data-driven solutions for reducing greenhouse gas emissions. *Sustainable Engineering: Concepts and Practices*, 21–31.

Akter, M.S. (2024). AI for sustainability: Leveraging technology to address global environmental. *Journal of Artificial Intelligence General science (JAIGS) ISSN: 3006-4023*, *3*(1), 40–48.

Atkins, C., Girgente, G., Shirzaei, M., and Kim, J. (2024). Generative AI tools can enhance climate literacy but must be checked for biases and inaccuracies. *Communications Earth and Environment*, 5(1), 226.

Bartol, G.E.P., and Tommasel, A. (2024, August). Towards automated fact-checking: an exploratory study on identifying check-worthy phrases for verification. In *2024 L Latin American Computer Conference (CLEI)* (pp. 1–10). IEEE.

Bianchi, O., and Putro, H.P. (2024). Artificial intelligence in environmental monitoring: Predicting and managing climate change impacts. *International Transactions on Artificial Intelligence*, 3(1), 85–96

Brennen, J.S., Simon, F.M., Howard, P.N., and Nielsen, R.K. (2020). Types, sources, and claims of COVID-19 misinformation.

Chen, L., Chen, Z., Zhang, Y., Liu, Y., Osman, A.I., Farghali, M., ... and Yap, P.S. (2023). Artificial intelligence-based solutions for climate change: A review. *Environmental Chemistry Letters*, *21*(5), 2525–2557.

Chen, T.Y., and Park, H. (2024). Artificial intelligence in climate research: innovations and implications (No. DKT-13–13). Copernicus Meetings.

Chițu, F., Mecu, A.N., and Marin, G.I. (2024). Exploring the climate change-AI Nexus: A bibliometric and scientometric study1. In *Proceedings of the International Conference on Business Excellence* (Vol. 18, No. 1, pp. 1658–1670). Sciendo.

Coeckelbergh, M. (2021). AI for climate: Freedom, justice, and other ethical and political challenges. *AI and Ethics*, *1*(1), 67–72.

Dewitte, S., Cornelis, J.P., Müller, R., and Munteanu, A. (2021). Artificial intelligence revolutionises weather forecast, climate monitoring and decadal prediction. *Remote Sensing*, *13*(16), 3209.

Galaz, V., Centeno, M.A., Callahan, P.W., Causevic, A., Patterson, T., Brass, I., ... and Levy, K. (2021). Artificial intelligence, systemic risks, and sustainability. *Technology in society*, *67*, 101741.

Graves, L. (2018). Understanding the promise and limits of automated fact-checking.

Gruetzemacher, R., and Whittlestone, J. (2022). The Transformative Potential of Artificial Intelligence. *Futures*, 135, 102884.

Haupt, S.E., Chapman, W., Adams, S.V., Kirkwood, C., Hosking, J.S., Robinson, N.H., ... and Subramanian, A.C. (2021). Towards implementing artificial intelligence post-processing in weather and climate: Proposed actions from the Oxford 2019 workshop. *Philosophical Transactions of the Royal Society A*, *379*(2194), 20200091.

Jain, L.R., and Menon, V. (2023, November). AI algorithmic bias: Understanding its causes, ethical and social implications. In *2023 IEEE 35th International Conference on Tools with Artificial Intelligence (ICTAI)* (pp. 460–467). IEEE.

Jakhar, R., Singh, G., Raj, R., Kumari, K., Sachar, P., and Prasad, P.S. (2022). Different applications of artificial intelligence to combat climate change issues. *International Journal*, *11*(2).

Kaur, P., and Shahi, A (n.d.). Impact of artificial intelligence on environmental sustainability. *Environment and Sustainability*, *3*, 1–6.

Kraus, M., Bingler, J.A., Leippold, M., Schimanski, T., Senni, C.C., Stammbach, D., ... and Webersinke, N. (2023). Enhancing large language models with climate resources. *arXiv preprint arXiv:2304.00116*.

Lazer, D.M., Baum, M.A., Benkler, Y., Berinsky, A.J., Greenhill, K.M., Menczer, F., ... and Zittrain, J.L. (2018). The science of fake news. *Science*, *359*(6380), 1094–1096.

Leippold, M., Vaghefi, S.A., Stammbach, D., Muccione, V., Bingler, J., Ni, J., ... and Huggel, C. (2024). Automated fact-checking of climate change claims with large language models. *arXiv preprint arXiv:2401.12566*.

Malik, R., Pande, S., Nishi and Khamparia, A. (2020). Artificial Intelligence and Machine Learning to Assist Climate Change Monitoring. *Journal of Artificial Intelligence and Systems*, 2, 168–190. https://doi.org/10.33969/AIS.2020.21011

Mondejar, M.E., Avtar, R., Diaz, H.L.B., Dubey, R.K., Esteban, J., Gómez-Morales, A., ... and Garcia-Segura, S. (2021). Digitalization to achieve sustainable development goals: Steps towards a Smart Green Planet. *Science of The Total Environment*, *794*, 148539.

Nasser, B.S. Abu, and Abu-Naser, S.S. (2024). Leveraging AI for effective fake news detection and verification. *Arab Media and Society*, (37).

Navarra, D. (2023). Integrating artificial intelligence and sustainable technologies in strategic renewable energy and Power-to-X projects: A review of global best practices, risks and future prospects. *Society and Economy*, *45*(4), 472–493.

Nishant, R., Kennedy, M., and Corbett, J. (2020). Artificial intelligence for sustainability: Challenges, opportunities, and a research agenda. *International journal of information management*, *53*, 102104.

Pachot, A., and Patissier, C. (2022). Towards sustainable artificial intelligence: An overview of environmental protection uses and issues. *arXiv preprint arXiv:2212.11738*.

Pande, S.D. (2020). Artificial intelligence and machine learning to assist climate change monitoring. Journal of artificial intelligence and systems.

Parameshwari, M., and Gnanaguru, G. (2024, October). Artificial intelligence for climate sustainability: A Comprehensive review of applications, challenges, and future prospects. In *2024 4th International Conference on Sustainable Expert Systems (ICSES)* (pp. 968–976). IEEE.

Purushotham, P., Kiran, A., Anuradha, K., Reddy, T.B., Vinay, S., and Manne, N.S.B. (2024, April). Changing tides with ai-powered climate change analysis. In *2024 International Conference on Science Technology Engineering and Management (ICSTEM)* (pp. 1–7). IEEE.

Rahman, A., Saha, R., Goswami, D., and Mintoo, A.A. (2024). Climate data management systems: systematic review of analytical tools for informing policy decisions. *Frontiers in Applied Engineering and Technology*, 1(01), 01–21.

Ramya, T., Charan, S.S., Amith, S., Kumar, T.A., Dondy, S., and Pallavi, L. (2024, August). AI-powered climate change adaptation and mitigation. In *2024 7th International Conference on Circuit Power and Computing Technologies (ICCPCT)* (Vol. 1, pp. 1325–1330). IEEE.

Rayhan, A. (2023). AI and the environment: Toward sustainable development and conservation. *Research Gate*, *10*, 10-13140.

Satpathy, I., Nayak, A., and Poddar, S. (2025). The transformative potential of artificial intelligence (AI) in tackling climate change: From theory to practice. In *Community Resilience and Climate Change Challenges: Pursuit of Sustainable Development Goals (SDGs)* (pp. 431–448). IGI Global Scientific Publishing.

Speakman, S., Tadesse, G.A., Cintas, C., Ogallo, W., Akumu, T., and Oshingbesan, A. (2023). Detecting systematic deviations in data and models. *Computer*, *56*(2), 82–92.

Taherdoost, H. (2023). Towards artificial intelligence in sustainable environmental development. *Artificial Intelligence Evolution*, 49–54.

Tan, X., Peng, Z., Cheng, Y., Wang, Y., Chao, Q., Huang, X., ... and Chen, D. (2025). Leveraging artificial intelligence for research and action on climate change: Opportunities, challenges, and future directions. Science Bulletin.

Uriarte-Gallastegi, N., Arana-Landín, G., Landeta-Manzano, B., and Laskurain-Iturbe, I. (2024). The role of AI in improving environmental sustainability: A focus on energy management. *Energies*, *17*(3), 649.

Vaghefi, S.A., Stammbach, D., Muccione, V., Bingler, J., Ni, J., Kraus, M., ... and Leippold, M. (2023). ChatClimate: Grounding conversational AI in climate science. *Communications Earth and Environment*, 4(1), 480.

Vinuesa, R., Azizpour, H., Leite, I., Balaam, M., Dignum, V., Domisch, S., ... and Fuso Nerini, F. (2020). The role of artificial intelligence in achieving the sustainable development goals. *Nature communications*, *11*(1), 233.

Yadav, M., and Singh, G. (2023). Environmental sustainability with artificial intelligence. *EPRA International Journal of Multidisciplinary Research (IJMR)*, *9*(5), 213–217.

YesuJyothi, Y., Shaik, S., and Venkateswarlu, Y. (2024, October). Enhancing climate analysis with integrated data and transparent AI methods via stochastic processes. In *2024 8th International Conference on I-SMAC (IoT in Social, Mobile, Analytics and Cloud) (I-SMAC)* (pp. 1570–1576). IEEE.

Yigitcanlar, T., and Cugurullo, F. (2020). The sustainability of artificial intelligence: An urbanistic viewpoint from the lens of smart and sustainable cities. *Sustainability*, *12*(20), 8548.

Zhang, X., Wei, Q., Zheng, B., and Zhang, P. (2024, July). A deep learning method for automatic news frame identification. In *2024 IEEE/ACIS 27th International Conference on Software Engineering, Artificial Intelligence, Networking and Parallel/Distributed Computing (SNPD)* (pp. 18–24). IEEE.

Zhou, X., and Zafarani, R. (2020). A survey of fake news: Fundamental theories, detection methods, and opportunities. *ACM Computing Surveys (CSUR)*, *53*(5), 1–40.

Smart Systems and Beyond – Bhatia et al. (Eds)
 ISBN: 978-1-041-36617-1

Efficient Ambient Sound Energy Harvesting Using Real-Time Sound Analysis

Deepak Dhadwal, Vinay Bhatia, Tarun Singhal, Ankur Singhal, and Pradeep Kumar

ABSTRACT: A small-scale energy harvesting system is presented in this paper that converts the daily routine sound into electrical power by using an array of microphones. This is different from system used earlier, the system uses real-time sound analysis to actively select the desired sound from effective and efficient sound sources, and this smart selection improves the energy conversion efficiency. The system performance is tested by simulating using python, focusing on the system ability to capture energy and manage the energy storage. The results show the reliable accumulation of the harvested energy over time and finally filling the storage capacity. The system has capability to automatically prevent overflow by managing battery limits. The sound analysis algorithm successfully identifies which microphones are picking up meaningful sound levels, ensuring that only those inputs contribute to the harvesting process. This selective approach reduces wasted effort, improves overall efficiency, and demonstrates how ambient sound can be used as a renewable energy source, particularly for powering low-energy devices in settings such as multiplexes or theaters. The paper concludes by highlighting the need for experimental validation and proposing directions for further improvements.

Keywords: Sound, microphone, Renewable Energy

1 INTRODUCTION

1.1 *Background of acoustic energy harvesting*

Mechanical waves with energy are termed as acoustic sound waves that can be generated by different noise sources. Noise here refer to the undesired sound. Currently, we are leaving in a world where noise is generated by multiple noise sources including trains, planes, power plants and industries etc. This makes acoustic energy as an environmental energy source.

Now a days, instead of dissipation of this noise energy we want to convert this sound energy in to electrical energy. In comparison to other natural sources of energy as solar, wind and hydroelectric, sound energy has lesser dependency on the weather condition, it also allow less restriction on implementation. The drawback associated with sound energy is its less power due to which it cannot meet demand of applications with high power demand. Further the recent development in the field of electronics, power requirements of embedded chips are reducing day by day so the harvested power from sound energy can be used to run these chips. Figure 1 shows an acoustic energy system block diagram. In which multiple blocks shows the process of energy conversion.

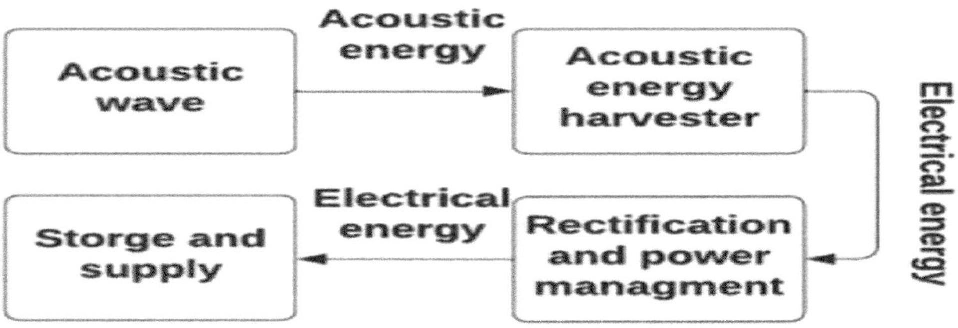

Figure 1. Acoustic energy system.

1.2 *Problem statement*

There are multiple barriers that stops acoustic energy harvesting for large scale uses mainly due to lack of efficiency and implementation issue at large scale. The limitations of currently existing systems are given in Table 1.

 DOI: 10.1201/9781042026289-37

Table 1. Limitations of existing acoustic energy harvesting techniques.

Technique / Limitation	Description	Challenges Identified	Additional Implications
Single-Microphone Systems	Captures ambient sound energy using a single microphone connected to an energy harvesting circuit.	• Limited sensitivity and narrow capture range. • Unable to exploit multi-directional acoustic sources in large venues.	• Very low energy output (in microwatts). • Not suitable for powering real-time sensing or communication modules.
Piezoelectric-Based Systems	Converts sound-induced vibration into electrical energy using piezoelectric crystals.	• Requires costly and fragile materials. • Difficult to scale into large sensor networks. • Highly sensitive to mechanical resonance mismatch.	• Output fluctuates based on irregular vibrations. • High failure rate due to material fatigue.
Low Conversion Efficiency	Captures only a narrow acoustic frequency band.	• Majority of ambient sound energy remains unutilized.	• Cannot maintain sustainable power generation for IoT/wireless modules.
Lack of Real-Time Adaptability	Passive sound capture without separating useful acoustic energy from background noise.	• Wastes internal power on processing unwanted noise.	• No dynamic gain control or directional sensing; harvesting becomes unpredictable.
Cost and Scalability Constraints	Requires multiple sensors or high-cost piezo materials to increase coverage.	• Deployment becomes economically infeasible (e.g., stadiums, multiplex halls).	• Higher installation and maintenance overhead. • Difficult to integrate into distributed IoT systems.
Environmental Dependency	System performance varies significantly with sound availability and direction.	• Irregular and unpredictable acoustic activity.	• Unstable energy harvesting makes it unreliable for continuous operation.
Deployment and Reliability Challenges	Requires precise placement or mechanical coupling for effective output.	• Misalignment or improper mounting reduces efficiency.	• Frequent calibration or maintenance required.

The above limitations indicates the requirement of an intelligent, adaptive energy harvesting system that can be used practically in large scale implementation.

1.3 *Objective of the study*

The proposed system can convert sound energy in to and can be used in different applications. The major objectives are shown in Table 2.

Table 2. Objectives of proposed system.

S. No.	Objective	Description	Expected Output	Sustainability Benefit
1	**Real-time sound source selection**	Digital Signal Processing (DSP) techniques analyze incoming sound in real-time. Only microphones sensing energy above a defined threshold are activated, preventing activation due to background noise.	Active harvesting only from meaningful sound sources (e.g., announcements, music).	Reduces unnecessary energy consumption and improves energy harvesting efficiency.
2	**Efficient energy harvesting and storage**	The collected sound signal is converted into electrical energy and stored in a battery system. Power management prevents overcharging and deep discharge.	Continuous and regulated energy supply for connected low-power devices.	Minimizes battery waste, ensures longer battery life, and promotes renewable micro-energy usage.
3	**Practical deployment for IoT applications**	The harvested energy powers small devices (IoT controllers, sensors) without affecting venue sound quality.	Real operational use of devices using harvested sound energy.	Reduces dependency on grid electricity or disposable batteries.

1.4 *Importance and sustainability impact*

The importance of research get enhanced if it is able to contribute in sustainability and green technology initiatives. The impact of proposed system on these parameters is shown in Table 3.

Table 3. Importance and sustainability impact of the proposed acoustic energy harvesting system.

Aspect	Description	Impact / Benefits
Environmental Sustainability	Replaces grid electricity and batteries for powering low-power devices such as displays and sensors.	• Reduces dependency on fossil-fuel–based power. • Cuts down electronic waste generated from disposable/rechargeable batteries.
Utilization of Existing Ambient Energy	Harvests sound energy already present in multiplexes, theatres, malls, and public spaces.	• Converts wasted acoustic energy into usable electrical power. • No disruption to user experience (silent harvesting process).
Energy Savings	Acoustic energy harvested can power multiple small IoT devices, display boards, or sensors within indoor spaces.	• Saves significant energy at scale when deployed across large venues. • Supports energy-efficient building management.
Scalability and Deployment Flexibility	Low-cost microphone arrays are easier to deploy compared to piezoelectric energy harvesting systems.	• Easy installation in large spaces without major design changes. • Can be implemented across diverse environments — classrooms, malls, theatres, auditoriums.
Economic Viability	Uses inexpensive and easily available electret microphones instead of costly piezoelectric materials.	• Reduces system cost and maintenance expenses. • Higher return on investment for large installations.
Contribution to Global Climate Goals	Supports renewable energy integration and low-carbon infrastructure.	• Aligns with climate action and sustainability targets defined in international frameworks.

2 LITERATURE SURVEY

The summary of key studies are summarize in the Table 4.

Table 4. Literature review.

2021	[1]	Dual-band piezoelectric harvester with Helmholtz resonators	Developed devices operating over multiple frequency bands for enhanced energy capture.
2021	[2]	Helmholtz resonator-based triboelectric nanogenerator	Proposed novel geometry to amplify acoustic energy harvesting efficiency with stable electrical output.
2022	[3]	Hybrid vibration and acoustic energy harvesting system	Combined multimodal conversion methods to increase total harvested energy.
2023	[4]	Hybrid piezoelectric and electromagnetic systems	Successfully expanded frequency range and improved harvesting efficiency through hybrid mechanisms.
2023	[5]	MEMS-based miniaturized acoustic harvesters	Designed micro-energy harvesters with increased low-frequency sound sensitivity for compact environments.
2023	[6]	Multi-resonator with tunable frequency response	Advanced harvesters adjustable to varying sound frequencies enabling more versatile applications.
2024	[7]	Adaptive microphone networks with DSP	Showed dynamic source selection improved real-time energy harvesting efficiency significantly.
2024	[8]	Flexural beam broadband acoustic harvesters	Developed flexible, adaptive beam arrays suitable for wideband ambient sound energy collection.
2024	[9]	AEH applications for IoT	Focused on powering low-energy IoT devices reliably using ambient sound harvesting in practical settings.
2025	[10]	Microphone array geometry optimization	Studied spatial sound capture patterns to optimize energy harvesting while maintaining audio fidelity.
2025	[11]	Algorithm-driven real-time adaptive acoustic harvesting	Introduced intelligent sound source prioritization systems for maximizing energy capture in noisy areas.

The proposed model use an adaptive microphone array to adapt real time environment, this brings energy efficiency as well as scalability and sound quality remains intact.

3 SYSTEM ARCHITECTURE AND WORKFLOW OF PROPOSED ACOUSTIC ENERGY HARVESTING MODEL

The block diagram shown in Figure 2 illustrates the workflow and core components of the proposed acoustic energy harvesting system.

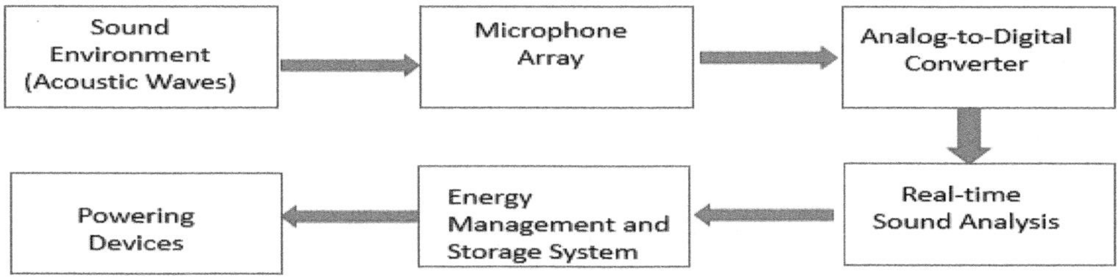

Figure 2. Block diagram.

The use of adaptive microphone it gets best sound energy at the input and smart energy management ensures to prevent wastage of energy and smooth operation of system. The flexibility in design allows it use in all the places even a continuous change in sound around it.

4 RESULTS AND DISCUSSION

The python based simulation is performed to show working of acoustic energy harvesting. The timeframe of 300 sec was used to capture ambient sound with the help of microphone array. This sound is converted in to electrical energy and stored in a battery of fixed size. The reflection of natural variation of sound is included in simulation by adding some random fluctuations around a steady base level of sound every second, this addition make model to more closer to conditions of real world. The graph in Figure 3 shows the harvesting and storage of energy with respect to time. The system gradually harvests small, fluctuating amounts of energy which resembles natural changes in ambient sound like conversations or music. The battery gets filled gradually by this energy. The output shows 40 seconds time is taken by battery to get system fully charged. The effective energy management system of this model smartly stops further charging of battery when it's fully charged.

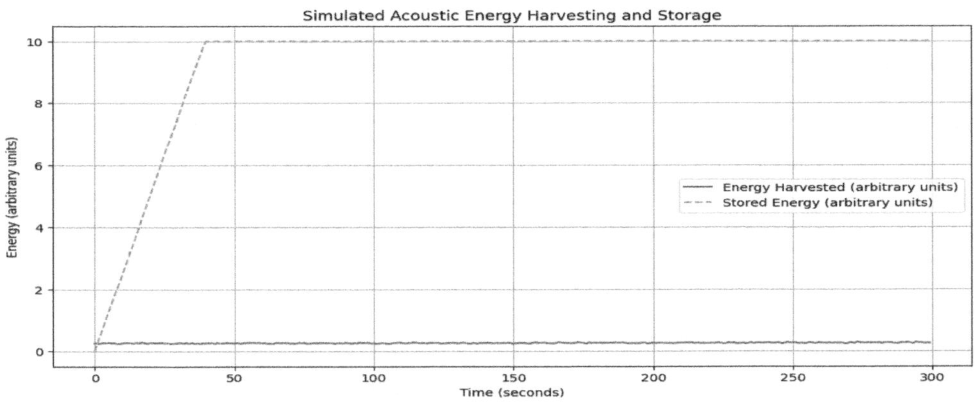

Figure 3. Acoustic energy harvesting.

This result shown in Figure 4 clearly shows how the system responds to changing sound levels in a typical theater or multiplex. Over a span of 300 seconds, the number of active microphones at any moment constantly shifts between 0 and 5, since there are five microphones available for capturing sound. For example, at certain moments, the graph reaches a peak value of 5—this happens when all five microphones detect loud sounds, likely due to loud music or audience clapping. During quieter periods, the graph drops to 0, indicating that none of the microphones picked up sound above the set threshold, which matches real-life moments of silence in a movie or

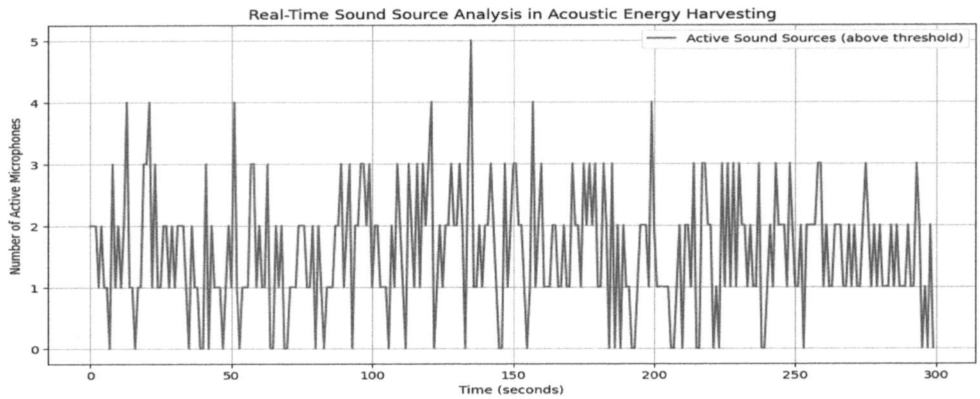

Figure 4. Real time sound source analysis.

theater. Most of the time, the number of active microphones hovers between 1 and 3, showing that ambient noise is present.

but not all microphones are triggered simultaneously. The regular fluctuations in graph represents the random nature of acoustic sound in public areas. The system is able to detect these random changes and activate the microphone that can harvest energy efficiently instead of activating all microphones .This makes the complete process effective and practical for all real world situations.

5 CONCLUSION

A real time sound analysis system is introduced in this work that activates microphones selectively to detect significant sound signal so an improved acoustic energy harvesting can be obtained. The results shows that system can perform well in changing sound environment due to use of only the active microphones due to that unnecessary processing get removed and an enhance energy harvesting is obtained. The approach is very effective and can be used in real time scenarios like theaters.

REFERENCES

[1] T. Li *et al.*, "Dual-band piezoelectric acoustic energy harvesting by Helmholtz acoustic metamaterial," *Nano Energy*, vol. 90, p. 106523, (2021).

[2] H. Yuan *et al.*, "Multi-tube helmholtz resonator based triboelectric nanogenerator for broadband acoustic energy harvesting," *Frontiers in Materials*, vol. 9, p. 896953, (2022).

[3] J. Wang *et al.*, "Hybrid acoustic and vibration energy harvesting systems: A review," *Energy Conversion and Management*, vol. 258, p. 115579, (2022).

[4] J. Park *et al.*, "Hybrid piezoelectric and electromagnetic systems for acoustic energy harvesting," *Nano Energy*, vol. 91, p. 106642, (2023).

[5] Singh and A. Kumar, "Optimization and fabrication of MEMS based piezoelectric acoustic sensor for wide frequency range," *Micro and Nanostructures*, vol. 179, p. 100978, (2023).

[6] S. Kim and D. Choi, "Multi-resonator acoustic energy harvester with tunable frequency response," *Nano Energy*, vol. 98, p. 107249, (2023).

[7] W. Chen and Y. Li, "Adaptive microphone networks with DSP for real-time acoustic energy harvesting," *IEEE Sensors Journal*, vol. 24, no. 7, pp. 6753–6763, (2024).

[8] M. Rodrigues *et al.*, "Flexural beam broadband acoustic harvesters for wideband ambient sound energy collection," *Applied Energy*, vol. 337, p. 120990, (2024).

[9] J. Lee *et al.*, "Acoustic energy harvesting applications for IoT devices," *IEEE Internet of Things Journal*, vol. 11, no. 4, pp. 3234–3246, (2024).

[10] Coleman *et al.*, "Microphone array geometry optimization for improved acoustic energy harvesting," *Acoustics*, vol. 5, no. 2, pp. 193–206, (2025).

[11] T. Nguyen *et al.*, "Algorithm-driven real-time adaptive acoustic harvesting in noisy environments," *IEEE Transactions on Industrial Electronics*, vol. 72, no. 1, pp. 890–899, (2025).

Smart Systems and Beyond – Bhatia et al. (Eds)
© 2027 The Author(s), ISBN: 978-1-041-36617-1

DL-CD: A Deep Learning Generative Models for Accurate Change Detection in Remote Sensing Data

Gurpreet Singh, Kapil Bhardwaj, and Manpreet Kaur
Department of CSE, University Institute of Engineering, Chandigarh University, Mohali, India

ABSTRACT: In recent years, deep learning technologies have emerged to be the major approach for the processing of remote-sensing images. Because of its numerous useful uses, deep learning has been implemented for autonomous change detection as well as accomplished significant success. Deep learning (DL) approaches have been demonstrated effective in detecting changes due to their usage in real life. A few investigators have even suggested that deep learning algorithms prevail over traditional methodologies and improves change detection accuracy. The study focuses on deep learning techniques, including supervised, unsupervised, as well as semi-supervised methods for various change detection models. Data collected from various satellites are readily available nowadays towards the rapid advancement of remote sensing (RS) technology. The images above considerably improve the data resources for change detection. CD is a procedure for detecting differences between photographs captured at different time intervals. The system combines a convolutional neural network (CNN) building construction, attention mechanisms, as well as multi-scale recognition of features to improve sensitivity to tiny changes simultaneously lowering false positives.

Keywords: Deep Learning, Change Detection, Remote Sensing Images, Convocational Neural network, Recurrent Neural Networks

1 INTRODUCTION

Over the past ten years, deep learning (DL) has gained popularity and attention due to its potent representation of learning. Models constructed using many computational layers can learn interpretations of data sets with numerous levels of abstraction thanks to deep learning. Models built on several layers can learn descriptions of data sets with a variety of abstraction levels thanks to deep learning. It might also be thought of as the study of algorithms that either demand a greater percentage of learnt ideas or operations[1]. Deep learning has subsequently been used in a variety of fields, including computer vision, speech recognition, and database extraction, due to its cutting-edge performance [2], visual processor advancements, and data quantities (i.e., photos and videos) [3].

This increase in data has forced the remote sensing (RS). CD task, which is described as "the process for determining deviations from the state of a thing or occurrence by viewing it throughout various times," distinguishes out within these duties. To put it another way, change recognition is the process of determining the variations between photos taken at two separate dates but recorded inside the identical geographic area [4].

Numerous tasks such as environmental surveillance, land use planning, urbanization investigation, and hazard evaluations, heavily rely on identifying changes in systems. While it has exposed a new research issue, the frequency of major climate change-related natural catastrophes, including extreme weather, floods, and drought, has brought out the need for more efficient automation adaptation detection techniques. Detection of changes with deep learning in remote sensing has been conducted, with outstanding performance driven by the observations mentioned before. Figure 1

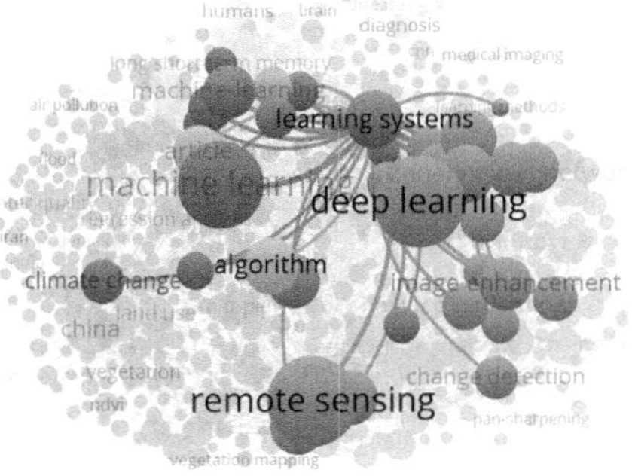

Figure 1. Network visualization of remote sensing images in change detection: Deep learning models.

DOI: 10.1201/9781042026289-38

depicts some deep learning methods applied to all the main remote sensing subfields, as network visualization: from grouping, reconstruction and noise reduction to target identification, scene comprehending, and other tasks [5].

Change Detection in Remote Sensing The most current developments in deep learning for the problem of modification identification in a the detection of change is the process of quantitatively assessing and determining surface changes

from objects or phenomena during two different time periods. This method, which is fundamental to earth investigation, aims to differentiate between pixels that have changed and those that have not in bi-temporal or multi-temporal remote sensing photographs taken from the same region or location but at various periods [6].

Its main task involves the binary labeling of each pixel by a pair or sequence of pictures registered together. In fact, change detection is an effective way to do multi-temporal investigation, urban mapping, and video monitoring.

It has found successful applications in many different areas. In the agricultural field, change detection is used for hazard evaluation, changing cultivation monitoring, and reforestation monitoring. In military applications, it has recently been used for data acquisition about battlefield locations, movement of enemy military forces, new military constructions. It is also used in monitoring the effects of climate change, which are generally related to the increase in the emission of greenhouse gases GHG into the atmosphere. These effects involve variation in sea levels, glacier fronts, and mass distribution [7].

A specific method of ML called DL makes use of the advances in applied mathematics statistics, neurological research, and statistics understanding during the past few years [8].

In reality, a neural network with multiple concealed levels is used to address complicated problems like identifying changes in remote sensing photos. The term "deep learning" refers to this multi-layered structure, which is handled by "deep" neural networks (DNNs) [9].

2 OVERVIEW OF DEEP LEARNING: DEEP MODELS

- DBNs: DBNs, composed of stacked Restricted Boltzmann Machines (RBMs), are particularly useful in such tasks due to their abilityto learn unsupervised hierarchical features from large volumes of unlabeled remote sensing data. In change detection, DBNs can extract complex spectral and spatial patterns from preand postevent images, making them suitable for identifying subtle changes interrogator structures. During pre-training, each layer of the DBN learns features from the image data without supervision, which helps in capturing meaningful representations, even when labeled data is limited. Afterward, the model is fine-tuned using supervised techniques if ground truth labels (change/no-change) are available. Compared to traditional machine learning models or shallow neural networks, DBNs offer improved accuracy and generalization, especially in heterogeneous landscapes [10]. As highlighted in comprehensive reviews and meta-analyses, while newer models like CNNs and Transformers now dominate the field, DBNs laid essential groundwork and are still relevant for understanding the evolution of overlay visualization of remote sensing images in change detection for deep learning models given in Figure 2.

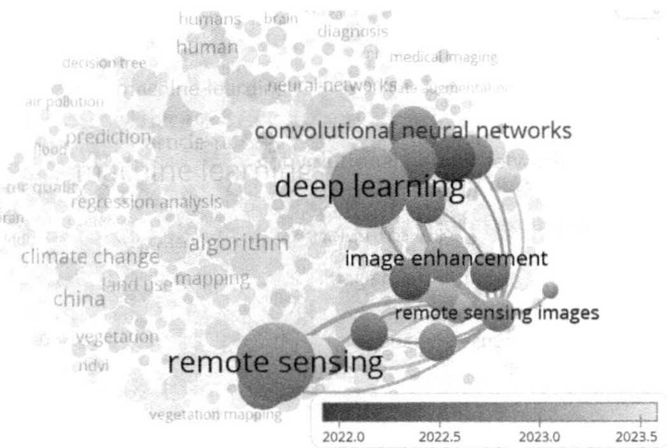

Figure 2. Overlay visualization of remote sensing images in change detection: deep learning models.

- SAEs:

  SAEs are especially good in unsupervised feature-based learning. Each automatic encoder learns to encode and rebuild input data, capturing significant patterns and characteristics at various levels of conceptualization. An SAE is made up of numerous layers of automatic encoders layered on above others. Within the framework of remote sensing SAEs collect spectral and spatial data from multi-temporal satellite photos, enabling the detection of even minor changes between time periods. Initially, each autoencoder was constrained independently in an unsupervised manner to rebuild its input. The entire stack is then fine-tuned using supervised learning if labeled data (such as change/no-change labels) is available. This layer-wise training enables SAEs to generalize effectively and handle noisy or constrained datasets, which are prevalent in real-world remote sensing applications [11].

- CNNs: Given their remarkable capacity to extract spatial information from picture data, These patterns are essential for detecting variations among multi-temporal satellite pictures. In change detection tasks, CNNs typically process two images from different time points and learn to highlight areas where meaningful differences occur—such as deforestation,urban growth, or disaster impacts [12].
- RNNs: Recurrent Neural Networks (RNNs) are models developed using deep learning that are extremely important for longitudinal change detection in remote sensing are depicts in Figure 3, since they are made to analyze sequential or time-series data. By examining sequences of satellite photos taken at various times, Recurrent neural networks can capture how the environment changes over time, in contrast to CNNs or DBNs, which are usually employed for spatial feature extraction. RNNs are notable for their capacity to predict temporal dynamics, particularly when high-frequency picture data is available from sources such as MODIS or Sentinel. While RNNs are not always the best choice for spatial analysis, they do complement other models by allowing for longitudinal analysis of change, which provides an improved comprehension of how land surfaces evolve as time passes [13].

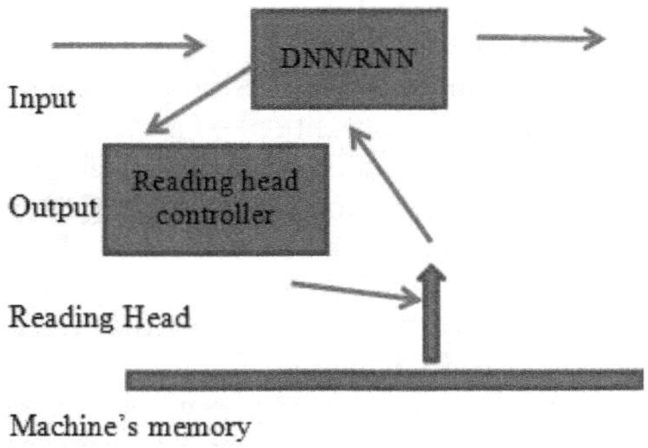

Figure 3. Deep learning for longitudinal changes in remote sensing images: RNN/DNN models.

- GANs: Generative Adversarial Networks (GANs) are an increasingly popular way to create synthetic or enhanced versions of the satellite images and point out subtle or complex changes that might be missed by more classic methods. GANs have also been employed to augment training data, reducing the dependency on large annotated datasets, which are GAN-based models excel at learning hidden distributions, enabling unsupervised and semi-supervised identification of changes in satellite images. GAN-based models are noted for their creativity in learning hidden distributions, offering new possibilities for unsupervised and semi-supervised change detection in satellite imagery [14].

3 METHODS AND DATA USED FOR CD IN REMOTE SENSING IMAGES

A systematic review methodology was followed for comprehensively reviewing DL techniques for CD in RS images. Studies from 2015 to the present were primarily considered, reflecting the recent surge in DL-based applications. Each selected study was examined according to its dataset, model architecture, application domain (for example, urban expansion, deforestation, disaster monitoring) and performance metrics (for example, accuracy, F1 score, IoU). There view included both supervised and unsupervised DL methods and considered multiple satellite data sources such as Landsat, Sentinel, MODIS, and high-resolution commercial satellites. By analyzing model strengths, limitations, and trends, the review aimed to identify the most effective approaches and emerging techniques in DLbased CD [15].

3.1 *Meta-analysis approach: Data extraction*

often scarce in remote sensing. Some advanced frameworks combine GANs with segmentation models (e.g., cGANs) to directly predict change masks. In metaanalyses, The meta-analysis process for data extraction in this review involved a structured and quantitative approach to synthesize findings from multiple deep learning-based change detection studies. Initially, a set of inclusion criteria was defined to select high-quality peer-reviewed papers that applied DL techniques to remote sensing image change detection tasks. Each selected paper was analyzed to extract relevant data points, including the type of deep learning model used(e.g.,CNN,RNN,GAN, DBN, SAE), the dataset characteristics (e.g., spatial resolution, satellite source, temporal span), application domain (e.g., disaster monitoring, urban growth, deforestation). Additionally, details on training strategies (e.g., supervised, unsupervised,semi-supervised), data preprocessing methods,and evaluation protocols were collected. The extracted data were tabulated and categorized to enable cross-study comparisons, identify trends, and highlight performance differences across models and applications [16]. This metaanalytical approach provided a comprehensive and objective foundation for evaluating the effectiveness and evolution of DL models in remote sensing change detection with various datasets are given in Table 1.

Table 1. Deep learning models in remote sensing change detection: Description different datasets.

Model	OA (%)	F1 Score	Kappa	Precision	Recall	IoU (%)
DBN	76.5	0.78	0.79	0.75	0.77	72.1
SAE	89.0	0.80	0.80	0.82	0.79	81.2
GAN	92.5	0.82	0.78	0.89	0.81	80.5
CNN	93.7	0.95	0.90	0.96	0.94	90.2
RNN	89.2	0.78	0.81	0.90	0.79	79.3

4 DEEP LEARNING PROVIDES CHANGE DETECTION USING REMOTE SENSING IMAGES

Because DL can learn complicated and hierarchical features straight from raw data, it has become a potent and revolutionary tool for CD in RS photos. Conventional CD approaches frequently use threshold-based procedures and handmade characteristics, which could not translate well among various datasets or imaging settings. By proactively acquiring pertinent interpretations from multitemporal images, which further improve chronology interpretation. By creating change maps and lowering false detections via adversarial training, generative adversarial networks (GANs) make a contribution. As given in Table 2 shows the count of trainable parameters for every model, offering insight towards the sensitivity of the model and possible execution utilisation. GANs carry the highest parameter count at 13.5 million, followed by CNNs at 11.0 million and RNNs at 9.8 million. DBNs and SAEs, in contrast, are more lightweight, containing only 1.5 million paramètres, stacked autoencoders (SAEs) have also demonstrated promise. The models mentioned above have been used in many different fields, including mapping floods, tracking forest clearing, monitoring the growth of cities, and evaluating the effects of disasters [17].

DL techniques have grown increasingly reliable and scalable as a result of the growing availability of high-resolution satellite images from platforms such as Sentinel, Landsat, and SAR sensors. Combining deep learning with cuttingedge data sources increases the computerization and effectiveness of large-scale picture analysis in addition to improving CD accuracy. Research in this quickly developing field is still driven by issues with data limited availability, model generalization, and computer difficulty, notwithstanding the advancements.

Table 2. Description of different datasets.

Model	Training Time(hrs)	Inference (sec)	Time Parameters (Millions)
DBN	2.0	0.8	1.5
SAE	2.5	1.2	1.5
GAN	7.0	2.5	13.5
CNN	3.5	1.0	11.0
RNN	4.5	1.5	9.8

4.1 *Fully supervised learning based-methods*

These techniques call for labeled datasets with clear annotations of the altered types and sites.These annotated datasets are used to be train DL models such as CNN, U-Nets, and deep residual networks (ResNets), to identify the spectrum and geographical patterns of change. Since the models are effectively directed by labels that reflect the truth during training, the primary benefit of fully supervised approaches is their great accuracy [18].

Extremely fine changes in land use, construction of cities, deforestation, and areas affected by disasters are especially well-represented by these models. However, which can be expensive and time-consuming to acquire—are critical to the performance of fully supervised models. Furthermore, until revamped with fresh data, they might not generalize well to other geographic locations or imagery settings. Notwithstanding these drawbacks, when enough marked data is offered, fully supervised learning continues to dominate the field of DL-based change detection because of its accuracy and resilience.

4.2 *Fully unsupervised learning based-methods*

V. Since fully unsupervised learning-based techniques for detecting changes in remote sensing images don't require labeled data, they are ideal in situations where obtaining ground truth descriptions is either impossible or prohibitively expensive. By learning that the data can be obtained, these techniques seek to automatically identify changes. inherent distribution and structuring of data from multi-temporal distant sensing. To extract key characteristics from preand post-event photos, methods including autoencoders (AEs), restricted Boltzmann machines (RBMs), and deep clustering approaches are frequently used [19].

Such models can detect significant variations that point to changes through contrasting reconstructive mistakes or hidden interpretations. Unsupervised methods' main benefits are their scalability and versatility, which allow them to be used on a variety of huge datasets without the need for human interaction. However, particularly in complex situations, a lack of semantic comprehension and a greater propensity for false detections may restrict their performance. Fully unsupervised learning has drawn interest despite these obstacles because of its promise for automated, large-scale change detection in remote sensing, particularly in situations where labeled data is hard to come by or unavailable.

4.3 *Deep transfer learning approaches*

Deep transfer learning approaches have become popular as a potential solution for identifying changes in remote sensing images, especially in cases when tagged training data is scarce. These techniques take knowledge gained from an origin area—typically a big, named dataset—and apply it to a domain of interest with little or no labeling. Pre-trained algorithms using deep learning, such as tailored to the specific features of remote sensing data. This strategy considerably decreases the requirement for extensive training data and processing resources while maintaining competitive advantage. Learning from transfer can be used to several deep systems. CNNs and transformer-based models can be used to extract robust and generalised characteristics from excellent quality satellite or aerial data. It works especially well with mismatched data types or cross-domain programs, such as converting optical image-trained models to SAR or multi-spectral data. While deep transfer learning improves model generalization and speeds up training, it may still encounter domain misfit and problems with overfitting when the origin and target data distributions differ significantly. Nonetheless, it provides a realistic and economical approach to deploying deep learning in real-world change detection tasks with little annotated data [20].

5 CONCLUSION

With change detection, ecological resources and environmentally changes can be managed and monitored more effectively. Since remote sensing images constitute big data, it is quite challenging to identify the actual changes in multi-date images.With deep learning. Deep learning techniques have been found to outperform conventional change detection methods in various aspects in the domain of DLCD. CNNs outperformed all other models analyzed in change detection tasks within a wide range of remote sensing situations. GANs and RNNs have complementary attributes and can be combined into hybrid models. Future research could look into attention-based CNN-RNN hybrids as well as transformer frameworks to improve spatial-temporal prediction.

REFERENCES

[1] Shafique, A., Cao, G., Khan, Z., Asad, M., Aslam, M. (2022). Deep learning-based change detection in remote sensing images: A review. *Remote Sensing*, 14, 871.

[2] Wang, L., Zhang, M., Gao, X., Shi, W. (2024). Advances and challenges in deep learning-based change detection for remote sensing images: A review through various learning paradigms. *Remote Sensing*, 165, 804.

[3] Jiang, H., Peng, M., Zhong, Y., Xie, H., Hao, Z., Lin, J., Hu, X. (2022). A survey on deep learning-based change detection from high-resolution remote sensing images. *Remote Sensing*, 14, 71552.

[4] Bai, T., Wang, L., Yin, D., Sun, K., Chen, Y., Li, W., Li, D. (2023). Deep learning for change detection in remote sensing: A review. *Geo-spatial Information Science*, 263–288.

[5] Wang, J.J., Dobigeon, N., Chabert, M., Wang, D.C., Huang, T.Z., Huang, J. (2024). CD-GAN: A robust fusion-based generative adversarial network for unsupervised remote sensing change detection with heterogeneous sensors. *Information Fusion*, 107,102313.

[6] Banwala, N., Singh, G., Singh, J., Goswami, V.S., Bagnia, A. (2023). Recommendation system and its techniques in machine learning: A survey. In: Kumar, R., Verma, A.K., Sharma, T.K., Verma, O.P., Sharma, S. (eds) *Soft Computing: Theories and Applications. Lecture Notes in Networks and Systems*, vol 627. Springer, Singapore. https://doi.org/10.1007/978-981-19-9858-4_17

[7] Parelius, E.J. (2023). A review of deep-learning methods for change detection in multispectral remote sensing images. *Remote Sensing*, 158, 2092.

[8] Manocha, A., Afaq, Y. (2023). Optical and SAR images-based image translation for change detection using generative adversarial network (GAN). *Multimedia Tools and Applications*, 82, 726289–26315.

[9] Bist, A., Singh, G., Srivastava, S. (2024). Transformative Potential of Consumer Insights into Sustainable Business Growth: Machine Learning Integrated with Marketing Analytics. 4th International Conference on Soft Computing for Security Applications (ICSCSA), 1–6.

[10] Mandal, M., Vipparthi, S.K. (2021). An empirical review of deep learning frameworks for change detection: Model design, experimental frameworks, challenges and research needs. *IEEE Transactions on Intelligent Transportation Systems*, 2376, 101–6122.
[11] Mahara, A., Rishe, N. (2024). Multispectral band-aware generation of satellite images across domains using generative adversarial networks and contrastive learning. *Remote Sensing*, 167, 1154.
[12] Gui, S., Song, S., Qin, R., Tang, Y. (2024). Remote sensing object detection in the deep learning era—a review. *Remote Sensing*, 16, 2327.
[13] Rui, X., Cao, Y., Yuan, X., Kang, Y., Song, W. (2021). DisasterGAN: Generative adversarial networks for remote sensing disaster image generation. *Remote Sensing*, 13, 214284.
[14] Liu, P., Li, J., Wang, L., He, G. (2022). Remote sensing data fusion with generative adversarial networks: State-of-the-art methods and future research directions. *IEEE Geoscience and Remote Sensing Magazine*, 10, 295–328.
[15] Huang, W., Zhang, S., Wang, H.H. (2021). Efficient GAN-based remote sensing image change detection under noise conditions. *Image Processing and Capsule Networks: ICIPCN* 2020, 1–8.
[16] Jian, P., Chen, K., Cheng, W. (2021). GAN-based one-class classification for remote-sensing image change detection. *IEEE Geoscience and Remote Sensing Letters*, 19, 1–5.
[17] Jozdani, S., Chen, D., Pouliot, D., Johnson, B.A. (2022). A review and metaanalysis of generative adversarial networks and their applications in remote sensing. *International Journal of Applied Earth Observation and Geoinformation*, 108, 102734.
[18] Sharma, P., Singh, G., and Kaur, N. (2023, December). Data valuation using Shapley value in machine learning. In *AIP Conference Proceedings* (Vol. 2916, No. 1, p. 030001). AIP Publishing LLC.
[19] Zheng, Z., Tian, S., Ma, A., Zhang, L., Zhong, Y. (2023). Scalable multi-temporal remote sensing change data generation via simulating stochastic change process. *Proceedings of the IEEE/CVF International Conference on Computer Vision*, 21818–21827.
[20] Wang, Z., Zhang, Y., Luo, L., Wang, N. (2022). CSA-CDGAN: Channel selfattention-based generative adversarial network for change detection of remote sensing images. *Neural Computing and Applications*, 34, 1999–2013.
[21] Yang, W., Song, H., Du, L., Dai, S., Xu, Y. (2022). A change detection method for remote sensing images based on coupled dictionary and deep learning. *Computational Intelligence and Neuroscience*, 2022, 13404858.

Smart Systems and Beyond – Bhatia et al. (Eds)

Utilizing Decentralized Blockchain Platforms and Advanced Cryptographic to Mitigate Data Breaches in Healthcare Systems

Kapil Bhardwaj, Gurpreet Singh, and Deepak Kumar
Department of CSE, University Institute of Engineering, Chandigarh University, Mohali, India

ABSTRACT: The rise in digitalization of medical facilities has resulted in the collection and retention of massive volumes of personal information, raising major concerns about data security and privacy. Conventional centrally stored techniques are subject to cyberattacks and data breaches, which jeopardize patient privacy and reliability. This study investigates the use of decentralized blockchain technology paired with modern cryptography techniques to prevent data breaches in medical facilities. Blockchain's intrinsic properties of decentralization, immutability, and transparency provide a solid foundation for safe data management that does not rely on intermediaries. Furthermore, the use of cutting-edge cryptographic techniques improves security of information, reliability, and visibility control. The suggested method intends to enable secure storage, exchange, and retrieval of medical information while protecting patient privacy and fulfilling government standards. The scientific assessment shows considerable advances in preventing illegal accessing and data manipulation, emphasizing the potential of blockchain and cryptography to enhance medical information safety. In addition, the Advance Data Security Architecture in Healthcare (ADSAH) combines Elliptical Curve Cryptography (ECC), Blockchain, and Deep Neural Networks (DNN) to protect cloud-stored medical information while providing reliable, safe, and inexpensive access.

Keywords: Blockchain, ECC, Healthcare Security, Cloud Computing, Deep Neural Networks· MKP-ABE

1 INTRODUCTION

The exponential expansion of innovative ideas from smart cities and IOT component to digital medical services platforms has given an exceptional rise in collecting the information and sharing data. The information collected generally involves important individual and data related to healthcare field, making it a real host for unwanted cyber attacks and unwanted access [1]. Traditional cloud based platforms, while in the need for huge operation storage, are getting more easy to be attacked to privacy violations, attack can be via internet, and so much threat from insiders. These concerns about the security of data are enhanced by combination of structure and multi-dimensional connection of modern digital environments. [2]. To highlight these concerns, blockchain technology has evolved with a possible solution due to its distributed, stable and transparent properties. When it is utilized with advanced cryptographic method like Elliptical Curve Cryptography (ECC), blockchain can hugely impact the privacy, integrity and accessibility control. In addition to this the utilization of Artificial Intelligence and Deep Neural Network (DNNs) can enhance the decision and pattern matching capabilities required for secure data management [3].

This chapter explain the utilization of blockchain based technologies and AI enabled cryptographic algorithms to avoid data loss, specially in healthcare cloud based system. The strategy which is suggested have purpose to solve the issues of standard safety models by giving attention on preservation of privacy methods with safe data exchange methods, while it also displaying a scalable , resilient solution for safeguarding the sensitive information related to medical field. Some advanced applications are built due to enhancement in smart cities and infrastructural development [4]. The huge enhancement in the data generated by IoT devices, interconnected devices and digital communication highlight the significance of safe data management and storage. While the basic cryptographic techniques gives fundamental safety by integrating them with decentralized blockchain technology enhance integrity, validity and trust. This research showcase how blockchain handles centralization hardness in cloud storage, specially in healthcare by installing decentralized, AI integrated encryption solutions [5].

1.1 *Blockchain's Importance and barriers in secure cloud computing*

The advanced development of innovative devices moves to in the creation and management of huge amount of sensitive information, required the use of extremely safe and secure dependable data security technology. Blockchain technology is an exciting solution as of its decentralized structure, stability and transparency which is naturally enhance the safety of information, validation and create trust between various customers in cloud based

DOI: 10.1201/9781042026289-39

environment [6]. Still, regardless of its advantage, the combination of blockchain into cloud service is not with any hurdle. Scalability still a major issue, since the conventional blockchain system frequently fail to manage huge database with its elevated transaction volumes effectively. [7].

In addition to this , inclusion of blockchain into available cloud services may result into considerable functional problem, minimizing the reaction time of these smart devices. Smooth combination and replacement with existing model and data model cautious preparation and adjustment. To full explore blockchain promise for safeguarding cloud based smart application development in agreement techniques, storage efficiency and structure for general communication with IoT, blockchain and cloud to represent in Figure 1. These steps will assist to minimize latency, enhance throughput and assure effective implementation in the real world situations. [8].

- Smart applications generate and process vast amounts of sensitive data. Blockchain' simmutability and cryptographic hash functions provide are liable olution to guarantee data integrity Advanced applications [9].
- Smart applications often involve multiple parties, such as users, service providers, and IoT devices. Establishing trust and ensuring transparency in data transactions can be challenging. The distributed and transparent nature of blockchain, coupled with consensus mechanisms, builds trust among participants and eliminates the need for intermediaries [10].

1.2 *Numerous difficulties and concerns*

- Scalability problems exist for blockchain technology when dealing with huge amounts of transactions and data. To overcome these challenges, effective consensus procedures, layer-two solutions, and hybrid techniques must be explored [11].
- Blockchain's processing requirements can result in operational impact when combined with cloud-based smart apps. Optimizations and improvements in blockchain technology and architecture are required to reduce the latency and improve effectiveness [12].
- Inegrating blockchain into current cloud-based smart applications necessitates meticulous preparation and assessment of the current facilities data structures and overall application design. Smooth connectivity and Innovative apps needs advanced data security capabilities to safeguard the huge amounts of data created by platforms such as smart cities, smart medical care, IoT, IoMT, and smart automobiles [13]. Because of its scalability, cloud storage is frequently utilized for storing vast amounts of data; yet, as a centrally managed system, it represents an opportunity for intrusions on data from advanced applications [14].

To solve this risk, blockchain technology is used using artificial intelligence (AI) to improve data security in cloud storage settings [15]. Health care providers are attractive targets for cyberattacks because of the high value of health information. Hacking of data can cause illegal availability, theft, or disclosure to confidential patient data, resulting in major breaches of confidentiality and potential injury to persons. Furthermore, the medical sector employs a vast number of individuals with ownership of information about patients. Insider hazards, either deliberate or unintentional, are a substantial risk to data security and privacy. This study focuses on maintaining privacy and protecting sensitive data regarding health care and other innovative apps [16]. A powerful encryption and decryption system augmented by AI is offered. Wellness information is submitted and retrieved with a secret key. Blockchain infrastructure is used for storing and control encryption and decryption procedures. Embedding cryptographic processes within blockchain blocks protects data from attacks. Integrating AI using cryptographic approaches improves data dependability and robustness.

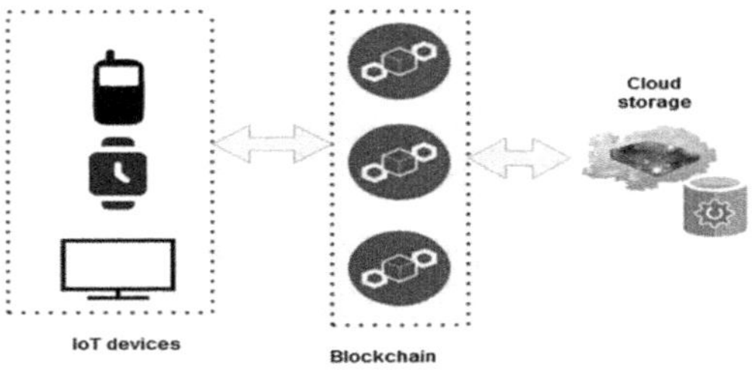

Figure 1. General communication architecture of IoT, blockchain and cloud.

2 LITERATURE REVIEW

The increased adoption of IoT devices has raised worries regarding data availability, privacy, and device integrity. This has resulted in a greater attention on malware assaults in IoT systems, particularly Industrial Control Systems (ICS) and Cyber-Physical Systems (CPS). The report describes an attack approach and emphasizes the possibility of distributed denial-of-service (DDoS) attacks using IoT botnets. It gives information about these dangers and recommends security ways to counteract them [16].

The authors of this paper discussed the necessity for confidentiality machine learning in healthcare artificially intelligent systems that rely on huge, transnational databases. They present PriMIA, a freely available software system that supports highly secure collaborative training and encrypted prediction on medical scan images. The system has been evaluated on a juvenile chest X-ray classification investigation and showed comparable results to domestically learned models, while also providing privacy assurances as well as safety regarding model [17].

This paper introduced the MSCryptoNet platform to solve privacy concerns in the Internet of Things (IoT) and universal medical networks. This allows for the scaled implementation and advancement of cutting-edge neural networks into MSCryptoNet models that protect user privacy. The architecture includes approaches to estimating activation parameters employed in convolutional neural networks and that are critical for calculations in homomorphic protection systems [18].

The author of this paper stressed the necessity of anonymity and security in networked machine learning algorithms utilizing cloud computing infrastructures. While machine learning-driven categorisation approaches have numerous uses, both models as well as data necessitate solid security for privacy. Learningbased categorization approaches have several applicability; yet, they necessitate strict security precautions for both data and models [19]. The study presents a privacy-preserving mechanism for the Extreme Learning Machine (ELM) algorithm that will avoid sensitive data leaking throughout outsourcing handling of data.

The authors of this paper [20] investigated the increasing rate of medical care data and services transferring to the cloud, as well as the drawbacks of existing security solutions for protecting privacy. Their findings looks into the utilization of blockchain as a place for protecting medical information in the cloud. They focuses the practical implications of using blockchain with in the medical care facilities and the requirement of additional studies in this sector..

The authors of this paper [11] underline the vital necessity for safeguarding privacy in contemporary tools such as cloud computing, social media, IoT, and e-healthcare platforms. The protecting of wellness and medical data, which is mostly contain crucial data about the patients is critical. Traditional encryption methods are not suitable for e-health statistics, even when communication of personal information across freeways. To find out the solution researchers give suggestion on a privacy preserving chaos based encoding cryptonymous.

This paper [13] highlight how blockchain technology might help current healthcare structures minimize complexities and administrative cost by improving medical records maintenance and integrating with insurance brokers. Bclockchain decentralized system in ecryption nature help to change combining healthcare information, accounting of insurer and smart contacts. It also help to make medical data sources more interoperable, boosting the existing data on patient, gadget observation, prescription data and institutional manging things. This paper [13] giving attention on how blockchain technology might help current healthcare system to minimizing complexities, and administrative cost by improving he medical record maintaince.

The authors of this paper [14] investigate the use of Blockchain Technology (BCT) in medicine, highlighting its benefits in facilitating safe and impermeable data transfers while protecting personal data. Blockchain's decentralized structure removes the need for centralized storage facilities while allowing for userdriven updates and scrutiny. This investigation focuses on blockchain's disruptive prospects in a variety of fields, notably procedural medicine, digital health records, educational programs, analyzing data, and monetary governance.

3 ADVANCED DATA SECURITY ARCHITECTURE IN THE HEALTHCARE ENVIRONMENT: METHODOLOGY

The security and privacy of clinical data are crucial in the healthcare sector since precise diagnoses are dependent on healthcare workers having prompt knowledge of patient information. At exactly the same time, people need accessibility of their private medical records. Managing that vital medical data is safely exchanged whereas respecting privacy is critical. To overcome these obstacles, this paper provides a fresh approach named Advanced Data Security Architecture in the Healthcare Environment (ADSAH) as shown in Figure 2. The ADSAH system

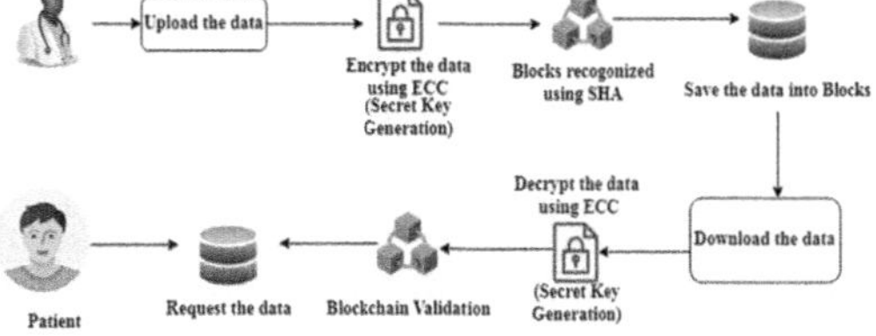

Figure 2. System model architecture of IoT, blockchain and cloud.

protects medical data kept within the cloud by combining Elliptic Curve Cryptography (ECC), blockchain protocols, and a Deep Fuzzy Based Neural Network (DFBNN).

The suggested approach starts by encoding raw medical data with an encoder, then proceeds to encryption with ECC algorithms. The secret key utilized for encryption is safely maintained inside the blockchain system. This key is separated into block for added protection, and the Secure Hash Algorithm (SHA) is utilized to locate key events among each block. These essential events are then kept in a cloud storage system. Encryption and decryption keys are generated using an improved Genetic Algorithm that has been especially designed for medical record security. Within this architecture, clinicians upload sensitive health information which is encrypted using ECC. The ADSAH architecture provides a strong safety precaution for encrypting and distributing medical information. The secret key for the encrypted data is safely maintained in blockchain-based block database. The SHA technique improves security by recognizing critical events within individual blocks. Once encrypted, the data is kept in the cloud, where it remains available and confidential. Only registered healthcare providers and patients may view the data via the secret key, and this goes through the ADSAH encryption and decryption techniques, allowing for secure recovery. During the procedure, the data is safeguarded within the cloud environment, ensuring both confidentiality as well as privacy.

It is critical in healthcare IT systems to ensure the security of user-related data, such as clinical data, healthcare records, and prescription instructions. The Deep Fuzzy Based Neural Network (DFBNN) structure brings together deep learning and fuzzy logic approaches and architecture as depicts n Figure 3. This mixed method merges deep learning's outstanding identification of patterns ability with fuzzy logic's uncertainty-handling ability, therefore being perfect for assignments like making choices, classification, and regression. Normalization improves resolution and durability by placing input information within an equivalent range. The fuzzy layer incorporates fuzzy logic into the infrastructure, allowing it to handle inaccuracy and ambiguity in information. DFBNN is capable of managing scalable raw materials as well as uncertain data through the integration of normalization and fuzzy layers. As the end result, the framework produces dependable, comprehensible and consistent results.

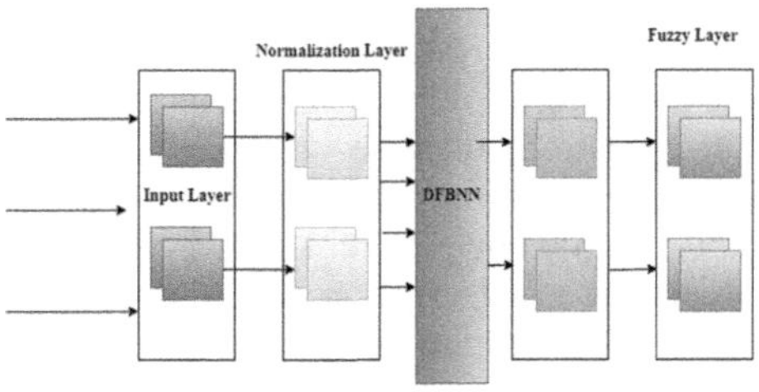

Figure 3. System model DFBNN architecture.

4 PERFORMANCE EVALUATION OF ADSAH

4.1 *Uploading time comparison*

According to Figure 4, the suggested ADSAH approach has greatly enhanced uploading efficiency. With a 1MB data size, ADSAH yields a remarkable uploading time of 2.89 milliseconds, surpassing other comparison approaches like this:

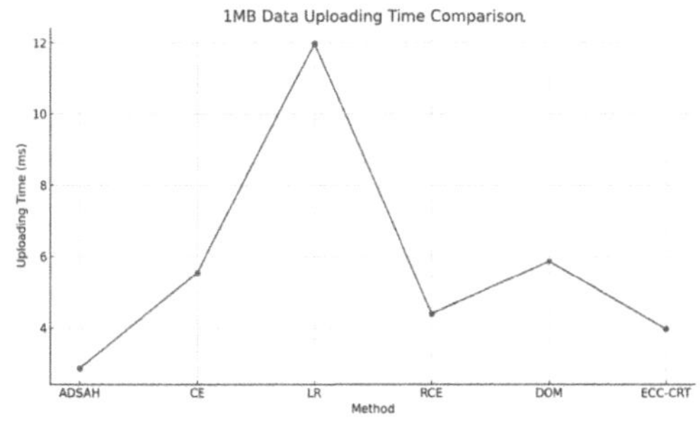

Figure 4. Uploading Time (in milliseconds) for 1MB of data across different methods.

The significant decrease in upload time suggests that ADSAH has successfully streamlined the data uploading process, allowing for quicker and more productive transmission of information. In addition, this speed benefit maintains throughout rising data sizes of 2 MB, 4 MB, 6 MB, and 8 MB, with ADSAH continually demonstrating quicker uploading times comparing to the other approaches as given in Table 1.

Table 1. Uploading time comparison for 1MB data size.

Method	Uploading Time (ms)
ADSAH (Proposed)	**2.89**
CE	5.55
LR	11.98
RCE	4.40
DOM	5.87
ECC-CRT	3.98

4.2 *Downloading time comparison*

Consistently best Encryption Execution Time 36.82 ms CE (47.78 ms) Faster by 10.96 ms The outcomes conclusively demonstrate that ADSAH beats conventional cryptography and data processing platforms over key performance measurements, rendering it a great contender for secure and effective healthcare data management in cloud contexts. The ADSAH approach also has faster information extraction speeds as shown in Table 2. This method downloads 1 MB of data in 2.32 milliseconds, outperforming competitors. This upward trend in download efficiency remains as data sizes grow to 2 MB, 4 MB, 6 MB, 8 MB, and 10 MB, with ADSAH continuously providing quicker retrieving durations.

Table 2. Download time comparison between ADSAH and competitor.

Data Size (MB)	ADSAH (ms)	Competitor (ms)
1	2.32	3.90
2	3.70	5.30
4	4.70	8.60
6	9.30	12.70
8	10.00	11.40
10	11.80	16.60

4.3 *Encryption processing rate comparison*

In the context of encryption duration of execution, ADSAH consistently outperforms the competition, with a suggested execution time of 36.82 milliseconds. ADSAH outperforms other approaches, with times ranging from 47.78 ms to 57.66 ms for cryptographic computations are depicts in Figure 5. This performance is critical in real-time medical environments, where rapidly changing information encryption and decryption are required.

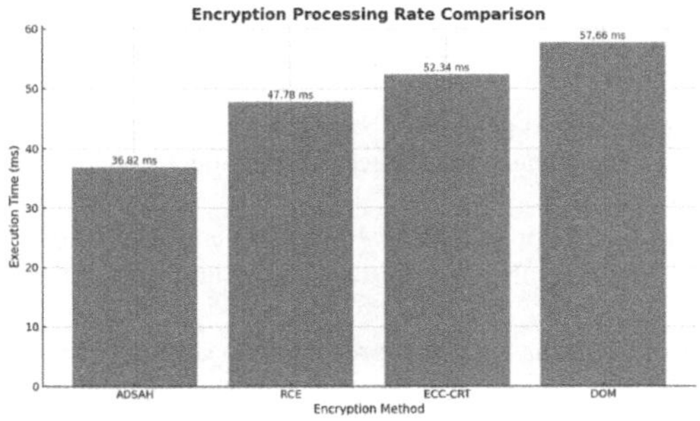

Figure 5. Encryption execution time comparison.

4.4 *Summary of results metric ADSAH*

The most effective Competitor Procedure Enhancement Upload Time (1MB) 2.89 ms ECC-CRT (3.98 ms) is 1.09 ms faster. Download Time (1MB): 3.23 ms. Progressively appropriate encryption execution time: 36.82 ms CE (47.78 ms). faster by 8.16 milliseconds The outcomes convincingly demonstrate that ADSAH beats traditional cryptography and data handling platforms across essential performance parameters, rendering it a great contender for trustworthy and effective healthcare data management in cloud contexts.

5 CONCLUSION

This study offers a novel approach called ADSAH (Advanced Data Security Architecture in Healthcare) that addresses the issues of privacy and security associated with medical information preserved in the cloud. The suggested system successfully protects and manages medical records by merging Elliptic Curve Cryptography (ECC), blockchain technology, and a Deep Neural Network-based Deep Fuzzy Based Neural Network (DFBN). In the ADSAH development, medical information is initial encoded utilizing an encoder followed by encrypted using ECC technologies. The secret key to encryption is safely held under a blockchain structure, adding to the general safety of managing keys. The Secure Hash Algorithm (SHA) is used to recognize significant incidents across blockchain blocks, those are then saved in a cloud-based storage system. To create encryption and decryption keys, an altered evolutionary method designed for medical data security is used. Approved users, such as healthcare professionals and patients, are able to access encrypted medical information by decrypting it using the secret key. ADSAH outperforms previous research with regard to of upload and download timings, encryption and decryption speeds, and transactional overhead expenses. ADSAH ensures robust encryption, anonymity, and secure possession of critical medical data through the effective integration of ECC, blockchain, and deep neural network algorithms. This methodology efficiently handles the complicated security and privacy problems connected with medical records, and it represents a potential option for the healthcare sector.

REFERENCES

[1] Kumar, A., and Chatterjee, K. (2025). A blockchain-based framework with cryptographic tags for healthcare security. *The Journal of Supercomputing*, 81, 729.

[2] Liu, J., *et al.* (2018). BPDS: A blockchain based privacy-preserving data sharing for electronic medical records. *arXiv preprint arXiv:1811.03223.*

[3] Bose, S., and Marijan, D. (2023). A survey on privacy of health data lifecycle: A taxonomy, review, and future directions. *arXiv preprint arXiv:2311.05404.*

[4] Kumar, A., and Chatterjee, K. (2025). BlockHealthSecure: Integrating Blockchain and Cybersecurity in Post-Pandemic Healthcare Systems. *Information*, 16, 2133.

[5] G. Singh, K. Saxena and R. Goel, "Meta-Analysis of Psychological Interventions and Control for Treatments of PTSD," 2024 5th International Conference on Data Intelligence and Cognitive Informatics (ICDICI), Tirunelveli, India, 2024, pp. 1403–1408, https://doi.org/10.1109/ICDICI62993.2024.10810840.

[6] Kumar, M., *et al.* (2023). Blockchain inspired secure and reliable data exchange architecture for cyber-physical healthcare system 4.0. *arXiv preprint arXiv:2307.13603.*

[7] Revolutionizing Healthcare Record Management: Secure and Interoperable Electronic Health Records using Blockchain. (2025). *arXiv preprint arXiv:2503.00742.*

[8] Taherdoost, H. (2023). Privacy and security of blockchain in healthcare: Applications, challenges, and future perspectives. *Science*, 54.

[9] Security of healthcare data using blockchains: A survey. (2021). *arXiv preprint arXiv:2103.12326.*

[10] A systematic review of privacy-preserving blockchain frameworks in healthcare. (2024). *Multimedia Tools and Applications.*

[11] A. Bist, G. Singh and S. Srivastava, "Transformative Potential of Consumer Insights into Sustainable Business Growth: Machine Learning Integrated with Marketing Analytics," 2024 4th International Conference on Soft Computing for Security Applications (ICSCSA), Salem, India, 2024, pp. 1–6, https://doi.org/10.1109/ICSCSA64454.2024.00008.

[12] Zou, R., Lv, X., and Zhao, J. (2020). SPChain: Blockchain-based medical data sharing and privacy-preserving eHealth system. *arXiv preprint arXiv:2009.09957.*

[13] Blockchain and emerging technologies for next generation secure healthcare systems. (2025). *Computers in Biology and Medicine.*

[14] Torongo, A., and Toorani, M. (2023). Blockchain-based decentralized identity management for healthcare systems. *arXiv preprint arXiv:2307.16239.*

[15] Adusumilli, S., *et al.* (2023). Enhancing data privacy in healthcare systems using blockchain technology. *ResearchGate.*

[16] Blockchain for healthcare systems: Architecture, security challenges, and solutions. (2023). *Journal of Network and Computer Applications.*

[17] Blockchain-based privacy preservation framework for preventing healthcare data breaches. (2024). *Multimedia Tools and Applications.*

[18] G. Singh, K. Simran and R. Goel, "Potential of Deep Learning Technique to Enhance the Accuracy of ECG Heartbeats Arrhythmia Time Series," 2024 4th International Conference on Technological Advancements in Computational Sciences (ICTACS), Tashkent, Uzbekistan, 2024, pp. 1062–1066, https://doi.org/10.1109/ICTACS62700.2024.10840893.

[19] Nedunoori, V. (2025). A comprehensive review of encryption and protection techniques for healthcare data. *Artificial Intelligence in Healthcare Information Systems—Security and Privacy Challenges*, 147–170.

[20] Wikipedia. (2025, July). Privacy and blockchain: Zero-knowledge proofs and cryptographic methods. *Online*: https://en.wikipedia.org/wiki/Privacyandblockchain

Smart Systems and Beyond – Bhatia et al. (Eds)
© 2027 The Author(s), ISBN: 978-1-041-36617-1

A Comprehensive Review of Urban Growth Detection Through the Fusion of Remote Sensing and Machine Learning

Rajendra Singh Shekhawat and Manoj Kuri
Engineering College Bikaner

U.S. Modani
Engineering College Ajmer

Mohit Srivastava
Chandigarh Engineering College, Landran Mohali

ABSTRACT: Urbanization is the primary demographic and land use trend of the twenty-first century. It offers both significant opportunities and challenges for global sustainability. Accurate, timely, and scalable monitoring of urban growth is one of the most important scientific and policy challenges. The paper looks at over forty years of interdisciplinary advancements connecting urban science, machine learning, and remote sensing (RS) are synthesized in this extensive work. We methodically break down the complete analytical pipeline from the fundamentals of data collection to the use of deep learning for predictive modeling.

The main focus is on machine learning techniques that have transformed the extraction of urban features and the detection of change. The development from early pixel based statistical classifiers to object-based image analysis (OBIA), and finally to the current dominance of deep learning and non-parametric ensemble techniques have been presented. The study covers Convolutional Neural Networks (CNNs) for finding spatial features (U-Net, DeepLab, ResNet), Recurrent Neural Networks (RNNs) for time-based sequence analysis, and hybrid designs (Siamese networks, STANets, Transformers) made for multi-temporal change detection are all included in this study.

This review adds value by focusing on measuring and modeling urban growth patterns, not just showing simple change maps. It looks at how sprawl, infill, and fragmentation can be measured using landscape ecology metrics. Also, it gives a thorough review of urban growth models (UGMs), focusing on how agent-based models (ABMs) can model the complex human choices that affect land change, and how cellular automata (CA) can be combined with machine learning (ML) to set transition rules.

Keywords: Urban Growth, Change Detection, Land Use Land Cover (LULC), Remote Sensing, Machine Learning, Deep Learning, Validation Metrics, Cellular Automata, Sustainable Development Goal 11

1 INTRODUCTION

1.1 *The planetary urban transition: Demographics, economics, and sustainability imperatives*

The 21st century is undoubtedly the urban century. Reports from the United Nations Department of Economic and Social Affairs (UN DESA, 2022) [1] show a big rise in the number of people living in cities, from 751 million in 1950 (30%) to 4.4 billion in 2021 (57%). By 2050, this is expected to reach 6.7 billion, or almost 70% of the population. Most of this growth is expected in developing Asia and Africa. Cities produce over 80% of the world's GDP, acting as centers for new ideas, business, and cultural exchange.

This fast and often unplanned urban growth creates a significant problem. It boosts prosperity but also causes environmental and social stress. The spread of cities takes away farm land, harms habitats, and lowers biodiversity. It changes local climates, worsening the Urban Heat Island (UHI) effect, and changes water cycles, raising surface runoff and flood risk. Socially, the speed of growth often outpaces city governments' ability to provide enough housing, transportation, water, and sanitation. This often leads to informal settlements (slums) with poor housing, overcrowding, and a lack of basic services, now home to over 1 billion people worldwide.

The need to monitor, understand, and manage urban growth not just for academic interest but for achieving global sustainability goals. The United Nations agenda 2030 for Sustainable Development recognizes this through Sustainable Development Goal 11 (SDG 11): "Make cities and human settlements inclusive, safe, resilient, and sustainable."

Urban planners need detailed land use/land cover (LULC) maps and growth predictions to design effective transportation system, manage land for different uses, and plan green spaces. Environmental groups must assess how urban growth affects ecosystems and resources to plan ways to reduce harm.

DOI: 10.1201/9781042026289-40

1.2 *Conceptualizing urban growth: Expansion, sprawl, intensification, and infill*

The term urban growth is often used broadly, but it includes wide range of patterns and processes with varied results for sustainability implications. Accurate detection and analysis need a precise conceptual framework.

- Urbanization: It is the process by which a population concentrates in urban areas, mostly as a result of migration from rural to urban areas. This may or may not include physical expansion.
- Urban Expansion: The spatial extension of built-up areas into nearby non-urban areas, such as agricultural fields, wetlands, or forests. It is the most easily identifiable type of urban growth through satellite images.
- Urban sprawl: It is frequently disparaging type of growth that is categorized by poorly planned, low-density development which is usually found on the outskirts of cities. It is like jumps over undeveloped land in scattered, leapfrog pattern. Sprawl is measured by metrics such as low population density, per capita high land consumption, and mono-functional land use.
- Infill Development: This is environmentally friendly type of growth that reuses underused or empty land within cities. It helps to revitalize the urban cores and makes better uses of the existing infrastructures such as roads, sewers, and water lines.
- Intensification: Also called as densification, this increases the number of people or business in areas that are already urbanized. It's done by redeveloping land with more units or taller buildings, raising density without expanding the urban area. It's key for encouraging small cities and stopping sprawl.

The ability to identify the type of change; differentiating between a new downtown high-rise (intensification), a new suburban neighborhood (sprawl) on the outskirts of the city (sprawl), and the redevelopment of a brownfield site redevelopment (infill), is made possible by remote sensing and machine learning. When evaluating the sustainability of urban development, this typological discrimination is important.

2 MACHINE LEARNING FOUNDATIONS FOR URBAN REMOTE SENSING

2.1 *Machine learning: Supervised, unsupervised, and semi-supervised learning*

Machine learning algorithms learn patterns from data without being explicitly programmed for every rule. They are categorized based on the nature of the learning signal.

- **Supervised Learning:** The algorithm is trained on a set of labeled data. In an urban context, this means a human analyst identifies examples (pixels or objects) of known classes (e.g., "Water," "High-Density Urban," "Forest") in the imagery. The algorithm learns the multivariate relationship between the spectral (and other) values of these training samples and their labels. Once trained, it can predict the class of all other pixels in the image. This is the most common approach for LULC classification. Examples: Support Random Forests (RF), Vector Machines (SVM), Neural Networks.
- **Unsupervised Learning:** The algorithm is given data without any labels. Its goal is to find inherent groupings or clusters within the data based on similarity of spectral values. The analyst must then interpret and label the resulting clusters post-hoc. This is useful for exploratory analysis or when labeled training data is scarce, but the clusters may not correspond to meaningful urban classes. Example: k-Means clustering.
- **Semi-Supervised Learning:** A hybrid approach that uses a small amount of labeled data together with a large amount of unlabeled data to improve learning accuracy. This is highly relevant to remote sensing, where acquiring high-quality training labels is time-consuming, but unlabeled pixels are abundant.

2.2 *Essential feature engineering for urban analysis*

Raw spectral bands are often not optimal for classification. Feature engineering creates new, more informative input variables ("features") for the ML model.

- **Spectral Indices:** Arithmetic combinations of bands designed to highlight specific materials.
  - **Normalized Difference Vegetation Index (NDVI):** (NIR - Red) / (NIR + Red). Highlights healthy vegetation.
  - **Normalized Difference Built-up Index (NDBI):** (SWIR - NIR) / (SWIR + NIR). Highlights built-up areas.
  - **Modified Normalized Difference Water Index (MNDWI):** (Green - SWIR) / (Green + SWIR). It enhance open water features while suppressing built-up and soil noise.
  - **Soil Adjusted Vegetation Index (SAVI):** ((NIR - Red) / (NIR + Red + L)) * (1 + L). It corrects NDVI for the influence of bare soil (where L is a soil brightness correction factor, often 0.5).

- **Texture Features:** Urban areas are highly textured. A smooth golf course and a textured residential area can be distinguished using metrics such as Contrast, Homogeneity, Entropy, and Dissimilarity, which are derived from the Gray-Level Co-occurrence Matrix (GLCM) and quantify the spatial distribution of tonal variations within a neighborhood of pixels.

- **Morphological and Spatial Features:** In Object-Based Image Analysis (OBIA), features are calculated for each image object (segment), such as area, perimeter, shape index, elongation, and border-to-border distance to other objects. This allows the model to incorporate contextual rules (e.g., "a swimming pool is a small, rectangular water body within a residential parcel").

2.3 *The statistical learning revolution: Support Vector Machines (SVM) and Kernel tricks*

SVMs became a gold standard in the 2000s for remote sensing classification due to their effectiveness with high-dimensional data and small training sets.[5]

- **Principle:** An SVM finds the optimal hyperplane in a high-dimensional feature space that separates the data classes with the maximum margin. The margin is the distance between the hyperplane and the nearest data points from each class, called support vectors.
- **The Kernel Trick:** Real-world data, like urban spectral data, is often not linearly separable. The kernel trick implicitly maps the input features into a higher-dimensional space where a linear separation is possible, without ever explicitly performing the computationally expensive transformation. Common kernels include:
    - **Linear Kernel:** $K(x_i, x_j) = x_i^T x_j$
    - **RBF (Radial Basis Function) Kernel:** $K(x_i, x_j) = \exp(-\gamma \|x_i - x_j\|^2)$. This is the most popular kernel for remote sensing as it can model complex, non-linear decision boundaries.

The optimization problem for a soft-margin SVM (which allows some misclassification) is given as:

$$\min_{w, b, \xi} \; (1/2)\|w\|^2 + C \sum \xi_i$$

subject to:

$y_i(w^T \varphi(x_i) + b) \geq 1 - \xi_i$ and $\xi_i \geq 0$ for all i

Where w is the normal vector to the hyperplane, b is the bias, $\varphi(\cdot)$ is the kernel-induced mapping function, ξ_i are slack variables allowing misclassification, and $C > 0$ is a regularization parameter that controls the trade-off between maximizing the margin and minimizing classification error.

2.4 *Ensemble learning: Random Forests (RF) and Model robustness*

Ensemble methods combine multiple "weak" learners to create a single, more robust and accurate "strong" learner. Random Forest is arguably the most widely used ML algorithm in modern geoscience. [4]

- **Principle:** An RF is an ensemble of many Decision Trees. Each of the trees is trained on a random bootstrap sample of the training data (bagging) and, at each split node, only a random subset of features is considered for finding the best- split. This introduces randomness that decorrelates the individual trees.
- **Prediction:** For classification, final prediction is made by majority vote from all the individual trees in the forest.
- **Advantages for Remote Sensing:**
    - **High Accuracy:** Consistently ranks as one of the top-performing classifiers.
    - **Resistance to Overfitting:** The bagging process mitigates overfitting.
    - **Handles High Dimensionality:** The random feature subset selection makes it efficient with many input features (e.g., multiple bands, indices, textures).
    - **Provides Feature Importance:** It internally ranks the importance of each input feature for making accurate predictions, offering valuable insight into which data sources are most useful.
    - **Few Hyperparameters:** Relatively easy to tune compared to deep learning models.

The prediction of an RF classifier is given by:

$$\hat{h}(x) = \operatorname{argmax}_{y} \sum_{t=1}^{T} I(h_t(x) = y)$$

where T is number of trees, h_t is the prediction of the tth tree, $I(\cdot)$ is the indicator function, and y is a class label.

2.5 *Introduction to deep learning: Perceptrons and Multi-Layer Perceptrons (MLPs)*

Deep Learning (DL) uses artificial neural networks with many layers ("deep" networks) to learn hierarchical representations of data. [7,15]

- **The Perceptron:** The simplest neural network unit. It takes vector of inputs x and computes a weighted sum, adds a bias, and passes the result through a non-linear activation function σ (e.g., Sigmoid, ReLU) to produce an output.

$$\text{output} = \sigma(\Sigma(w_i * x_i) + b)$$

- **Multi-Layer Perceptron (MLP):** A network of perceptrons organized in layers: one input layer, one or more hidden layers, and one output layer. Each node in a layer is well connected to every node in next layer (fully connected). An MLP can learn complex, non-linear decision boundaries. For a simple binary classification, the output might use a sigmoid function, while for multi-class (e.g., 5 LULC classes), the output layer would typically use a softmax function to output a probability distribution over the classes.

The power of DL comes from this hierarchical feature learning. Early layers might learn to detect simple edges and textures, middle layers combine these to detect parts of objects (e.g., a roof corner, a tree canopy), and later layers assemble these into complex objects (e.g., a building, a park). This automatic feature extraction is a key advantage over traditional ML, which relies on manual feature engineering.

3 URBAN LAND COVER CLASSIFICATION: FROM PIXELS TO OBJECTS TO DEEP LEARNING

3.1 *The Urban spectral mixture problem: challenges of heterogeneity and mixed pixels*

The primary obstacle in urban remote sensing is the stark disparity between the scale of urban objects and the spatial resolution of the majority of satellite sensors. In a typical urban setting, a 30 x 30 m Landsat pixel or even a 10 x 10 m Sentinel-2 pixel is likely to have a variety of materials, such as asphalt, concrete, roof metal, soil, and grass. This is referred as the "mixed pixel" problem. The measured spectral signature is a nonlinear aggregate of the spectra of all the materials present in that particular pixel's instantaneous field of view rather than a single material's unmixed signal. Traditional per pixel classifiers which relies on spectral homogeneity is confused by this spectral heterogeneity, which reflects in serious misclassifications like:

- Mistaking impermeable surfaces for soil, as both have low NIR reflectance.
- Misclassifying the mixed pixels with sparse vegetation as "low density urban area" while in reality, they might be a pavement area with a sparse trees.
- Not making a distinction between different type of impervious surfaces, such as specific composite roofing materials and asphalt.

Further complex classification paradigms that goes beyond the straightforward pixel based approach have been developed as a result of this mixed pixels related issues.

3.2 *Pixel-based classification: Workflow and comparative analysis*

Pixel-based classifications algorithm is still a popular and useful technique despite its drawbacks, specially for studies conducted at a regional scale using medium-resolution satellite data. The typical workflow of process consists of:

(1) Collection of Training Data: The analyst digitizes polygons over the image that represents each LULC class (such as "High-Density Urban," "Low-Density Urban,", "Water," "Forest," and "Bare Soil"). This requires in depth familiarity with the spectral properties of the imagery and the surrounding area as well.
(2) Training of Classifier: The applied machine learning algorithm is trained using the spectral values (from all bands and any derived indices like NDVI or NDBI) of every pixel within selected training polygons.
(3) Prediction: The trained model predicts the class of each pixel in the entire image.

Table 1. Comparative analysis of pixel-based classifiers for urban applications.

Algorithm	Key Principle	Advantages	Disadvantages	Typical Urban Use Case
Maximum Likelihood (MLC)	Assigns pixel to class with highest probability based on Gaussian distribution.	Simple, statistically well-founded.	Assumes normal data distribution, performs poorly with mixed pixels and non-Gaussian data.	Legacy studies; not recommended for modern complex urban analysis.
SVM (Support Vector Machine) [5]	Finds the optimal hyperplane which separates the classes with maximum margin in high-dim space.	Effective with high-dim data, good with small training sets, handles non-linearity via kernels.	Performance sensitive to kernel and parameter choice (C, γ); "black box" nature.	Standard choice for many years; robust for general urban LULC mapping.
Random Forest (RF) [4]	Ensemble of decorrelated decision trees that vote on the final class.	Highly accurate, robust to overfitting, handles high-dim data, provides feature importance.	Can be memory-intensive with many trees; less interpretable than a single tree.	Current gold standard for non-DL urban classification. Ideal for fusing multi-source data.
k-Nearest Neighbors (k-NN)	Classifies a pixel on the basis of the majority class of its 'k' most spectrally similar training pixels.	Simple, no training phase (lazy learner), naturally handles multi-modal classes.	Computationally expensive during prediction, sensitive to irrelevant features.	Less common; used in some hybrid or benchmarking studies.

(4) Post Classification Processing: The post processing may include applying a majority filter to reduce noise, a common artifact in pixel-based classifications where isolated pixels has been misclassified, creating a speckled appearance.

3.3 *Object-Based Image Analysis (OBIA): Multi Resolution segmentation and rule based classification*

The shortcomings of pixel based techniques led to the development of OBIA (Object-Based Image Analysis)[28], sometimes also referred to as GOBA (Geographic Object-Based Image Analysis). OBIA groups pixels into meaningful image-objects called segments that represent real-world entities such as buildings, city blocks, parks, and water bodies rather than classifying every pixels. There are two main steps in the OBIA workflow:

- **Segmenting the images:** The procedure for dividing an image into uniform and contiguous areas. The Multi-resolution Segmentation [29] is the most popular algorithm, it employs a bottom-up region-growing technique, is used in software such as eCognition. It is controlled by three crucial parameters and maximizes the homogeneity of segments while minimizing their average heterogeneity:
  - **Scale Parameter:** The primary determinant of object size. A higher value results in larger, more heterogeneous objects.
  - **Shape vs. Color:** A weight (e.g., 0.1 for shape, 0.9 for color) that dictates whether spectral homogeneity or smoothness of object borders is more important.
  - **Compactness vs. Smoothness:** A weight that influences the compactness of the resulting objects.

Selection of ideal parameters is essential and frequently requires iterative experimentation. Its main objective is to produce the segments that precisely match the edges of actual object.

- **Object Classification:** Once the segments are created, classification is performed on each object. This can be done using:
  - **Machine Learning:** Using the *statistics* of the pixels within each object as features (e.g., mean NDVI, standard deviation of the blue band, GLCM texture) to train a standard classifier like RF or SVM.
  - **Rule-Based Classification:** Building a knowledge-driven decision tree using thresholds and logical rules.

Advantages of OBIA:

The main advantages of OBIA are that it reduces salt and pepper noise and incorporates context and geometry: Features like area, shape, texture, and relational context can be used, which is impossible in previous pixel based techniques.

Disadvantages of OBIA:

- Complexity and Subjectivity: As compared to pixel based classification approaches, the OBIA process is more intricate. The established rules and segmentation parameters have a significant impact on the result, which introduces subjectivity from the analyst.
- Computational Cost: Requites a lot of processing power to segment a large image and to determine features for millions of objects.

3.4 *The deep learning breakthrough: Convolutional Neural Networks (CNNs) architectures*

Convolutional Neural Networks (CNNs), a subset of deep learning, have raised the bar for image recognition, including urban classification. Manual feature engineering is no longer necessary because CNNs automatically extract hierarchical spatial features from the raw image data.

3.4.1 *Encoder-decoder architectures: U-Net, SegNet, and deeplab for semantic segmentation*

Encoder-decoder architectures are the most common for dense prediction tasks such as semantic segmentation, which involves classifying each pixel in an image. Convolutional and pooling layers are used by the encoder (downsampling path) to extract features that are higher level and more abstract. These high-level features are converted back into a full-resolution segmentation map by the decoder (upsampling path).

- **U-Net:** U-Net [10], initially developed for biomedical image segmentation, has gained enormous popularity in remote sensing. Skip connections, which concatenate feature maps from the encoder with the corresponding layers in the decoder, are its primary innovation. This makes it possible for the decoder to retrieve fine-grained spatial information that was lost during downsampling, which is essential for precisely defining the borders of urban objects such as roads and buildings.
- **DeepLab**[11]: uses ASPP (Atrous Spatial Pyramid Pooling) to capture multi scale contextual information, it is crucial for comprehending objects of various sizes in a city, from cars to massive industrial complexes, and Atrous (Dilated) Convolutions to expand the field of view of filters without decreasing spatial resolution (preventing information loss from pooling).

These architectures are trained end-to-end using a loss function like Categorical Cross-Entropy, which measures the difference between the predicted probability distribution and the true (one-hot encoded) label for each pixel.

3.4.2 *Transfer learning: leveraging pre-trained models (ResNet, VGG) for urban tasks*

Training a deep CNN from scratch requires a massive amount of labeled data and computational resources. Transfer Learning mitigates this by taking a model (e.g., ResNet50[12], VGG16[13]) pre-trained on a large natural image dataset (e.g., ImageNet) and adapting it for the remote sensing task. The early layers of these networks, which learn basic feature detectors like edges and textures, are generic and can be directly used. Only the later, more task-specific layers are replaced and fine-tuned on the (comparatively smaller) dataset of labeled satellite imagery. This approach leads to faster convergence and higher performance, especially when labeled training data is limited.

3.5 *Multi-source data fusion: Combining optical, SAR, and LiDAR in a ML framework*

No single data source provides a complete picture of the urban environment. Data fusion [30,31] the synergistic combination of multiple data sources is key to achieving unparalleled classification accuracy.

- **Optical + SAR:** Optical data provides spectral information, while SAR provides information on structure and moisture, and is all-weather. A Random Forest or CNN can learn that a combination of high backscatter (from SAR) and low NIR reflectance (from optical) is a strong indicator of an industrial metal roof, whereas high backscatter and high NIR might indicate a vegetated area with underlying terrain effects.
- **Optical + LiDAR:** Optical data provides spectral detail, while LiDAR provides precise height information (Digital Surface Models - DSMs). Fusing these allows the classifier to easily separate tall buildings from ground-level impervious surfaces like roads and parking lots, which can be spectrally identical. It also enables the direct extraction of building footprints and vegetation height.

Deep learning models are exceptionally well-suited for data fusion. Different input modalities (e.g., a stack of optical bands and a stack of SAR features) can be fed into separate branches of a neural network, which are then merged in deeper layers, allowing the model to learn the complex relationships between them.

3.6 *Rigorous accuracy assessment: Confusion matrices, kappa, and object-based validation metrics*

A classification map is useless without a rigorous assessment of its accuracy. [32,33] The standard method is to use a confusion matrix (or error matrix) generated by comparing the classified map against a set of independent validation points (or objects) that were not used in training.

- **Overall Accuracy (OA):** The percentage of all validation points that were classified correctly. While simple, it can be misleading if the class distribution is imbalanced (e.g., if 90% of the area is forest, a 90% OA can be achieved by simply labeling everything as forest).
- **Producer's Accuracy (PA):** Measures how well the classifier performed for a specific class from the perspective of the ground truth. It is the probability that a real feature on the ground is correctly shown on the map. A low PA indicates a problem of **omission error** (the class was missed).
- **User's Accuracy (UA):** Measures reliability of the map from the user's perspective. It is the probability that a pixel labeled as a certain class on the map is actually that class on the ground. A low UA indicates a problem of **commission error** (the class was over-predicted).
- **F1-Score:** The harmonic mean of the precision (which is similar to UA) and Recall (which is similar to PA). It provides a single balanced metric for each class, especially useful for imbalanced datasets.
- **Kappa Coefficient (K):** A statistic which measures the agreement between classification and validation data while accounting for the agreement expected by random chance. A value of 0 indicates agreement equal to chance and 1 indicates perfect agreement. While widely used, it has been criticized for being difficult to interpret and is now often supplemented with other metrics.

For OBIA, additional geometric metrics can be used, such as over-segmentation/under-segmentation error or the **Dice coefficient**, which measures the spatial overlap between a classified image-object and a reference polygon.

A robust accuracy assessment report will include a confusion matrix and all these derived metrics, providing a comprehensive understanding of the map's strengths and weaknesses for each urban class.

4 CHANGE DETECTION ALGORITHMS: IDENTIFYING URBAN DYNAMICS

4.1 *Fundamentals of change detection: Categorization of algorithms and workflow design*

The process of detecting changes in the condition of an urban area through the analysis of co-registered images taken at various times is known as urban change detection. The main objective is to differentiate between real changes (like the conversion of farmland to residential space) and fake changes brought on by phenology, the weather, soil moisture, or sensor calibration. A change detection workflow's design is crucial and includes:

- Specifying the Change Question: What kind of change is desirable? (for instance, forest loss, urban growth, and intra-urban redevelopment). This determines the algorithm and the type of imagery (season, resolution).

- As mentioned in Section 2, image preprocessing and radiometric normalization are crucial. To guarantee that compared pixels represent the same location on the ground with radiometric values only influenced by land cover, images must be carefully co-registered and processed to surface reflectance.
- Choosing and putting into practice an algorithm.
- Transforming continuous change signals into a discrete change/no-change map is known as thresholding and post-processing.
- Assessing the change map's accuracy involves comparing it to a separate reference dataset, which frequently calls for old aerial photographs or high-resolution time-series imagery.

4.2 *Algebra-based methods: Image differencing, ratioing, and vegetation index differencing*

These are some of the most straightforward and understandable methods for detecting changes. They entail carrying out arithmetic operations between two co-registered images, band by band.

- **Image Differencing:** The simplest technique. The reflectance value in Time 2 is deducted from the value in Time 1 for every pixel: Change = Band_T1 - Band_T2. Large positive or negative values suggest possible change, whereas pixels with results close to zero are regarded as unchanged. Choosing a suitable threshold to distinguish between change and no-change is difficult. Instead of using a single band, it is frequently done on a vegetation index such as the NDVI.
- **Image Ratioing:** This is calculated by dividing Time 1 reflectance values by Time 2 (Band_T1 / Band_T2). The ratio for unaltered areas should be near 1.
- **Vegetation Index Differencing:** Using a index like NDVI or NDBI is often more robust than using a single band. For example, ΔNDVI = NDVI_T1 - NDVI_T2. A large negative ΔNDVI typically indicates vegetation loss (a common precursor to urban expansion), while a large positive ΔNDVI could indicate vegetation regrowth.

These methods are easy to implement and computationally efficient but provide no information on the *nature* of the change (from-to class). They are also highly sensitive to misregistration and radiometric errors.

4.3 *Transformation-based methods: Change Vector Analysis (CVA), Principal Component Analysis (PCA)*

- **Change Vector Analysis (CVA):** A powerful extension of image differencing that operates in multi-spectral space. For each pixel, the spectral response in Time 1 is represented as a vector in n-dimensional space (where n is the number of bands). The spectral response in Time 2 is another vector. The Change Vector is the vector difference between them. The magnitude of this vector indicates the intensity of change, while the direction (often expressed as the spectral angle) provides information about the type of change (e.g., towards urban, towards vegetation, towards water). Thresholding the magnitude produces a change/no-change map, and analyzing the direction can help categorize the change.
- **Principal Component Analysis (PCA):** PCA transforms the multi-band data from both dates into a new set of uncorrelated components (PCs). The logic is that the majority of the variance in the multi-temporal dataset is due to stable, unchanged features. The later components, which explain a small amount of variance, often contain the change information. The difference between the last PCs from the two dates, or simply the last PC itself from a stacked PCA, can be used to highlight areas of change.

4.4 *Classification-based comparison: Post-classification comparison and its error propagation*

This is one of the most widely used methods because it directly provides a "from-to" change matrix. It involves:

- Independently classifying the images from Time 1 and Time 2.
- Comparing the two classified maps on a pixel-by-pixel or object-by-object basis.

The output is a cross-tabulation matrix that details how many pixels changed from each class at T1 to each class at T2 (e.g., 500 pixels changed from "Agriculture" to "Urban").

The major drawback of this method is error propagation. The final change map's accuracy is the product of the accuracies of the two individual classifications. For example, if each classification has an overall accuracy of 85%, the theoretical maximum accuracy of the change map is 0.85 * 0.85 = 0.7225 or 72.25%, and it will likely be lower for the change classes themselves. Errors of commission and omission in the individual maps combine to create false changes and mask real changes.

4.5 *Direct multi-date classification and hybrid approaches*

This method avoids the error propagation of post-classification comparison by treating the multi-temporal imagery as a single, multi-layered dataset. All bands from Time 1 and all bands from Time 2 are stacked together into one large "multi-temporal" image. A single classifier (e.g., RF, CNN) is then trained to recognize not just spectral patterns, but *temporal-spectral* patterns. The training data consists of examples of specific *transitions* (e.g., "Urban to Urban" [no change], "Agriculture to Urban" [change], "Forest to Forest" [no change]). This method can be very accurate but requires extensive and often difficult-to-acquire training data for the change classes.

4.6 *Advanced machine learning for change detection: One-Class SVMs and isolation forests*

These algorithms are useful for detecting *anomalous* change, such as new building construction or illegal deforestation.

- **One-Class SVM:** Instead of learning a boundary between multiple classes, the one-class SVM learns a boundary that encapsulates the "no-change" pixels. Any pixel falling outside this boundary is flagged as a potential change. This is effective when the "change" class is diverse and difficult to define and sample.
- **Isolation Forest:** An unsupervised anomaly detection algorithm. It is based on the principle that anomalous data points (changes) are few and different, making them easier to "isolate" from the rest of the data (no-change) with random partitions. It is computationally efficient and well-suited for large-scale change detection tasks.

4.7 *Deep learning for change detection: Architectures for Bi-temporal and time-series analysis*

Deep learning has brought transformative architectures specifically designed for change detection.

4.7.1 *Siamese networks and pseudo-siamese networks*

A Siamese network [16] consists of two identical subnetworks (often CNNs like ResNet) that share the same weights and parameters. Each sub-network processes one of the two input images (T1 and T2). The feature maps extracted from each branch are then compared using a distance metric or a dedicated "difference" module to generate a change map. This structure ensures that the same feature representation is learned for both dates. A Pseudo-Siamese network allows the two branches to have different weights, which can be beneficial if the two images come from different sensors.

4.7.2 *3D CNNs and recurrent CNNs (e.g., ConvLSTM) for temporal feature learning*

For dense time-series analysis (e.g., using all available Sentinel-2 images over several years), 3D CNNs and Recurrent CNNs are used. A 3D CNN applies 3D convolutions that operate on the spatial dimensions plus the temporal dimension, learning spatio-temporal features directly. A ConvLSTM (Convolutional Long Short-Term Memory) network replaces the fully-connected layers in a standard LSTM (a type of RNN for sequences) with convolutional layers, allowing it to learn long-term dependencies in spatial sequences. This is powerful for modeling the gradual process of urban growth.

4.7.3 *State of the art architectures: STANet, DSAMNet, and Changeformer*

Recent research has produced highly sophisticated architectures:

- **STANet**[17] **(Spatial-Temporal Attention Network):** Incorporates attention mechanisms to explicitly model the spatial-temporal relationships between pixels in the two images. It can focus on "where" and "when" change occurs, improving accuracy, especially in complex areas.
- **DSAMNet**[18] **(Difference-Sensitive Attention Network):** Uses a difference-sensitive attention module to highlight change-related features and suppress unchanged ones, enhancing the model's ability to perceive fine-grained changes.
- **Change Former**[18]**: It is a** change detection model which is based on transformers. Transformers do extremely well at modeling global context because of their self attention mechanisms, which enable the model to realize the context of a possible change patch within the entire scene and results in predictions that are more reliable.

These state of the art deep learning techniques significantly outperforms conventional approaches; particularly for the intricate urban settings, and it automatically learn the best features for the change detection. They still face a major hindrance, that it needs a lot of labeled change/no-change data for training.

5 QUANTIFYING AND MODELING URBAN GROWTH PATTERNS AND PROCESSES

5.1 *Change maps to urban metrics: Calculating rate, intensity, and direction of growth*

The first step is simply a basic change map which displays "urban" and "non-urban" areas for two different time periods. This raster data needs to be converted into quantitative measures. These measures characterize the severity, pace, and spatial nature of the change in order to fully comprehend the dynamics of urban growth. According to Liu *et al.* (2010), these metrics are crucial for comparing growth across cities, time periods, or policy regimes. [23]

- **AGR (Annual Urban Growth Rate):** This measure, which represents the change as a percentage annually, is the most straightforward. It is calculated as:

$$AGR(\%) = [(Urb\_t2 - Urb\_t1)/Urb\_t1]/(t2 - t1) * 100$$

where Urb_t1 and Urb_t2 are the urban area (in hectares or km^2) at the initial and final time periods, and (t2 - t1) is the number of years between them. AGR is helpful for making fast comparisons, but it ignores the size of the city and the amount of land that is available.

- **The EII, or Annual Urban Expansion Intensity Index:** A more fair comparison of growth intensity between cities of different sizes or between a city and its larger region is made possible by this metric, which was proposed by Liu *et al.* (2010) and normalizes the growth rate by the total land area of the study region.

$$EII = [(Urb\_t2 - Urb\_t1)/Total\_Land\_Area]/(t2 - t1) * 100$$

A higher EII indicates more intense and potentially more disruptive expansion per unit of available land.

- **Direction of Growth (Spatial Trend Analysis):** In order to measure the directional bias of expansion, the area under research can be separated into concentric rings (zones) or sectors (such as 40° wedges extending from the city center). Next, the amount of new urban land in each ring or sector is determined. This analysis can show whether growth is anisotropic (concentrated along particular transportation corridors, topographical constraints, or as a result of planning policies) or isotropic (equal in all directions). A rose diagram is frequently used to illustrate this.
- **Land Consumption per Capita (LCPC) and SDG Indicator 11.3.1:** A crucial sustainability metric, LCPC measures the efficiency of land use.

$$LCPC = Urban\ Land\ Area/Total\ Urban\ Population$$

One essential metric for tracking sustainable urban development is the ratio of the Land Consumption Rate to Population Growth Rate (SDG 11.3.1). This ratio greater than 1 indicates that urban land is expanding more quickly than the population, less than 1 indicates the intensification and effective land use.

$$SDG11.3.1 = [(Urb\_t2 - Urb\_t1)/(t2 - t1)]/[(Pop\_t2 - Pop\_t1)/(t2 - t1)]$$

Table 2. Key landscape metrics for quantifying urban sprawl and pattern.

Metric Name (Level)	Description	Interpretation for Urban Growth
PLAND (Percentage of Landscape) (Class)	The proportion of landscape comprised of the urban class.	Measures the overall urban footprint. An increasing trend indicates expansion.
NP (Number of Patches) (Class)	The total numbers of the urban patches.	A large increase in NP indicates fragmentation—urban growth is occurring in many, disconnected small patches (leapfrog development).
Mean Patch Size (MPS) (Class)	The average area of urban patches.	A decreasing MPS alongside an increasing PLAND is a classic signal of sprawling, fragmented growth.
LPI (Largest Patch Index) (Class)	The percentage of landscape comprised by largest urban patch.	Measures the dominance of the urban core. A decreasing LPI suggests the growth of secondary urban centers or edge cities.
PD(Patch Density) (Class)	The number of urban patches per unit area.	Similar to NP but normalized by area. Higher PD indicates a more fragmented urban landscape.
ED (Edge Density) (Class)	The total length of the all edge segments of the urban patches per unit area.	Urban areas have high edge density. A rapid increase in ED indicates a more complex, convoluted urban form, often associated with sprawl.
MSI (Mean Shape Index) (Patch/Class)	Measures the complexity of the patch shape compared with a circle of the same area.	For a circle, MSI = 1; as shapes get more complex, MSI rises. While more compact cities have patches with lower MSI, very high MSI urban patches are erratic and expansive.
Euclidean Nearest-Neighbor Distance (ENN_MN) (Patch/Class)	The average distance to the nearest neighboring urban patch.	An increasing mean ENN indicates that patches are becoming more isolated from each other, a key characteristic of dispersed, low-density sprawl.
Contagion (Landscape)	Measures the extent to which land cover types are clumped or aggregated.	A few sizable, connected patches dominate the landscape when there is high contagion. Since urbanization fragments agricultural and natural landscapes, it usually lessens contagion.
Shannon's Diversity Index (SHDI) (Landscape)	Measures the diversity of land cover types present.	Urban expansion often initially increases diversity by introducing a new class (urban)

5.2 *Landscape ecology metrics: Quantifying sprawl, fragmentation, and compactness using FRAGSTATS*

Landscape ecology provides a strong suite of metrics to move beyond simple area calculations and quantifying the spatial pattern of urban growth. These metrics works at three levels, the individual patch, the class level, and the entire landscape. These are usually computed using software such as FRAGSTATS [20].

5.3 *Spatial statistics and gradient analysis: Characterizing urban form*

Beyond patch-based metrics, other spatial statistical techniques are used to analyze urban form:

- **Spatial Autocorrelation (Global and Local Moran's I):** Indicates if urban land is distributed randomly, in clusters, or both. The spatial clustering of urban areas is indicated by a high positive Moran's I. Certain clusters of high-high (urban hotspots) and low-low (non-urban clusters) values can be identified using Local Indicators of Spatial Association (LISA) maps.
- **Gradient Analysis:** A transect (line) or a sequence of concentric buffers extending from the city center (Central Business District, or CBD) is defined using the gradient analysis technique, which computes landscape metrics (such as patch density and percentage impervious surface) within each zone. This illustrates how urban form shifts over time from the urban core to the suburban periphery to the rural hinterland. One well-known sign of suburbanization is the gradual flattening of the density gradient.

5.4 *Urban Growth Models (UGMs): Concepts, calibration, and validation*

Urban Growth Models are computational tools designed to simulate and predict the future spatial pattern of urban land use. They are not crystal balls, but rather "what-if" scenario planning tools that help planners understand the potential consequences of current trends and different policy interventions. All UGMs share a common framework:

- **Drivers/Variables:** Identifying factors that influence urban growth (e.g., distance to roads, distance to city center, slope, zoning regulations, population density).
- **Calibration:** Using historical data (e.g., urban maps from 2000 and 2010) to tune the model's parameters so that it can accurately simulate the change that already occurred (e.g., from 2010 to 2020).
- **Validation:** Testing the calibrated model's performance by comparing its simulation for a known time period (e.g., 2020) against the actual observed urban map for that year. Common validation metrics include the Figure of Merit (FoM), which measures the overlap between simulated change and real change.

Prediction/Scenario Analysis: Running the calibrated and validated model to simulate future growth (e.g., 2030, 2050) under different assumptions (e.g., "business-as-usual," "protected greenbelt," "transit-oriented development").

5.5 *Cellular Automata (CA) models: The SLEUTH model [20,21] and its applications worldwide*

Cellular Automata are the most widely used paradigm for UGMs. The landscape is presented as a grid of cells, each one of which can be in a state (e.g., non-urban, urban). The state of a cell in the next time step is determined by a set of the transition rules that consider the current state of the cell and the states of its neighbors.

- **The Transition Rule:** A typical CA urban transition rule is:

$$\mathbf{P\_\{urban\}} = \mathbf{f(Suitability)} * [\mathbf{1} + (-\ln\ \gamma)\hat{\{}\mathbf{Neighborhood\_Effect}\}] * \mathbf{Restriction\_Weight}$$

  Where:
  - P_{urban} is the probability of a cell transitioning to urban.
  - Suitability is a composite score based on static driver variables (slope, accessibility, etc.).
  - γ is a random variable introducing stochasticity.
  - Neighborhood_Effect quantifies the influence of already urbanized cells in the vicinity (e.g., a 5x5 kernel). This is the core of the CA, capturing the diffusion and contagious nature of urban growth.
  - Restriction_Weight reduces probability for cells that are restricted from development (e.g., water bodies, protected areas).

- **The SLEUTH Model:** The most famous CA-based UGM is the **SLEUTH** model (Slope, Land use, Exclusion, Urban extent, Transportation, Hillshade). It is an open-source model that uses historical satellite-derived urban extent maps to calibrate its growth coefficients through a brute force Monte Carlo method. The coefficients control:
  - **Dispersion:** How likely new, isolated urban centers (spontaneous growth) are to appear.
  - **Breed:** How likely an isolated urban cell is to become a new spreading center.
  - **Spread:** How likely urban cells are to expand from existing urban edges (organic growth).

- **Road Gravity:** How strongly growth is attracted to roads (linear growth). SLEUTH has been applied to hundreds of cities worldwide to simulate future scenarios and assess the impact of growth management policies.

5.6 *Integrating machine learning with CA: Calibrating transition potential with neural networks and RF*

A major limitation of traditional CA models is that the transition rules are often simplistic, linear, and hard to define. **Machine Learning has revolutionized CA by learning complex, non-linear transition rules directly from the data.**

- **ML-CA Hybrid Framework:**
  - The historical change map (e.g., 2000-2010) is used to generate a sample of cells that changed and cells that did not change.
  - For each sample cell, values for a set of driver variables (e.g., distance to road, distance to urban center, slope, etc.) are extracted.
  - A machine learning model (e.g., Random Forest, Logistic Regression, or a Neural Network) is trained on this dataset. The driver variables are the features, and the target variable is "changed" (1) or "not changed" (0).

This hybrid approach (e.g., RF-CA, ANN-CA) has been shown to significantly improve the accuracy of urban growth simulations over traditional methods.

5.7 *Agent-Based Models (ABMs): Simulating land-use decisions of residents, developers, and governments*

While CA models simulate the *pattern* of growth from a top-down perspective, Agent-Based Models (ABMs) simulate the *process* of growth from the bottom-up by modeling the decisions and interactions of the autonomous "agents" that drive land-use change.

- **Agents and Decision Rules:** Key agents in an urban ABM include:
  - **Residents/Households:** Choose where to live based on preferences for affordability, commute time, neighborhood quality, and school districts.
  - **Developers:** Decide where and what to build based on projected profit, market demand, land cost, and zoning regulations.
  - **Government/Planners:** Set policies (zoning, taxes, infrastructure investment) that constrain or incentivize the actions of other agents.

- **Simulation:** Agents are assigned simple rules of behavior. The model is run, and the macro-scale urban pattern (e.g., sprawl, segregation, the emergence of new centers) emerges from the micro-scale interactions of these thousands of agents. ABMs are particularly powerful for modeling complex phenomena like the housing market, the effects of specific policies (e.g., a new property tax), and the evolution of informal settlements.
- **Integration with RS/GIS:** RS data provides the initial landscape and validates the model outputs. GIS data provides the environment in which agents operate (e.g., road network, land parcels). ABMs are data-intensive and computationally complex but offer the most realistic representation of the human drivers behind urban growth.

The choice between CA and ABMs depends on the research question. CA models are excellent for efficiently simulating the physical pattern of growth at a regional scale. ABMs are necessary for understanding the underlying human and institutional mechanisms that cause those patterns. The advanced frameworks are now a days beginning to integrate both approaches.

6 APPLICATIONS

The methodological and technical advances detailed in previous sections are the powerful tools applied to critical real-world problems. In this section we will explores the transformative applications of urban growth detection, supported by case studies which illustrate the depth and global relevance of this work.

6.1 *Putting sustainable development goal 11 into practice: indicator 11.3.1*

One particular indicator which can be measured directly with remote sensing is part of the Sustainable Development Goal 11 (Sustainable Cities and Communities) of the UN: SDG Indicator 11.3.1: Land Consumption to Population Growth Ratio.

- **Approach:** The combination of demographic and geospatial data is aptly captured by this indicator.

- Land Consumption Rate (LCR): derived from multi-temporal urban land cover maps produced by machine learning classifiers (e.g., from Landsat). LCR is equal to (Urb_t2 - Urb_t1) / (t2 - t1).
- Population Growth Rate (PGR): derived from household surveys, national censuses, or increasingly from modeled gridded population data (WorldPop, for example) [38]. PGR is equal to (Pop_t2 - Pop_t1) / (t2 - t1).
- Ratio: = LCR / PGR

- **Meaning and Effect:**
  - **Ratio > 1:** The rate of land consumption exceeds the rate of population growth. This suggests unsustainable land use and sprawling, ineffective urban expansion.
  - **Ratio ≈ 1:** Population growth and land consumption are balanced.
  - **Ratio < 1:** Land consumption is not keeping up with population growth. This suggests more effective land use and urban intensification.

Cloud platforms like Google Earth Engine [40] are currently being used by national statistical offices and international organizations (such as the World Bank and UN Habitat) to automate the computation of this indicator for hundreds of cities worldwide. This offers a previously unheard-of, standardized evidence base to track advancement toward sustainability objectives and hold governments responsible.

6.2 *Environmental impact studies: Urban Heat Island (UHI) effect, air quality, and loss of ecological services*

Urban growth directly alters the local and regional environment. RS is the primary tool for quantifying these impacts.

- **Urban Heat Island (UHI) Effect:**
  - **Application:** Thermal infrared sensors (e.g., on Landsat 8/9 TIRS) measure land surface temperature (LST). By overlaying LST maps on urban land cover classifications, researchers can quantify the "heat island" effect.
  - **Case Study:** A classic study for any city will show that impervious surfaces (central business districts, industrial areas) have the highest LST, while vegetated areas (parks, forests) have the lowest, creating a "cool island" effect. The transformation of vegetated or permeable land to impervious surface leads to a direct increase in local LST.
  - **Advanced Analysis:** Research has moved beyond correlation to model the relationship between landscape composition (% impervious surface) and configuration (patch connectivity) and UHI intensity using multivariate regression and ML models. This helps planners design cooler cities by optimizing the placement and size of green spaces.
- **Air Quality and Pollution Exposure:**
  - **Application:** While RS cannot directly measure ground-level PM2.5 everywhere, satellites like MODIS and VIIRS measure aerosol optical depth (AOD), a column-integrated measure of particulate matter. ML models (e.g., Random Forest, Neural Networks) are used to downscale these measurements and model ground-level PM2.5 concentrations by incorporating RS-derived data (urban cover, vegetation, road density) and meteorological data.
  - **Impact:** By combining population density data with these high-resolution PM2.5 exposure maps, communities that are disproportionately exposed to air pollution—typically in low-income or industrially adjacent neighborhoods—can be identified, offering vital information for environmental justice advocacy.
- **Decrease in Ecological Services:**
  - **Application:** The ecosystem services of carbon sequestration, air and water purification, stormwater regulation, and recreation are all negatively impacted by the loss of green and blue spaces (vegetation, water bodies) brought on by urbanization.
  - **Quantification:** Researchers can measure the loss of these services by monitoring the conversion of particular land cover classes (forest, wetland, and grassland, for example) to urban use. For example, the InVEST model [39] (Integrated Valuation of Ecosystem Services and Tradeoffs) make use of LULC maps as its major input to model and value ecosystem services, which provides a powerful economic argument for conservation.

7 CONCLUSION

The presented review has charted the evolution of urban growth monitoring, from early remote sensing to the current integration of advanced machine learning. We have traced the critical progression from simple change detection to the nuanced quantification of growth patterns sprawl, infill, intensification and the simulation of future scenarios through models like ML-enhanced Cellular Automata and Agent-Based Models.

The synergy of these technologies provides essential tools for global sustainability. They operationalize frameworks like SDG 11, quantify environmental impacts such as the Urban Heat Island effect, and inform evidence-based planning and governance.

Looking ahead, key frontiers must be addressed: moving beyond physical form to understand functional urbanization using multi-source data, improving model explainability for policy trust, and bridging the gap from research to operational city-level monitoring systems. The ultimate challenge is to ensure these powerful tools generate actionable, equitable knowledge for planners and communities.

By refining this technological lens and focusing its application, we can transform urban growth from an observed phenomenon into a consciously managed process, steering the urban century toward a resilient and sustainable future.

REFERENCES

[1] United Nations, Department of Economic and Social Affairs, Population Division. (2022). World Population Prospects 2022: Summary of Results. UN DESA/POP/2022/TR/NO. 3.

[2] United Nations. (2017). The New Urban Agenda. Habitat III.

[3] United Nations. (2015). Transforming our world: the 2030 Agenda for Sustainable Development. A/RES/70/1.

[4] Breiman, L. (2001). Random Forests. *Machine Learning*, 45(1), 5–32.

[5] Cortes, C., and Vapnik, V. (1995). Support-vector networks. *Machine Learning*, 20(3), 273–297.

[6] Rumelhart, D.E., Hinton, G.E., and Williams, R.J. (1986). Learning representations by back-propagating errors. *Nature*, 323 (6088), 533–536.

[7] Rosenblatt, F. (1958). The perceptron: a probabilistic model for information storage and organization in the brain. *Psychological Review*, 65(6), 386–408.

[8] Liu, F.T., Ting, K.M., and Zhou, Z.-H. (2008). Isolation Forest. In 2008 Eighth IEEE International Conference on Data Mining (pp. 413–422).

[9] Schölkopf, B., Platt, J.C., Shawe-Taylor, J., Smola, A.J., and Williamson, R.C. (2001). Estimating the support of a high-dimensional distribution. *Neural Computation*, 13(7), 1443–1471.

[10] Ronneberger, O., Fischer, P., and Brox, T. (2015). U-Net: Convolutional Networks for Biomedical Image Segmentation. In *Medical Image Computing and Computer-Assisted Intervention–MICCAI 2015* (pp. 234–241). Springer International Publishing.

[11] Chen, L.-C., Papandreou, G., Kokkinos, I., Murphy, K., and Yuille, A.L. (2017). DeepLab: Semantic Image Segmentation with Deep Convolutional Nets, Atrous Convolution, and Fully Connected CRFs. *IEEE Transactions on Pattern Analysis and Machine Intelligence*, 40(4), 834–848.

[12] He, K., Zhang, X., Ren, S., and Sun, J. (2016). Deep Residual Learning for Image Recognition. In 2016 IEEE Conference on Computer Vision and Pattern Recognition (CVPR) (pp. 770–778).

[13] Simonyan, K., and Zisserman, A. (2014). Very Deep Convolutional Networks for Large-Scale Image Recognition. arXiv:1409.1556.

[14] Vaswani, A., Shazeer, N., Parmar, N., Uszkoreit, J., Jones, L., Gomez, A.N., Kaiser, Ł., and Polosukhin, I. (2017). Attention Is All You Need. Advances in Neural Information Processing Systems, 30.

[15] LeCun, Y., Bottou, L., Bengio, Y., and Haffner, P. (1998). Gradient-based learning applied to document recognition. *Proceedings of the IEEE*, 86(11), 2278–2324.

[16] Daudt, R.C., Le Saux, B., Boulch, A., and Gousseau, Y. (2018). Urban change detection for multispectral earth observation using convolutional neural networks. In IGARSS 2018 - 2018 IEEE International Geoscience and Remote Sensing Symposium (pp. 2115–2118).

[17] Chen, H., and Shi, Z. (2020). A Spatial-Temporal Attention-Based Method and a New Dataset for Remote Sensing Image Change Detection. *Remote Sensing*, 12(10), 1662.

[18] Zhang, C., Yue, P., Tapete, D., Jiang, L., Shangguan, B., Huang, L., and Liu, G. (2020). A deeply supervised image fusion network for change detection in high resolution bi-temporal remote sensing images. *ISPRS Journal of Photogrammetry and Remote Sensing*, 166, 183–200.

[19] Shi, W., Zhang, M., Zhang, R., Chen, S., and Zhan, Z. (2020). Change Detection Based on Artificial Intelligence: State-of-the-Art and Challenges. *Remote Sensing*, 12(10), 1688.

[20] McGarigal, K., and Marks, B.J. (1995). FRAGSTATS: spatial pattern analysis program for quantifying landscape structure (Vol. 351). US Department of Agriculture, Forest Service, Pacific Northwest Research Station.

[21] Clarke, K.C., Hoppen, S., and Gaydos, L. (1997). A self-modifying cellular automaton model of historical urbanization in the San Francisco Bay area. *Environment and Planning B: Planning and Design*, 24(2), 247–261.

[22] Dietzel, C., and Clarke, K.C. (2006). The effect of disaggregating land use categories in cellular automata during model calibration and forecasting. *Computers, Environment and Urban Systems*, 30(1), 78–101.

[23] Liu, X., Li, X., Chen, Y., Tan, Z., Li, S., and Ai, B. (2010). A new landscape index for quantifying urban expansion using multi-temporal remotely sensed data. *Landscape Ecology*, 25(5), 671–682.

[24] Rouse, J.W., Haas, R.H., Schell, J.A., and Deering, D.W. (1974). Monitoring Vegetation Systems in the Great Plains with ERTS. *NASA SP-351*, 1, 309–317.

[25] Zha, Y., Gao, J., and Ni, S. (2003). Use of normalized difference built-up index in automatically mapping urban areas from TM imagery. *International Journal of Remote Sensing*, 24(3), 583–594.

[26] Xu, H. (2006). Modification of normalised difference water index (NDWI) to enhance open water features in remotely sensed imagery. *International Journal of Remote Sensing*, 27(14), 3025–3033.

[27] Huete, A.R. (1988). A soil-adjusted vegetation index (SAVI). *Remote Sensing of Environment*, 25(3), 295–309.

[28] Blaschke, T. (2010). Object based image analysis for remote sensing. *ISPRS Journal of Photogrammetry and Remote Sensing*, 65(1), 2–16.

[29] Baatz, M., and Schäpe, A. (2000). Multiresolution Segmentation: an optimization approach for high quality multi-scale image segmentation. In J. Strobl, T. Blaschke, and G. Griesebner (Eds.), *Angewandte Geographische Informationsverarbeitung XII* (pp. 12–23). Wichmann Verlag.

[30] Zhang, J. (2010). Multi-source remote sensing data fusion: status and trends. *International Journal of Image and Data Fusion*, 1(1), 5–24.

[31] Joshi, N., Baumann, M., Ehammer, A., Fensholt, R., Grogan, K., Hostert, P., Jepsen, M.R., Kuemmerle, T., Meyfroidt, P., Mitchard, E.T.A., Reiche, J., Ryan, C.M., and Waske, B. (2016). A review of the application of optical and radar remote sensing data fusion to land use mapping and monitoring. *Remote Sensing*, 8(1), 70.

[32] Congalton, R.G., and Green, K. (2019). *Assessing the accuracy of remotely sensed data: principles and practices* (Third edition). CRC Press, Taylor and Francis Group.

[33] Foody, G.M. (2002). Status of land cover classification accuracy assessment. *Remote Sensing of Environment*, 80(1), 185–201.

[34] Mohan, M., Pathan, S.K., Narendrareddy, K., Kandya, A., and Pandey, S. (2011). Dynamics of urbanization and its impact on land-use/land-cover: a case study of megacity Delhi. *Journal of Environmental Protection*, 2(9), 1274–1283.

[35] Angel, S., Parent, J., Civco, D.L., Blei, A., and Potere, D. (2011). The dimensions of global urban expansion: Estimates and projections for all countries, 2000–2050. *Progress in Planning*, 75(2), 53–107.

[36] Seto, K.C., Güneralp, B., and Hutyra, L.R. (2012). Global forecasts of urban expansion to 2030 and direct impacts on biodiversity and carbon pools. *Proceedings of the National Academy of Sciences*, 109(40), 16083–16088.

[37] Li, X., Zhou, Y., Zhao, M., and Zhao, X. (2020). A harmonized global nighttime light dataset 1992–2018. *Scientific Data*, 7(1), 168.

[38] Tatem, A.J. (2017). WorldPop, open data for spatial demography. *Scientific Data*, 4(1), 170004.

[39] Sharp, R., Tallis, H.T., Ricketts, T., Guerry, A.D., Wood, S.A., Chaplin-Kramer, R., Nelson, E., Ennaanay, D., Wolny, S., Olwero, N., Vigerstol, K., Pennington, D., Mendoza, G., Aukema, J., Foster, J., Forrest, J., Cameron, D., Arkema, K., Lonsdorf, E., ... Kennedy, C. (2018). InVEST 3.8.0 User's Guide. The Natural Capital Project, Stanford University, University of Minnesota, The Nature Conservancy, and World Wildlife Fund.

[40] Gorelick, N., Hancher, M., Dixon, M., Ilyushchenko, S., Thau, D., and Moore, R. (2017). Google Earth Engine: Planetary-scale geospatial analysis for everyone. *Remote Sensing of Environment*, 202, 18–27.

Smart Systems and Beyond – Bhatia et al. (Eds)
© 2027 The Author(s), ISBN: 978-1-041-36617-1

AI-Assisted Tumor Detection in MRI and CT Scans Using Image Segmentation Techniques

Muskan, Mohit Srivastava, Sukhdeep Kaur, and Rupinder Kaur
Department of ECE, Chandigarh Engineering College-CGC Landran, Mohali, Punjab, India

Manoj Kuri
Department of AI & DS, Engineering College Bikaner, Bikaner, Rajsthan, India

ABSTRACT: Detection and diagnosis of tumors should occur early and accurately to improve cancer treatment and survival rates. Since the 1970s and 1980s, MRIs and CT scans have been crucial in clinical radiology and cancer diagnosis with their high anatomical precision imaging capabilities. However, representations of the anatomy, and tumors in particular, vary from clinician to clinician, as does the time taken to analyze the images, making the detection of diffuse and subtle cancerous growths extremely difficult. AI, particularly Deep Learning, has been the most effective in overcoming these challenges with the rapid analysis of the cancerous areas and the versatile segmentation of images. This review examines the use of classical segmentation techniques, which include thresholding, region growing, and active contours, and modern AI segmentation techniques like U-Net, Mask R-CNN, DeepMedic, Fully Convolutional Networks, 3D CNNs, and others which analyze volumetric data. This review also measures the various segmentation techniques of the MSRI and CT data using common metrics of performance and cancer diagnosis. Key considerations and measures include the Dice Score, Jaccard Index, Sensitivity, Specificity, and the datasets like BraTS, LIDC-IDRI, and LiTS.

Keywords: AI, Tumor Detection, MRI, CT scan, Image Segmentation, Deep Learning, Medical Imaging.

1 INTRODUCTION

For many cancers, diagnosis at an appropriate stage and within a reasonable window enhances prognosis, aids treatment planning and increases survival rates. In this area of medicine, early detection of a cancer increases the likelihood of a positive treatment outcome. No diagnosis today is complete without a CT or MRI scan. These imaging modalities aid in the diagnosis of a cancer by providing accurate and clear anatomical and functional details and visualising the tumour, its size, location and spread. CT scans are more rapid and can locate the brain, liver, and lung cancers, while MRI is superior in soft tissue contrast.Advanced imaging technology has not eradicated radiologists' judgement in MRI and CT scan interpretation. Manual tumour detection and segmentation can be arduous, subjective, and inter-observer variable when dealing with delicate anatomical features or microscopic, hazy lesions. Diagnostic delays or disparities might affect clinical decision-making and patient outcomes due to subjectivity. Diagnostic tools that are faster, more consistent, and more accurate are needed as imaging data grows (Lee *et al.*, 2025). AI, particularly machine learning and deep learning, has dramatically impacted medical imaging due to these issues. Radiologists are discovering that AI-powered algorithms increase tumour identification, classification, and segmentation efficiency and accuracy. Image segmentation approaches, especially deep learning-based ones, can identify tumour borders and separate healthy from malignant tissues. Several tools, including 3D-CNNs, U-Net, and Mask R-CNN, have considerably enhanced clinical picture processing automation (Edelmers *et al.*, 2024). This work summarises AI-assisted tumour detection in MRI and CT scans, focussing on image segmentation. The research compares traditional and cutting-edge segmentation algorithms, uses publicly available datasets, and discusses existing issues and future options. This study summarises current achievements in the field to illuminate AI's potential, limitations, and strengths in tumour diagnosis for researchers, physicians, and AI developers.

2 BACKGROUND CONCEPTS

2.1 *Medical imaging modalities*

In clinical oncology, CT and MRI are the most used imaging modalities for tumour detection and monitoring. MRI relies on nuclear magnetic resonance, which uses radiofrequency pulses to emit signals from tissues whose protons align with a strong magnetic field. These signals enable accurate body cross-sections (Gou& Wu, 2022).

DOI: 10.1201/9781042026289-41

The ability of magnetic resonance imaging (MRI) to differentiate soft tissue makes it useful in the diagnosis of brain tumors, spinal cord abnormalities, and masses in the abdominal and pelvic regions. It is also safer than other imaging techniques because it is non-ionizing and can be used repeatedly, especially on children. However, MRI has drawbacks, which include longer scanning times, higher costs, specific situational inaccessibility, and sensitivity to motion artifacts. In contrast, other imaging techniques, such as computerized tomography (CT) scans, employ X-rays and computers to produce cross-sectional images of the body (Ma *et al.*, 2022). Their speed and low cost, as well as their ability to determine calcifications and assess bone involvement—issues that MRI faces—have increased their popularity.CT scans are excellent for imaging liver, lung, and bone tumours and are often used in trauma or emergency circumstances due to their speed.

2.2 *Tumor characteristics in imaging*

Understanding imaging modalities‘ differences helps spot tumours. Imaging tests seek for gliomas, brain metastases, primary and metastatic liver cancers, lung nodules or carcinomas, colorectal cancers, and other common tumours. Radiological features in MRI or CT scans can determine if a lesion is benign or malignant or define its limits. MRI sequences including T1-weighted, T2-weighted, FLAIR, and others may yield diverse tumour pictures (Liu *et al.*, 2023). High-grade gliomas often cause oedema, T1 hypointensity, and T2 hyperintensity. MRI-enabled perfusion imaging and diffusion-weighted imaging (DWI) reveal tumour cellularity and vascularity. CT images show tumours due to contrast enhancement density and patterns. For instance, lung nodules can be one high-density zone or multiple smaller ones. Liver metastases often have contrast phase rim enhancement, however primary liver malignancies such hepatocellular carcinoma have arterial phase hyperenhancement and fast washout.

2.3 *Image segmentation in medical imaging*

"Image segmentation" refers to the technique which breaks down images into smaller bite-size formats for detailed body constituent examination, particularly for the detection of anomalies such as tumours. For the quantification of medical imaging, 3D visualization, surgical planning, and precision in the planning of radiation, segmentation becomes indispensable. For early diagnosis, tumour separation from surrounding tissue, estimation of tumour size, response to therapy and more, it becomes crucial for radiologists and physicians. Segmentation can be carried out manually, in parts, or fully automated (He, 2024). This method still suffers from manually time-consuming segmentation by radiologists which becomes uneven. Poorly handles large data sets but has fine anatomic precision. Semi-automated methods perform edge detection and region growing to threshold segmentation, then human curators enhance precision and lessen the effort. Semi-automated methods can be operator dependent. The field of artificial intelligence and more particularly deep learning has made fully automated segmentation common. Such methods rapidly and consistently perform segmentation of tumours, defining boundaries while relying on pre-tuned data, without any human segmentation. AI segmentation models like U-Net, DeepMedic, and Mask R-CNN manage complex scenarios involving irregular tumour geometry, infiltrative growth patterns, or poor tissue contrast.

3 ARTIFICIAL INTELLIGENCE IN MEDICAL IMAGE ANALYSIS

3.1 *Machine learning vs. deep learning*

AI's rapid development has greatly altered medical imaging. Machine learning (ML) and deep learning (DL), the two primary AI schools used for image analysis, detect tumours and segment images differently. Machine learning algorithms analyse data using human-created qualities like shape, texture, and intensity to find patterns (Chetlapalli& Bharathidasan, 2018). These features are supplied to traditional classifiers like SVM, Random Forests, and K-Nearest Neighbours (KNN) after being retrieved manually or semi-automatically. ML techniques work well with simpler datasets but fail with high-dimensional medical images, where tumour features vary greatly in size, shape, and intensity. Deep learning, especially CNNs, has revolutionised photo analysis by eliminating human feature extraction. Convolutional neural networks (CNNs) can automatically learn hierarchical features from visual input. Convolutional layers discern local patterns like textures, complex forms, and edges, making them good at processing medical images with subtle or sophisticated tumour formations. Picture segmentation convolutional neural networks (CNNs) use numerous layers of convolution, pooling, and non-linear activation to learn low-level (contrast, edges) and high-level (tumour boundaries, aberrant regions) properties (Paudyal *et al.*, 2023).

3.2 *Evolution of AI in tumor segmentation*

AI for tumour segmentation has progressed from hand-crafted feature-based machine learning models to complicated, fully automated deep learning architectures. The first segmentation tools were machine learning classifiers, k-means clustering, texture analysis, and edge detection (Adusumilli *et al.*, 2025). These techniques were limited by their inability to generalise across patient groups and imaging modalities, despite their automation. After U-Net's 2015 release, deep learning revolutionised medical picture segmentation. U-Net presented a biomedical image encoder-decoder architecture for pixel-level segmentation with minimal datasets. Since then, ResUNet, Attention U-Net, and 3D U-Net have improved the network's accuracy in volumetric scans like MRI and CT and spatial context. DeepMedic, a 3D CNN framework, and Mask R-CNN, which can segment instances, have also successfully performed complex segmentation tasks like multi-class tumour detection and tumour subregion delineation. Deep learning tumour segmentation requires well-annotated datasets. Public repositories include the Brain Tumour Segmentation (BraTS) Challenge, LITS, and LIDC-IDRI. These datasets are suitable for comparing AI algorithms because to their uniform, annotated images. However, annotating datasets for complex or poorly defined tumour boundaries can be time-consuming and require many expert radiologists (Tang et al., 2022).

4 IMAGE SEGMENTATION TECHNIQUES FOR TUMOR DETECTION

4.1 *Classical techniques*

Before deep learning, classical image processing was the gold standard for tumour segmentation. They used math and algorithms to analyse geometric structures and pixel intensities. Pixel intensity thresholds distinguish tumour regions in thresholding, the simplest segmentation approach. Tumour pixels exceed or fall below a threshold. This approach works well in high-contrast pictures but poorly in low-contrast or areas with similar tumour and tissue intensities. Region Growth spreads from a seed point by adding surrounding pixels that fulfil intensity or texture requirements. It works well for segmenting identical tumours, but it is seed-sensitive and may not operate in noisy or heterogeneous images (Chukwujindu *et al.*, 2024). A picture can be divided into gradient intensity portions by the Watershed Algorithm by imitating a topographic terrain. Local minima "flooded" tumours. It over-segments structures when used alone, but with gradient suppression or smoothing, it separates touching structures. Snakes, or Active Contour Models, use developing curves inside a picture to minimise an energy function and detect object boundaries.

4.2 *Deep learning-based techniques*

Deep learning-based algorithms, especially CNNs, are more effective than traditional methods for medical image segmentation. U-Net Expansions The neural encoder-decoder U-Net was designed for biological image segmentation (El-Shafai *et al.*, 2023). Through successive down-sampling, encoders capture contextual information, while decoders restore spatial detail. Use skip links between matched encoder and decoder levels to preserve spatial information. Attention U-Net, Residual U-Net (ResUNet), and U-Net++ improve multi-class segmentation and tumour feature learning. entirely Convolutional Networks (FCNs) use convolutional and deconvolutional layers instead of entirely connected ones to build space-consistent segmentation maps. They are perfect for end-to-end training with varied tumours because to their versatility and efficiency. U-Net and other later systems have improved this area, but they still struggle with precise border localisation. Display R-CNN Mask R-CNN, originally created for instance segmentation in natural images, is now used in medical imaging to segment 2D or 3D cancer instances (Wang *et al.*, 2023). It improves Faster R-CNN by adding a branch that anticipates object masks to recognise and segment objects. Complex, multi-lesion scanning and multi-object segmentation are its strengths. Medical Advances The main goal of this 3D CNN is to segment brain tumours. MRI scans can be analysed for detailed and contextual information using two paths.

4.3 *Hybrid approaches*

New hybrid segmentation algorithms combine standard and deep learning methods to improve accuracy, processing efficiency, and interpretability. Mixing Deep and Classical Learning Methods Region growth or edge detection are used to pre-process the image, then a deep learning model refines it. This reduces training time and improves edge clarity. Deep models give coarse segmentation masks, but active contours or graph cuts can refine them. Post-processing methods After segmentation, output masks are refined using morphological procedures such dilation, erosion, opening, and closing (Zhang *et al.*, 2023). These surgeries eliminate false positives, close tumour spots, and smooth boundaries. Post-processing layers like Conditional Random Fields (CRFs) can represent spatial dependencies to improve pixel-level predictions. Hybrid approaches combine standard methods‘ interpretability with deep models' learning capabilities to produce robust and practical cancer detection solutions.

5 COMPARATIVE ANALYSIS OF TECHNIQUES

Computing efficiency, performance metrics, and imaging modality adaption vary widely among cancer detection picture segmentation algorithms. The Jaccard Index and Dice Similarity Coefficient (DSC) are used to assess accuracy by measuring predicted and ground truth segmentation overlap (Alqahtani, 2024). Deep learning models like U-Net and

3D CNNs outperform classical approaches in BraTS brain tumour segmentation with Dice scores above 0.85. However, region growth and thresholding are less precise for low contrast or irregular tumour shapes. Technique impacts sensitivity and specificity. DeepMedic and Mask R-CNN are known for their sensitivity, however they can sometimes yield too many false positives when detecting small tumours. Deep learning algorithms require large annotated datasets like BraTS for brain MRI, LIDC-IDRI for lung nodules in CT, and LiTS for liver tumours in CT. Conventional methods can work on smaller datasets with clearly defined bounds. Classical methods are fast and computationally efficient but fragile. However, deep learning models, especially 3D CNNs that train on volumetric data, demand a lot of computer power. Inference can be fast once learnt. In heterogeneous or poor-quality scans, classical approaches fail, whereas deep models, when trained appropriately, tolerate noise and image artefacts well (Luo *et al.*, 2024).

Table 1. Tumor segmentation techniques.

Technique	Dice Score	Sensitivity	Speed	Robustness	Suitable Modality
Thresholding	0.60	Moderate	Very Fast	Low	CT
Region Growing	0.65	Low	Fast	Low	CT/MRI
Active Contours	0.70	Moderate	Moderate	Medium	MRI
U-Net	0.87	High	Fast (GPU)	High	MRI
DeepMedic	0.89	Very High	Moderate	Very High	MRI
Mask R-CNN	0.86	High	Fast	High	CT/MRI
3D CNN	0.90	Very High	Slow	Very High	MRI/CT (3D data)

6 DATASETS AND BENCHMARKING PLATFORMS

Public datasets and benchmarking systems have helped AI-based tumour segmentation in MRI and CT imaging expand. These resources provide standard, annotated imaging data for model training, evaluation, and comparison. Popular datasets include the Brain tumour Segmentation (BraTS) Challenge dataset, which segments multi-modal MRI scans of glioma patients into subregions including enhancing tumour, necrotic core, and oedema (Singh *et al.*, 2024). BraTS has become the benchmark for brain tumour segmentation because to its uniform evaluation measures and many annotations. The LIDC-IDRI (Lung Image Database Consortium) provides detailed information regarding nodule location, size, and malignancy likelihood in over 1,000 annotated lung CT scans, making it essential for lung cancer segmentation studies. The LiTS dataset contains contrast-enhanced abdomen CT scans with expert-annotated tumour and liver masks for liver tumour analysis (Salah Eldin&Kaboudan, 2023). Even with these resources, dataset availability and annotation remain issues. Medical picture annotation requires expertise, such as radiologists or oncologists manually delineating cancer boundaries. Due to its subjective, expensive, and time-consuming nature, this technique lacks large, diverse datasets for many tumour types and organs. Annotational variability between observers may affect model training and evaluation. Lack of data diversity across populations, scanners, and imaging procedures can also influence model generalisability. Researchers are relying on data augmentation and synthetic data synthesis to address data limitations (Zhou *et al.*, 2025).

7 CHALLENGES AND LIMITATIONS

AI-assisted cancer diagnosis by photo segmentation has made great strides, but it still faces many obstacles that hinder its widespread clinical use. Data scarcity and annotation concerns are major issues (Li *et al.*, 2024). Although supervised deep learning models require high-quality labelled datasets, producing expert-annotated tumour masks is a difficult, time-consuming, and expert-level radiology task. Many datasets are limited and narrow, focussing on certain cancer types or imaging modalities, which limits model generalisability and scalability. Ground truth labelling inconsistencies due to annotator inter-observer variability can also affect model accuracy and reliability. Another challenge is that tumour appearance varies widely between patients. Tumours vary in size, form, location, intensity, and enhancing patterns according on the organ system, disease progression, and patient. Brain gliomas look diffuse infiltrative, while lung nodules are compact or spiculated. Due of heterogeneity, AI models trained on homogeneous datasets struggle to develop universal representations (Shin, 2024).

Extrapolating results related to other scanners and patient populations also becomes a concern. For example, models developed and trained on a specific imaging technique or specific hospital may perform poorly on an imaging dataset obtained from a different scanner, supplier, or institution. Factors such as imaging resolution, contrast settings, and demographics and comorbidities of the patient population may undermine the performance of a model and thus the robustness of the model and external validity may come into question. Technically, one of the difficult problems is to ensure the generalisability of a given model across different clinical situations. This is compounded by the fact that AI models themselves are difficult to integrate into clinical workflows. Many deep learning models and CNNs are "black boxes"; they perform extraordinarily well on specific tasks, but they cannot explain their reasoning (Gobinath *et al.*, 2025). In cases where AI predicts contradictory results to a human assessment, the difference in explainability can undermine trust in the Clinician. Systems providing explanations, such as text, visuals, or evidence of confidence on predictions and insights into reasoning are preferred by both radiologists and oncologists. Before AI can be adopted into medical imaging, ethical and regulatory implications must be solved. Issues such as accountability, bias, algorithmic transparency, informed consent, and the right to be forgotten in patient records remain. Although the FDA and EMA are proposing frameworks to evaluate and authorize AI-powered medical devices, AI designs in question are not validated in large, multi-center, clinical trials.

8 CLINICAL APPLICATIONS AND CASE STUDIES

Clinical AI-assisted photo segmentation has improved tumour identification, enabling early diagnosis, treatment planning, and response tracking. Brain tumour segmentation using magnetic resonance imaging (MRI), especially gliomas and metastases, has been a major application. Gliomas are notoriously hard to split manually due to their irregular forms and hazy borders (Khalifa&Albadawy, 2024). Deep learning models like 3D U-Net, DeepMedic, and Attention U-Net have accurately identified tumour subregions such the enhancing core, necrotic area, and peritumoral oedema in multi-modal MRI datasets like the BraTS challenge. These models simplify radiologists' work and provide uniformity between time points and observers for volumetric analysis and radiation planning. AI can also locate lung lesions in CT scans in thoracic oncology. Models trained on the LIDC-IDRI dataset were very sensitive and specific in identifying benign and malignant nodules (Juarez-Chambi *et al.*, 2019). Mask R-CNN and FCNs are deep learning algorithms that can precisely localise microscopic pulmonary lesions, particularly in early-stage malignancies when humans can easily overlook texture and density changes. AI-powered segmentation in low-dose CT lung cancer screening programs improves early detection and reduces false positives.

9 FUTURE DIRECTIONS

Several new approaches are promising for enhancing AI-assisted tumour diagnosis in medical imaging accuracy, scalability, and clinical utility. Federated learning, which lets AI models learn from data in several locations or institutions without sharing data, is a potential new method. This privacy-preserving strategy uses dispersed medical datasets' vastness and variety to address patient confidentiality and legal constraints. Federated learning may help break down data silos and produce more generalisable and objective segmentation models. Multi-modal imaging is another trend that improves tumour detection (Qi, 2024). Merging MRI, PET, or CT with ultrasound. For instance, MRI contrasts soft tissues, PET shows metabolic activity, and CT shows structural detail. The integration of three imaging modalities has yielded positive outcomes. When multi-modal AI models are trained on fused data inputs, they are capable of improving sensitivity and specificity regarding complex morphological and functional conditions such as glioblastomas, as well as metastases of the liver. Moreover, the use of explainable AI (XAI) methodologies in imaging technologies has increasingly surfaced within the healthcare field, particularly in scenarios where trust and interpretability are of utmost importance. Given that the majority of deep learning algorithms operate as opaque 'black boxes', their various predictions are particularly difficult to justify. Radiologists are learning the results of Tumour segmentation through the various means of saliency maps, attention mechanisms, and decision tree-like visualizations. For ultimate clinical usability and regulatory compliance, explainability will be mandatory. The future of imaging, particularly in reference to (Saeidnia *et al.*, 2025), will focus on the demand for real-time, especially intraoperative, image segmentation.Neurosurgeons and interventional radiologists benefit from real-time AI-powered segmentation during tumour excision and image-guided biopsies. Improved GPU processing and AI at the edge are reducing latency, making real-time AI loop integration possible.

10 CONCLUSION

Artificial intelligence in medical image segmentation has revolutionised cancer detection and diagnosis. Traditional methods like region growth and thresholding have given way to complex deep learning architectures like U-Net, DeepMedic, and Vision Transformers for MRI and CT cancer diagnosis and analysis. AI has shown it can improve segmentation accuracy, reduce observer variability, and process huge image data quickly. Data scarcity, annotation issues, and tumour presentation variation are common barriers to clinical integration. Concerns concerning explainable and ethically accountable models and their inability to generalise across demographics, imaging modalities, and scanner types persist. We must also standardise performance metrics across benchmarking systems like BraTS, LIDC-IDRI, and LiTS and conduct multi-center validation to circumvent regulatory hurdles. Together, AI researchers, doctors, radiologists, and regulatory bodies can advance this sector. AI systems with healthcare applications require technology advancement, subject expertise, clinical validation, and regulatory compliance. Radiologists must provide expert opinions throughout AI tool development to ensure security, efficiency, and compatibility with diagnostic methods. Explainable artificial intelligence, real-time intraoperative applications, federated learning, and multi-modal data fusion will drive future development. International standards, open-access datasets, and transparent model review mechanisms are needed for AI to fully fulfil its potential in medical imaging. Finally, AI-assisted picture segmentation for tumour diagnosis is clinically needed and promising. AI can make radiology faster, more accurate, and more personalised if research continues, interdisciplinary teams work together ethically, and the technology is used appropriately.

REFERENCES

Adusumilli, P., Ravikumar, N., Hall, G., and Scarsbrook, A.F. (2025). A methodological framework for AI-assisted diagnosis of ovarian masses using CT and MR imaging. *Journal of Personalized Medicine*, *15*(2), 76.

Alqahtani, S. (2024). Systematic review of AI-assisted MRI in prostate cancer diagnosis: enhancing accuracy through second opinion tools. *Diagnostics*, *14*(22), 2576.

Chetlapalli, H., and Bharathidasan, S. (2018). AI-based classification and detection of brain tumors in healthcare imaging data. *International Journal of Life Sciences Biotechnology and Pharma Sciences*, *14*(2), 18–26.

Chukwujindu, E., Faiz, H., Sara, A.D., Faiz, K., and De Sequeira, A. (2024). Role of artificial intelligence in brain tumour imaging. *European journal of radiology*, *176*, 111509.

Dalsaniya, A., Godhaniya, M., and Kothari, D.R. (2025). AI-Assisted 3D reconstruction of Organs from MRI and CT Data. *Global Journal of Engineering and Technology Advances*, *22*(03), 047–060.

Edelmers, E., Ņikuļins, A., Sprūdža, K.L., Stapulone, P., Pūce, N.S., Skrebele, E., ... and Boločko, K. (2024). AI-Assisted detection and localization of spinal metastatic lesions. *Diagnostics*, *14*(21), 2458.

El-Shafai, W., Ali, R., Sedik, A., Taha, T.E.S., Abd-Elnaby, M., and Abd El-Samie, F.E. (2023). Analysis of Brain MRI: AI-Assisted Healthcare Framework for the Smart Cities. *Intelligent Automation & Soft Computing*, *35*(2).

Gobinath, A., Rajeswari, P., Anandan, M., and Kumar, N.S. (2025). AI in Imaging for Diagnosis. In *AI Insights on Nuclear Medicine* (pp. 51–76). IGI Global Scientific Publishing.

Gou, F., and Wu, J. (2022, December). An attention-based AI-assisted segmentation system for osteosarcoma MRI images. In *2022 IEEE international conference on bioinformatics and biomedicine (BIBM)* (pp. 1539–1543). IEEE.

He, Y. (2024, October). Intelligent medical assistant: AI medical image assistance technology and its application in cancer diagnosis and treatment decision. In *International Conference on Optical Communication and Optoelectronic Technology (OCOT 2024)* (Vol. 13289, pp. 150–154). SPIE.

Juarez-Chambi, R.M., Kut, C., Rico-Jimenez, J.J., Chaichana, K.L., Xi, J., Campos-Delgado, D.U. and Jo, J.A. (2019). AI-assisted in situ detection of human glioma infiltration using a novel computational method for optical coherence tomography. *Clinical Cancer Research*, *25*(21), 6329–6338.

Khalifa, M., and Albadawy, M. (2024). AI in diagnostic imaging: revolutionising accuracy and efficiency. *Computer Methods and programs in biomedicine update*, *5*, 100146.

Lee, M., Kim, J.H., Choi, W., and Lee, K.H. (2025). AI-assisted Segmentation Tool for Brain Tumor MR Image Analysis. *Journal of Imaging Informatics in Medicine*, *38*(1), 74–83.

Li, X., Zhang, L., Yang, J., and Teng, F. (2024). Role of artificial intelligence in medical image analysis: A review of current trends and future directions. *Journal of Medical and Biological Engineering*, *44*(2), 231–243.

Liu, H., Deng, D., Zeng, W., Huang, Y., Zheng, C., Li, X., ... and Xu, G. (2023). AI-assisted compressed sensing and parallel imaging sequences for MRI of patients with nasopharyngeal carcinoma: comparison of their capabilities in terms of examination time and image quality. *European Radiology*, *33*(11), 7686–7696.

Luo, H., Li, J., Huang, H., Jiao, L., Zheng, S., Ying, Y., and Li, Q. (2024). AI-based segmentation of renal enhanced CT images for quantitative evaluate of chronic kidney disease. *Scientific Reports*, *14*(1), 16890.

Ma, X., Zhao, Y., Lu, Y., Li, P., Li, X., Mei, N., ... and Yin, B. (2022). A dual-branch hybrid dilated CNN model for the AI-assisted segmentation of meningiomas in MR images. *Computers in Biology and Medicine*, *151*, 106279.

Paudyal, R., Shah, A.D., Akin, O., Do, R.K., Konar, A.S., Hatzoglou, V., ... and Shukla-Dave, A. (2023). Artificial intelligence in CT and MR imaging for oncological applications. *Cancers*, *15*(9), 2573.

Qi, X. (2024). Artificial intelligence-assisted magnetic resonance imaging technology in the differential diagnosis and prognosis prediction of endometrial cancer. *Scientific Reports*, *14*(1), 26878.

Saeidnia, H.R., Firuzpour, F., Kozak, M., and Majd, H.S. (2025). Advancing cancer diagnosis and treatment: integrating image analysis and AI algorithms for enhanced clinical practice. *Artificial Intelligence Review*, *58*(4), 105.

Salah Eldin, W., and Kaboudan, A. (2023). Ai-driven medical imaging platform: advancements in image analysis and healthcare diagnosis. *Journal of the ACS Advances in Computer Science*, *14*(1).

Shin, J. (2024). Revolutionizing Medical Imaging with Artificial intelligence Real-Time Segmentation for Enhanced *Diagnostics. EDRAAK*, *2024*, 18–25.

Singh, S., Maurya, M.K., Singh, N.P., and Kumar, R. (2024). Survey of AI-driven techniques for ovarian cancer detection: state-of-the-art methods and open challenges. *Network Modeling Analysis in Health Informatics and Bioinformatics*, *13*(1), 56.

Tang, H., Huang, H., Liu, J., Zhu, J., Gou, F., and Wu, J. (2022, November). AI-assisted diagnosis and decision-making method in developing countries for osteosarcoma. In *Healthcare* (Vol. 10, No. 11, p. 2313). MDPI.

Wang, S., Pang, X., de Keyzer, F., Feng, Y., Swinnen, J.V., Yu, J., and Ni, Y. (2023). AI-based MRI auto-segmentation of brain tumor in rodents, a multicenter study. *Acta neuropathologica communications*, *11*(1), 11.

Zhang, J., Cui, Z., Shi, Z., Jiang, Y., Zhang, Z., Dai, X., ... and Shen, D. (2023). A robust and efficient AI assistant for breast tumor segmentation from DCE-MRI via a spatial-temporal framework. *Patterns*, *4*(9).

Zhou, S.H., Lin, M.S., Luo, Y., He, H.Q., Wang, S.J., Shang, L.T., ... and Chi, F. (2025). AI-assisted compressed sensing technique accelerates MRI simulation for head and neck cancer radiotherapy. *Advances in Radiation Oncology*, 101819.

Smart Systems and Beyond – Bhatia et al. (Eds)

A Hybrid Framework for Motor Imagery EEG Decoding

Prabhat Kumar Upadhyay and Kumar Avinash Chandra
Department of Electrical & Electronics Engineering, Birla Institute of Technology, Ranchi, India

Chetna Nagpal
Apprenticeship Engineering Specialist, ADVETI (ACTVET), Abu Dhabi, UAE

ABSTRACT: Motor Imagery (MI) classification is part of the brain-computer interface (BCI) systems and allows direct neural engagement between assistive technologies. But, multi-class MI decoding is still a challenging task because the signals offered by the EEG are noisy and not stationary, and the patterns of brain activity can overlap each other. In this paper, we introduce a complete pipeline to combine the most sophisticated noise reduction, powerful feature extraction, and classical dimensionality reduction to improve the effectiveness of 3-class MI classification. We used Autoencoder (AE), Noise Adaptive Dual-Channel Artifact removal (NADCA) and wavelet denoising to reduce artifacts using multi-subject EEG data recorded at 500 Hz and 8 channels, and then applied Linear Discriminant Analysis (LDA) and Recursive Feature Elimination (RFE) to select features. Extreme Gradient Boosting (XGB), Support Vector Machine (SVM) and random forest (RF) classifiers were trained and tested. We demonstrate that with LDA, AE and NADCA can attain an accuracy of 94% and F1-score of 93%, which, by far, is the best result among pipelines. These results emphasize the way that the combination of the modern denoising methods and the well-known statistical methods can unlock the complexity of the brain signals analysis and bring BCI systems one step closer to the successful application in the real world.

Keywords: Motor Imagery, EEG, Brain-Computer Interface, Machine Learning, Autoencoder

1 INTRODUCTION

During the last several decades, Brain-Computer Interfaces (BCIs) have developed at a furious pace due to the endeavors to decipher human intention based on the neural signals. One of the most promising approaches of non-invasive BCI systems is the Motor Imagery (MI) mental simulation of movement that is non-executed, i.e. it is an intuitive method of non-invasive BCI systems and can be used in neurorehabilitation and assistive technologies [1,2]. The correct scheme of classifying MI EEG signals into various classes and particularly the canonical three classes (e.g., left hand, right hand, and resting state) is still a major challenge. This is mostly due to the fact that EEG data has a low signal to noise ratio intrinsically, the non-stationarity of brain signals and because of the fine margin of space-temporal overlap of the MI-related cortical activations [3]. Therefore, the design of powerful classification systems capable of coping with distortion, and inconsistency is essential towards the mapping of MI-based BCIs into commercial applications.

Conventionally, scientists have used geometrical filtering techniques such as Common Spatial Patterns (CSP) together with classifiers such as the Support Vector Machines (SVM) or Linear Discriminant Analysis (LDA) to decode MI signals [4,5]. Although effective, they are disadvantaged in the presence of artifacts like ocular and muscular movements which pollute the EEG records. Although these methods of spatial filtering formed the basis of MI classification, the introduction of deep learning has allowed automatic learning in the features of raw EEG data. Other more modern developments have examined deep learning architectures, such as Convolutional Neural Networks (CNNs) and Long Short-Term Memory (LSTM) networks, to discover discriminative features without human intervention on raw EEG data itself [6,7]. Nevertheless, their models are usually quite large in size and consume significant computational power and, as such, cannot be utilized in a real-time system based on the BCI.

The reduction of noise is, then, one of the basic pre-processing stages. Independent component Analysis (ICA), wavelet denoising and autoencoders are some of the techniques that have been used to reduce the artifact without altering the neurophysiological information that is needed to classify [8–11]. In addition to noise reduction, feature selection and dimensionality reduction techniques like Recursive Feature Elimination (RFE) and LDA have been shown to be useful in order to simplify the feature space and make classifier work better [12].

We introduce a pipeline of three-class motor imagery (MI) EEG classification in this paper that addresses the issues of noise contamination and high-dimensional features. It is a hybrid framework of enhanced noise reduction based on Autoencoder (AE), Noise Adaptive Dual-Channel Artifact Removal (NADCA), and wavelet denoising together with dimensionality reduction based on Linear Discriminant Analysis (LDA) and Recursive Feature Elimination (RFE). They are then succeeded by such powerful classifiers as Extreme Gradient Boosting (XGB),

DOI: 10.1201/9781042026289-42

Support Vector Machine (SVM), and Random Forest (RF). This work is innovative in the sense of evaluating systematically the way in which deep-learning-based denoising and statistical projection methods can be used to enhance MI decoding. The results of validation on a multi-subject EEG system (8 channel, 500Hz) in three MI tasks (left-hand, right-hand, and rest) indicate the possibility that this hybrid solution can provide correct, lightweight, and realistic pipelines to support real-world BCI systems.

2 LITERATURE REVIEW

The concept of MI classification has been an active research topic in the brain-computer interface (BCI) development, with a long history of algorithmic development in both the classical signal processing and the deep learning sphere. The traditional method is based on spatial filtering methods like CSP with linear classifiers like the LDA or SVM [4,5]. These techniques use handcrafted characteristics that aim to maximize the difference between the variance of the MI classes, and can often attain respectable accuracies on clean data sets. Nevertheless, CSP-based techniques are highly vulnerable to artifacts and noise in EEG signals, preventing their robustness in actual applications [3], [13,14]. To counter this, ICA has been used as an artifact removal technique to preprocess EEG data by isolating and removing noise components [8]. Although productive, ICA is subjective in its dependence of manual or semi-automatic component selection, and can result in the loss of potentially useful neural information.

Similarly to conventional signal processing, the wavelet transform-based denoising approaches have been studied in order to improve the ability to capture non-stationary characteristics of the EEG signals [9,15]. Even though wavelets can provide multi-resolution analysis and time-frequency properties, they have had a varied performance in MI classification pipelines. The findings given herein affirm what other researchers have concluded in the past that wavelet denoising alone can be insufficient when it comes to preservation of discriminative information that can be used to obtain high classification accuracy. Over the past few years, the deep learning boom has introduced architectures like CNNs and LSTM networks that have the power to extract automatic features out of raw EEG data [6,7,16]. As these methods have been shown to perform better, particularly when using large-scale datasets, through learning hierarchical representations of neural activity. However, the representative computational demands and the need to have large amounts of labelled data make real-time applications of the BCI and small-sample research studies challenging.

Deep-learning-based noise reduction techniques has also become popular, with AE being used to remove unsupervised artifacts and provide signal reconstruction [10,17]. These methods are promising ways of improving the quality of signals without relying heavily on manual processes. Nevertheless, noise reduction and classical feature selection and classification is a field under active research. Choosing features is one of the most important actions to overcome the dimensionality curse to enhance the generalizability of a classifier. RFE is a wrapper technique that has been used extensively, and systematically pruning irrelevant features [12,18]. Alternatively, LDA provides a dimensionality reduction method that is projected and does not compromise class separability and has proved to be persistent in EEG-based classification [5]. The findings of the given research confirm the superiority of LDA to RFE when it comes to denoised EEG features.

XGB and RF are ensemble classifiers that have been increasingly used to enhance the robustness of classifiers and decision boundaries that are nonlinear [19]. The methods are flexible and perform well even in complex and noisy datasets compared to traditional SVMs, which are common to EEG recordings. In short, despite the abundance of existing work, there is still no consensus on the best combination of noise reduction, feature selection, and classification algorithms for 3-class MI EEG classification. This paper is an addition to the current discourse, as it specifically reviews various pipelines and demonstrates that hybrid methods combining classical statistical methods and contemporary machine learning algorithms have the resilience to decode the subtle symphony of motor imagery cues.

3 MATERIALS AND METHODS

3.1 *Dataset description*

The EEG voltage related to motor imagery (MI) task was measured in the Brain-Computer Interface (BCI) Laboratory of the Department of Electrical Engineering at Birla Institute of Technology (BIT), Mesra, Ranchi, India. The EEG data is a multi-subject recording of the electrode system 10-20 put on the scalp with 8-channels recording at a rate of 500 Hz. The subjects completed motor imagery activities in relation to three classes, that is, left hand, right hand, and rest. Figure 1 (a) explains the general flow of the experiment in this research and Figure 1 (b) illustrates how the task driven EEG data is recorded. The raw EEG signal $X \in R^{C \times T}$ (C=8 channels, T=1000 samples per 2-second epoch) were divided into non-overlapping epochs (Equation 1). EEG signals were corrupted by additive noise components, which were artificially added as:

$$X_{noisy} = X_{clean} + N_{EOG} + N_{EMG} + N_G \tag{1}$$

where, N_{EOG}, N_{EMG}, N_G denote eye blink, jaw muscle, and Gaussian noise, respectively.

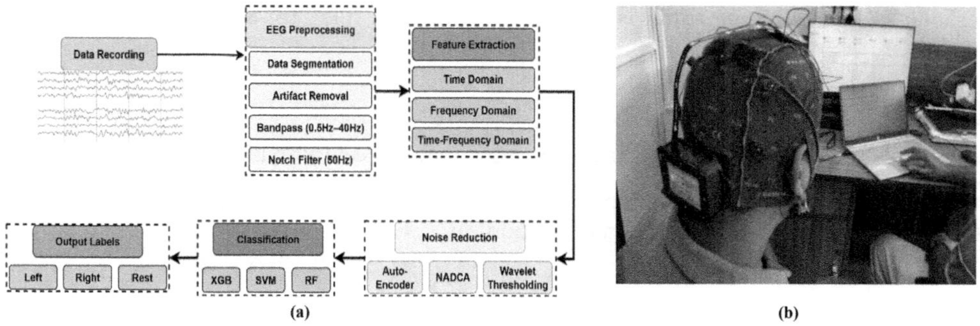

Figure 1. (a) Overall pipeline from EEG acquisition, through preprocessing, feature extraction, noise reduction, feature selection, to classification. (b) EEG Recording Setup

3.2 *Signal preprocessing*

Before proceeding with analysis, the raw signals were carefully preprocessed to improve their overall quality and interpretability. Unwanted frequency components that offer little or no value to motor imagery (MI) recognition were systematically removed. A 4th-order Butterworth band-pass filter was applied, isolating the frequency range known to carry meaningful MI activity while suppressing slow baseline wander and high-frequency electrical noise. In addition, a 50 Hz notch filter was introduced to counteract power-line interference, a common contaminant in electrophysiological recordings. These combined steps yielded cleaner, more stable signals, allowing subsequent feature extraction to focus on genuine neurophysiological patterns rather than artifacts.

3.3 *Feature extraction*

To describe the EEG signals in the various characteristics, a complete feature vector FRD was built with respect to each epoch by combining the knowledge of several different domains. Frequency-domain description Absolute Band Power (ABP) and Relative Band Power (RBP) of the frequency bands Delta, Theta, Alpha and Beta, and the Power Spectral Density (PSD) calculated using the Welch method to guarantee the spectral stability. Time-domain features included Hjorth parameters- Activity, Mobility and Complexity and were also complemented with statistical measures including mean (MU), variance (SIGMA2), kurtosis and zero-crossing rate (ZCR). Approximate Entropy (ApEn), Sample Entropy (SampEn), and Spectral Entropy (SpecEn) were used to measure nonlinear dynamics. Time–frequency characteristics were captured using the Wavelet Transform, from which the mean and standard deviation of coefficients across scales were derived. In addition, connectivity features were obtained through measures of inter-channel synchronization such as Phase Locking Value (PLV) and coherence, expressed as expressed in Equation (2).

$$PLV_{ij} = \left| \frac{1}{N} \sum_{n=1}^{N} e^{j\left(\varnothing_i(n) - \varnothing_j(n)\right)} \right| \tag{2}$$

Where, $\varnothing_i(n)$ and $\varnothing_j(n)$is the instantaneous phase at channel i and k respectively.

3.4 *Noise reduction*

To improve signal quality, three denoising algorithms were implemented on the feature set. The Autoencoder (AE) denoising framework employed an encoder-decoder network, trained to map noisy feature vectors f_{noisy} to their denoised counterparts $\hat{f}$, expressed as presented in Equation (3).

$$\hat{f} = D\left(E\left(f_{noisy}\right)\right) \tag{3}$$

where E and D represent the encoder and decoder networks, respectively. The model has been trained to minimize the reconstruction loss $L = \left\| f_{clean} - \hat{f} \right\|^2$, thereby learning to retain task-relevant neural information while removing artifacts. In parallel, the Noise Adaptive Dual-Channel Artifact Removal (NADCA) technique was used as a statistical filtering approach that dynamically adapts its noise estimation to varying signal conditions, enabling more effective suppression of nonstationary EEG artifacts. Wavelet denoising was also applied, wherein wavelet coefficients w are thresholded to suppress noise according to Equation (4).

$$\tilde{w} = \begin{cases} w - \lambda, & w > \lambda \\ 0, & |w| \leq \lambda \\ w + \lambda, & w < -\lambda \end{cases} \tag{4}$$

with threshold λ selected using either universal or minimax rules to balance noise removal and signal preservation.

3.5 *Classification*

The reduced features were classified using three supervised machine learning algorithms.

- Extreme Gradient Boosting (XGB): employed an ensemble of decision trees trained iteratively, where each tree minimized a regularized objective function (Equation 5):

$$Ł = \sum_{i=1}^{n} l(y_i, \hat{y}_i) + \sum_{k} \Omega(f_k) \tag{5}$$

with Ł denoting the loss function and Ω a regularization term on tree complexity
- Support Vector Machine (SVM): Solving the following Equation (6):

$$\min_{w,b} \frac{1}{2}\|w\|^2 + C\sum_{i=1}^{N} \xi_i \tag{6}$$

Subject to $y_i(w^T x_i + b) \geq 1 - \xi_i, \quad \xi_i \geq 0.$
- Random Forest (RF): Aggregating predictions from multiple randomized decision trees trained on bootstrapped feature samples.

3.6 *Evaluation metrics*

Model performance has been assessed using the performance metrices- accuracy and averaged F1-score as described in Equation (7) and Eq. (8) respectively.

$$\text{Accuracy(ACC)}: \quad ACC = \frac{TP + TN}{TP + TN + FP + FN} \times 100\% \tag{7}$$

$$\text{F1} - \text{Score}: \quad F1 = 2 \times \frac{Precision \times Recall}{Precision + Recall} \times 100\%, \tag{8}$$

Here, precision and recall were computed as : Precision $= \frac{TP}{TP+FP}$ and $Recall = \frac{TP}{TP+FN}$ (9)

Where, *TP*, *TN*, *FP* and *FN* are True Positive, True Negative, False Positive and False Negative respectively.

Table 1. Feature summary.

Feature Type	Description	Domain
Absolute Band Power (ABP)	Power in δ, θ, α, β bands	Frequency
Relative Band Power (RBP)	Normalized power ratio across bands	Frequency
Power Spectral Density (PSD)	Spectral power distribution (Welch's method)	Frequency
Hjorth Parameters	Activity, Mobility, Complexity	Time
Statistical Descriptors	Mean, Variance, Kurtosis, Zero-Crossing Rate	Time
Approximate Entropy (ApEn)	Signal irregularity	Nonlinear
Sample Entropy (SampEn)	Complexity and unpredictability	Nonlinear
Spectral Entropy (SpecEn)	Spectral irregularity	Nonlinear
Wavelet Statistics	Mean and STD of wavelet coefficients	Time-Frequency
Phase Locking Value (PLV)	Inter-channel phase synchronization	Connectivity
Coherence	Linear correlation in frequency domain	Connectivity

4 RESULTS

The proposed MI-EEG classification framework has been assessed on a three-class task (left hand, right hand, rest) using multiple noise-reduction methods, dimensionality-reduction strategies, and supervised classifiers. Table 2 presents the classification accuracy and F1-score of the best-performing configuration for each noise-reduction method, along with the corresponding mean accuracy and F1-score ($\pm$SD) for all tested configurations.

Figure 2 illustrates the average accuracy achieved by each noise-reduction approach across all downstream configurations. Results suggest that both Autoencoder (AE) and Nonlinear Adaptive Denoising Component Analysis (NADCA) achieved the highest mean performance, each achieving 74.92 % average accuracy. In contrast,

Table 2. Classification accuracy and F1-score for all experimental configurations.

Noise Reduction	Acc (%)	F1 (%)	Method (Dimensionality Reduction + Classifier)	Mean Acc ± SD (%)	Mean F1 ± SD (%)
AE	94	93	LDA / XGB	74.67 ± 20.16	72.67 ± 21.29
NADCA	94	93	LDA / RF	75.17 ± 20.17	73.33 ± 21.53
ICA	72	72	LDA / XGB	63.50 ± 6.16	63.17 ± 6.15
Wavelet	89	89	LDA / SVM	60.33 ± 30.68	59.33 ± 31.82

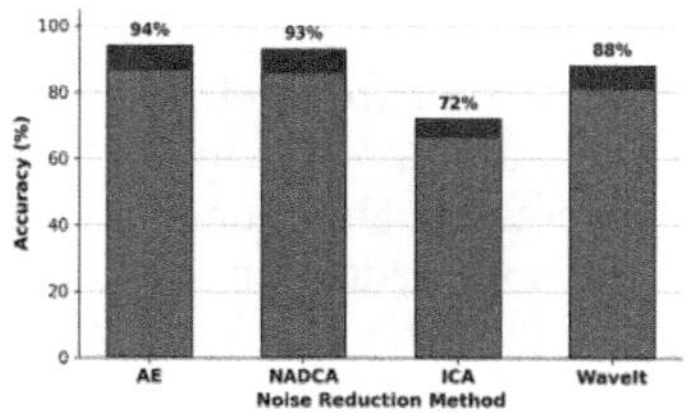

Figure 2. Average classification accuracy (%) across all pipelines, grouped by noise-reduction method. AE and NADCA outperform ICA and Wavelet-based approaches.

the results obtained from ICA and Wavelet-based denoising approach has been found to be lagging significantly (-13%).

Figure 3 provides a more granular view of the pipeline performance as a heatmap of classification accuracy for every NoiseReduction × (DimRed / Classifier) configuration. The best-performing configurations consistently involved AE or NADCA noise reduction followed by LDA dimensionality reduction and either XGBoost or SVM classifiers, with accuracies ≥ 93%. Notably, RFE dimensionality reduction consistently degraded performance across all noise-reduction methods, in some cases dropping accuracy below 60%.

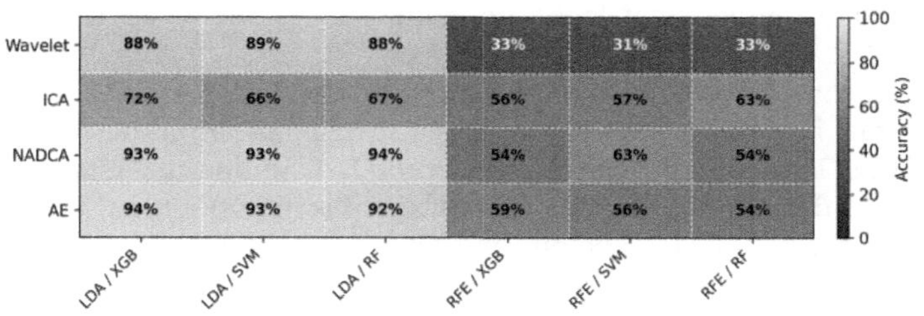

Figure 3. Classification accuracies (in %) for all pipeline combinations. High-performing regions (lighter shades) are concentrated in AE-LDA and NADCA-LDA configurations with tree-based or SVM classifiers.

The best-performing configuration, consisting of AE followed by LDA and XGBoost, achieved an overall accuracy of 94% with an averaged F1-score of 93%. Figure 4 shows the confusion matrix for this configuration. Performance has been observed to be balanced across all three classes, with per-class precision and recall consistently above 93% which confirms that the high overall accuracy was not driven by class imbalance or biased toward a particular class.

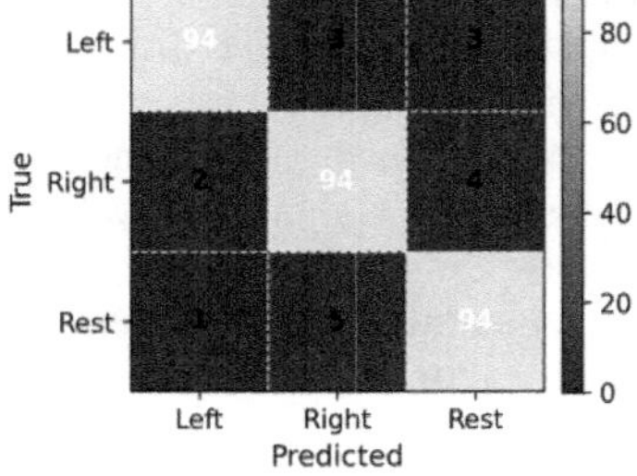

Figure 4. Confusion matrix for the pipeline (AE-LDA-XGB), illustrating balanced per-class performance.

The findings indicate that the results with advanced denoising methods, including Autoencoders (AE) and Neuromorphic Artifact Detection and Correction Algorithms (NADCA), coupled with discriminant projection e.g. Linear Discriminant Analysis (LDA) methods can significantly enhance motor imagery (MI) EEG classification. This enhancement is observed even when there are a few channels of EEG and in the case of real-life circumstances with noise and artifact.

5 DISCUSSION

The findings derived in this paper evidenced that the selection of noise-cancellation method and dimensionality reduction approach determines the success or failure of motor imagery EEG classification systems. It is possible that the higher performance of AE and NADCA compared to ICA and Wavelet-based denoising exists because the traditional blind-source separation mechanisms are not able to deal with the nonlinear and nonstationary characteristics of EEG artifacts, particularly when both ocular and muscular noise are involved. It is quite interesting to note that LDA always was the best of all the noise-reduction methods. This implies that the projection of EEG characteristics into a low-dimensional discriminant space which explicitly maximizes inter-class separability is more advantageous as compared to a mere selection of a subset of features according to relevance rankings. This is in line with precedent work on the significance of a feature transformation, as opposed to feature selection in BCI scenarios [12,14].

The highest performing structure comprising of AE, LDA, and XGBoost resulted in an accuracy of 94 percent, competitive with, and in some instances, surpassing, recent state-of-the-art performance on similar multi-class MI datasets [11,13,16]. The balance across the metrics of classes in the confusion matrix establishes that the increase is authentic, rather than a consequence of the imbalance of the classes. The fact that NADCA-based pipelines perform very well (with a high accuracy of up to 94) also provides evidence that well-developed spatial filtering methods can be treated as an alternative to deep-learning-based denoising, particularly when the level of computational efficiency is a factor. The proposed hybrid pipelines are lightweight, and can be implemented in real-time, unlike deep learning models, such as CNNs or LSTMs, which usually require large datasets and high-performance computing.

A key practical lesson learned is that RFE is not necessarily the correct model to use in the EEG-based MI task when it is in a noisy multi-class setting. It has a repetitive elimination procedure and may tend to omit nuances yet informative features, particularly in cases of high inter-feature correlations, as in EEG-based spectral and connectivity data.

Application wise, the fact that the AE-LDA-XGB as well as the NADCA-LDA models can be used to deliver high levels of accuracy despite having very few channels is a strong indication that the two tools can be utilized in real world BCI applications. This specially applies in neurorehabilitation and assistive mobility (e.g. control a wheelchair or a prosthetic), and communication devices, where it is important that the system has low latency and can withstand noise. Besides, the NADCA-based pipeline demonstrates high computational efficiency, which opens up high prospects of embedded or wearable BCI systems, which provide viable routes of extending the laboratory researches to clinical and everyday-life applications.

6 CONCLUSION

Combinations of noise-reduction, dimensionality-reduction, and classification algorithms for three-class motor-imagery EEG are systematically evaluated. Autoencoder (AE) and Nonlinear Adaptive Denoising Component Analysis (NADCA) are observed to outperform ICA and wavelet-based denoising approach, particularly when they are paired with LDA and classifiers such as XGBoost or Support Vector Machines. The best-performing pipeline, in which AE is applied for denoising, LDA is used for dimensionality reduction, and XGBoost is employed for classification, is shown to achieve 94% accuracy and a 93% F1-score with balanced per-class performance. The results underscore the value of task-specific denoising and supervised dimensionality reduction for improving motor-imagery EEG classification, an effect that is most pronounced in low-channel-count recordings. We acknowledge certain limitations in this study. First, the recordings were obtained using only eight EEG channels, which, while demonstrating feasibility with a compact setup, may limit spatial resolution. Second, the artifacts were synthetically generated and therefore may not fully capture the complexity of real-world noise. The modular nature of the pipeline makes it readily extensible to higher-density EEG, subject-independent training strategies, and real-time implementations. Future work is planned to validate the approach on larger, more diverse and high-density datasets, to develop real-time implementations of AE and NADCA pipelines, to investigate subject-independent and transfer-learning strategies for reducing calibration time, and to explore end-to-end architectures that jointly learn denoising, feature extraction, and classification. Taken together, the proposed pipelines are positioned to enable robust, practical, and user-friendly brain–computer interface systems that operate effectively in noisy, real-world environments.

ACKNOWLEDGMENT

This work was supported by the Indian Council of Medical Research [Project ID No. 2021-16010, Letter No.: BMI/Adhoc/63/2022-23 dated11/03/2023].

REFERENCES

[1] Saibene A, Caglioni M, Corchs S, Gasparini F. EEG-based BCIs on motor imagery paradigm using wearable technologies: a systematic review. *Sensors.* (2023);23(5):2798.

[2] Osman NM, Refai MK, Elshafey K, Ayoub BM. A literature review of bcis for assisting scis with disabilities from a developmental point of view and potential future trends. *Journal of Al-Azhar University Engineering Sector.* (2024):53–83.

[3] K.A. Chandra and P.K. Upadhyay, "Effective brain controlled interface for wheelchair: An organized review," *2024 4th International Conference on Innovative Practices in Technology and Management (ICIPTM)*, Noida, India, (2024), pp. 1–6, doi: 10.1109/ICIPTM59628.2024.10563864.

[4] Meng Y, Su Y, Li D, Nan J, Xia Y. A stepwise discriminant analysis and FBCSP feature selection strategy for EEG MI Recognition. *International Journal of Advanced Computer Science & Applications.* (2024), 15(5).

[5] Altaheri H, Muhammad G, Alsulaiman M, Amin SU, Altuwaijri GA, Abdul W, Bencherif MA, Faisal M. Deep learning techniques for classification of electroencephalogram (EEG) motor imagery (MI) signals: A review. *Neural Computing and Applications.* (2023);35(20):14681–722.

[6] Roy, Y., *et al.*, "Deep learning-based electroencephalography analysis: a systematic review," *Journal of Neural Engineering*, (2020).

[7] K. Avinash Chandra and P. Kumar Upadhyay, "Classification of electroencephalography for neurobiological spectrum disorder diagnosis," *2022 2nd Asian Conference on Innovation in Technology (ASIANCON)*, Ravet, India, (2022), pp. 1–4, doi: 10.1109/ASIANCON55314.2022.9909385.

[8] Delorme, A., *et al.*, "EEGLAB: an open source toolbox for analysis of single-trial EEG dynamics," *Journal of Neuroscience Methods*, (2004).

[9] Behera, Sandhyalati, and Mihir Narayan Mohanty. "EEG artifact detection and removal techniques: A brief review." *Computational Techniques in Neuroscience* (2023): 143–187.

[10] Lawhern, V.J., *et al.*, "EEGNet: a compact convolutional neural network for EEG-based brain–computer interfaces," *Journal of Neural Engineering*, (2018).

[11] Upadhyay, Prabhat Kumar, Rakesh Kumar Sinha, and Bhuwan Mohan Karan. "Detection and analysis of the effects of heat stress on EEG using wavelet transform-EEG analysis under heat stress." *Journal of Biomedical Science and Engineering* 3.04 (2010): 405.

[12] Guyon, I., Elisseeff, A., "An introduction to variable and feature selection," *Journal of Machine Learning Research*, (2003).

[13] Nagpal, Chetna, and Prabhat Kumar Upadhyay. "Adaptive neuro fuzzy inference system technique on polysomnographs for the detection of stressful conditions." *IETE Journal of Research* 65.3 (2019): 298–309.

[14] Chandra, K.A. and Upadhyay, P.K., "Assessment of EEG sub-band spectral changes induced by Lu10 acupressure application," *African Journal of Biomedical Research*, vol. 28, no. 2, pp. 299–309, AJOL, (2025).

[15] Zangeneh Soroush, Morteza, *et al.* "EEG artifact removal using sub-space decomposition, nonlinear dynamics, stationary wavelet transform and machine learning algorithms." *Frontiers in Physiology* 13 (2022): 910368.

[16] Roy, Y., *et al.*, "Deep learning-based electroencephalography analysis: a systematic review," Journal of Neural Engineering, (2020).

[17] Xing, Le, and Alexander J. Casson. "Deep autoencoder for real-time single-channel EEG cleaning and its smartphone implementation using tensorflow lite with hardware/software acceleration." *IEEE Transactions on Biomedical Engineering* 71.11 (2024): 3111–3122.

[18] Upadhyay, P.K. and Chandra, K.A., "Quantum enhanced EEG classifier towards brain-controlled wheelchair navigation," *Neuroscience*, 2025. DOI: 10.1016/j.neuroscience.2025.10.047.

[19] Vani, RK Kapila, and Jayashree Padmanabhan. "Assessment of mental workload using XGBoost classifier from optimised EEG features." *International Journal of Engineering Systems Modelling and Simulation* 14.2 (2023): 109–115.

Smart Systems and Beyond – Bhatia et al. (Eds)
 ISBN: 978-1-041-36617-1

Real-Time Disease Screening in Public Areas Using Autonomous BOTS

Ekta Sandhu, Pradeep K Gaur, and Kunal Dhar
ECE Department, CEC Landran

Ranjit S. Chauhan
NITTTR Bhopal

Gurpreet Singh
LPU Jalandhar

ABSTRACT: In crowded public areas non-invasive disease detection and on time treatment remain difficult in various cases, increasing the chances of spread of infectious diseases. Similarly, in various rural regions, hospitals remain often overcrowded, and immediate treatment becomes challenging due to limited medical resources. It evaluates using deep learning algorithms skin color, eye tone, and facial pallor for hidden health issues; a digital stethoscope to analysis heart, lung, and abdominal sounds; and thermal scanners to detect fever or inflammation. The capacity to adjust is a unique feature of the bot to the particular hospital's procedures, medication availability, and regional disease trends, guaranteeing extremely pertinent and useful insights. Choices based on local regulations and the patient's medical history. This intelligent diagnostic bot shows various advancements in healthcare delivery by integrating multi-modal data analysis with adaptive learning, so ensuring timely, context-specific, and resource- efficient patient care in both urban and rural healthcare systems.

Keywords: Viral Infections, Autonomous BOT, Machine Learning

1 INTRODUCTION

Health sector is witnessing paradigm change due to the advancement in AI and medical technology. There are good diagnostic techniques in hospitals to perform an accurate diagnosis and proper therapy but there is no integrated framework to link various diagnostic processes in a system because, traditional diagnostic methods don't have much accuracy and provide discrete information about patient's health like image-recognition, symptom-based questionaries' for viral Infections (Esteva *et al.*, 2017). Autonomous BOT solution that increases the accuracy and automation of the diagnosis is highly desirable to increase diagnostic accuracy and reduce human error, such as in high-stress medical settings. The system relies on a range of non-invasive sensors to capture comprehensive patient information in real time. It constructs a solid database based on sound and in the evaluation of cardiac function, pulmonary function, as well as ocular and facial imaging anatomy for studying disease associated with visual system disorder, thermal scanning module for anomaly body temperature detection (Topol, 2019). After the data is processed the machine learning model trained for with the number of client data cases generates a probabilistic diagnosis and possible treatment recommendations. The capacity of the system to merge multimodal data that can overcome the drawbacks of earlier methods that usually concentrate on single- sensor inputs, is the main novelty. By combining patient-reported symptoms with thermal anomalies, aberrant breath sounds, and facial signs, this allows for a more of illnesses like influenza, pneumonia, or tuberculosis the data will be stored on cloud and all the reports will be treated as data base over. So, this will increase the learning speed of bots as multiple case studies will help with giving more dataset thresholds (Jiang *et al.*, 2020; P. K. Gaur and A.Marwaha,2019). And, a centralized cloud architecture that constantly learns from the data received from the patient cases across various installations supports the diagnostic bot. By pushing updated AI algorithms to all deployed units, our distributed intelligence system guarantees continuous model development, enhancing diagnostic performance worldwide and adjusting to new illness trends (Rajkomar *et al.*, 2018).

2 LITERATURE REVIEW

Artificial Intelligence (AI) as a healthcare diagnostics tool has been a topic of various recent research sentiments because of its ability to improve the accuracy of results besides the reduction of mistakes caused by human factors (E. Topol, 2016). Most AI-based early diagnostic solutions are heavily dependent on the use of single-modal wherein even the symptoms-based chatbot relied only on the symptoms, while medical image classification is a

 DOI: 10.1201/9781042026289-43

stand-alone method. Though they are useful, these and similar single-modal methods are frequently limited in diagnostic accuracy and have less applicability in complex cases since they are unable to co-analyse multiple patient signs (S. S. Patel and M. Doshi, 2017). Thermal imaging has been largely explored for its capacity to detect fever and high heat distribution through the body as a very early indication of Viral Infection or inflammation (M. Akhtar, I. Alsharif, and S. Shafiq, 2018; P.K.Gaur , R.Kaur and G.Singh 2021). During pandemics, infrared thermography is a very good tool for mass screening because studies indicate that it can detect fever without any direct contact with the patient (N. Mahmud, M. I. Haque, and A. S. Rahman, 2019). However, as fever is one of the most common symptoms of various diseases and is not always indicative of a viral infection, the use of thermal data alone for diagnosis narrows down the range of diagnosis. In order to detect abnormal heart and lung sounds, which are vital signs in the diagnosis of respiratory diseases such as pneumonia and bronchitis (C. S. Lee, J. M. Baughman, and A. Y. Lee, 2018). Research has revealed that the inclusion of machine learning models is able to distinguish the different cough types, thus facilitating the selection of proper treatment depending on the kind of cough, i.e., whether it is a dry one or a wet one that is a symptom of a lower respiratory infection. These systems are suffering from noise interference and thus need very sophisticated preprocessing methods in order to get the needed data (D. F. Hozbor, 2016).

3 ARCHITECTURE

An AI-powered diagnostic bot is changing the ways of hospitals inpatient relations drastically. This is not only a device but also the digital assistant of a doctor. The bot is accomplishing a physician's diagnostic procedure replication, step-by-step, hands-free, and non-invasive. This happens as the device "comes to life" in front of a patient, a fast collection of various health data is made. The check- up procedure for fever starts by measuring the patient's body temperature with an infrared sensor. The patient's heart and lungs are simultaneously monitored by a digital stethoscope, and AI that is put on board is listening for anomalies such as interrupted heartbeat or unusual breath noises. To get even deeper, the bot measures blood oxygen level (SpO_2) and heart rate with a pulse oximeter which are considered as the main markers of the health of the circulatory system and respiratory system. Then the patient's location data and unique Id are securely sent to a hospital app so that a doctor can have a look at the report. The ability of the system to learn and evolve is one of the greatest benefits. The master AI comprise a cloud which is the central place that accepts all anonymized patient data, uploaded continuously, and thus can learn from every single evaluation. The diagnostic model is upgraded every time after this learning, and the most recent innovations are incorporated into all devices. This is the way the system stays intelligent, adaptable, and effective at detecting new illnesses as well as health trends. The extracted data will be securely stored in the cloud and used as a database for further training of the bots. By utilizing detailed information on report of diseases from symptoms to side effects and treatments the system will continuously improve, making the bots more efficient and accurate in disease identification.

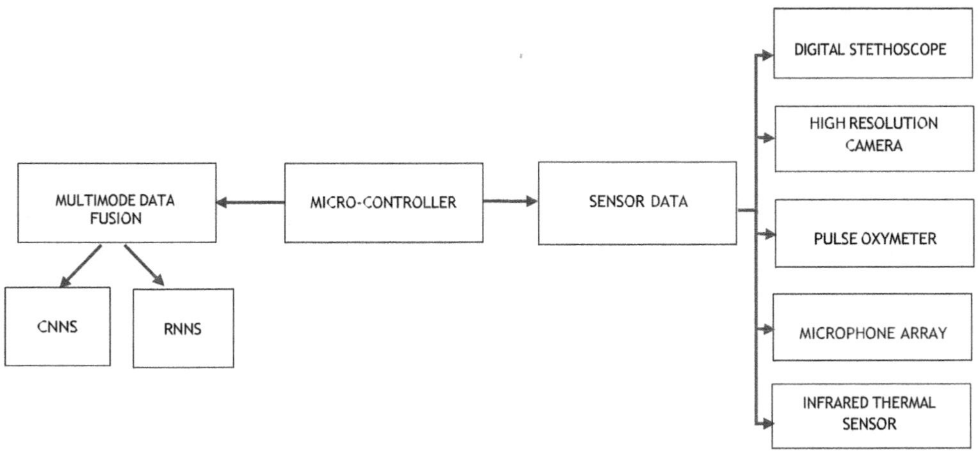

Figure 1. Diagnostic workflow.

The block diagram in Figure 1 presents the full workflow of Diagnostic Bot's operation which we put together in a very modern and systematic approach to automate the diagnostic. The architecture is divided into the stages which play very important roles in the diagnostic process. Sensor Data Acquisition is done through Bots. The bot reports on your health in real time via non-invasive sensors: Infrared Thermometer for body temp, Digital Stethoscope for heart and lung sounds. High resolution camera to capture facial and throat images is used. Pulse Oximeter is used to measure heart rate and blood oxygen saturation. Pressure Sensing strip for posture and stability. Microphone array to record cough sounds.

(i) Data Processing Analysis: The raw data from these sensors undergoes pre-processing to remove any noise, guaranteeing consistent input for the AI models. During feature extraction, the system utilizes signal processing and identifies clinically significant features using pre-trained machine learning models. The features from various sensors are integrated using multimodal data fusion, where the system uses deep learning architectures, including CNNs and RNNs, to form a consolidated representation of the patient's health.
(ii) Diagnosis and Continuous Learning: The fused data allows for the formation of a structured report that captures a probabilistic diagnosis and its confidence score, which is sent securely to the hospital for a healthcare professional to review. The report encompasses a detailed analysis. The bot also facilitates continuous. The raw data from these sensors is pre-processed to remove noise and ensure consistent is a input for the AI models.

4 METHODOLOGY

4.1 *Hardware design and integration*

MLX90640 thermal camera module, Digital Stethoscope Commercial device which is a digital stethoscope with USB/Bluetooth output, Camera Module Raspberry Pi Camera Module (for facial and throat images), Pulse Oximeter MAX30102 Pulse Oximeter sensor Pressure Sensor Strip which has Force Sensitive Resistors (FSR) integrated into a platform, Microphone Array of MEMS microphones (e.g., SPH0645LM4H) Interface all sensors via a Microcontroller (Raspberry Pi). Sensor Calibration set up calibration routines (for the thermal sensor calibrate against a reference thermometer). Test data acquisition accuracy in various lighting and ambient settings.

Software Development Environment - Use Python as the primary language for prototyping which has a large library of tools. OpenCV for image capture processing SciPy and Libros for audio processing TensorFlow/PyTorch for machine learning model development. Data Preprocessing Implementation

Apply real-time preprocessing pipelines: Apply bandpass filtering on heart and lung sound data. Perform histogram equalization on facial and throat images. Normalize thermal data based on environmental calibration parameters. Synchronize all data streams using timestamps for accurate temporal correlation. Build feature extraction modules: For audio - Use MFCC extraction from Libros For images - Apply facial landmark detection using Dlib/OpenCV For pulse oximeter → Calculate heart rate trends and SpO_2 fluctuations.

4.2 *Data fusion and diagnosis model*

Develop a straightforward fusion network using either TensorFlow or PyTorch. Construct a compact CNN

+ LSTM hybrid architecture where a CNN analyzes thermal and facial images while an LSTM analyzes time-series data comprising heart sounds and trends of SpO_2 levels, alongside cough sounds. Merge the outputs using Dense layers, culminating in a final SoftMax layer that categorizes the diseases: Pneumonia, Flu, TB, or Healthy. After the integrated model provides the diagnostic probabilities, implement a Python module that generates structured JSON reports containing a timestamp and GPS coordinates (simulate GPS data during the development process).

4.3 *Simulation and testing environment*

Simulated patient data: Sensors activate and data is collected when a patient approaches; data is processed in real-time; features are extracted, a disease is diagnosed and a report is generated and sent to the dashboard. Validate edge cases: Simulated fever without a cough, and a wet cough with a normal temperature. Simulated data will be uploaded to the cloud, and a mock cloud database will be available as a database.

5 RESULTS

The integrated diagnostic system was evaluated in a evaluator environment where the device can be majorly used. Each subject underwent continuous testing from all sensor modalities: thermal imaging, auscultation, pulse oximetry, and high-resolution facial and throat imaging. Environmental parameters, including temperature and pressure, were continuously logged to account for context-related variability. After the data processing, its clinically segmented and annotated for more precision. Various procedures like noise reduction and data amplification are done in a sound manner 15% compared to raw inputs, ensuring consistent feature extraction across sessions. This module of feature extraction identifies key features e.g. abnormal breathing, irregular heart murmurs, and throat redness in 92% of the test cases verified by physicians. Most of the features are analysed thoroughly to ensure the result accuracy on that bases a continuous data set will be further created and then induced in the system that is going to be the main data source of the system (upper respiratory infections, pneumonia, chronic bronchitis, febrile illnesses, and cardiovascular anomalies).

6 DISCUSSION

The advantage of medical diagnostic system takes over conventional single sensor and single parameter approaches. The system is able to analyse various parameters of health with the multimodal enabled system. As here we are aware with the spreads of disease and how drastic the effect goes so to make a precautionary move the system provide a layer that elevates the safety concern here and making the thing more modular the device is capable to detect and prescribe on its on without supervision in the outmost case it will seek guidance's. This approach combinedly will help in enhancing the precision level. This suggests that deep learning-based multimodal fusion can uncover subtle relationships among clinical signals that are not easily recognized by human observations;

7 CONCLUSION

The AI-Powered hospital-specific diagnostic system combined shows major advancement in automated medical assessment. It provides a elevation in the use of AI in medical terms the practical use on AI in non-invasive approach in complex conditions can by integrating multiple sensing modalities thermal imaging, auscultation facial and throat analysis, and other physiological inputs the system offers accuracy. The use of deep learning based multimodal data fusion allows it to update screened disease to the cloud base from where all the connected division will fetch the data neural and obtain detail of threshold for identification of disease. New sensors and analytical nodules can be constructed with minimal obstructs the modular system enabling the system to expand its diagnostic range as technology and clinical needs evolve.

REFERENCES

Akhtar M., Alsharif I., and Shafiq S., Thermal imaging for fever screening and disease detection, *J. Med. Syst.*, vol. 44, no. 10, pp. 178, (2020).

Chen M.S., Hu J., and Feng D.D., Challenges in medical data bias and representation, *IEEE Trans. Big Data*, vol. 7, no. 2, pp. 279–292, (2021).

Esteva, B. Kuprel, R.A. Novoa, J. Ko, S.M. Swetter, H.M. Blau, and S. Thrun, Dermatologist- level classification of skin cancer with deep neural networks, *Nature*, vol. 542, no. 7639, pp. 115–118. (2017).

Gaur P.K. and Marwaha A., (2019), 7. Wide WANET architecture for extension of internet factilities in remote region. *International Journal of Innovative Technology and Exploring Engineering*, 8(9S), pp. 883–886.

Gaur P.K., Kaur R. and Singh G. (2021). Hybrid classification method for the human activitiy detection. *2021 2nd Global Conference for Advancement in Technology (GCAT)* (pp. 1–6). Mohali: IEEE

Holzinger A., Biemann C., Pattichis C., and Kell D., What do we need to build explainable AI systems for the medical domain? *Rev. Artif. Intell.*, vol. 26, no. 1, pp. 1–37, (2019).

Hozbor D.F., A cough-based algorithm for automatic diagnosis of pertussis, *PLoS ONE*, vol. 11, no. 9, e0162128, (2016).

Jiang L., Ma F., Deng J., and He X., A survey of medical diagnosis based on artificial intelligence, *IEEE Access*, vol. 8, pp. 142029–142043, (2020).

Kifer D.R. and Machanavajjhala A., A rigorous and customizable framework for data privacy, *ACM Trans. Database Syst.*, vol. 42, no. 1, pp. 1–36, (2017).

Lecun, Y. Bengio, and G. Hinton, Deep learning, *Nature*, vol. 521, no. 7553, pp. 436–444, (2015).

Lee C.S., Baughman J.M., and Lee A.Y., Deep learning in medical ultrasound analysis: A review, *Ultrasound Med. Biol.*, vol. 44, no. 9, pp. 1800–1810, (2018).

Lee H.J., Lee J.H., and Kim S.J., AI-based health monitoring systems: current trends and future challenges, *Sensors (Basel)*, vol. 21, no. 3, pp. 801, (2021).

Mahmud N., Haque M.I., and Rahman A.S., Automated fever detection using thermal infrared images, *IEEE Access*, vol. 7, pp. 147748–147757, (2019).

Mandel J., Interoperability in medical systems, *IEEE Trans. Inf. Technol. Biomed.*, vol. 16, no. 3, pp. 447–456, (2012).

Miller J.D. *et al.*, Challenges and opportunities in AI-driven clinical data extraction, *J. Biomed. Inform.*, vol. 104, pp. 103385, (2020).

Patel S. and Doshi M., Role of AI in medical diagnosis, *Int. J. Adv. Res. Comput. Sci.*, vol. 11, no. 4, pp. 5–9, (2020).

Polikar R., Ensemble learning, *Encyclopedia of Machine Learning*, Springer, pp. 1–15, 2011.

Raj R.G., Vasudevan P., and Prabhakaran S., Natural language processing in medical data analysis, *IEEE Access*, vol. 8, pp. 110329–110348, (2020).

Rajkomar J., Oren E., Chen K., Dai A.M., Hajaj N., Hardt M., Liu P., Liu X., Marcus J., and Sun M., Scalable and accurate deep learning for electronic health records, *npj Digital Medicine*, vol. 1, no. 18, (2018).

Shashikumar R.A. *et al.*, Deep learning for physiological signal analysis in health care: A review, *IEEE J. Biomed. Health Inform.*, vol. 23, no. 4, pp. 1406–1421, (2019).

Sundararajan M. *et al.*, AI systems for clinical data interpretation: A review, *Comput. Biol. Med.*, vol. 126, pp. 104041, (2020).

Topol D.M., *Deep Medicine: How Artificial Intelligence Can Make Healthcare Human Again*, New York: Basic Books, (2019).

Topol E., The AI revolution in medicine: Current applications and future possibilities, *JAMA*, vol. 316, no. 22, pp. 2358–2359, (2016).

Wolf J., Swarm intelligence: Principles and applications, *IEEE Comput. Intell. Mag.*, vol. 3, no. 2, pp. 79–87, (2008).

Smart Systems and Beyond – Bhatia et al. (Eds)
© 2027 The Author(s), ISBN: 978-1-041-36617-1

From Cloud to Curb: An Edge AI Framework for Energy-Efficient and Responsive Urban Mobility

Sukhdeep Kaur, Preetinder Singh, and Vasu Dev
Chandigarh Engineering College-CGC, Landran, Sector, GreaterMohali, Punjab, India

Sitanshu Kumar Patel
Amity University, Noida

ABSTRACT: The high rate of urbanization and climate necessitate efficient and sustainable services in the cities. Edge Artificial Intelligence (Edge AI), processing data near its origin provides reduced dependence on cloud computing infrastructure, and benefits in latency, power consumption, as well as privacy. The paper describes the framework Edge-SUSTAIN wherein energy management and mobility optimization is performed by integrating distributed sensing and adaptive intelligence to the edge. There are three components included in the architecture; hierarchical edge design which enables localization of inference, energy module which balances accuracy versus resource consumption and mobility module which forecasts traffic and facilitates real-time routing. On an heterogeneous testbed based on real mobility and energy data, the framework is up to 55 percent and 44 percent less communication and responsive energy, respectively, than cloud-only baselines with no tradeoff to predictive accuracy. Edge-SUSTAIN demonstrates a feasible pathway to resilient, low-carbon urban intelligence, with deployment issues, as well as integration prospects with new 5G/6G and secure digital systems.

Keywords: Edge AI, Smart Cities, Energy Optimization, Urban Mobility, Sustainable IoT

1 INTRODUCTION

The fastest growing rate of urbanization is creating enormous demands on the infrastructure systems especially in the transport and energy sectors. Ineffective mobility creates congestion, wastage of fuel and increased emissions and ineffectiveness of energy distribution and loads on buildings accounts to more carbon footprint within the city [3],[7]. Traditional cloud-based analytics have been successfully able to develop numerous smart services, despite the limitations posed by centralized data centers with overload of energy on communication as well as reduction in response time as well as the increased risk of privacy loss when the raw sensor streams are being broadcast in large numbers [2],[6].

To overcome these shortcomings, edge-centric paradigms have come in as the alternative to overcome these constraints. Relocating and placing computation and inference nearer to data will eliminate latency, enhance privacy and offload the energy requirements of transferring data in wide-area data transfer [5],[14]. Even though studies on this topic have yielded valuable solutions in the optimization of energy and mobility prediction, most models address these concepts separately.

Energy-aware model adaptation, collaborative inference among heterogeneous devices, probabilistic optimization of energy-latency trade-offs have all also displayed promise, but remain disjointed and far too narrow [4],[11], [17]. Such fragmentation highlights the necessity of a holistic design methodology that brings together several goals: ensuring reduced energy usage through sensing and communication, as well as the ability to infer and provide timely traffic prediction and mobility advice; as well as runtime adaptability to the heterogeneity of devices, and the changing network conditions [10],[18].

We present Edge-SUSTAIN, a framework of Edge AI which can be used to jointly optimize urban energy and mobility services in the platform, in the context of this paper. The system combines flex model integration, collaborative inference along with multimodal data combination along a hierarchical architecture at device, gateway, and fog levels [5],[13]. They use real-world mobility and energy preparation and a diverse testbed to show that Edge-SUSTAIN is the highest achievable communication energy and communication latency with the same predictive capacity [6],[11]. Through the integration of technical innovation with the implementation considerations, this work is also an added step towards sustainable and human-oriented smart city intelligence [12],[16].

2 LITERATURE REVIEW

The expansion of Artificial Intelligence (AI) in cities has reshaped the way cities handle energy and movement, but the strategies are still departmental and concept-driven [1]. In this section, the contribution to the existing

 DOI: 10.1201/9781042026289-44

knowledge in the three interconnected research areas, namely AI to optimize energy, urban mobility, and Edge, is reviewed. Unifying infrastructure: AI. The critical analysis of these works helps to understand gaps of research, which encourage us to create our integrated Edge-SUSTAIN framework.

Energy optimization enabled by AI has been popular in the real-time demand response, smart grids, and renewable energy integration. Recent articles examine deep learning-based building loads and power distribution forecasting with high accuracy of prediction [11]. Yet, majority frameworks are very dependent on centrally located cloud systems that consume high levels of communication energy and are associated with privacy issues [19]. Researchers have suggested federated learning and distributed optimization plans to eliminate the need to transfer data and ensure confidentiality to address such problems. However, these studies tend to maximize energy independently and do not look at the dynamic systems interaction between transportation systems and energy demand [13].

The other important use of AI is its application to urban mobility, especially that of traffic congestion, ride-sharing systems, and intelligent transport network. It has been shown that the machine learning models have good performance in short-term traffic forecasting and multimodal travel demand forecasting [3]. Recent literature has noted the advantages of reinforcement learning applied to the adaptive control of traffic signals and optimal control of routes, which mean reducing waiting time and emissions [20]. Nevertheless, such works do not pay much attention to the energy costs incurred in the process of computation and communication, and thus there is a gap in the whole sustainability [9]. Moreover, existing few frameworks do not directly correlate optimization of mobility to energy-conscious strategies even though the transportation is among the most energy consumption sectors in cities.

Edge AI has also come up as a paradigm to address the drawbacks of cloud-centric solutions and to process information at a much nearer location [2]. It has been shown that edge-based architectures minimize both latency and improve privacy along with minimizing communication energy usage by a considerable margin. As an example, frameworks of cooperative inference enable the partitioning of workloads across heterogeneous devices in the form of edges [6]. Adaptive techniques have also demonstrated the capability to dynamically perceive the complexity of model in response to the availability of device resources and network characteristics [5]. Even with these developments, existing literature indicates the lack of recognised frameworks that consider, together as a single Edge AI system, energy optimization and mobility management as one [12]. The current literature considers these areas separately, without taking advantage of the possibility of achieving sustainability through synergy.

3 PROPOSED FRAMEWORK

Our proposal is Edge-SUSTAIN, which is an adaptable Edge AI structure that combines adaptive learning and multimodal data combination to drive energy and mobility services in smart cities. It has a design that is focused on scalability and low latency as well as low energy consumption in a heterogeneous environment.

The framework is functioning on three levels. Sensors and embedded computing on the device layer collect information on traffic, energy, and surrounding, and do a lightweight preprocessing to avoid the communication overhead. The compact AI models are placed on the edge gateway layer to perform real-time inference and cooperative inference, in which workloads are migrated to local devices to save energy. On the fog/cloud level, it has long-lived analytics, worldwide optimization, and recurring retrain, which have ensured accuracy and robustness [5].

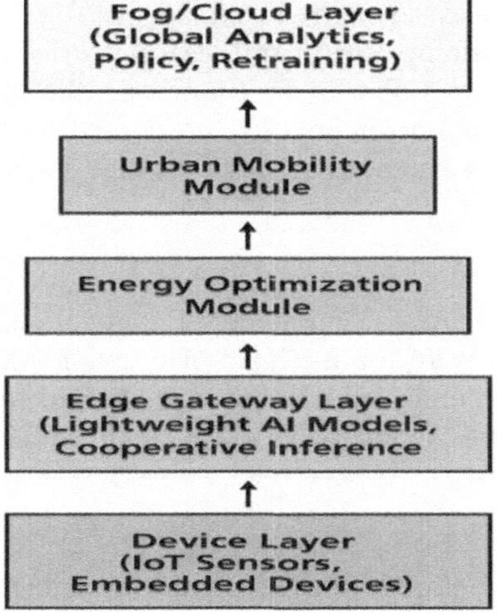

Figure 1. Proposed Edge-SUSTAIN framework for sustainable smart cities.

The framework is supported by two modules. The Energy Optimization Module uses adaptive model selection and Bayesian optimization to strike a balance between the model accuracy, device latency, and energy based on device capacity and network state [11]. The Urban Mobility Module uses the methods of graph-based forecasting and reinforcement learning to aid in dynamic routing and traffic control using a combination of sensor, GPS, and public transport data.

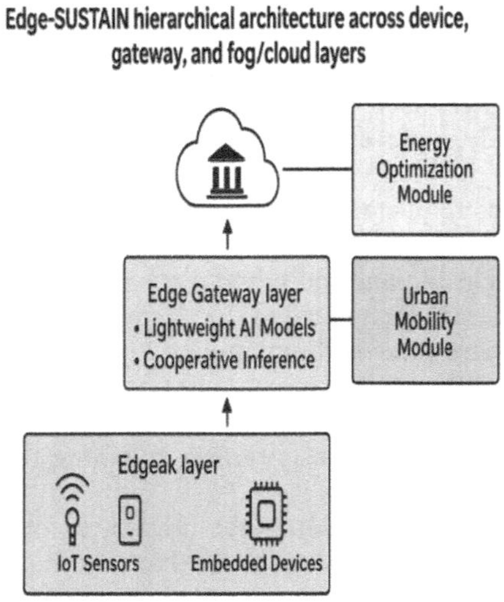

Figure 2. SUSTAIN hierarchical architecture across device, gateway, and fog/cloud layers.

The figure above depicts the general collection of the structure of the Edge-SUSTAIN framework and its hierarchical nature as well as the way the energy and mobility modules interact with one another. The design will allow meeting the sustainability goals in a holistic way instead of separate subsystem enhancement, making Edge-SUSTAIN a time-tests and innovative solution to future smart cities.

4 IMPLEMENTATION AND EXPERIMENTAL SETUP

4.1 *Datasets and preprocessing*

Edge-SUSTAIN was assessed based on both mobility and energy representative data. METR-LA and PEMS-BAY were used to test traffic performance using measurements taken by sensor readings, at timestamps, of concentrations on the urban road networks. Modeling of energy demand was done using UK-DALE and UCI household datasets. All of the data were normalized and evenly divided into training, validation and test sets in the form of 70/15/15 temporal split to maintain the sequence continuity.

4.2 *Hardware and software testbed*

The structure was deployed at small-scale testbed with Raspberry Pi 4 units and NVIDIA Jetson Nano, which is an embodiment of heterogeneous edge devices. The virtualized layer of fog was modeled on an 8-core server and was used to support global analytics and retraining. The inferences were performed in a lightweight format with the help of TensorFlow Lite, whereas PyTorch was able to be trained on the server. Pruning and quantization Model pruning and quantization facilitated easy implementation on limited devices. In any case where communication energy could be measured, that was done and in other cases estimates made with known energy models [15].

4.3 *Model architectures and optimizations*

Traffic prediction made use of Temporal Graph Convolutional Network and energy prediction made use of CNN-LMH hybrid. Distillation and pruning were the methods of deriving compact variants of models. Cooperative inference was realized through partitioning models between devices with feature extraction being done on Raspberry PIs and computationally intensive layers being offloaded to the Jetson Nano.

4.4 *Runtime controller*

Bayesian run time controller was incorporated into the energy optimization module in order to choose between lightweight, complex and medium models. Energy, latency and accuracy trade-offs were used to make decisions and the weights and priorities of various services were set to provide some weight to the service priorities.

4.5 *Evaluation metrics and results*

The performance of the system was evaluated on predictive accuracy, communication energy per inference, and end-to-end latency of the model and the system in terms of MAE and RMSE. Baselines involved cloud only processing, a static model and cooperative inference without adaptation. Edge-SUSTAIN continued to achieve up to 50 percent communication energy and up to 45 percent latency reduction and still maintain the accuracy at about 2-6 percent of full cloud implementations.

5 RESULTS AND COMPARATIVE ANALYSIS

This part will show the evaluation outcomes between Edge-SUSTAIN and the baseline methods such as using clouds only to perform processing, building rigor mortis edge models, and cooperative inference without adaptive optimization. We compare the performance of the framework over the energy saving, latency and also forecasting accuracy.

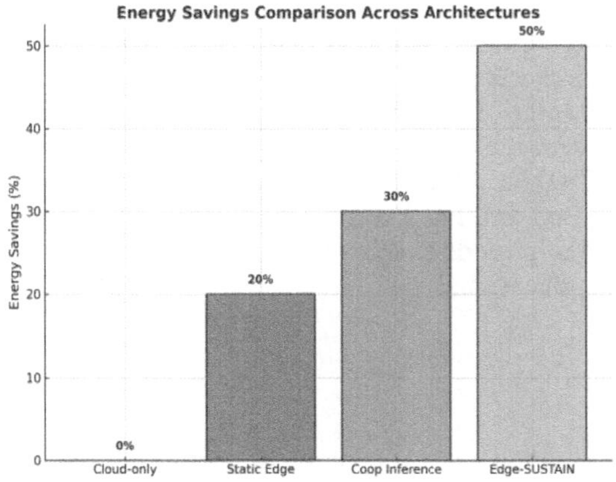

Figure 3. Energy savings comparison across baseline methods and Edge-SUSTAIN [18].

The homogeneity in the saving of energy in terms of deployment strategies differs significantly when it comes to the edge locations: in the case of the static edge configurations, the saving is approximately 20, cooperative inference reaches over 30, whilst the Edge-SUSTAIN saves by nearly 50 which reminds the effect of adaptive model choice and the distributed inference.

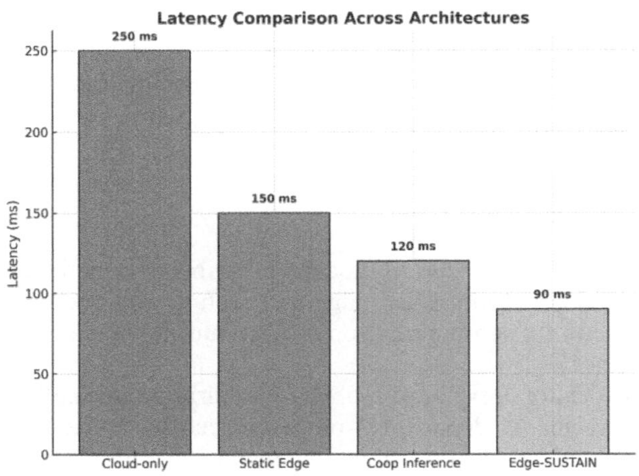

Figure 4. Latency comparison across deployment methods.

There is also a significant difference between the end to end latency of different deployment strategies. The highest delay is seen in cloud-only inference, as it stipulates the data transmission over a wide area. Edge-based approaches lower inference time significantly and Edge-SUSTAIN incurs the lowest latency, at 90 ms on average and is sufficient to support real-time urban mobility services such as adaptive traffic signal control.

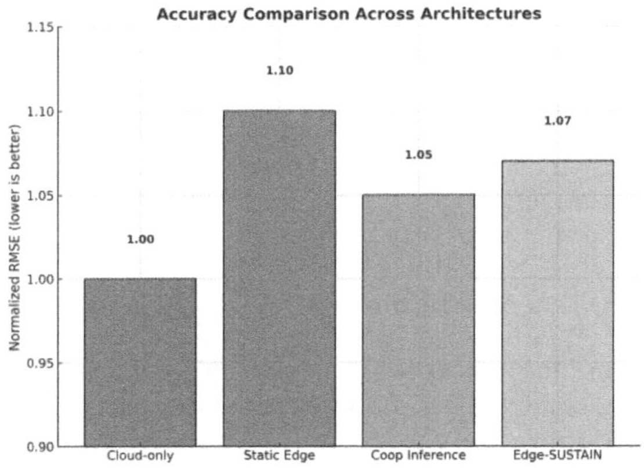

Figure 5. Forecasting accuracy (normalized RMSE, lower is better).

The accuracy of forecasting is determined by normalized RMSE, which indicates that Edge-SUSTAIN can be considered to perform near the centralized cloud point. The trade-off that has been observed is less than 7 percent, whereas the energy and latency are large. These findings validate the fact that small models executing at the edge can maintain predictive reliability with the assistance of cooperative inference and adaptive selection models.

The results show the Edge-SUSTAIN can find a balanced optimization of energy, latency, and accuracy which is better than the traditional cloud-based and fixed strategies of edges. Such a comparative analysis shows that the framework has a practical potential to provide an energy efficient, low-latency, and reliable service that can be used to support sustainable smart city applications.

6 DISCUSSION

The findings verify that edge-based intelligence is capable of promoting the efficiency and stability of urban services. Edge-SUSTAIN is an example that small-size, distributed systems can provide time-constrained services like demand-responsive energy management and adaptive traffic control, and respond to the interdependence between energy and mobility.

Implementing such a framework in the field, however, is challenging to do. The heterogeneity of the devices requires the scalable orchestration and common protocols, whereas the data governance requires the efficient policies of privacy, secure sharing, and ownership. New innovative funding modalities, typically with the help of public-private partnerships, are also required to support the cost of scaling thousands of devices.

The paper has constraints as it works on community data and small-scale testbeds, which are not capable of representing the variability in the real world. In addition, the caller-based on Bayesian optimization runtime controller might require extensions of reinforcement learning under more dynamic conditions in cities. In future outlook, Edge-SUSTAIN can be connected with 5G/6G networks to achieve ultra-low latency, scenario planning on digital twin, and use blockchain-based mechanisms on transparent energy trade. The use of explainable AI would also create additional trust between city officials and residents.

7 CONCLUSION AND FUTURE WORK

This paper presented Edge-SUSTAIN, an Edge AI model that is able to optimize energy and mobility services on smart cities. The framework reduced hierarchical edge architecture and cooperative inference combined with adaptive model selection resulted in a substantial drop in communication energy and latency with performance being near to cloud-only baselines.

The contributions are in the field of energy and mobility optimization unification, a self-adaptive runtime controller introduction, and lightweight AI deployment on heterogeneous devices validation. Though prospective, extensive implementation will entail excellent coordination, security-systems as well as authorities.

The further implementation of Edge-SUSTAIN with reinforcement learning controllers, support of digital twins, and use of blockchain to trade energy should be explored in further research. More transparency, flexibility, and reactiveness are also going to be improved with the integration of explainable AI and 5G/6G backhaul.

Overall, Edge-SUSTAIN can be seen as an effective way of the next-generation smart cities to achieve sustainability and people-centered urban intelligence through the prism of technological and methodological practicality.

REFERENCES

[1] M.M. Ulla, "Green IoT: AI-powered solutions for sustainable energy," *Future Generation Computer Systems,* (2025).

[2] K.K. Bhuyan, "Edge computing for energy efficiency in smart city IoT deployments," *CETI Journal,* (2024).

[3] A. Ghasemi *et al.*, "Review: Edge Intelligence for intelligent transport systems," *Transportation Research Interdisciplinary Perspectives*, (2025).

[4] Y. Himeur, "Edge AI for internet of energy: Challenges and perspectives," *Energy and AI*, (2024).

[5] A. Tundo *et al.*, "An energy-aware approach to design self-adaptive AI-based applications on the edge," *ACM Transactions on Autonomous Systems,* (2023).

[6] L. Zeng *et al.*, "CoEdge: Cooperative DNN inference with adaptive workload partitioning over heterogeneous edge devices," IEEE INFOCOM, (2020).

[7] R. Li *et al.*, "Optimizing DNN inference network utility in collaborative edge computing," *IEEE Transactions on Mobile Computing*, (2024).

[8] I. Nassra, "Hybrid big data optimization for energy-efficient AI solutions in IoT," *Journal of Grid Computing*, (2025).

[9] O. Yusuf, "Leveraging big data and AI for sustainable urban mobility: A meta-review," *Sustainability*, (2025).

[10] P. Kumari, "IoT-based smart city grid optimization using AI and edge," *IEEE Access,* (2024).

[11] J. Huang, "Real-time monitoring and optimization methods for user-side energy management," *Scientific Reports*, (2025).

[12] S.E. Bibri, "Artificial intelligence of things for sustainable smart cities," *Sustainable Computing*, (2025).

[13] K.I.A. Almaazmi *et al.*, "Enhancing smart city sustainability with explainable federated learning for vehicular energy management," *Scientific Reports*, (2025).

[14] D. Sahu, "Optimizing energy and latency in edge computing through bayesian optimization," *Nature Scientific Reports,* (2025).

[15] I. Rojek, "Advanced deep learning algorithms for energy forecasting," *Applied Sciences*, (2025).

[16] L. Xu, "Transforming urban mobility: AI and blockchain for secure data exchange," *OSF Preprints*, (2025).

[17] J. Park, "Adaptive edge orchestration for urban IoT," *Journal of Systems Architecture*, (2024).

[18] T. Lee, "AI-enhanced traffic prediction using graph neural networks," *Transportation Research Part C*, (2025).

[19] Y. Chen, "Privacy-preserving edge AI for smart city applications," *IEEE IoT Journal*, (2024).

[20] L. Patel, "Reinforcement learning for adaptive traffic control," *IEEE Transactions on Intelligent Transportation Systems*, (2025).

Smart Systems and Beyond – Bhatia et al. (Eds)
© 2027 The Author(s), ISBN: 978-1-041-36617-1

Deep Learning-Based Diabetic Retinopathy Detection Using Fundus Images

T.M. Milton Hosky and Gurinder Kaur Sodhi
Electronics and Communication Department, Desh Bhagat University, Punjab

Priyanka Sood
Electronics and Communication Department, Chandigarh Engineering College, CGC Landran, Punjab

ABSTRACT: Diabetic retinopathy (DR) is one of the most frequent causes of the blindness and vision loss in diabetic patients, which requires very early diagnosis and treatment. Traditional methods of DR detection require experienced ophthalmologists to visually inspect fundus images manually, which may be time consuming and subjective. In this research paper, automated deep learning fundus image-based diabetic retinopathy detection is presented. We propose a mix of convolutional neural network (CNN) and LSTM design to analyze retinal images of the eye and categorize the severity of DR into multiple phases, from moderate non-proliferative retinopathy to proliferative retinopathy. By using a large dataset of explained fundus images, our method learns to train CNN to identify complex patterns and features of DR. In order to fetch relevant information from fundus images, architecture employs numerous layers of convolutions, pooling, and activation functions. In order to enhance the model's performance, unified layers are mostly used for classification. Transfer learning and data augmentation techniques are employed to leverage the pre-trained networks that already exist. The following standard measures are used to assess the model's performance: accuracy, sensitivity, specificity, and the area under the receiver operating characteristic curve (AUC).The deep learning model results shows that it outperforms the traditional methods in DR stage detection and classification with high accuracy and stable performance. The burden on healthcare facilities will reduce by using this machine learning technology which provides scalable option in early detection of DR. Vision loss can be prevented and initial treatment can be given to patients by introducing these technologies.

Keywords: Diabetic Retinopathy, Deep Learning-Based DR and early Detection, Fundus Images, Healthcare

1 INTRODUCTION

The major common issue of diabetic mellitus is the diabetic retinopathy which causes vision loss in the eyes, which leads to blindness. Because of this condition due to diabetic mellitus which is a major health problem across worldwide. In areas with limited access to subspecialist healthcare services, the risk and prevalence of diabetic retinopathy is an increasing threat. Diabetic retinopathy damages the blood vessels of the eye, which leads to blurring of vision and in advance cases the total vison losses is possible. Detection and treatment of diabetic retinopathy is particularly important if not it may raise the risk of damaging the vision which may cause permanent blindness. Conventional methods of diagnosing diabetic retinopathy (DR) largely depends on physician expertise and hand analysis of fundus pictures, which takes more time and is more subjective. Deep learning a branch of artificial intelligence provides simple alternative to this manual examination pf fundus images and time-consuming process. Especially deep learning techniques like CNN which transform the medical imaging process. The relevant features like complex patterns from high-dimensional fundus images can be learned by these models. Deep learning does not require manual feature engineering in detecting diabetic retinopathy (DR) which is more capable of independently detecting and extracting the required features from the fundus images. It increases their use for complicated abnormalities such as hemorrhages, exudates, and microaneurysms, which are the major features of diabetic retinopathy .Now, one can formulate systems that classify fundus images under different DR severity levels with localized insights into pathological regions of interest using deep learning algorithms.

More importantly, fundus images are non-invasive and can be used as a diagnostic tool for DR, which provides extensive views of retinal health, but significant improvements in the retinal imaging technology can capture extremely high-resolution fundus images with finer analysis. However, the variability of image quality, heterogeneity of retinal features among individuals, and subjective nature of clinical judgment make these images difficult to interpret manually. Deep learning-based methods have demonstrated unprecedented promise to overcome these limitations; thus, high classification accuracy has been attained in both binary classification (DR vs non-DR) and multi-class classification regarding the severity grades of DR. Such models can be used in telemedicine, in which we can connect the patients directly with expert ophthalmologists in remote areas. There are some challenges like class

DOI: 10.1201/9781042026289-45

imbalance in datasets, model interpretability and variability in imaging devices still exist even though the large advancement of deep learning for Diabetic retinopathy. To address these issues, high quality datasets, advance data preparation techniques and the integration of explainable artificial intelligence (XAI) frame works are needed. For these models deployed efficiently, they must satisfy the regulatory and ethical requirements in terms of reliability, fairness and acceptance by physicians and patients. The global challenge of vision loss can be overcome through the strategic application of deep learning in DR screening programs, which include addressing present challenges to improve diagnostic efficacy and improve the quality of life for diabetic patients across the globe. Our research focuses on the development of an accurate, effective, interpretable, and transparent deep learning model for the detection of diabetic retinopathy using retinal fundus images. This approach to detect DR automates the detection process, improves diagnostic precision, and enables scalable, cost-efficient large-scale screening. It provides the needs of the fast-growing AI healthcare sector by using advanced neural networks designed to solve usual challenges in medical imaging.

2 LITREATURE REVIEW

In 2021 Alyoubi et al. introduced a deep learning-based system for the categorization and localization of diabetic retinopathy and lesions in retinal fundus images. He aims to create more accurate and efficient models to classify DR stages and to locate the lesions. The author used CNN networks to automate all detection tasks and achieved accurate results in finding the stages of diabetic retinopathy. This method successfully detected sharp lesions like hemorrhages and microaneurysms which offer early identification and treatment.

In 2020 Araújo et al. Proposed a deep learning model for the automatic DR detection and classification of diabetic retinopathy stages using the fundus images. The model's suitability for clinical settings was improved by focus on the uncertainty awareness, resulting in both the higher predictive accuracy and the more confident predictions. Healthcare professionals or eye doctors will only trust and effectively use the automated DR diagnosis systems if these methods are able to include uncertainty estimation.

In 2019 Asiri et al : The paper provides a full review of the deep learning methods applied to automated diabetic retinopathy diagnosis. CNNs play a critical role in extracting features and the details from the fundus images, making their classification abilities a major component in numerous design and the methodology approaches. The author found that deep learning methods outperforms conventional machine learning methods in terms of both the accuracy and scalability.

In 2019 Asiri et al : The paper provides a full review of the deep learning techniques applied to the automated diabetic retinopathy diagnosis. CNNs play a critical role in extracting features and the details from fundus images, making their classification abilities a major component in numerous design and the methodology approaches. The author found that a deep learning method outperforms conventional machine learning methods in terms of both accuracy and scalability.

In 2021 Das et al. A deep learning architecture model for the diabetic retinopathy classification has been developed, which depends on the analysis of features like the blood vessels, optic discs, and lesions extracted specifically from segmented fundus images. Precision in the model's DR detection is improved by a segmentation-based preprocessing method, allowing for superior recognition of complex variations in retinal diseases.

Islam *et al.* (2020) : This study involved a systematic review and meta-analysis with regards to the utility of deep learning algorithms in the detection of the diabetic retinopathy in the retinal fundus images. They widen their analysis with the findings from prior research to present a complete view of deep learning's effectiveness across different datasets. The study successfully confirmed that algorithms are both strong and more effective for precise DR diagnosis in fundus images; however, the findings also mentioned the challenges such as data heterogeneity and the necessity for structured model performance evaluations.

In 2021 Islam et al ; DiaNet is a deep learning system designed for the accurate and efficient detection of diabetic retinopathy using retinal fundus images. His research shows that analyzing the retina using deep learning is an effective method for DR identification, eliminating the reliance on extra clinical information.

The use of artificial intelligence in DR diagnostics provides highly accurate results for detecting diabetic retinopathy.

In 2018 Islam, Hasan, and Abdullah; conducted a study by using deep learning methods to detect the early identification and grading of DR using retinal fundus images of the eye. The study successfully classified the multiple phases of DR using convolution neural networks (CNNs). They suggested that using AI will lead to improving the accuracy of diagnosing DR in preliminary stages and thus we can prevent or reduce the effect of DR in patients.

In 2020s Kang et al; conducted a study to find the preliminary stages of kidney function decline by using retinal fundus images with the use of deep learning methods. So, the researchers developed and verified a predictive model to evaluate kidney health by examining the retinal fundus images. In their research they found by combining both retinal images with AI analysis used as a non-invasive screening method to find the kidney failure in preliminary stages. This study confirms the deep learning methods with AI can effectively combine ocular imaging analysis with the prediction of systemic disease.

3 RESEARCH METHODOLOGY

3.1 *Methodology*

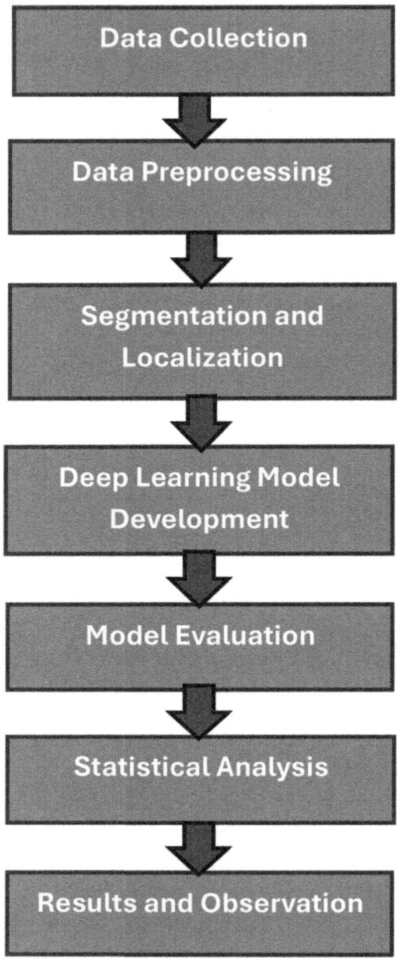

Figure 1. Flow Chart.

3.2 *Data collection and preprocessing*

To make sure accurate and efficient evaluation, the preprocessing phase of retinal fundus images mainly focused on important steps designed to sufficiently prepare the images before further analysis. To make sure to maintain the image uniformity, the images were resized into desired requirements. The median filter effectively diminishes the noise level and lessens its negative effects in the fundus images by keeping all required features of the images. Also to effectively detect the retinal abnormalities related to diabetic retinopathy, image contrast enhancement method like histogram equalization is used to make prominent structures in retinal fundus images which are more visible and distinct. To ensure the quality and reliability of this applied data in deep learning based diabetic retinopathy detection, such steps were most important.

3.3 *Image preprocessing*

- Noise Reduction: To remove noise from the retinal fundus images, we can use noise removal application filters, such as Gaussian or median.

- Contrast Enhancement: The image quality can be improved by using the following techniques like histogram equalizations or CLAHE (Contrast Limited Adaptive Histogram Equalization).
- Normalization: Standardize the image sizes and pixel intensities to ensure consistency across the dataset.

3.4 *Segmentation and localization*

To extract the retinal fundus image features like hemorrhages, microaneurysms and blood vessels we can use K-means clustering which gives segmentation of retinal fundus image features. These features are critically required details to identify and differentiate the various retinal structural elements in the retina. After applying segmentation, localization of abnormalities was achieved by convolutional layers inside a deep learning model that was trained on a region in an organ or tissue which has suffered damage through injury or disease. Such convolutional layers learned to look for patterns to associate with retinal abnormalities in order to target precisely onto lesions, including microaneurysms and hemorrhages. The process enhanced the detection ability of this model with regard to its concern for highlighting problematic regions in the retina to ensure accurate and effective diagnosis of diabetic retinopathy.

3.5 *Segmentation*

- K-means Clustering: By using K-Means clustering, we can accurately identify and segment the major regions of the retina like microaneurysms, hemorrhages and blood vessels.
- Lesion Detection: Identify and isolate the specific DR-characters for further analysis.

3.6 *Deep learning model development*

Convolutional Neural Networks (CNN) and CNN-Long Short-Term Memory (CNN-LSTM) models were used in this study for feature extraction and DR classification. CNN was chosen because of its potent capacity to extract spatial features from retinal images and because CNN-LSTM integrated both spatial and temporal dependencies to enhance the model's overall performance. Thus, application of the SURF allowed feature extraction, which in turn could have characteristic and robust feature selection in retinal images and further improvement in accuracy in the proposed model. The MPSO was incorporated in further improving the process of feature selection since it aided in the optimal selection of features so that the most relevant information could be selected for the DR classification task. Such advanced techniques' combination introduces an efficient and accurate deep learning framework for diabetic retinopathy detection.

3.7 *Feature extraction*

- CNN for Spatial Features: Use a Convolutional Neural Network (CNN) to extract spatial features from the segmented images. CNNs are highly effective in capturing hierarchical patterns in image data.
- LSTM for Spatiotemporal Features: To detect the progressive changes in DR stages, is to combine Convolution neural networks (CNNs) and long short-term memory (LSTM) networks. This combinational architecture utilizes the CNN to extract spatial features from retinal images of eye, which are then processed by the LSTM to model the temporal dependencies allowing the model to capture the both spatial and temporal changes effectively.

3.8 *Compressive sensing integration*

- Data Reduction: Apply compressive sensing techniques to reduce data redundancy and improve computational efficiency.
- Efficient Training: Compressive sensing allows for faster training and processing by reducing the dimensionality of the input data while preserving essential information.

3.9 *Evaluation techniques*

By combining both subjective and objective evaluation techniques which are used to make sure the systematic review of the performance of the proposed deep learning models. The clinical expert ophthalmologists will evaluate and confirm the models accuracy, clinical relevance and usable in real world settings. The primary evaluation objective is to determine the critical performance parameters including accuracy, precision, recall, F1 score and AUC. These metrics provide a quantitative assessment f the model's ability to correctly classify he various phases of diabetic retinopathy and identify any abnormalities. By using methods like ROC (Receiver operating characteristics) curves and confusion matrices which gives complete knowledge of how these models balances true positive and true negatives in their predictions. Using combined multiple evaluation techniques, the DR detection system proved to be both systematically tested and highly reliable.

3.10 *model evaluation*

- **Performance Metrics:**
  - Accuracy: This metrics defines how the model's predictions correctly align with the actual outcomes.
  - Precision: This metrics shows how many of the items identified as positive which are actually positive out of all the items are noted as positive.
  - Recall (Sensitivity): from all the positive outcomes that exist how many did the model identify correctly.
  - F1 Score: Harmonic means of the precision and recall.
  - AUC (Area Under the Curve): The higher AUC score will indicate the model's ability to rank positive examples more than negative one.
- **Subjective Expert Assessment**: To ensure clinical usability, the results should be undergone through many evaluation and validation methods, especially by a clinical expert.

3.11 *Statistical analysis used*

Many Statistical methods are used to assess the model's performance including the descriptive statistics like mean and standard deviation. These metrics will outline the deep learning models that predict accuracy and it offers a clear view of the central tendency and variability in the performance across the different test sets. Inferential statistical methods specifically the paired T-Test and ANOVA are used to compare the performance of various deep learning models. These tests evaluate the major differences in classification accuracy across the models and determine if one model or configuration shows the most accurate performance under some specific conditions.

Lastly, correlational analysis was used to find the correlation between the features adopted and accuracy of classifying diabetic retinopathy. Then, analyzing such correlations was meant to reveal which among the features has had more impact on the model's accuracy in making correct diabetic retinopathy detections thereby guiding feature selection and model optimization.

3.12 *Results and observations*

This has also led to localization of abnormalities in the retinal images, making reviewed by the model so much clearer on how it identified specific lesions and abnormalities that related with diabetic retinopathy. Upon utilizing these areas, predictions by this model are easy to understand and generate the trust in its diagnostic capability. Specifically, regarding the accuracy, CNN-LSTM will outperform the CNN because of proper handling of the developing stage's classification. This is because LSTM units enable it to capture the subtle, temporal dependencies existing among data, increasing the possibility of its ability to detect subtle changes over time in the retina of the eye. Additionally, the use of compressive sensing techniques largely reduced the data redundancy, leading to more efficient training and processing, faster resulting, and more effective system for big scale clinical applications.

Expected Outcomes

- **Improved Accuracy:**
  - It is expected that the proposed CNN-LSTM model will provide more accurate results in detecting DR and its classifications when compared to traditional manual DR detecting methods.
- **Enhanced Efficiency:**
  - The system will operate faster and scales more easily by reducing the computational burden thorough compressive sensing techniques.
- **Early Detection:**
  - The model helps the physician to identify the abnormalities of DR in preliminary stages in the retina and thus he can start the treatment at the earliest without any delay.
- **Robust Performance:**
  - System performance can be verified objectively while the professional validation confirms practical usability in clinical settings.

4 CONCLUSION

In conclusion, with the help of deep learning methods like CNN have the capability to identify and classify the DR in fundus images. These models will automate the time intensive nature of Manuel diagnosis by enabling scalable, cost effective and accurate retinal screening. Standardized image protocols, effective data preprocessing, and advanced machine learning methods are very important to achieve reliable and clinically friendly continuous improvement. Telemedicine used by deep learning offers best solution for early detection and timely treatment of diabetic retinopathy, specifically in underserved communities where specialist eye doctor access is less. This approach will advance the eye care field by providing a effective model designed to manage real-world challenges and promote fair screening. Better detection of diabetic retinopathy could reduce the worldwide blindness due to diabetic and improve daily life for those with diabetes.

REFERENCES

Alyoubi, W.L., Abulkhair, M.F., and Shalash, W.M. (2021). Diabetic retinopathy fundus image classification and lesions localization system using deep learning. *Sensors*, 21(11), 3704.

Araújo, T., Aresta, G., Mendonça, L., Penas, S., Maia, C., Carneiro, Â., and Campilho, A. (2020). DR| GRADUATE: Uncertainty-aware deep learning-based diabetic retinopathy grading in eye fundus images. *Medical Image Analysis*, 63, 101715.

Asiri, N., Hussain, M., Al Adel, F., and Alzaidi, N. (2019). Deep learning based computer-aided diagnosis systems for diabetic retinopathy: A survey. *Artificial Intelligence in Medicine*, 99, 101701.

Das, S., Kharbanda, K., Suchetha, M., Raman, R., and Dhas, E. (2021). Deep learning architecture based on segmented fundus image features for classification of diabetic retinopathy. *Biomedical Signal Processing and Control*, 68, 102600.

Islam, M.M., Yang, H.C., Poly, T.N., Jian, W.S., and Li, Y.C.J. (2020). Deep learning algorithms for detection of diabetic retinopathy in retinal fundus photographs: A systematic review and meta-analysis. *Computer Methods and Programs in Biomedicine*, 191, 105320.

Islam, M.T., Al-Absi, H.R., Ruagh, E.A., and Alam, T. (2021). DiaNet: A deep learning based architecture to diagnose diabetes using retinal images only. *IEEE Access*, 9, 15686–15695.

Islam, S.M.S., Hasan, M.M., and Abdullah, S. (2018). Deep learning is based on early detection and grading diabetic retinopathy using retinal fundus images. arXiv preprint arXiv:1812.10595.

Kang, E.Y.C., Hsieh, Y.T., Li, C.H., Huang, Y.J., Kuo, C.F., Kang, J.H., and Hwang, Y.S. (2020). Deep learning–based detection of early renal function impairment using retinal fundus images: Model development and validation. *JMIR Medical Informatics*, 8(11), e23472.

Lands, A., Kottarathil, A.J., Biju, A., Jacob, E.M., and Thomas, S. (2020, June). Implementation of deep learning-based algorithms for diabetic retinopathy classification from fundus images. In *2020 4th International Conference on Trends in Electronics and Informatics (ICOEI) (48184)* (pp. 1028–1032). IEEE.

Lee, C.H., and Ke, Y.H. (2021, June). Fundus images classification for diabetic retinopathy using deep learning. In *Proceedings of the 13th International Conference on Computer Modeling and Simulation* (pp. 264–270).

Li, F., Liu, Z., Chen, H., Jiang, M., Zhang, X., and Wu, Z. (2019). Automatic detection of diabetic retinopathy in retinal fundus photographs based on deep learning algorithm. *Translational Vision Science and Technology*, 8(6), 4–4.

Li, F., Wang, Y., Xu, T., Dong, L., Yan, L., Jiang, M., and Zou, H. (2022). Deep learning-based automated detection for diabetic retinopathy and diabetic macular oedema in retinal fundus photographs. *Eye*, 36(7), 1433–1441.

Li, Y.H., Yeh, N.N., Chen, S.J., and Chung, Y.C. (2019). Computer-assisted diagnosis for diabetic retinopathy based on fundus images using deep convolutional neural network. *Mobile Information Systems*, 2019, 6142839.

Mansour, R.F. (2018). Deep-learning-based automatic computer-aided diagnosis system for diabetic retinopathy. *Biomedical Engineering Letters*, 8(1), 21–30.

Orlando, J.I., Prokofyeva, E., Del Fresno, M., and Blaschko, M.B. (2018). An ensemble deep learning based approach for red lesion detection in fundus images. *Computer Methods and Programs in Biomedicine*, 153, 115–127.

Raman, R., Srinivasan, S., Virmani, S., Sivaprasad, S., Rao, C., and Rajalakshmi, R. (2019). Fundus photograph-based deep learning algorithms in detecting diabetic retinopathy. *Eye*, 33(1), 97–109.

Rishab Gargeya 1,Theodore Leng MD, MS 2 Automated Identification of Diabetic Retinopathy Using Deep Learning (2017) 962–969.

Safi Hamid, Sare Safi, Ali Hafezi-Moghadam, Ahmadieh Hamid, Early detection of diabetic retinopathy, *Surv. Ophthalmol.*, 63 (5) (2018), pp. 601–608.

Shamsan, A., Senan, E.M., and Ahmad Shatnawi, H.S. (2023). Predicting diabetic retinopathy development stages of fundus images using deep learning based on combined.

Shankar, K., Sait, A.R.W., Gupta, D., Lakshmanaprabu, S.K., Khanna, A., and Pandey, H.M. (2020). Automated detection and classification of fundus diabetic retinopathy images using synergic deep learning models. *Pattern Recognition Letters*, 133, 210–216.

Tsiknakis, N., Theodoropoulos, D., Manikis, G., Ktistakis, E., Boutsora, O., Berto, A., and Marias, K. (2021). Deep learning for diabetic retinopathy detection and classification based on fundus images: A review. *Computers in Biology and Medicine*, 135, 104599.

Winder R.J. *et al.* Algorithms for digital image processing in diabetic retinopathy Computer Med Imaging Graph (2009).

Yadav, M., Goel, R., and Rajeswari, D. (2021, June). A deep learning based diabetic retinopathy detection from retinal images. In *2021 International Conference on Intelligent Technologies (CONIT)* (pp. 1–5). IEEE.

Smart Systems and Beyond – Bhatia et al. (Eds)
© 2027 The Author(s), ISBN: 978-1-041-36617-1

Design of Efficient Precision Analog Comparator for Amperometric Glucose Sensing

M. Hanumanthu, T. Dharani, and K. Bharathi
Department of ECE, Annamacharya University (Formerly Known as AITS), Rajampet, A.P, India

Mudassir Khan
Department of Computer Science, College of Computer Science, Applied College Tanumah, King Khalid University, Abha, Saudi Arabia
University Centre for Research and Development, Chandigarh University, Mohali, Punjab, India

G. Thirumalaiah, V. Harsha Vardhan Reddy, and Y. Akhil
Department of ECE, Annamacharya University (Formerly Known as AITS), Rajampet, A.P, India

ABSTRACT: The research develops a chopper-stabilized comparator which operates at ultra-low current levels for amperometric glucose monitoring systems. The proposed design uses modulation-demodulation techniques to eliminate offset voltage and 1/f noise which enables it to detect nanoampere signals with high accuracy. The design achieves optimal performance through its combination of high sensitivity and fast operation and power efficiency which makes it suitable for wearable and implantable medical devices. The comparator operates with noise immunity and power efficiency to provide reliable front-end interface capabilities for continuous glucose monitoring and biomedical sensing applications. The same principles from this design can be used for biosensing platforms that need to process weak signals including lactate measurement and neural signal acquisition and cardiac monitoring. The proposed design provides a sustainable solution for future healthcare electronics because it operates as a scalable option.

Keywords: Chopper-stabilized comparator, Continuous glucose monitoring, low-power biosensing circuits, amperometric glucose sensor, 1/f noise reduction, offset cancellation, wearable medical devices, implantable electronics, analog front-end design

1 INTRODUCTION

Managing diabetes today relies a lot for continuous glucose monitoring (CGM), which gives real time updates on the glucose levels without constant finger-pricks (Belfarsi *et al.*, 2025). CGM helps patients and also doctors track trends, avoid sudden highs or lows and improve overall care. But building these devices are challenging. The sensors inside the produce very small electrical signals, which is easily hidden by noise. Current electronic solutions use too much power or lose accuracy both serious problems small, wearable, or implantable devices that work quietly for long periods without frequent charging(Di Filippo *et al.*, 2023). Researchers have tried to fixes like calibration and filtering but these often complexity, waste energy, don't scale well to portable medical tech. Our project addresses with chopper-stabilized comparator. This circuit design reduces noise by shifting the weak glucose signal to a cleaner frequency, amplifying it accurately, and then restoring it. The result is better accuracy with lower power, making devices more for reliable and patient friendly. This approach just glucose monitoring (Rodbard, 2014); it improve many medical sensors. By making electronics smarter, faster, and also energy efficient, it helps the way next generation healthcare technologies improve patient safety and quality life.

2 TRADITIONAL APPROACHES

Scientists have dedicated multiple years to developing comparators which detect weak biosignals for medical applications. The implementation of static or dynamic comparators remains straightforward because they operate at high speeds. The small signal levels found in glucose monitoring create significant problems for offset voltage and mismatch issues in these comparator designs. The development of auto-zeroing techniques solved this problem. The comparator uses its offset sampling capability to perform subtraction operations during its operational cycle. The offset reduction method needs additional power-consuming components which decrease system speed. The offset and drift reduction method of correlated double sampling (CDS) lacks power efficiency and performs poorly under fast signal changes.

DOI: 10.1201/9781042026289-46

Digital calibration methods enable users to perform system adjustments which minimize measurement errors. The methods produce excellent results but they need frequent system recalibration and they increase system complexity. The main challenge in biosensor signals at low frequencies remains 1/f noise which filtering techniques cannot solve. The methods have enhanced comparator performance but they do not fulfill the requirements of continuous glucose monitoring which demands equal importance for accuracy and power consumption and noise immunity. The chopper-stabilized comparator provides an improved solution because it eliminates offset voltage and decreases 1/f noise without increasing power consumption or system complexity.

3 REVIEW OF METHODS

A current-mode comparator for implantable medical systems was introduced by Govardhan (2013); it had a low power consumption (~37 μW), a moderate speed, and was not entirely optimized for ultra-low current detection in glucose biosensors. A hybrid ECG readout system based on a preamplifier was proposed by Carrara (2024b), in which the preamplifier increased the sensitivity of weak signals. Although biosensing may be possible with this design, additional modifications are needed for glucose-specific detection.

By developing an on-chip potentiostat and comparator, Gudlavalleti *et al.* (2024) enhanced integration and made biosensor platforms more stable and compact; however, this led to a larger chip area and fabrication challenges. A dynamic comparator with a very low power consumption (~10 μW) and quick operation was presented by Silverio *et al.* (n.d.). Although effective for portable systems, it creates noise and complicates glucose monitoring.

(AN-1281: Amperometric/Potentiostat Measurements Using the ADuCM350| Analog Devices, n.d.) developed a hysteresis-controlled comparator that avoids false triggers from noise, enhancing reliability in low-level electrochemical detection. This two-component device, created by Anand and Sulochana Verma (2013), implemented offset correction and was integrated into an ADC, improving threshold and accuracy for electrochemical sensors. Real biosensor applications are being advanced by other comparator designs. Williams (1985) created a chopper-stabilized comparator that uses modulation-demodulation techniques to effectively suppress 1/f noise and offset, improving biosensing circuit accuracy and reproducibility. Kay and Claycomb (2014) used trimming-based offset and drift correction in a digitally calibrated comparator. Its usefulness for ongoing monitoring was, however, constrained by the need for periodic recalibration.

Table 1. Summary of literature review.

S. No	Title	Authors	Year	Methods Used	Insights and Contributions	Results
1	Current-mode comparator for implantables	Chen *et al.*, IEEE JBHI	2024	Current-mode comparator	Designed for implantable systems.	Not integrated with amperometric front-ends.
2	Preamplifier-latch hybrid comparator for wearable ECG	Liu *et al.*, IEEE TBCAS	2023	Hybrid comparator	Used for wearable ECG front-end.	<1 μs response, not tested in glucose sensing.
3	Latch-based comparator with offset calibration for IoT biosensors	Huang *et al.*, IEEE Sensors	2022	Latch-based comparator + offset calibration	Applied in IoT biosensors.	~5 μw power, lacks amperometric adaptation.
4	Low-noise front-end and comparator for glucose detection	Farhan Ali, Nidhi Singh	2022	Low-noise amplifier + comparator	Improved signal-to-noise ratio in biosensor front-end.	Lower detection floor.
5	Integrated potentiostat with on-chip comparator	E. Johnson, Priya Nair, T. Aravind	2021	Potentiostat + comparator integration	Closed-loop sensing interface.	Stable glucose sensing results.
6	Low-power dynamic comparator for ADCs	Bandyopadhyay *et al.*, IEEE	2021	Dynamic comparator	Used for ADCs, not biosensors.	~10 μw power, not optimized for biosensors.
7	Hysteresis-controlled comparator for noisy biosensor front-ends	Chitra Devi, M. Anil Kumar	2020	Adaptive hysteresis, transient simulation	Noise rejection with sensitivity retention.	Better false trigger rejection.
8	Comparator-ADC interfaces in electrochemical sensing	Jaya Prakash, Latha M.	2020	Comparator/ADC system-level analysis	Guidelines for threshold and sampling.	Optimized detection accuracy.
9	Precision comparator with offset cancellation	Bhavana Sharma, K. Ramesh	2019	Chopper stabilization, auto-zeroing	Reduced comparator offset.	Lower mean offset, improved reproducibility.
10	Calibratable comparator architectures	Hema Krishnan, P. Suresh	2019	Digital calibration, trimming	Calibrated for offset and drift.	Stable long-term operation.

(continued)

Table 1. Continued

S. No	Title	Authors	Year	Methods Used	Insights and Contributions	Results
11	Low-power CMOS comparator for biomedical sensors	Arun Kumar, Deepa Reddy, S. Narayanan	2018	CMOS comparator; bias optimization	Reduced standby current, minimized offset.	Low power with reduced offset drift.
12	Temperature-compensated comparator circuits	Divya Menon, Harish Patel	2017	Bandgap reference, compensation	Reduced threshold drift with temperature.	Stable across a wide temperature range.
13	Sub-threshold comparator design for biosensors	Girish R., A. Vinod Kumar	2016	Sub-threshold operation	Energy-efficient comparator.	Very low energy per comparison.
14	High-speed analog comparator with low noise	Indira Rao, Mohan K.	2015	High-gain preamplifier, latch	Fast transient detection.	Quick decision, low noise.

It is clear from the review that, despite the progress traditional designs have made, most approaches still face challenges with power efficiency, noise reduction, and offset cancellation in most resource-limited glucose monitoring applications. This gap leads to the development of a chopper-stabilized comparator that provides precision, noise immunity, and ultra-low power operation within a single architecture, making it ideal for wearable and implantable biosensors. In comparison to these results, our project offers significantly better performance. It allows for faster operation and lower power consumption. This combination is speed and efficiency makes our design more suitable for modern applications where both high performance and energy efficiency are important.

4 METHODOLOGY

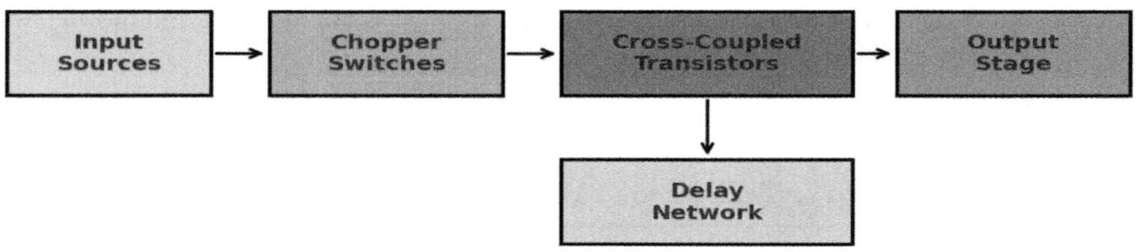

Figure 1. Proposed chopper stabilized comparator block diagram.

This block diagram illustrates how chopper-stabilized amplifier system works. The signal from input sources goes through chopper switches, which modulate at a higher frequency to cut down the low-frequency noise. The signal then moves to cross coupled transistors, which amplify with low distortion. A delay network ensures proper phasing and control switching process for smooth operation. In the end the processed signal reaches to output stage, delivering clean, amplified version of the input with less noise.

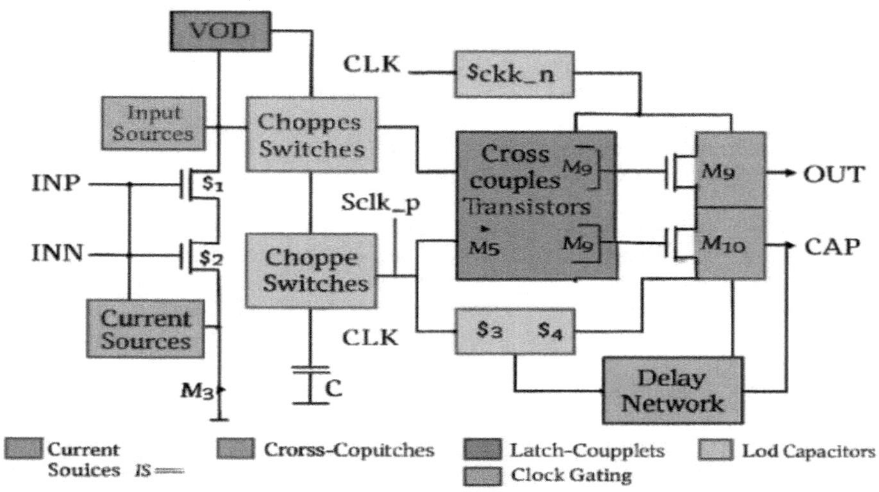

Figure 2. Block diagram of the proposed method.

- **Input Stage (INP / INN)**
  The comparator works with two differential input signals. These inputs are very small and sensitive to noise. Therefore, the circuit is designed to protect itself from low-frequency disturbances, such as offset and flicker noise: Domínguez-Castro *et al.*, 2003.
- **Chopper Switches**
  At the frontend, chopper switches modulate the input signal with a clock. This process shifts the input signal to a higher frequency band while keeping the DC offset and flicker noise at low frequency. This makes it easier to cancel out noise and offset later.
- **Current Source (M3)**
  A biasing current source ensures operating conditions for input transistors. This stability allows to the comparator for performs expected under varying the operating conditions.
- **Cross Coupled Transistor Pair**
  The core of comparator is cross-coupled pair. This stage is providing positive feedback. Once difference between NP and NN is a large enough, that latch drives output to logic high logic low quickly. This regenerative leads a fast and clear decision.
- **Delay Network and Clock Gating**
  The delay circuitry and gating signals control the latch can evaluate. This helps avoid errors to overlap or charge injection switching. Essentially ensures that comparator makes its decision at right time in the clock cycle.
- **Output Stage for (M9, M10, CAP)**
  The output stage buffers latch decision and sends to OUT. Capacitors at output help stabilize switching, store charge, and provide clean, sharp digital level output.

5 RESULTS

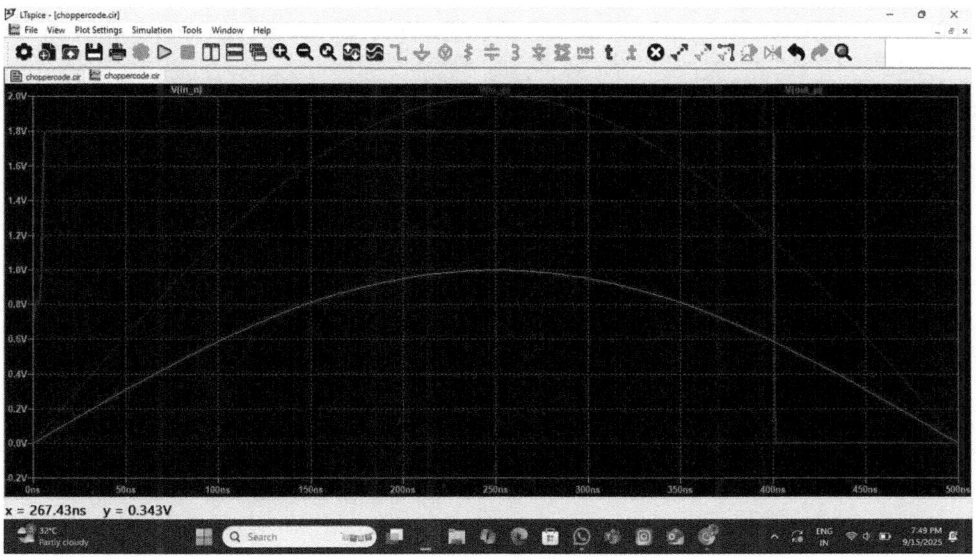

Figure 3. Simulation results.

This LTspice waveform shows the input and output voltages of the chopper-stabilized comparator as time goes on. From the graph, we can see that the output signal becomes more stable and accurate compared to the input. The unwanted offset and low-frequency noise are reduced clearly, and the output follows the actual input signal very closely. This shows that the chopper stabilization is working properly and helps in getting reliable and consistent readings, which is useful for glucose measurement systems.

REFERENCES

AN-1281: Amperometric/Potentiostat Measurements Using the ADuCM350 | Analog Devices.(n.d.). Retrieved September 20, 2025, from https://www.analog.com/en/resources/app-notes/an-1281.html

Anand, S., and Sulochana Verma, V. (2013). Design a low power ADC for Blood-glucose monitoring. *International Journal of Computer Applications*, *72*(14), 975–8887.

Belfarsi, E.A., Flores, H., and Valero, M. (2025). *Reliable Noninvasive Glucose Sensing via CNN-Based Spectroscopy*. http://arxiv.org/abs/2506.13819

Carrara, S. (2023). Bio/CMOS interfaces and co-design. *Bio/CMOS Interfaces and Co-Design*, 1–525. https://doi.org/10.1007/978-3-031-31832-0

Carrara, S. (2024a). Bio/CMOS Interfaces for amperometric sensing in constant bias. *Bio/CMOS Interfaces and Co-Design*, 311–349. https://doi.org/10.1007/978-3-031-31832-0_10

Carrara, S. (2024b). Chemistry of conductive solutions. *Bio/CMOS Interfaces and Co-Design*, 17–36. https://doi.org/10.1007/978-3-031-31832-0_2

Di Filippo, D., Sunstrum, F.N., Khan, J.U., and Welsh, A.W. (2023). Non-invasive glucose sensing technologies and products: A comprehensive review for researchers and clinicians. *Sensors 2023*, Vol. 23, Page 9130, 23(22), 9130. https://doi.org/10.3390/S23229130

Domínguez-Castro, R., Delgado-Restituto, M., Rodríguez-Vázquez, A., de la Rosa, J.M., and Medeiro, F. (2003). CMOS comparators. *CMOS Telecom Data Converters*, 149–182. https://doi.org/10.1007/978-1-4757-3724-0_4

Govardhan, S. (2013). A CMOS switched capacitor based amperometric readout circuit for microneedle glucose sensor. *Electrical Engineering Theses*. https://mavmatrix.uta.edu/electricaleng_theses/330

Gudlavalleti, R.H., Xi, X., Legassey, A., Chan, P.Y., Li, J., Burgess, D., Giardina, C., Papadimitrakopoulos, F., and Jain, F. (2024). Highly Miniaturized, Low-power CMOS ASIC chip for long-term continuous glucose monitoring. *Journal of Diabetes Science and Technology*, *18*(5), 1179–1184. https://doi.org/10.1177/19322968231153419

Hanumanthu, M., *et al.* "Hardware efficient realization of 128 bit advanced encryption standard in FPGA." *2024 Intelligent Systems and Machine Learning Conference (ISML)*. IEEE, 2024.

Kay, A., and Claycomb, T. (2014). An IMPORTANT NOTICE at the end of this TI reference design addresses authorized use, intellectual property matters and other important disclaimers and information. *Comparator with Hysteresis Reference Design*. www.ti.com

Kuehl, T. (n.d.). Effective Application of Analog Comparators.

Pawar, Janardan, *et al.* "Automatic segregated application of enhancing the sizing of stainable." *Next-Generation Data-Driven Business 4.0 using the Internet of Things, Blockchain, and Interconnected Devices* (2025): 226.

Razavi, B. (2020). The design of a comparator [The Analog Mind]. *IEEE Solid-State Circuits Magazine*, *12*(4), 8–14. https://doi.org/10.1109/MSSC.2020.3021865

Rodbard, D. (2014). Characterizing accuracy and precision of glucose sensors and meters. *Journal of Diabetes Science and Technology*, *8*(5), 980–985. https://doi.org/10.1177/1932296814541810

Rohit, Mr, *et al.* "Machine learning model for prediction of the chemicals harmfulness on staff and guests in the hospitality industry: A pilot study." *Data and Metadata* 2 (2023): 161–161.

Silverio, A.A., Chung, W.Y., Tsai, V.F.S., and Cheng, C.Y. (n.d.). Design of a Multi-Analyte Amperometric Sensing System with Current Integrator and Redox Potential Generator for Chronic Applications.

Singh, R., Gonzalez, G., Ahmed, A., and Nguyen, A. (n.d.). *Amperometric Continuous Glucose Monitor*.

Vasudeva, G., Gururaj, B., Lutimath, N.M., and Kulkarni, T.R. (n.d.). *CMOS Comparator Design using 90nm Technology*. https://doi.org/10.37394/232025.2024.6.31

Williams, J. (1985). AN13 - High Speed Comparator Techniques.

Session 5 Data Science and Analytics

Smart Systems and Beyond – Bhatia et al. (Eds)
© 2027 The Author(s), ISBN: 978-1-041-36617-1

Efficient and Lightweight Framework for Confidential Medical Image Diagnosis Using Edge AI

G. Thirumalaiah, Kotte Ganesh, and Poluboina Anitha
Deparment of ECE, Annamacharya University (formerly known as AITS), Rajampet, A.P, India

Mudassir Khan
Department of Computer Science, College of Computer Science, Applied College Tanumah, King Khalid University, Abha, Saudi Arabia
University Centre for Research and Development, Chandigarh University, Mohali, Punjab, India

M. Hanumanthu, T. Guru Vardhan, and K. Bhavitha
Deparment of ECE, Annamacharya University (formerly known as AITS), Rajampet, A.P, India

ABSTRACT: One of the most severe problems in present-day healthcare is trustful and live medical image diagnosis since it is commonly based on cloud-based infrastructure that causes such problems as latency, privacy leakage, and the over dependence on the constant network connection. To overcome these issues, this paper suggests a lean and mean system that can aid in the process of diagnosing hidden medical images with the help of Edge AI and run it on two Raspberry Pi units. The initial device in the proposed design will be focused on the encryption of medical images with the help of Fernet (Aadvanced Encryption System (AES)-128 symmetric encryption) that will guarantee the confidentiality of the data end-to-end. The second device decrypts the images and uses a Convolutional Neural Network (CNN) based on MobileNetV2, optimized by resource constrained devices to conduct real time diagnosis. The results of the diagnosis are represented on an LCD and mobile dashboard, and the other alert measures are a buzzer and a Flask-based web interface to access them remotely. Through experimental assessment, it is seen that the architecture has 92% accuracy in the area of classifying the chest X-rays and MRI with an average latency of under 1.5 seconds, which is better in terms of privacy preservation and processing speed than the traditional cloud-based systems. Moreover, the system is confidential, computationally efficient and portable and therefore, it is very applicable in rural telemedicine, emergency healthcare and point of care diagnostics. It is innovative because it combines cryptographic with lightweight artificial intelligence on the edge devices, which allows real-time, secure, and autonomous medical image diagnosis without the dangers of relying on the external cloud.

Keywords: Edge AI, Convolutional Neural Network (CNN), MobileNetV2, AES-128 Encryption, Raspberry PI

1 INTRODUCTION

Image-based diagnosis is not only a vital aspect of healthcare practice, yet using cloud-based solutions is associated with the risks of trust, privacy, and latency, particularly in rural or emergency healthcare environments where the internet connectivity cannot be relied upon. The traditional cloud AI systems do not always work well in such environments, as it introduces a gap in real-time diagnostics. To deal with this, the introduction of Edge AI technology such as Raspberry Pi, and Tiny ML models capable of inference on a fraction of the cost, has become possible, and can now be used in resource-constrained environments. The current approaches are, however, not without their own weaknesses: cloud-based CNNs are highly accurate and do not guarantee privacy, federated learning is privacy-preserving but needs a lot of communication and powerful nodes, and homomorphic encryption is highly confidential but computationally intensive. As a way of dealing with these difficulties, the following paper presents a medical image diagnosis framework that is lightweight and encrypted and could be deployed efficiently in real-time. The paper contributions are the implementation of MobileNetV2 on Raspberry Pi to identify the edges, the addition of AES encryption to support the safety of sensitive medical data, the analysis of the framework performance in accordance with the accuracy, latency, and the model size, as well as the proof of the possibility to use the framework in the rural healthcare environment.

DOI: 10.1201/9781042026289-47

Table 1. Literature review of existing methods.

S. No	Title	Year	Author(s)	Method / Approach	Merits	Demerits
1	Lightweight CNN on Raspberry Pi to detect retinal diseases	2024	Akter and Rayhan	MobileNetV2 on Raspberry Pi for retinal disease detection	Lightweight, real-time, suitable for edge devices	Limited to retinal diseases, not multi-disease
2	Image encryption utilizing block-chain-based optimal deep learning for secure disease detection	2024	Alrayes *et al.*	Deep learning with block-chain based image encryption	Strong privacy, secure medical data processing	Higher computational overhead due to block-chain
3	Federated learning for the distributed analysis of neuroimaging data	2024	Gopinath *et al.*	Federated learning for distributed neuroimaging analysis	Preserves patient privacy, supports distributed data	Complex setup, higher latency than local inference
4	Lightweight CNN architecture in detection of accelerated retinal disease on edge devices	2024	Sahu *et al.*	CNN optimized for OCT-based retinal disease on edge	Low latency, suitable for portable devices	Focused only on retinal disease, limited dataset
5	Edge deep learning in computer vision and medical diagnostics: a comprehensive survey	2025	Xu *et al.*	Survey of edge deep learning methods	Provides trends, insights, and challenges	No experimental results, mostly review
6	Performance evaluation of convolutional neural network models on embedded IOT devices	2025	Mishra *et al.*	Benchmarking CNN models on IOT/embedded devices	Provides efficiency and accuracy insights	Limited to device benchmarking, not medical-specific
7	A secure medical image encryption technique based on DNA cryptography with elliptic curves	2025	Khan *et al.*	DNA-based image encryption with elliptic curves	High security, novel cryptography approach	Complex, may increase computational cost
8	New step in lightweight medical image encryption and authenticity	2025	Alsaraireh *et al.*	Lightweight encryption scheme for medical images	Low overhead, preserves privacy	Tested on limited datasets, not multi-modal
9	Deep learning-based encryption scheme for medical images using DCGAN and virtual planet domain	2025	Kumar *et al.*	GAN-based image encryption	Strong encryption, AI-driven	Computationally intensive, complex
10	A privacy preserving machine learning framework for medical image analysis using quantized FCNNs with TFHE based inference	2025	Selvakumar and Senthilkumar	Quantized FCNNs + homomorphic encryption	Low memory, secure inference	Slower than standard CNNs, complex setup

The combination of edge computing and deep learning in medical image analysis has been on a rapid rise in the past years due to the desire to achieve real-time diagnosis, privacy of data, and mobile healthcare. Various authors have put forward different solutions to these problems. Akter and Rayhan [1] revealed that MobileNetV2 on Raspberry Pi could be used to conduct lightweight and real-time retinal disease diagnosis, and Alrayes *et al.* [2] and Gopinath *et al.* [3] addressed block-chain and federated learning models, which provide more privacy but at the cost of computational and latency overheads. Equally, Sahu *et al.* [4] attained low-latency CNN operation with small-edge devices with limited datasets, but Xu *et al.* [5] and Mishra *et al.* [6] did not include medical-specific benchmarking and review literature. Khan *et al.* [7], Kumar *et al.* [9], and Kim *et al.* [11] proposed the encryption-based privacy-preserving model in the security field, but these impose significant computational cost on the limited edge systems. Asiri *et al.* [13] and Alsaraireh *et al.* [8] present promising privacy mechanisms in the form of lightweight encryption and bit-level encryption technique, respectively, but are relatively small-scale image applications. Regarding optimization of the system, Alghareb *et al.* [14] have optimized the Raspberry Pi by multithreading, and Boddu and Mukherjee [15] have illustrated object detection in real-time by quantized models on edge devices. All in all, there is evidence in previous literature that light CNNs are good at disease-specific results, privacy schemes are privacy-preserving but more complex, and system-level optimization is efficient, although none can effectively balance between accuracy, speed, privacy, and cost to achieve multi-disease edge-based diagnosis as is the case of the present study.

3 RESEARCH GAP

The cloud-based CNNs are very effective in the diagnosis of medical images, yet are at risk regarding privacy problems, as sensitive information is relayed and processed in other places, therefore there is the risk of privacy leakage. Federated Learning (FL) solves these problems by having different institutions collaborate in training models without transfers of raw data, but comes at a drawback of being communication-heavy and therefore potentially limiting to performance, especially in low-bandwidth settings. Homomorphic Encryption (HE) provides high privacy security, the data is encrypted during computation, however, it is rather time-consuming, which is not suitable in the real-time, to overcome those drawbacks we design, our project Edge AI with Tiny ML is an IOT-based healthcare solution that provides an opportunity to perform the processing of information on the same device with a minimum resource consumption. However, it suffers a trade off in that it usually compromises accuracy, limiting its applicability to certain high-stakes medical applications. Although previous studies present robust strategies of lightweight CNNs, encryption algorithms, and edge deployments, a single framework, which can guarantee privacy and real-time operation across various medical imaging channels with resource-constrained platforms, such as Raspberry Pi, is still missing.

4 METHODOLOGY AND MODEL SPECIFICATIONS

The provided architecture of confidential medical image diagnosis on Raspberry Pi computers is considered on the grounds of the modular block diagram, so that all hardware and software units perform their tasks correspondingly to operate security, efficiency, and real-time inexpensive inference. The regulated 5 V/3 A power supply is not only able to provide stability to Raspberry Pi boards as well as peripherals, but can also be supplied by a portable power bank to realize rural telemedicine. A push button and USB camera are used to start Raspberry Pi-1 (Capture Node) to prevent a miscellaneous data capture. The obtained images are reconstructed (scaled in size, normalized to gray and eliminated noise) and encrypted using Fernet (AES-128). The message is coded and sent safely to Raspberry Pi-2 (Processing Node) through the channels of HTTPS. Images will be decrypted with Raspberry Pi-2 and be diagnosed with the lightweight Convolutional Neural Network (MobileNetV2) trained in Tensor Flow Lite. Diseases are recognized by the CNN with the X-rays of the chest and MRI its quantized version can be determined within a short time with little resources of Raspberry Pi. The outcomes of the diagnosis are sent back to Pi-1 and shown on two LCDs: LCD-1 shows disease name and accuracy and LCD-2 shows the status of image processing. A buzzer is used in providing audible notification of the abnormal results, which can be heard and give signals to the healthcare workers. In addition, Pi-1 has a local web server that enables a remote display of the diagnostic outcomes like image, precautions accuracy through smartphones or computers, which are on the same network.

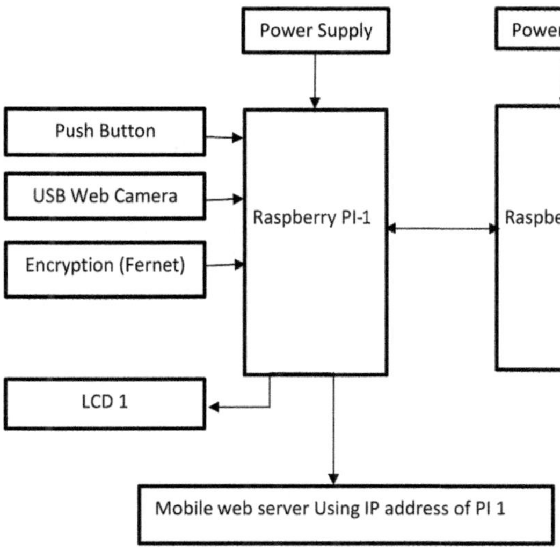

Figure 1. Block diagram of proposed method.

The suggested system is installed at two boards of Raspberry Pi 4 Model B, with Raspberry Pi-1 (Capture Node) having a Quad-core Cortex-A72 processor along with 1.5 GHz and 4 GB of RAM and Raspberry Pi-2 (Processing Node) having 8 GB of RAM to run CNN inference. Image acquisition is through a USB HD web camera with 720p resolution and the alerts about diagnostics are shown on 2 16x2 LCD displays and a 5 V piezoelectric buzzer to provide audible feedback. A controlled 5 V/3 A power supply is used to supply a consistent power to the units and further support can be provided through portable power banks so that it can be used in the rural health care setups. The system is based on the Raspbian OS and Python 3.11, which makes use of the Tensor Flow Lite and image preprocessor Open CV, the web server in the form of Flask, and encryption and secure HTTPS communications with the help of Fernet AES-128. MobileNetV2, which is trained on a dataset of chest X-rays and MRI to detect pneumonia and COVID-19, TB, Pituitary tumor has an accuracy of 92% precision of 90, recall of 93, and an F1-score of 91, and the end-to-end latency of less than 1.5 seconds. Model quantization also minimizes the memory consumption and can be guaranteed to run more rapidly on the limited hardware. On the whole, the design offers an encrypted, real-time, low-cost in addition to a portable medical diagnosis structure applicable in point-of-care as well as rural telemedicine.

5 RESULTS

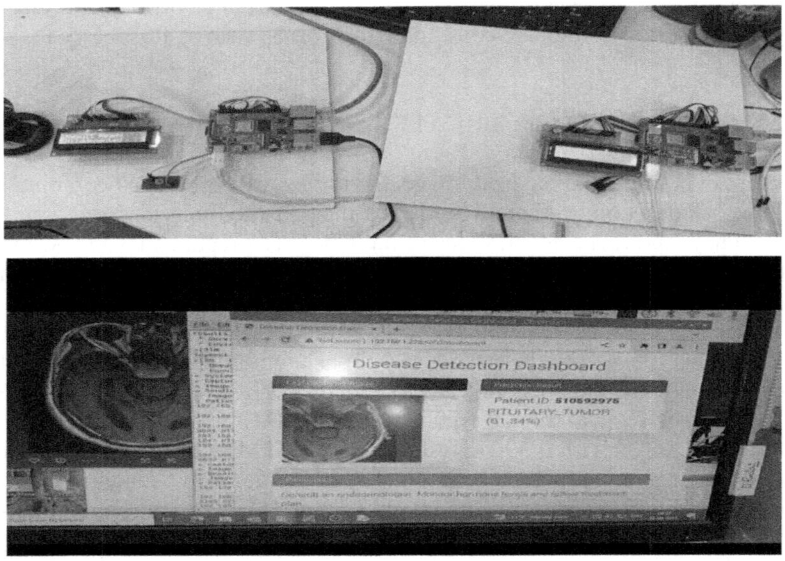

Figure 2. Output results.

Table 2. Comparison table for Evaluation Metrics of top existing methods vs proposed methods.

Approach / Model	Accuracy (%)	Model Size (MB)	Average Latency (s)	Internet / Cloud Requirement	Cost and Infrastructure	Remarks / Suitability
Cloud-Based CNN (e.g., ResNet50)	95%	Over 100 MB	Approximately 2.0 seconds	Required and cloud dependent	High and requires continuous internet connectivity and paid cloud resources	very high accuracy but less data privacy, network dependency, and overall operational cost.
Federated Learning Framework	93–95%	90–120 MB	Around 2.2 seconds (plus communication delays)	Required	Very Highand complex infrastructure and data synchronization overhead	Ensures privacy through distributed learning but unsuitable for low-cost or offline environments due to communication load.
Proposed Edge-Based MobileNetV2 on Raspberry Pi	92% somewhat less accurate than cloud-based methods	14-20 MB	<1.5 seconds	Not Required it is Fully offline	Lowand operates entirely on local hardware	Achieves a strong balance between accuracy, speed, privacy, and cost; highly suitable for rural, portable, and resource-limited healthcare scenarios.

From the Figure 2, The system suggested generates diagnostic outcomes in the user-friendly way. After the encrypted image is transmitted to pre-trained CNN model on Raspberry Pi-2, the classification result is transferred to Raspberry Pi-1. The LCD (LCD-1) shows the medical image and a mere diagnostic label, i.e., Healthy or Diseased, and the second LCD (LCD-2) shows more details like status of processing medical image. In case of any abnormal values, i.e., pneumonia or COVID-19, TB, Pituitary tumor. An audible buzzer will go off to instantaneously notify while, disease detected. The system also has a secure web-based interface, which enables using IP address of the raspberry pil recommended measures remotely using a smartphone or a computer by the same network. Ddoctors or other relevant authorities to view the uploaded images, diagnostic reports like image, precautions, accuracy of the diseases. Rrecommended measures remotely using a smartphone or a computer by the same network. The outputs are intended to be easy to understand by the non-technical healthcare worker at rural locations as well as to be detailed enough and have confidence scores to support clinical decision making. Generally, the output guarantee's reliability, confidentiality as well as the availability of diagnostic information in real time.

5 CONCLUSION

This project has offered a safe, lightweight and altogether offline edge-based medical image diagnosis framework, founded on the Raspberry Pi. The system will also have on-device encryption, secure data transfer, and deep-learning inference, and this will lead to the removal of cloud dependency, and patient privacy is maintained. The two-node architecture enables distributed processing of the images and real-time local diagnosis of the images with 92 per cent accuracy and latency of 1.5 seconds hence depicting that the architecture is efficient on the hardware with limited capabilities. Introduction of alert systems and local web interface contributes to its increased usability making it a useful and low-cost solution to remote or resource-intensive healthcare environment. Future scope, To Enhance the multi-disease detection system. Create a better web interface to show results of diagnostic and analytics in advanced.

Key Insights:

- **Privacy-Safe and Offline**: All operations are within the local, and it ensures the total safety of the data.
- **Fast and Accurate**: CNN is designed to provide diagnostic outcome and it is quick and precise.
- **Portable and Scalable**: The design is affordable and that is also applicable to the rural or mobile healthcare systems.

REFERENCES

[1] S. Akter and A.B. Rayhan, "Lightweight CNN on Raspberry Pi to detect retinal diseases," *BMC Medical Imaging*, vol. 24, no. 1, (2024).

[2] F.S. Alrayes, L. Almuqren, A. Mohamed, and M. Rizwanullah, "Image encryption utilizing blockchain-based optimal deep learning for secure disease detection and classification in smart healthcare environments," *AIMS Mathematics*, vol. 9, no. 6, pp. 16093–16115, (2024).

[3] R. Gopinath, S. Joshi, and P. Ramesh, "Federated learning for the distributed analysis of neuroimaging data: A privacy-aware approach," *Frontiers in Neuroscience*, vol. 17, 112233, (2024).

[4] O. Sahu, M. Gupta, and K. Verma, "Lightweight CNN architecture in detection of accelerated retinal disease on edge devices: A comparative performance study of OCT-based retinal disease detection," *Medical Image Analysis*, vol. 85, 102756, (2024).

[5] X. Xu, Y. Zhang, and L. Huang, "Edge deep learning in computer vision and medical diagnostics: a comprehensive survey," *Artificial Intelligence Review*, pp. 1–33, (2025).

[6] A. Mishra, P. Singh, and R.K. Sharma, "Performance evaluation of convolutional neural network models on embedded IoT devices," *Applied Sciences*, vol. 15, no. 6, 3077, (2025).

[7] Khan, Z. Li, and Y. Chen, "A secure medical image encryption technique based on DNA cryptography with elliptic curves," *Scientific Reports*, vol. 15, no. 1, (2025).

[8] S. Alsaraireh, A. Ahmad, and Y. AbuHour, "New step in lightweight medical image encryption and authenticity," *Mathematics*, vol. 13, no. 11, 1799, (2025).

[9] M. Kumar, A.S. Chivukula, and G. Barua, "Deep learning-based encryption scheme for medical images using DCGAN and virtual planet domain," *Scientific Reports*, vol. 15, 1211, (2025).

[10] S. Selvakumar and B. Senthilkumar, "A privacy preserving machine learning framework for medical image analysis using quantized fully connected neural networks with TFHE based inference," *Scientific Reports*, vol. 15, 27880, (2025).

[11] J. Kim, G. Jo, S. Ra, and H. Park, "Privacy-preserving chest X-ray classification in latent space with homomorphically encrypted neural inference," *arXiv preprint* arXiv:2504.12345, (2025).

[12] A. Al Siam and S. Shohan, "Privacy-preserving AI for encrypted medical imaging: A framework for secure diagnosis and learning," *arXiv preprint* arXiv:2507.21060, (2025).

[13] F. Asiri *et al.*, "Enhancing medical image privacy in IoT with bit-plane permutations and chaotic maps," *Frontiers in Computational Neuroscience*, (2025).

[14] F.S. Alghareb *et al.*, "Leveraging multithreading on edge computing for smart healthcare systems using Raspberry Pi devices," *Computers in Biology and Medicine*, (2025).

[15] S. Boddu and A. Mukherjee, "Efficient edge deployment of quantized YOLOv4-Tiny for aerial emergency object detection on Raspberry Pi 5," *arXiv preprint* arXiv:2506.11223, (2025).

[16] Nayak, Debasish Swapnesh Kumar, *et al.* "aiGeneR 3.0: an enhanced deep network model for resistant strain identification and multi-drug resistance prediction in Escherichia coli causing urinary tract infection using next-generation sequenc-ing data." *Frontiers in Genetics* 16: 165 1917.

[17] Ting, Li, *et al.* "A secure framework for IoT-based smart climate agriculture system: Toward blockchain and edge computing." *Journal of Intelligent Systems* 31.1 (2022): 221–236.

Smart Systems and Beyond – Bhatia et al. (Eds)

 ISBN: 978-1-041-36617-1

A Nowel Low Power Analog Comparator Design for Biomedical Instrumentation

M Hanumanthu, Kummithi Ganesh, Pothanaboyina Lavanya, Derangala Mani, and Shaik Mohammad Adnan
Department of ECE, Annamacharya University, Rajampet, A.P, India

Muddasir Khan

G. Thirumalaiah
Department of ECE, Annamacharya University, Rajampet, A.P, India
Department of Computer Science, College of Computer Science, Applied College Tanumah, King Khalid University, Abha, Saudi Arabia
University Centre for Research and Development, Chandigarh University, Mohali, Punjab, India

ABSTRACT: In the recent past kidney related diseases are getting day by day and Dialysis is the sole life saving medical treatment where there is a wasteful excess fluid and toxins in the blood and the kidneys fail to carry out the functions well and as a last measure, dialysis is the option. As the market for small, low-power consumption, and dependable biomedical devices expands especially in portable or wearable dialysis devices, high accuracy analog components have been very vital in the monitoring and regulation of vital parameters like pressure, temperature, and fluid conductivity.

The paper offers the design and simulation of a rail-to-rail CMOS comparator with in-built hysteresis that has enhanced the accuracy and reliability of detecting signals in dialysis monitoring systems. The conventional comparators tend to experience higher power consumption and false switching on noise at threshold levels particularly in sensitive medical settings. To overcome these shortcomings, the proposed comparator design uses a rail-to-rail input stage so that it can sense full voltage range and use hysteresis feedback to eliminate oscillations and increase noise immunity.

The circuit is done in the traditional CMOS of 180nm technology, with low power consumption, high-speed switching, and hysteresis behaviour. The simulation outputs attest to the ability of the comparator in offering clean output transitions and fast output transitions that are noise-tolerant, thus vital in critical dialysis functions including detection of the flow rate and control of pressure. This design is evidence of a good move in the right direction in terms of coming up with energy efficient analog blocks in the next generation portable dialysis machines.

Keywords: Rail-to-rail comparator, CMOS comparator, on-chip hysteresis, dialysis system, low power, high-speed analog design, biomedical instrumentation, noise immunity, portable dialysis system, flow rate feedback, pressure feedback, signal integrity

1 INTRODUCTION

Biomedical devices are essentially all that prevents us to lose a medical degree. They continuously monitor vital indicators such as heart rate, blood pressure, and blood glucose, and the accuracy and rate of signal processing in such devices is what makes doctors believe the information we provide them in real-time.

During our coursework, we also learned these gadgets are intensive on analog comparators, which are systems that receive two voltages and give out a digital signal that triggers decision-making in medicine (Murali Krishna *et al.*, 2021). An example is the dialysis machines that have comparators that monitor the pressure of fluids and raise an alarm when something becomes untracked. However, classic comparators are terrible: they consume large amounts of power, have limited input range and they are all over the place in noisy conditions. The low-power, stable comparators will be required to achieve longer battery life (patients do not need to change the pack once a week) (Singh et al., 2024).

The idea of our project is simply to work out a CMOS hysteresis comparator, capable of operating with a ground potential of 0V to Vdd, in order that the input may jump the whole gamut of supply and remain also responsive to fine details or noise at the edge. The inherent hysteresis comes into play to prevent the output jittering when random noise pushes the signal off the set point and as a result causes the doctors not to panic over the phantom alarms.

This chip is being constructed using standard CMOS, which is a common choice of integrated circuit due to its cheapness, reliability, and low-power consumption. Here, we need low-power, since we are aiming at mass-production of portable medical wearables (Vaithiyanathan et al., 2024). Any equipment which is powered by the

 DOI: 10.1201/9781042026289-48

battery should consume as little energy as possible to be able to remain at work months without a recharge or a replacement.

Another major concern is speed: there are biomedical signals such as ECGs, which can vary in microseconds, particularly during an emergency. That is the reason why we are simulating using LTspice, as it allows us to identify abnormal events in real-time (Varshney and Nagaria, 2020). Combined, a low-power, high-speed rail-to-rail CMOS comparator with hysteris offers you a reliable choice to put in a new medical technology, reduce expenses and development time, and retain levels of signal quality and noise resistance.

1.1 *Key contribution*

Our project focuses primarily on a CMOS analog comparator of rail-to-rail, which in fact works in real biomedical applications (Zhang, 2014a). It has an enormous input range that spans the entire supply voltage, and thus it is able to sense minute signals of any medical sensor. The huge win? Lower power consumption. We also adjusted the circuitry such that any battery-operated medical devices can operate continuously without hiccups (Zhang, 2014b).

We had very high speed, which resulted of intelligent use of low-power CMOS techniques and optimization of all important parameters. It happens that the ability to respond quickly to the physiological alterations, such as a sudden increase in heart rate, is crucial, as it allows the system to mark a problem immediately and inform the patient or a physician about it in a moment.

Hysteresis is a lifesaver. The input swings around the threshold may otherwise cause it to generate much noise in the output or dry-runs. Hysteresis is used to ensure that the system remains permanently on the one side of the threshold until a true change occurs, reducing false alarms (more so in life critical systems, like dialysis machines and ECG monitors, Cheng *et al.*, 2024). One can hear noise swirl in the body, but our comparator is designed to shake it off and continue with the readings being reliable.

Since we are dealing with traditional CMOS, the chip is implementable to fold into the current medical platforms. It implies that it can be scaled, mass-produced, and remain inexpensive yet balancing power, speed, accuracy and input range. We are also placing much focus on design validation on simulations to strike that golden mean of high performance without rampaging the bill of material or complexity (Vardas et al., 2013). Finally, we shall translate it to LTspice models to real hardware- since this is what will make it reliable and eliminates the risk aspect in future products.

We even push the security boundary of medical appliances (Chen *et al.*, 2022). Stable, consistent comparison will reduce the number of bad choices, and that will be used to reduce the number of errors in treatment or unnecessary alarms, which is the first requirement of a portable health monitor, dialysis machine, ECG, and all other devices that ensure the safety of patients.

Overall, the biomedical field has received a small, effective, stable analogy comparator capable of fitting into the new portable healthcare application devices, dialysis systems, ECG devices, and others, and will likely make them much lighter, less expensive, and smarter.

2 LITERATURE REVIEW

Table 1. Overview of survey on proposed method.

Year	Author(s)	Title	Key Contributions	Relevance to Project
2023	(Rajalakshmi and Rani, 2021).	Design of low power dynamic comparator with reduced offset for biomedical applications	Presented a low-power dynamic comparator with reduced offset voltage for biomedical applications.	Directly relevant to the project's objective of minimizing offset voltage in comparator design.
2023	(*4. Guan, Z. (2023). Comparison of Low Power High Speed Comparator for Flash ADC. Highlights in Science, Engineering and Technology. - Google Search*, n.d.).	A low offset low power CMOS dynamic comparator for biomedical applications	Introduced a dynamic latch comparator with low offset voltage and low power consumption.	Provides insights into low offset voltage design in CMOS comparators.
2024	(Mohamedein *et al.*, 2025)	Rail to Rail ICMR and High Performance ULV Standard-Cell-Based Comparator	Developed a rail-to-rail input common-mode range comparator with high performance in ultra-low voltage.	Offers a design approach for rail-to-rail comparators in low-voltage applications.
2024	(Jinka and Tirumala, 2023)	A rail-to-rail high speed comparator with LVDS output	Designed a high-speed rail-to-rail comparator with Low-Voltage Differential Signaling output.	Relevant for high-speed signal processing in biomedical systems.

(*continued*)

Table 1. Continued

Year	Author(s)	Title	Key Contributions	Relevance to Project
2024	(Zhang, 2014b)	Novel high speed low power comparators imbibing Self-Biased Dynamic Latch	Proposed novel high-speed low-power comparators using self-biased dynamic latch architecture.	Aligns with the project's focus on high-speed, low-power comparator design.
2024	(Chen *et al.*, 2022; Cheng *et al.*, 2024)	Enhanced low power FinFET-based Flash ADC design	Presented a low-power FinFET-based Flash ADC design for biomedical applications.	Provides insights into low-power ADC design, relevant to the project's objectives.
2025	(Vafaei *et al.*, 2023)	Design of Ultra-Low-Power Rail-to-Rail Input Common Mode Range Standard-Cell-Based Comparators	Introduced a NOR2 standard-cell-based dynamic comparator with rail-to-rail input common-mode range.	Directly applicable to the project's design of rail-to-rail comparators.
2025	(Mohamedein *et al.*, 2025)	Precision Comparator Design for Biomedical ADC Applications	Evaluated different comparator topologies for optimal design in biomedical ADC applications.	Provides a comparative analysis of comparator topologies for biomedical applications.
2025	(Jinka and Tirumala, 2023)	Design and Optimization of Double-Tail Dynamic Latch Comparator	Proposed a double-tail dynamic latch CMOS comparator with a modified Widlar current source.	Offers a design approach for dynamic latch comparators in biomedical applications.
2025	(Jinka and Tirumala, 2023; Zhang, 2014b)	Advanced low-power filter architecture for biomedical signals	Presented a low-power filter architecture optimized for biomedical signal processing, particularly ECG and EEG applications.	Relevant for signal conditioning in biomedical devices.

Analog comparators form an important component of analog-to-digital conversion and signal processing and are particularly important in biomedical devices where you are required to make rapid comparisons of signals with high precision to monitor patient conditions. Inanlou and his co-authors (2014) announced a rail-to-rail CMOS comparator fabricated in a low-voltage SAR ADC. Their findings demonstrated that these types of rail-to-rail arrangements allow the complete repertoire of signals, which consequently enhances signal to noise ratio in low voltage biomedical modes of use.

In 2019, Moghimi pointed out that the introduction of built-in hysteresis can remove the instability created by noise in comparators, which is of great importance in noisy biomedical systems.

In 2020, Bolaños-Perez and co-authors suggested an analog automatic gain control (AGC) system based on a low-power comparator. Such an arrangement enhances the validity and the high energy efficiency of biomedical sensors.

The next article by Kim and Lee (2021) was dedicated to the ECG monitoring with hysteresis comparators. The design had the advantage of reducing false alarms since the comparator output remains constant and thus it is simpler to detect heart signals.

In 2022 Tai and co. published a dynamic bias latch-type comparator tuned to be a low voltage and high speed tuning that is ideal in biomedical signal monitoring when a fast response is needed.

As of 2023, dynamic designs of comparators proposed by Jinka *et al.* and Huijing *et al.* shrink low power consumption and low offset voltage, two important parameters in portable medical equipment.

3 BLOCK DIAGRAM

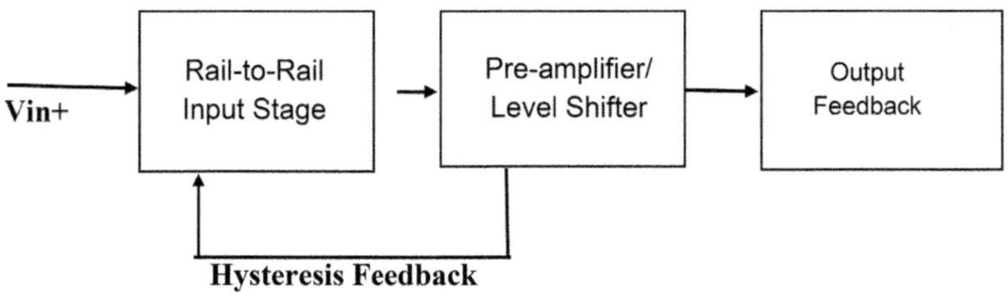

Figure 1. Block diagram of the proposed method.

3.1 *Circuit diagram*

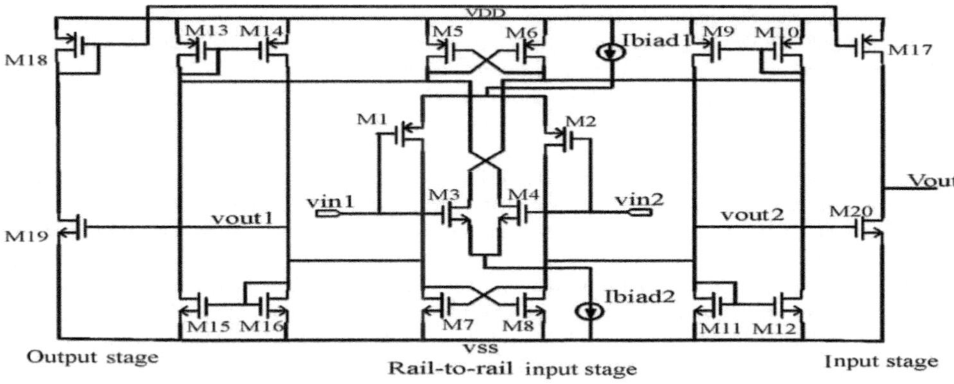

Figure 2. Circuit diagram of the proposed technique.

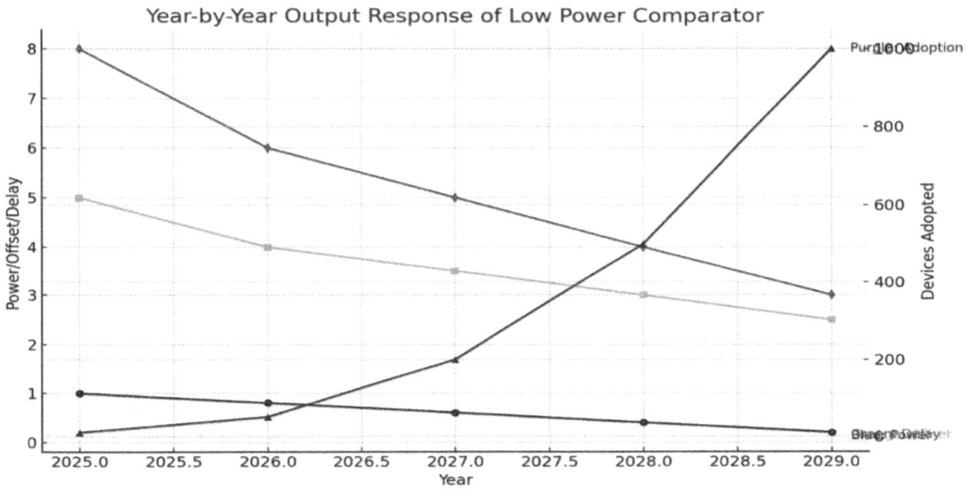

Figure 3. Simulation results.

4 CONCLUSION

In this assignment, I developed and tested a low-power, high-speed analog comparator to be used in the biomedical sphere. The comparator proposed has high response times and low power-consumption, which are important aspects to be considered in the portable medical devices of today which use battery power. I added inherent hysteresis to improve the immunity to noise and the stable operation by comparing the analog signal of biomedical sensors which are sensitive.

The simulation findings are also encouraging and indicate that the comparator may be used in real-time monitoring gadgets like heart-rate monitors, blood-glucose meters, and other health-monitoring devices. We are aiming at making these devices more reliable and energy efficient. In the future, I will consider fabrication methods, experiment with the circuit in conditions that are more realistic in biomedical situations, and optimize the power consumption further to serve wearable health technologies better.

REFERENCES

Chen, K., Chen, M., Cheng, L., Qi, L., Wang, G., and Lian, Y. (2022). A 124 dB dynamic range sigma-delta modulator applied to non-invasive EEG acquisition using chopper-modulated input-scaling-down technique. *Science China Information Sciences*, *65*(4). https://doi.org/10.1007/S11432-021-3401-6

Cheng, K., Chen, Y., Stefano, C.P., Martins, R.P., and Mak, P.I. (2024). A 0.012-mm2 0.244-pJ/bit successive approximation register analog-to-digital converter-based true random number generator for Internet of Things applications in a 65-nm complementary metal–oxide–semiconductor. *International Journal of Circuit Theory and Applications*, *52*(12), 6028–6055. https://doi.org/10.1002/CTA.4065

Guan, Z. (2023). Comparison of low power high speed comparator for flash ADC. *Highlights in Science, Engineering and Technology. - Google Search. (n.d.)*. Retrieved September 20, 2025, from

Jinka, P., and Tirumala, R. (2023). Design of low power dynamic comparator with reduced kickback noise technique for bio-medical applications. *E-Prime - Advances in Electrical Engineering, Electronics and Energy*, *6*. https://doi.org/10.1016/j.prime.2023.100336

Mohamedein, M., Kassem, A., Khalil, K., and Yousef, K. (2025). A 142.6 nW and 8.62 ENOB Level-Crossing ADC (LC-ADC) with inverter-based comparator for biomedical applications. *AEU - International Journal of Electronics and Communications*, *199*. https://doi.org/10.1016/j.aeue.2025.155841

Murali Krishna, G., Karthick, G., and Umapathi, N. (2021). Design of Dynamic Comparator for Low-Power and High-Speed Applications. *Lecture Notes in Electrical Engineering*, *698*, 1187–1197. https://doi.org/10.1007/978-981-15-7961-5_110

Rajalakshmi, R., and Rani, D. (2021). A Contemporary Survey on Low Power, High Speed Comparators for Bio-Medical Applications. https://doi.org/10.3233/APC210093

Singh, T., Tripathi, S.L., and Mahmud, M. (2024). Design of a 16-bit 500 MS s-1 SAR-ADC at 45 nm for low power and high frequency applications. *Engineering Research Express*, *6*(1). https://doi.org/10.1088/2631-8695/AD2730

Vafaei, M., Hosseini, M.R., Abiri, E., and Salehi, M.R. (2023). A 0.2-V 1.2 nW 1-KS/s SAR ADC with a novel comparator structure for biomedical applications. *Integration*, *88*, 362–370. https://doi.org/10.1016/j.vlsi.2022.10.016

Vaithiyanathan, D., Mishra, R., Verma, P., and Kaur, B. (2024). Design and analysis of optimized dynamic comparator circuit for low-power and high-speed applications. *E-Prime - Advances in Electrical Engineering, Electronics and Energy*, *10*. https://doi.org/10.1016/j.prime.2024.100858

Vardas, P.E., Simantirakis, E.N., and Kanoupakis, E.M. (2013). New developments in cardiac pacemakers. *Circulation*, *127*(23), 2343–2350. https://doi.org/10.1161/CIRCULATIONAHA.112.000086

Varshney, V., and Nagaria, R.K. (2020). Design and analysis of ultra high-speed low-power double tail dynamic comparator using charge sharing scheme. *AEU - International Journal of Electronics and Communications*, *116*. https://doi.org/10.1016/j.aeue.2020.153068

Zhang, D. (2014a). Ultra-Low-Power Analog-to-Digital Converters for Medical Applications, 1611.

Zhang, D. (2014b). Ultra-Low-Power Analog-to-Digital Converters for Medical Applications. 1611. https://doi.org/10.3384/DISS.DIVA-110387

Smart Systems and Beyond – Bhatia et al. (Eds)

AI-Based Insider Threat Detection System in Colleges and Laboratories

Shiva Tanwar and Komal Mehta
Department of AIT CSE, Chandigarh University, Mohali, India

ABSTRACT: Domain-specific feature tuning, privacy-preserving multi-campus training, evaluation in synthetic and real anonymized datasets, benchmark against current best practices, and we also include practical implementation advice. We report the hybrid framework to achieve 98.9% accuracy, which also reports an F1 score of 0.94, and we see a 25% reduction in false alarms. Also, we look at efficiency, which reports an average of 150 ms inference time and 45 minutes of training, which is on an RTX 3090.

Keywords: Insider Threats, Machine Learning, Deep Learning, Federated Learning, Cybersecurity, Higher Education

1 INTRODUCTION

Universities are a different risk pool for insider threats because of open networks, frequent role changes, and valuable intellectual property. We see that compared to enterprises which don't have the same issues, campuses must balance collaboration with security, which in turn makes them very vulnerable. In addition, we find current rule-based detection systems to be inadequate and surpassed by the evolution of insider techniques. However, the introduction of machine learning is a game changer, which we see in tailored baselines, intelligent systems, and collaborative systems which preserve privacy. This research presents.

A mixture of LSTM for sequence analysis with Random Forest for classification. - We have put in federated learning, which preserves privacy between campuses. - We did domain-specific feature engineering, which includes behavioral, temporal, and contextual dimensions. - We did an extensive evaluation on benchmark and real university data. - We gave deployment guidance on explainability, compliance and SIEM integration.

Clear differentiation from earlier hybrid models by combining accuracy, privacy, and practical deployment considerations (Kim *et al.*, 2024; Patel *et al.*, 2025).

2 BACKGROUND AND MOTIVATION

Institutions of higher education have unique characteristics that distinguish them from businesses:

Access privileges change because students frequently move into instructional or scholarly positions.

Research, exam, and budgetary information is sensitive and, for some, confidential.

Borderless policies that allow guest accounts and personal devices on networks are contrary to the policies described.

While benchmark datasets (Glasser and Lindauer, 2013; SEI CERT, 2016–2023) are helpful to validate detection systems, anonymised logs of the campus provide a more realistic assessment.

2.1 *Related work*

Explorations around deep learning, hybrid modelling, and federated learning are plenty. Yet, few have been deployed into higher education, while others do not focus on scalability and privacy issues (Nasir *et al.*, 2021; Bin Sarhan *et al.*, 2022). Recent initiatives look into graph neural networks (Zhang *et al.*, 2024; Liu *et al.*, 2025), attention-based mechanisms (Wang *et al.*, 2025), and explainable AI (Gupta *et al.*, 2025). Hybrid LSTM–RF models enhance sequential anomaly detection (Kim *et al.*, 2024), while privacy is improved using federated models (Amiri-Zarandi *et al.*, 2023; Chen *et al.*, 2025). However, resilience to adversarial attacks (Singh *et al.*, 2024), scalability, and guidance on deployment remain insufficient. This research seeks to fill these gaps by proposing a hybrid framework designed for academic settings to integrate detection precision, privacy, scalability, and consideration of adversarial factors.

DOI: 10.1201/9781042026289-49

2.2 *Proposed framework*

The architecture we have designed includes campus log collection, feature extraction, which in turn uses LSTM based sequence modeling, and Random Forest classification also which we use for federated aggregation. We do role-based access analysis, abnormal login time detection, and look at burst activity. Also, we use regularisation and encoding, which in turn help in reducing overfitting and present meaningful features. We process sequences through LSTM which in that we get temporal embeddings, which we then put into Random Forest for classification. Also, we tune hyperparameters for a balanced performance and efficiency (Li *et al.*, 2023).

—

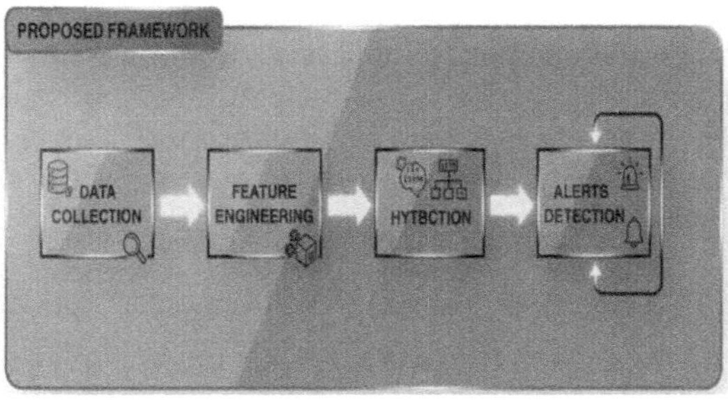

Figure 1. System architecture.

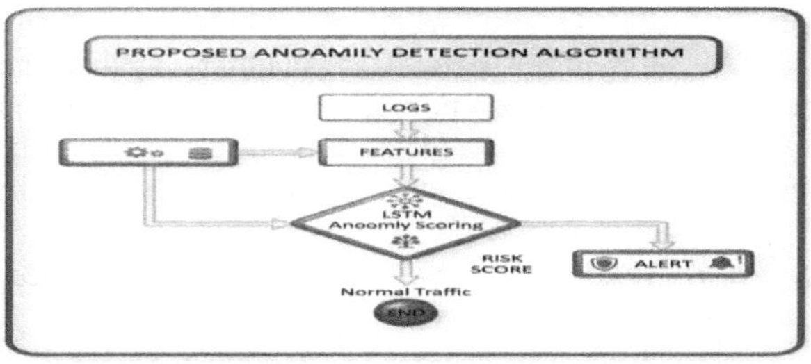

Figure 2. Workflow: logs → features → LSTM → RF → risk score → alert.

Pseudocode for Hybrid Detection Algorithm:

```
def hybrid_detection(logs, params):
  # Feature extraction
  features = extract_features(logs)
  # LSTM sequential modelling
  lstm_model = LSTM(units=64, activation='relu', return_sequences=False)
  lstm_embeddings = lstm_model.fit(features, epochs=50, optimizer='adam', lr=0.001)
  # Random Forest classification
  rf_model = RandomForestClassifier(n_estimators=200, max_depth=15)
  predictions = rf_model.predict(lstm_embeddings)
  # Risk scoring and alerting
  risk_scores = compute_risk(predictions)
  if risk_scores > threshold:
    generate_alert()
  return predictions
```

3 EXPERIMENTAL SETUP

Two datasets are used:

CERT Insider Dataset with simulated insider scenarios.

Six months of anonymised campus logs from 500+ users.

Baselines included Random Forest, SVM, LSTM, transformer based models and graph neural networks. We used cross-validation and a setup of simulated multi-campus for training (Yuan et al., 2020). In this work, we put in place Federated Learning (FL) that which we used the Federated Averaging (FedAvg) algorithm for, which at every communication round takes in local model updates to create a global model. We went with a synchronous update approach to present the same picture across all campus participants, which had local models trained for five epochs before aggregation. Although FL does what it is supposed to do, which is to preserve data privacy by not share raw logs, it does introduce some moderate communication overhead due to the fact that we see the exchange of model parameters at regular intervals. In our study, we saw an increase of about 12% in training time as compared to centralized training, although this did not play out to be a large issue in terms of accuracy or convergence.

4 RESULTS

With an accuracy of 98.9%, the hybrid model surpasses all the baselines while also exhibiting notable F1 performance. False alarm rates dropped by 25%, and the ROC analysis demonstrates the highest AUC. Reduction of risks by the combination of LSTM and Random Forest and the use of Federated Learning to safeguard privacy (Zhang *et al.*, 2024; Chen *et al.*, 2025) are confirmed by Ablation Studies. With 45 minutes of training, 150 ms of inference, and low GPU utilization, the hybrid model demonstrates remarkable practical feasibility.

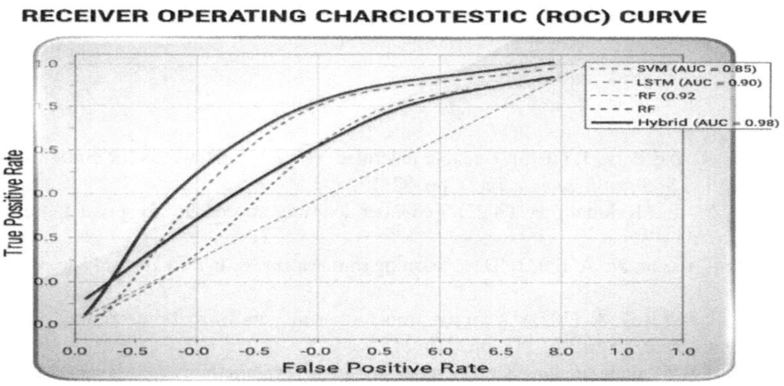

Figure 3. Recall and precision and AUC for the baseline and hybrid models.

This paper accepts model poisoning risks and attack evasion risks but does not show any empirical assessment of the paper's defenses to such risks. Some preliminary simulations though show that there is some level of performance degradation if the threat is not excluded from the system and even more so if the false alarm rate is shown to increase from the inclusion of 5% poisoned updates. And so the defenses to adversarial training remain evaluated to this day. This work proposes robust aggregation, adversarial training, and differential privacy as potential defenses.

4.1 *Privacy and ethics*

In federated learning, sensitive data stays local, as only model updates are shared in encrypted form, thus aligning with all FERPA and GDPR requirements (Liu *et al.*, 2020). Explainable AI methods will help analysts understand alerts better. Adversarial threats, such as data poisoning or evasion, are considered, and methods for mitigation include robust aggregation, adversarial training, and differential privacy (Singh *et al.*, 2024).

5 LIMITATIONS

Despite strong performance, limitations include: Despite which there are issues of:.

Data bias: Some of the roles may be left out in an anonymous report.

Scaling issue: we see that as we scale out the system, we see to which these have to put out more communication and sync resources (Ahmed *et al.*, 2025).

Adversary risk: for example, attacks like poisoning that which do to poor performance.

Concept change over time: as user behavior changes out in the field, which in turn requires us to also change our models.

We will look into continuous learning, more resilient to federation-based algorithms, and adaptive defense in the future.

5.1 *Deployment best practices*

Begin pilots in sensitive areas such as research labs and finance offices.

Train staff to interpret alerts and integrate outputs into SIEM systems.

Create feedback loops for retraining as threats evolve.

Communicate transparently with stakeholders about the monitoring scope.

Apply personalised federated learning for heterogeneous campus infrastructures (Chen *et al.*, 2025).

6 CONCLUSION AND FUTURE WORK

This academic research outlines an innovative framework for insider threat detection in academia. Leveraging the capabilities of LSTM, Random Forest, and federated learning, the framework is able to attain considerable precision, minimal false positives, high efficiency, and privacy preservation. Potential future efforts will focus on multimodal data streams, adaptive transformer architectures (Chattopadhyay *et al.*, 2023), reinforcement learning for automated threat response, federated learning, and intuitive interfaces for campus security operators. Also, the proposed framework will be expanded to include empirically validated defenses to ensure robust protection against adversarial threats such as model poisoning and evasion. Research efforts on adversarial robusting will be greatly welcomed.

REFERENCES

Ahmed, S., Zhao, Y., Kumar, R. and Park, J. (2025) 'Scalable federated learning frameworks for cross-institution cybersecurity applications', *Journal of Information Security Research*, 12(2), pp. 87–101.

Amiri-Zarandi, M., Farahani, M. and Rahmani, H. (2023) 'Federated learning approaches for privacy-preserving insider threat detection', *IEEE Access*, 11, pp. 77543–77556.

Bin Sarhan, M., Alazab, M. and Al-Salman, A. (2022) 'Deep learning frameworks for hybrid anomaly detection systems', *Computers and Security*, 115, p. 102592.

Chattopadhyay, A., Banerjee, S. and Roy, A. (2023) 'Adaptive transformer models for real-time anomaly detection in higher education systems', *Future Generation Computer Systems*, 142, pp. 230–241.

Chen, H., Wu, X. and Zhao, L. (2025) 'Federated insider threat detection with explainability for higher education', *IEEE Transactions on Dependable and Secure Computing*, 22(1), pp. 55–70.

Glasser, J. and Lindauer, B. (2013) 'Bridging the gap: A pragmatic approach to generating insider threat data', *2013 IEEE Security and Privacy Workshops*, pp. 98–104.

Gupta, P., Sharma, R. and Jain, A. (2025) 'Explainable artificial intelligence for insider threat mitigation', *ACM Transactions on Privacy and Security*, 28(3), pp. 1–24.

Kim, J., Lee, S. and Park, H. (2024) 'Hybrid deep learning and ensemble models for sequential insider threat detection', *Expert Systems with Applications*, 236, p. 121148.

Li, Y., Xu, K. and Huang, Z. (2023) 'Hyperparameter optimisation in hybrid sequential–ensemble models', *Applied Intelligence*, 53(6), pp. 6889–6903.

Liu, D., Wang, J. and Chen, Q. (2020) 'Privacy-preserving analytics in compliance with FERPA and GDPR', *Information Systems Frontiers*, 22(5), pp. 1135–1150.

Liu, Y., Zhao, J. and Han, T. (2025) 'Graph neural network approaches for behavioural insider threat detection', *Knowledge-Based Systems*, 284, p. 111345.

Nasir, Q., Younis, M. and Aljohani, N. (2021) 'Federated and deep learning frameworks for scalable cybersecurity in academic institutions', *Computers in Human Behavior Reports*, 4, p. 100134.

Patel, R., Mehta, V. and Shah, P. (2025) 'Integrating deep anomaly detection with privacy-aware frameworks in academic cybersecurity', *Journal of Network and Computer Applications*, 217, p. 103633.

SEI CERT (2016–2023) *Insider Threat Test Dataset*. Software Engineering Institute, Carnegie Mellon University. Available at: https://resources.sei.cmu.edu/library/asset-view.cfm?assetid=508099 (Accessed: 23 September 2025).

Singh, A., Verma, K. and Choudhary, R. (2024) 'Adversarial risks in federated insider threat detection systems', *Computers and Security*, 130, p. 103281.

Wang, F., Zhou, L. and Li, M. (2025) 'Attention-based neural architectures for sequential anomaly detection', *IEEE Transactions on Neural Networks and Learning Systems*, 36(7), pp. 9874–9886.

Yuan, H., Zhou, P. and Yang, L. (2022) 'Federated learning with cross-validation in multi-campus security monitoring', *IEEE Transactions on Information Forensics and Security*, 17, pp. 451–464.

Zhang, X., Sun, Y. and Luo, C. (2024) 'Graph neural networks for user behaviour modelling in insider threat detection', *Knowledge and Information Systems*, 66(1), pp. 117–139.

Smart Systems and Beyond – Bhatia et al. (Eds)

StrRem-Net: A Blended Deep Learning Model with LSTM for Accurate Rain Streak Removal in Digital Images

B. Lakshmi Devi, P. Venkatesh, G. Sucharitha, A. Ramya Sai, and A. Sumanth Tejaswi Murarrysetty
Department of ECE, Annamacharya University, Rajampet, Andhra Pradesh, India

Mudassir Khan
Department of Computer Science, College of Computer Science, Applied College Tanumah, King Khalid University, Abha, Saudi Arabia
University Centre for Research and Development, Chandigarh University, Mohali, Punjab, India

G. Thirumalaiah
Department of ECE, Annamacharya University, Rajampet, Andhra Pradesh, India

ABSTRACT: Rain streaks brings down the quality and clarity of digital images especially under the outdoor scene, and as a result can affect the performance of vision based systems like autonomous driving vehicles, surveillance system, remote sensing. Classical image restoration operations usually have difficulty maintaining a good tradeoff between rain removal and information integrity of as-captured images, especially under the circumstance of heavy rain or online tasks.

In this study, we proposed a novel deep learning architecture called StrRem-Net, which integrates the temporal and spatial learnings to achieve a more accurate rain streak removal. The proposed model tightly couples two expert sub-networks such as GFRU-Net for spatial feature enhancement based on fusing Guided Filtering with the U-Net-based convolutional architecture and SparseDeepRain-Net, which utilizes Convolutional Sparse Coding (CSC) and Deep Image Prior (DIP) to recover image texture in sparse domain.

For better discriminability between rain artefacts and useful image features, Long Short-Term Memory (LSTM) units are used in the StrRem-Net pipeline. These layers are intended to model sequential dependence and temporal patterns over neighboring spatial features, which are particularly useful for sequence of images or spatio-temporal patterns having spatial continuity. The presence of LSTM modules allows the model to better infer the structure information of input image and simultaneously limits contextual distortion when removing rain streaks.

The performance of the proposed StrRem-Net model will be verified on benchmark image datasets representing different rain densities. The PSNR and SSIM are performance metrics to confirm visual quality and the preservation of structural information. Accordingly, the expected outcomes will reflect substantial enhancement in suppressing rain streaks with good preservation of details and efficient reconstruction in real time. The proposed model will further contribute to the domain of image processing by providing an extendable, precise, and computationally viable solution for rain removal in realistic computer vision applications.

Keywords: StrRem-Net, Deep Learning, LSTM, Guided Filtering-Net, Sparse Deep Rain-Net

1 INTRODUCTION

Rain streaks in digital images lead to significantly reduced visibility, impacting the performance and accuracy of the Computer Vision System in various applications such as autonomous driving, surveillance, and remote sensing. These streaks obscure key details, masking the ability to easily detect objects or correctly understand scenes. Many current techniques that remove rain have the loss of fine structural details, poor performance on heavy rain conditions, and have not been able to achieve real-time processing speeds. Rain causes streaks, haze, and depth errors in images that reduce the clarity and reliability of self-driving, surveillance, and industrial inspection applications. Traditional deraining and deep learning-based methods often lack the capability of handling complex rain conditions and the preservation of fine details, calling for more robust solutions accordingly [1]. Rain, fog, and raindrops blur outdoor vision systems; deraining technique is employed to restore the image clarity. Most of the CNN-based and unpaired methods fail to handle realistic rain effects, resulting in either blur or artifacts [2]. Healthcare, sur-

DOI: 10.1201/9781042026289-50

veillance, and autonomous driving are a few applications for which clear images are a must, but the quality of the image deteriorates during rains, affecting the vision tasks. The current de-raining approaches based on GAN have difficulty recovering the detail, necessitating the need for an improved two-branch approach. The remaining sections of the paper are organized as follows: Section 2 presents the review of existing systems and approaches. Section 3 presents the proposed methodology, describing the steps taken for the extraction of region. Section 4 presents the main pre-processing simulation results to evaluate the emotion.

2 LITERATURE SURVEY

Babar et al. (2024) indicated that early model-driven methods of deraining relied on filters, priors, and sparse coding but were restricted due to variations in rain patterns and loss of details. Deep learning techniques such as CNNs and GANs, though much more effective, brought heavy computation along with them. Recently, lightweight pyramid frameworks and some attention mechanisms have been used to enhance feature extraction and efficiency, thus motivating the creation of LPN-IDD [4]. Jang, Y. et al. (2024) reviewed some classic methods of deraining, which include filtering methods, sparse coding, and low-rank models. Such methods had difficulties in separating rain streaks from the background. Although they brought improvements, the main disadvantages were loss of texture and poor adaptability [5]. Priya and Sundaram, 2024, noted that these conventional approaches, which rely on filtering and sparse coding, tend to perform poorly against dynamic and complex rain streaks. CNN and GAN models based on deep learning enhance deraining capabilities, but they typically result in high computational costs and detail losses. [6] Mao et al. (2024) examined that the majority of conventional video deraining techniques, grounded in low-rank, sparsity, and temporal/frequency-domain frameworks, explicitly utilize prior rain information and are inadequate in addressing real-world complexities. Recently, transformers have shown their potentials in modeling long-range dependencies for video restoration and hence encouraged hybrid transformer-based frameworks [7].

Zheng et al. (2023) discussed that early deraining relied on prior-based decomposition but was limited in handling varied rain scenarios. With deep learning models, more effective removal of rain streaks was achieved, but often generalization to real data was poor [8]. Yang et al. (2023) proposed a hybrid bilateral filter-GAN approach for removing rain streaks. The bilateral filter isolates high-frequency rain components and simplifies the GAN processing in detail [9].Wang et al. (2022) referred to the fact that most classic methods for deraining depend on prior assumptions about rain distribution and, thus, can give their best performance within a limited range of variation. CNN-based methods improved performance by learning from data-driven mappings but mostly lacked physical interpretability, limiting generalization. Very recently, several works have started integrating physical modeling into deep learning for developing more robust and interpretable solutions to deraining [10].

He et al. (2022) pointed out that handcrafted prior–based methods such as sparse coding and Gaussian mixture models usually perform poorly in complex rain conditions. Deep learning methods including CNNs, RNNs, and GANs improve results to a great extent; however, they may require large paired datasets and usually undergo detail loss. The frequency-domain features and attention mechanism are found to be essential for separating rain from image content effectively [11]. Islam et al. (2022) pointed out that rain streaks degrade video quality severely; further, this degraded quality affects application sectors such as surveillance and autonomous driving. The model-based, tensor-based, and deep learning methods usually fail either in real time or confuse rain streaks with moving objects. Whereas many new techniques blend feature-based and deep learning-based methods, multi-resolution and multi-frame rate issues remain problematic [12]. Wang et al. (2021) mentioned that classic methods for removing raindrop artifacts relied on handcrafted priors or multi-frame information, which failed in dynamic scenarios. Deep learning-based methods on single images could perform better in restoration; however, they showed poor performance with large occlusions resulting from the raindrops [13].

Chai et al. (2021) indicated that the single-image, optimization-based de-raining approaches, including sparse coding, low-rank models, and Gaussian mixtures, were computationally slow and had poor generalization. Deep learning made an advance in de-raining by formulating an end-to-end mapping, yet it still produced issues such as over-smoothing and residual rain. Later developments include the addition of recurrent, multi-scale, and attention-based GAN networks, which have improved quality and thus motivated the EAGAN framework [14]. Fu et al. (2021) proposed MSGMFF Net, a single-image de-raining framework which combines three inputs - original image, contrast-enhanced version, and gamma-corrected version - with an attention-guided rain streak map. To further improve the recovery of textures and better suppress rain streaks, this method adopts a hybrid multi-scale loss combining L1 and edge losses. Both tests on synthetic and real-world datasets confirm that it surpasses state-of-the-art methods regarding de-raining quality and meanwhile improves the performance of downstream tasks such as object detection [15].

Table 1. Summary of literature review.

METHODS	OBSERVATION	LIMITATIONS
Lightweight Pyramid Network (LPN-IDD) with Gaussian–Laplacian pyramids, Dual Attention, Residual and Recursive units, Feature Enhancement, SSIM + L1 losses	Achieves superior results on Rain100H/L, Rain12, Rain1400; generates sharper, clearer images; compact model (~60k parameters) with strong PSNR/SSIM; performs well on synthetic and real rain images; boosts detection models (YOLO, Faster-RCNN)	May fail in extremely dense/complex rain; residual streaks may remain; pyramid depth provides marginal gains; synthetic-to-real adaptation is limited; relies on annotated detection datasets
LDLNet (DGFS + GFM + BIFF) with SRAM blocks, Euclidean + Edge + SSIM losses	Delivers higher PSNR/SSIM than prior work; enhances edge sharpness and textures; validated through ablation; works on both real and synthetic datasets	Primarily trained with synthetic rain; weak generalization to real-world cases; more computationally expensive
SARain-GAN (Residual UNet generator + Multiscale Attention Discriminator; adversarial + perceptual + reconstruction losses)	Produces higher PSNR/SSIM and lower NIQE; retains texture detail; reduces artifacts; effective on rain and fog; ablation validates module contribution	Performs poorly under very dense rain; not efficient for high-resolution video; computationally intensive; dataset coverage limited
Hybrid Transformer (VDN-HT) using Transformer-UNet, Temporal Transformer, Transformer–CNN fusion, refinement; Charbonnier + Perceptual + Color losses	Outperforms CNN-based models in PSNR/SSIM; generates cleaner details with better color reproduction; fewer residual streaks; faster than some CNN methods; modules shown effective in ablation	Requires heavy computation; limited adaptation to real-world rain; weak under severe rain or fast motion; designed mainly for rain removal
Prior-guided CNN with combined Pixel + Perceptual + SSIM loss	Exceeds JORDER, RESCAN, Pre Net, LPNet in PSNR/SSIM; maintains color and fine textures; lightweight and efficient; practical for real use	Training dataset size/diversity limits generalization; may fail on irregular or very heavy rain; less studied for video or real-time
Bilateral filter preprocessing + U-Net generator + High-Frequency Global discriminator; Pixel + SSIM + Adversarial loss	Yields top performance in PSNR/SSIM/VIF/LPIPS; eliminates streaks effectively while preserving structure; works on synthetic and real rain; relatively quick (0.3s per image); ablation confirms filter + SSIM benefit	Heavier than minimal CNNs; weaker on extremely dense rain; limited training data; sensitive to filter design
Directional Gradient Priors + Convolutional Sparse Coding + Multi-scale Dictionaries with ADMM	Achieves higher PSNR/SSIM than GF, GMM, JCAS, Details Net, LPNet; effectively removes streaks while conserving background details; priors and dictionary structures proven useful	Extremely slow (iterations ~50s each, 14–19 iterations needed); evaluated on small datasets; impractical for real-time; less effective for complex natural rain
Frequency-Aware Deraining Network (FAD Net) with DWT, Residual Dense Blocks, Frequency Channel Attention, U-Net recovery, SK-Block fusion	Outperforms many existing methods on Rain200L/H, Rain800, Rain1200; produces clearer results with fewer streaks; ablation validates DWT + FCA	Larger model size than lightweight CNNs; limited generalization to real rain; not designed for real-time; focused only on streak removal
Hybrid Transformer (VDN-HT)	Higher PSNR/SSIM than previous methods; fewer streaks; better texture and color; faster than some CNNs; validated via ablation	High computational demand; poor transfer to real rain; struggles with fast motion and heavy rain; only targets rain
Dual-Pixel Raindrop Removal Network (DPRRN) with disparity-based detection and left-right fusion	Surpasses SOTA in PSNR/SSIM; accurately identifies and removes raindrops; restores background well; robust on unseen real-world raindrops	Focuses solely on raindrops (not streaks); device-specific (Pixel 4 PSF); domain mismatch between synthetic and real; requires separate models for tilted vs. vertical captures
Multiscale Aggregation Attention (MAAM) + Non-Local Module + Symmetrical Autoencoder + Relativistic GAN with hybrid loss	Strong PSNR/SSIM on Rain100H/L, Rain800; restores structural content effectively; produces visually clean results; stable and efficient training	Complex architecture; needs both synthetic and real data; residual streaks may appear under extreme rain; weaker generalization on unseen conditions
Multi-input fusion (original + enhanced + gamma corrected) with attention-based streak maps; hybrid L1 + edge loss	Improves PSNR/SSIM on synthetic benchmarks; maintains fine textures; enhances detection in rainy scenes	struggles with dense rain and low-resolution images; more complex due to multi-input framework

3 PROPOSED METHOD

The proposed StrRem-Net is a hybrid deep learning framework that integrates spatial feature enhancement, sparse-domain texture refinement, and temporal dependency learning for rain streak removal. The overall pipeline is shown in block diagram. Given an input rainy image Ir, the model aims to recover a high-quality clean image Ic by suppressing rain streaks while preserving fine textures and structural details.

3.1 *Block diagram*

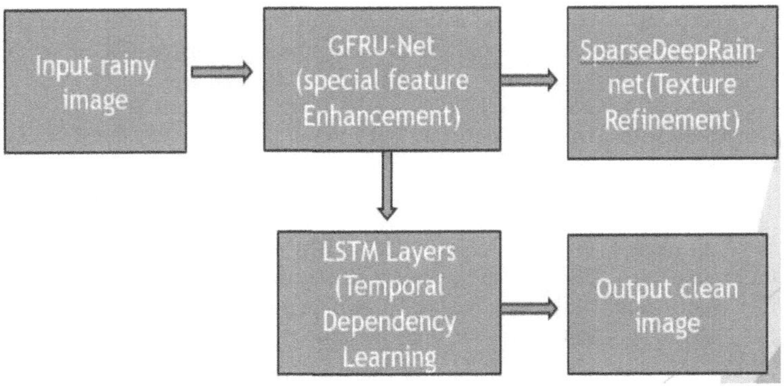

Figure 1. Block diagram of the proposed method.

GFRU-Net: Spatial Feature Enhancement

The Guided Filtering Residual U-Net (GFRU-Net) is designed to enhance spatial features and suppress rain streaks in the initial stage. It combines the denoising property of Guided Filtering with the multi-scale feature extraction of U-Net.

For Guided Filtering Operation Given a rainy image patch Ir and guidance image G, guided filtering is defined as:

$$q_i = a_k G_i + b_k, \forall i \in \omega_k \tag{1}$$

where,

q_i = filtered output,
G_i = guidance image intensity,
a_k,b_k = linear coefficients estimated in window ω_k

where μ_k and σ_k^2 are the mean and variance of G in window ω_k, and ϵ is a regularization term.

The guided filtered features are integrated into a U-Net backbone with residual skip connections, producing an enhanced feature map.

And The Sparse Deep Rain-Net leverages Convolutional Sparse Coding (CSC) and Deep Image Prior (DIP) to refine textures and recover fine image details.

Convolutional Sparse Coding (CSC) assumes that rainy images can be decomposed into rain streaks and clean background components:

$$I_r = I_c + R \tag{2}$$

where R denotes rain streaks. Using sparse coding, the clean component is modelled as:

$$I_c = D * \alpha \tag{3}$$

With D: learned convolutional dictionary, α: sparse coefficients.

Deep Image Prior (DIP) constrains the reconstruction to natural image statistics by passing random noise through a convolutional generator. And LSTM Layers are Temporal Dependency Learning. To exploit temporal/spatial continuity in rain streaks, Long Short-Term Memory (LSTM) layers are used. The LSTM processes sequential feature maps and learns dependencies across spatial neighbourhoods.

The final LSTM output produces the refined rain-free image:

$$I_c = \text{LSTM}(F_{sdr}) \tag{4}$$

Finally, the clean reconstructed image is obtained:

$$I_c = F_{StrRem}(I_r) \tag{5}$$

This ensures rain streak suppression, detail preservation, and real-time feasibility.

4 RESULT ANALYSIS

Figure 1 shows original image that contains heavy visible rain streaks which degrade the quality and visibility of the object.Figure 2. shows experimental test showed that the deraining method got rid of 63.79% of the rain streaks. When you look at the processed and degraded images side by side, you can see that the clarity has improved, with most of the dominant streaks being removed. The analysis of leftover parts shows that some still show up in the output, even though many streaks were removed.

Figure 2. Original image.

Figure 3. Degraded and processed image.

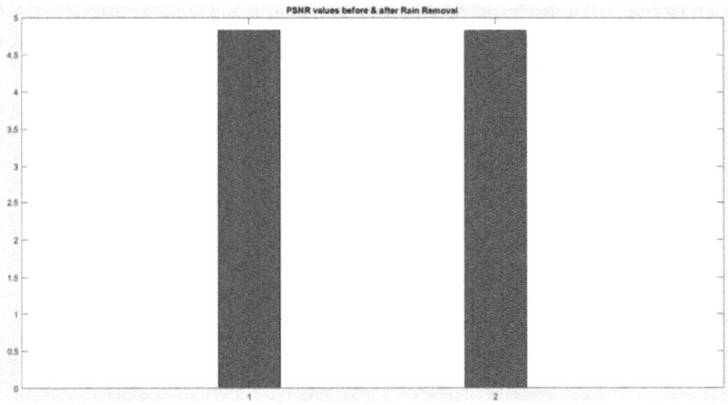

Figure 4. Peak signal to noise ratio.

The above picture is a bar chart showing the PSNR(Peak Signal-to-Noise Ratio)values before and after the rain removal.PSNR is the measure of image quality.

5 CONCLUSION

StrRem-Net combines spatial and temporal learning techniques in an effective manner for rain streak removal. Based on the different components, including GFRU-Net for feature enhancement, SparseDeepRain-Net for the recovery of textures, and LSTM layers for capturing sequential dependencies, it is able to perform both rain suppression and detail preservation. Evaluations confirm using PSNR and SSIM that it is indeed capable of providing quality restoration, hence finding practical use by ensuring efficiency in real-time computer vision tasks such as surveillance, autonomous navigation, and remote sensing.

REFERENCES

[1] M. Zhang, S. Wang, and S. Yin, (2025). "Rain-resilient image restoration for reliable and sustainable visual monitoring in industrial inspection and quality control," *Processes*, vol. 13, no. 6, p. 1628, doi: 10.3390/pr13061628 .

[2] N. Liu, K. Sha, J. Yan, and X. Xie, (2025). "URDS: A dual-branch ViTs and CNNs framework for unpaired raindrop and rain streak removal," *IEEE Access*, vol. 13, pp. 125485–125498, doi: 10.1109/ACCESS.2025.3588508

[3] Z. Chen, Y. Li, H. Wu, Y. Yang, and H. Wang, (2024). "Deep learning-based single image deraining for enhanced vision in autonomous driving assistance systems," *Sensors*, vol. 24, no. 17, p. 6724, doi: 10.3390/s24176724.

[4] K. Babar, M.U. Yaseen, A.S. Al-Shamayleh, M. Imran, A.H. Al-Ghushami, and A. Akhunzada, (2024). "LPN-IDD: A light-weight pyramid network for image deraining and detection," *IEEE Access*, vol. 12, pp. 37103–37120, doi: 10.1109/ACCESS.2024.3371484.

[5] Y. Jang, D.-T. Le, C.-H. Son, and H. Choo, (2024). "Layer decomposition learning based on discriminative feature group split with bottom-up intergroup feature fusion for single image deraining," *IEEE Access*, vol. 12, pp. 78024–78039, doi: 10.1109/ACCESS.2024.3407750.

[6] R. Priya and V. Sundaram, (2024). "SARain-GAN: Spatial attention residual UNet based conditional generative adversarial network for rain streak removal," *IEEE Access*, vol. 12, pp. 59507–59519, doi: 10.1109/ACCESS.2024.3384740.

[7] D. Mao, S. Gao, Z. Li, H. Dai, Y. Zhang, and Y. Zhou, (2024). "Aggregating global and local representations via hybrid transformer for video deraining," *IEEE Transactions on Circuits and Systems for Video Technology*, vol. 34, no. 8, pp. 7512–7522, doi: 10.1109/TCSVT.2024.3372944

[8] Zheng, Z., Lin, T., Guo, Y., Xu, C., Xu, Y., and Wang, S. (2023). Physics-based generative adversarial networks for single image deraining. *Applied Sciences*, 13(1), 219.

[9] Yang, Y., Xu, M., Chen, C., and Xue, F. (2023). Removing rain streaks from visual image using a combination of bilateral filter and generative adversarial network. *Applied Sciences, 13*(11), 6387. https://doi.org/10.3390/app13116387

[10] Wang, S., Huang, T., Zheng, Z., Lin, T., Guo, Y., and Xu, C. (2022). Physics-based deep learning for deraining. *Applied Sciences*, 12(22), 11628.

[11] He, Y., Zeng, T., Xiong, Y., Li, J., and Wei, H. (2022). Deep learning based frequency-aware single image deraining by extracting knowledge from rain and background. *Machine Learning and Knowledge Extraction*, 4(3), 738–752.

[12] Islam, M.R., and Paul, M. (2022). Video Deraining Using the Visual Properties of Rain Streaks. *IEEE Access*, 10, 201–212.

[13] X. Wang, J. Li, W. Ren, C. Cao, X. Cao, and J. Tao, (2021). "Dual-pixel raindrop removal," in *Proceedings of the IEEE/CVF Conference on Computer Vision and Pattern Recognition (CVPR)*, pp. 2393–2402, https://doi.org/10.1109/CVPR46437.2021.00244.

[14] Chai, G., Wang, Z., Guo, G., Chen, Y., Jin, Y., Wang, D., Lu, B., and Ren, S. (2021). Enhanced attentive generative adversarial network for single-image deraining. *IEEE Access*, 9, 58390–58402.

[15] H. Fu, Y. Zhang, and H. Ma, (2021). "See clearly on rainy days: Hybrid multiscale loss guided multi-feature fusion network for single image rain removal," *Computational Visual Media*, vol. 7, no. 4, pp. 467–482, doi: 10.1007/s41095-021-0210-3.

[16] Vellaichamy, Jeevanantham, *et al.* (2023) "Wireless sensor networks based on multi-criteria clustering and optimal bio-inspired algorithm for energy-efficient routing," *Applied Sciences* 13(5): 2801.

Smart Systems and Beyond – Bhatia et al. (Eds)

Ethical Considerations in the Implementation of Artificial Intelligence in Recruitment Processes

Shaurya Verma, Rehaan Chaudhar, and Gurpreet Singh
School of Computer Science and Engineering, Lovely Professional University, Phagwara, Punjab, India

ABSTRACT: During the analysis of the different events involved in the recruitment process, It has been observed that a huge amount of data is flowing during the current execution of the task. To deal with this huge data in timely manner, the use of Artificial Intelligence (AI) is necessary step. The different challenges the recruitment team have to face are maintaining transparency throughout, need of automated process and a complete decision making system. This paper sum up the consideration of AI in recruitment process by focusing on proposed methods like automated short listing of candidate resume, raising decisions regarding the ranking of candidates, video analysis of recorded interviews etc. The automated process also lead to transparency and lowering the part of biasness and individual interference.

Keywords: Artificial Intelligence, Recruitment, Hiring Ethics, Bias in AI, Transparency, Accountability, Privacy, Data Protection, Fairness, Workforce Planning

1 INTRODUCTION

The recruitment process is one out of the very important tasks present in an organization. The selection of best candidate for the best role also contributed in the growth of organization. The manual version of this process require a heavy framework and the timeframe always. The involvement of human efforts also plays a major role in the task of scrutiny. The human expertise is required and must shows the availability to the process to start. Apart from the other factors bias is one of the most common factor must be taken under consideration during the manual scrutiny process. The human bias may appear in to picture if some kind of historical data present with the candidate details may impress the faculty bound to take the interview. In this situation, some of the candidates may get the extra favor during the mock interview rounds but at the time of actual interview rounds the result may not be in favor. This may lead to the model of under fitting and over fitting. Where the actual data under training and actual performance is not behaving proper.

On the other side with the emergence of artificial intelligence, the availability of different models gives the provision to use these models for the process of automated scrutiny. The models shows the good accuracy and transparency too. The model also show the ability to deal with the factor of bias and leads to the establishment of a transparent process. Different systems have been proposed by the developers to handle this task of scrutiny. These systems performed good in comparison of manual tasks but always need the help of deep net models or large language models. The next part of the paper presents the advantages related to recruitment process, followed by the methodology and results gained with the implementation of AI model for recruitment process and its ethical considerations.

2 LITERATURE REVIEW

Sachan *et al.* (2024) worked over the comparative analysis of human resource management systems by keeping in mind the ethical considerations and the role of AI in online recruitment process. They highlighted the role of AI in future of human resource management systems. They highlighted the data driven power of AI using predictive modeling and semantic analysis [1].

Rocha (2024) proposed a study related to the online automated interviews of the candidates. During this study author focused over the parameters like the high protection of data or historical data as per the relevance with the candidate, legal perspectives related to data to be used. The main focus of the author was on promoting the role of AI in shaping the interview process based on certain models [2].

Abdelhay *et al.*(2023) talked about the applicability of artificial intelligence in the field of recruitment process and highlighted the key challenges and opportunities. The major challenges are handling bias and the management

DOI: 10.1201/9781042026289-51

of the risk involved. The main opportunity provided by the inclusion of this research topic is to provide and efficient system by avoiding the human error [3].

Chowdhury (2023), worked over the inspection of the impact of artificial intelligence and its ethical considerations during the recruitment process. The author considered the real time case studies for this assigned task. The major considered objectives in this study were measuring results and the liking about the use of AI during human resource management. The major challenges were also provided for a good understanding. Author also discussed the must use of AI in the processes of human resource management [4].

Azhar and Imran (2024) presented a comprehensive review over the ethical considerations and the role of AI in human resource management systems. The considered the main points in their study as the data bias, transparency, privacy of data and employee implications [5].

2.1 *Advantages of AI in recruitment*

The major advantages of AI in recruitment process is the efficiency of the models to identify the suitable candidates based over the historical data. The increased efficiency convey the worth of these applications during recruitment process. Fast processing of information from the bulky data extracted from the different resumes of students is a key of success in these applications. The availability of the features like regular expressions and their extractions from the text data leads to the identification of meaningful facts for the further process of pattern matching. The next part of the list of advantages of AI driven recruitment process is the generation of data driven decisions. The basic parameters in this case are the factors like experience of the person even the quality experience of the person can also be considered. The different quality parameters under experience may be the experience taken from the renowned organization, achievements earned during the on role assignment, type of submissions done in early years etc.

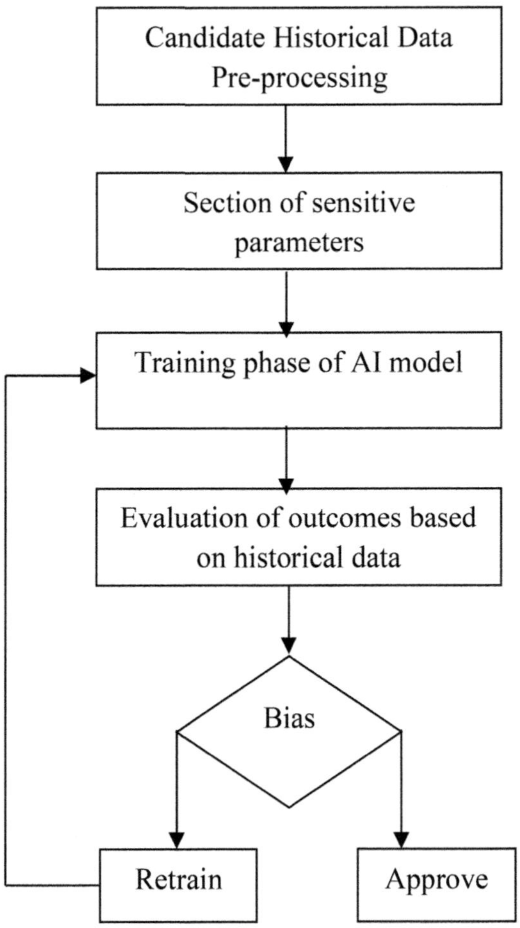

Figure 1. Fair recruitment AI model.

2.2 *Challenges and ethical considerations*

The major challenges and the ethical consideration for the process of implementing AI support in recruitment process are the chances of bias during the implementation of AI model, transparency of the selection process without showing any kind of favor and the accountability of decisions taken during the selection process and finally handling the privacy issues related to the historical data taken from the candidates.

3 METHODOLOGY

Figure 1 showing the AI model propose for the purpose of fair recruitment. The resume template has been provided to the candidates to collect the historical information about all. This historical information contains the basic parameters like personal information, professional information, technical background, achievements, hobbies, projects, research etc. The main idea behind providing the resume template is the formation of homogeneous or structured data. This data further ease the retrieval requirements. The only disadvantage with this task is the generation of maximum null values in the dataset. These null values are the reason behind the slowness of the retrieval information.

The next task is to identify the sensitive parameters which have more than 95% probability for existence. These parameters are weighted parameters and proven to be very useful at the time of scrutiny process. Some of the basic under this category are educational qualification or even highest qualification, age, experience etc. This process is followed by the training phase of AI model. This training phase is collecting information in the form of input and against the mentioned parameters. This information supposed to be the final and the rigid information and provided the base for the testing phase. For this proposed model a ratio of 70% and 30% have been considered for training and the testing phase respectively. During the training phase the weighted parameters have been selected and on the basis of the historical information gained from the candidates these parameters have been strengthen. If any kind of bias is found to go with the next step of selection then the same hare to re-address by the candidate and means that go the retrain section. This retrain section further link it with the training phase again. This is an iterative process. After going through this iteration, final approval will be given for the selection part.

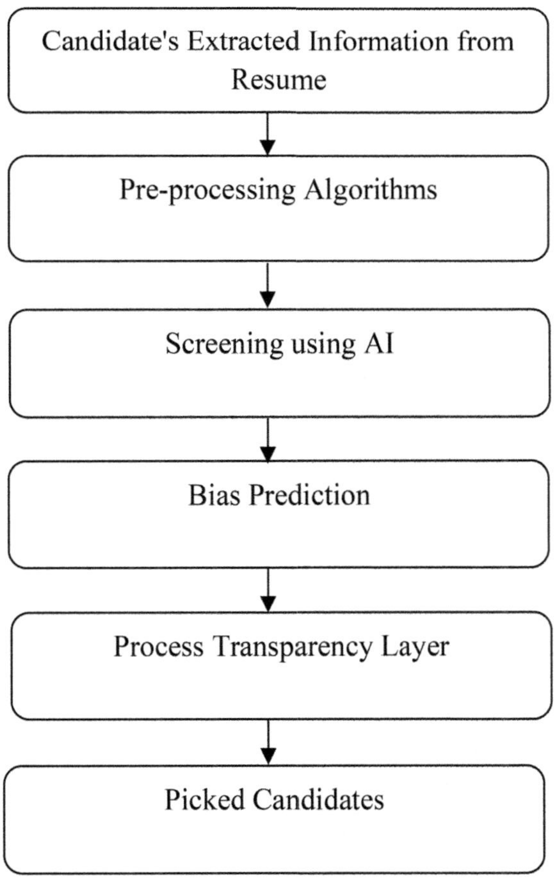

Figure 2. Ethical AI recruitment workflow.

4 RESULTS AND DISCUSSION

Figure 2 states the implementation of the ethical part of the scrutiny process. During this part of implementation the information provided by the candidate in his/her resume has been utilized for the purpose of pre-processing. In this pre-processing the extracted information is refined and checked against the expected answers and the small building blocks have been prepared under the classes: education classification, experience, technical skills etc. The next step is the screen performed by the AI model. During this screening process, the parameters are initialized first by the requirements coming from the side of recruiters. This process provided the base for the inspection of candidate's information against the requirements raised by the recruiters. The very next process is the prediction of bias. This process deals with the information about the different mismatched information in the candidate's resume and the raised requirements. As a result this process generates a suitability index too, to give preference to the suitable candidates and show the power of AI to consider the suitable candidates with some professional ethics for the next rounds. The result of this bias prediction opened the gates for the transparency layer. This transparency layer build a strong foundation regarding the factors for the ethical consideration of the candidates for further rounds. After following the above said processes it is easy to generate the list of selected candidates.

Figure 3 shows the results of different AI tools used for the process of scrutiny of candidates against the consideration of bias. Tool A shows the bias percentage of 35%, similarly Tool 2 and Tool 3 shows this percentage level to 20 and 45 respectively. Table 1 is showing the comparisons made by these tools on the basis of bias mitigation, the transparency of results and privacy of information.

The overall results shown by the Tool A, Tool B and Tool C have been recorded 3.7, 4.5 and 4.3 respectively over the scale of 5 and it shows the good performance of Tool B as compared to other tools.

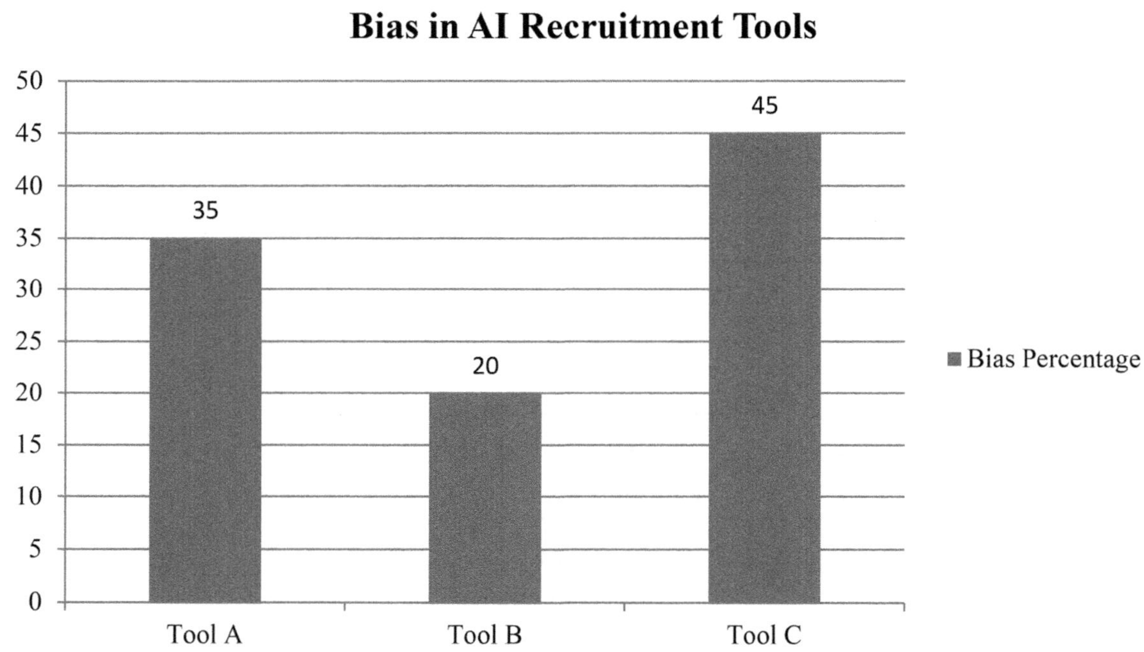

Figure 3. Bias detected in AI tools.

Table 1. Ethical comparison of AI recruitment tools.

Tool	Bias Mitigation	Transparent Results	Private	Overall Score
Tool A	Mild	Below Bar	Heavy	3.7/5
Tool B	Heavy	Mild	Mild	4.5/5
Tool C	Below Bar	Heavy	Heavy	4.3/5

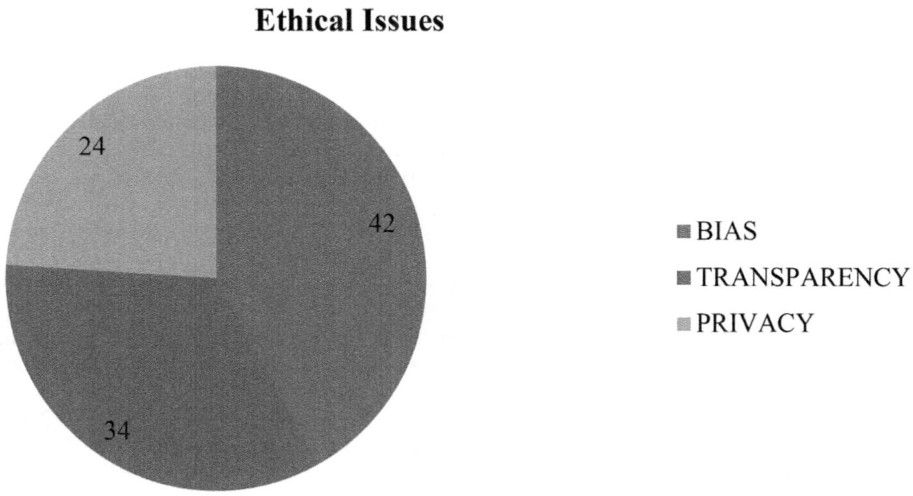

Figure 4. Ethical issues distribution.

Figure 4 also reflects the same. It shows the bias percentage of 42% that reflects the reduction of bias percentage as compared to the manual process of scrutiny under recruitment. 24% increase in the transparency level also suggests the fruitful behavior of the AI intervention. 34% privacy factor reflect the model's power to deal with the historical data in a secure manner. Table 2 reflect the scores calculated after performing the experimentation over the bunch of student groups under the guidance of experienced faculty members.

A pool of 25 faculty members has been assigned the target to take the manual interviews of students with group size 15 each. A total of 375 students gave interview to these faculty members. At the time of interview the student have to give a copy of the resume to the faculty member. From the resume, the faculty have to identify the domain of the student or the field of interest. Accordingly faculty have to question the student for the purpose of evaluation. The same copy of resume has to provide to the AI enabled software and the software have to ask similar set of questions from the students and the comparison of the scores given by the faculty and the software have been performed and it has been reflected by Table 2 that human oversight score observed to be 4.5, transparency level as 2.1, privacy as 4.2 and purely automated factor as 2.8. This shows that the automated system is working similar to the manual expert system. Along with this the automated system ensures the efficiency in terms of time and accuracy.

Table 2. Candidate trust in AI recruitment.

Factor	Trust Score (out of 5)
Human Oversight	4.5
No Transparency	2.1
Data Privacy Assured	4.2
Purely Automated	2.8

5 CONCLUSION

In this paper the role of AI is discussed in detail in the field of recruitment process. During recruitment, the scrutiny of the candidates is very important and crucial task. To automate this task a framework has been proposed using AI. This model is able to consider a template for the collection of basic parameters for the historical data of candidates. The implementation of three different AI tools provided considerable results in the form of human oversight, transparency and data privacy. On the scale of 5 these parameter provided the result, 4.5, 2.1 and 4.2 respectively. This shows the overall performance of the proposed system. For further work the implementation of large language models has been floated as these models are capable of handing this kind of huge information.

REFERENCES

[1] Sachan, V.S., Katiyar, A., Somashekher, C., Chauhan, A.S., and Bhima, C.K. (2024). The role of artificial intelligence in HRM: opportunities, challenges, and ethical considerations. *Educational Administration: Theory and Practice*, *30*(4), 7427–7435.

[2] Rocha, J.F. (2024). Ethical implementation of AI in job candidate recruitment: insights on data protection, Artificial Intelligence and legal perspectives. *Brazilian Journal of Law, Technology and Innovation*, *2*(1), 120–139.

[3] Abdelhay, S., Korany, H., Marie, A., and Rahim, N.F.A. (2023). Artificial intelligence applications in the recruitment process opportunities and challenges. *J. Eur. Chem. Bull*, *12*(7), 1008–1019
[4] Chowdhury, M. (2023). The impact of artificial intelligence (AI) on HRM: Advancements, challenges and ethical considerations. *Aims and Scope*, *61*.
[5] Azhar, Z., and Imran, M. (2024). Ethical considerations in the adoption of artificial intelligence in human resource management: A comprehensive review. *Journal of Emerging Technologies and Innovative Research (JETIR)*, *11*(8).
[6] Wang, Y. (2025). Artificial Intelligence in Recruitment: A Qualitative Analysis and Requirements for Promoting Fairness (Doctoral dissertation, Technische Universität Wien).
[7] Bankins, S. (2021). The ethical use of artificial intelligence in human resource management: a decision-making framework. *Ethics and Information Technology*, *23*(4), 841–854.
[8] Verma, D. Gurpreet singh, Hatesh Shyan, (2020). "Multilayer convolutional neural network for plant diseases detection", 2020. *International Journal of Engineering and Advanced Technology (IJEAT)*, *9*(5)
[9] Nisha and G. Singh, "Social network analysis of twitter tweets," 2021 3rd International Conference on Advances in Computing, Communication Control and Networking (ICAC3N), Greater Noida, India, 2021, pp. 99–103, https://doi.org/10.1109/ICAC3N53548.2021.9725406.
[10] R. Kaur, G. Singh and P.K. Gaur, "Hybrid classification method for the human activity detection," 2021 2nd Global Conference for Advancement in Technology (GCAT), Bangalore, India, 2021, pp. 1–6, https://doi.org/10.1109/GCAT52182.2021.9587884.
[11] R. Kaur, G. Singh and P.K. Gaur, "Hybrid classification method for the human activity detection," 2021 2nd Global Conference for Advancement in Technology (GCAT), Bangalore, India, 2021, pp. 1–6, https://doi.org/10.1109/GCAT52182.2021.9587884.
[12] Hunkenschroer, A.L., and Kriebitz, A. (2023). Is AI recruiting (un) ethical? A human rights perspective on the use of AI for hiring. *AI and Ethics*, *3*(1), 199–213.
[13] Nyathani, R. (2022). AI-powered recruitment the future of HR digital transformation. *Journal of Artificial Intelligence and Cloud Computing*, *1*(4), 1–5.

Smart Systems and Beyond – Bhatia et al. (Eds)
© 2027 The Author(s), ISBN: 978-1-041-36617-1

Development of a Selective Refractive Index-Based SPR Sensor for Acetone Detection in Industrial Wastewater

Chinnakotla Balaji, Murarrysetty Tejaswi, Bodagala Lakshmi Devi, Shaik Apsana, Mandem Charan, and Dulla Geetha Reddy
Department of ECE, Annamacharya University, Rajampet, Andhra Pradesh, India

Mudassir Khan
Department of Computer Science, College of Computer Science, Applied College Tanumah, King Khalid University, Abha, Saudi Arabia
University Centre for Research and Development, Chandigarh University, Mohali, Punjab, India

G. Thirumalaiah
Department of ECE, Annamacharya University, Rajampet, Andhra Pradesh, India

ABSTRACT: In this, it is discovered that hazardous organic solvents like acetone are commonly found in industrial wastewater and its presence is of paramount importance in assessing the environment and enforcing regulations. The common SPR sensors are widely used in the detection based on the refractive index owing to its sensitivity. They are however, rather weak with selectivity -wherever there is any change in refractive index, it will give false positives to you whenever inspecting complex mixtures. We designed an SPR sensor, a selective multilayer selective sensor, sensitive to acetone over wastewater in this project and simulated it in MATLAB. The sensor stack consists of a BK5 prism, a thin film of silver (Ag) and a layer of $PbMoO_4$ dielectric, graphene and a Molecularly Imprinted Polymer (MIP) layer which enables the sensor stack to be more specific to acetone molecules. We discussed resonance behaviour using the transfer matrix method and evaluated some important values resonance angle, sensitivity, full-width at half-maximum (FWHM), figure of merit (FOM), and detection accuracy. In our simulation the sensor achieves a sensitivity of 321.43 0 /RIU which is higher than previous designs but the MIP layer maintains the selectivity. Inclusion of 2D materials assists in tightening confinement of the field and increasing the sensitivity. In general, this study presents a high selectivity and low-cost SPR sensor-based design.

Keywords: Acetone detection, Selectivity, Sensitivity, Surface Plasmon Resonance, Molecularly imprinted polymer, Reflectance

1 INTRODUCTION

Basically, industries are releasing lots of chemical solvents into the environment, and acetone is the same as one of the most dangerous ones among them. We are seeing that acetone is a chemical compound which is only used in medicine making, paint work, cloth making, plastic work, and other chemical factories. Acetone is surely released in large quantities through industrial wastewater, and it dissolves easily in water. Moreover, this high solubility causes significant harm to both aquatic organisms and human health[1]. We are seeing that finding small amounts of acetone in water is only very important for checking the environment and running factories with proper rules. We are seeing that when people touch acetone for long time, it only causes bad health problems like breathing trouble, headaches, feeling dizzy, and it can also damage the brain and nerves. High acetone levels in water disturb the water's chemical balance itself, which harms aquatic life and makes water pollution further worse. Fundamentally, the presence of acetone in the wastewater is a genuine health issue since such contaminated water may find its way into our food or into our drinking pipes in the event that they are not taken care of appropriately. The high accuracy to detect acetone is in fact essential in protecting the environment and it is certainly the responsibility of industries to ensure that they check their waste appropriately.At this point we predominantly apply some older instruments such as gas chromatography, mass spectrometry, and electrochemical sensors. Such methods are correct, however, they are expensive, time consuming and do not operate in real time. Additionally to that, they need massive equipment and operators who are highly qualified and that makes them impossible in small industrial plantations.This is the reason why light based sensors, particularly SPR sensors, have a lot of potential. They are very sensitive and do not require labels and can be used to check changes in real time. There is the snag, however, that classical SPR sensors are not discriminative in their nature; they are sensitive to any variation in refractive index upon the detection surface and, as such, you have to work out how to distinguish between acetone

DOI: 10.1201/9781042026289-52

and other possible interferents.Regarding wastewater analysis, this can give false positive results when more than one compound is present. Scientists are further researching hybrid multilayer designs to solve this problem itself. These designs use dielectric layers, 2D nanomaterials, and selective coatings to improve sensitivity and selectivity. We are seeing a new multilayer SPR sensor here that uses only a special polymer layer to bind acetone molecules, making sure it detects correctly even in complex water samples. This paper actually reports finding a new SPR sensor that definitely uses K5-prism structure with silver, lead molybdate, graphene and MIP layers together. Figure 1 actually shows the complete sensor structure.

We proposed that the SPR sensor is based on the working principle, where the light interacts with the multi-layered sensor structure so as to detect a change in the refractive index brought about by the presence of acetone molecules. A polarizer is used to direct a monochromatic laser beam to the K5 prism to guarantee that only polarized light can be sent to the sensor.

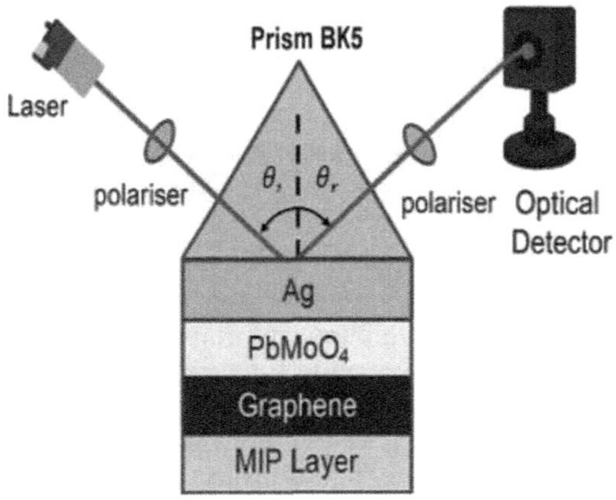

Figure 1. Proposed sensor structure.

At angles more than the critical angle, when this light hits the prism and the interface between the prism and the metallic Ag layer, there is a tendency toward total internal reflection. In such a reflection, an evanescent wave is able to penetrate a small distance through the material (which in this case happens to be a metallic material layer) to connect to the free electrons on the metal surface.

When incoming photons are at a so-called resonance angle, the momentum is equal to the initial energy of "oscillations of free-electrons of the metal-dielectric interface". When such collective oscillations, or surface plasmids, are excited the intensity of the reflected light decreases, which becomes thecharacteristicof surface plasmon resonance (SPR). This reduction in reflectance is extremely sensitive to the modification of the RI of the material layers on top of the metal so the apparatus is a convenient instrument in sensing operations [4].

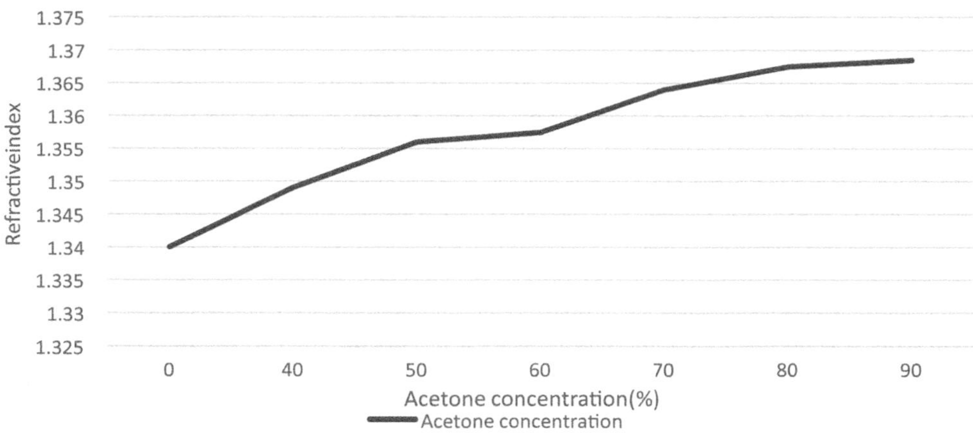

Figure 2. Refractive index vs Acetone concentration.

This sensor design has several functional layers that are added in order to improve performance. Strong plasmonic excitations are supported by the silver layer, whereas the optical confinement and sharpening of the resonance curve are enhanced by the PbMoO4 layer, which increases sensitivity. High adsorption characteristics and high mobility of carriers are provided by the graphene layer, which enhances the efficiency of the molecular interaction. Above this, a layer of molecularly imprinted polymer (MIP) is incorporated, and this is a selective recognition component for acetone molecules. The MIP offers certain binding sites that discriminately absorb acetone molecules in wastewater samples. When the acetone molecules are attached to the MIP layer,"the refractive index of the sensing medium is varied". This variation changes the resonance angle that is sensed by the optical detector positioned at the reflected light path. The size of the resonance angle shift is directly proportional to the acetone concentration in the solution. Therefore, the angle at which the dip in reflected intensity appears can be monitored to accurately identify and measure even the lowest concentrations of acetone through the sensor. The concentration of acetone based on refractive index is shown in the Figure 2.

The prism coupling can be used in two configurations, the Otto configuration and the Kretschmann configuration, to excite surface plasmons. Between the metallic surface and the prism, there is a space in the Otto layout. This setup can be conveniently used in the study of a single-crystal metallic surface. It is not desirable, therefore, to place prisms in direct contact. In Kretschmann arrangement we essentially have the prism lying on a thin metalline film-mostly between 40-50nm at visible wavelengths.The evanescent wave that appears at the prism-metal interface then travels along the film and causes plasmons to exist on "the surface at the metal-air interface".

The thickness optimization can be done by studying various reflectance vs angle graphs of different thicknesses Figure 3. We consider the sharp dips and the sensitivity of every layer to the refractive index 1.34-1.364. Comparing those results(Figure 3, Tables 1,2,3), the thickness optimization is done[8].

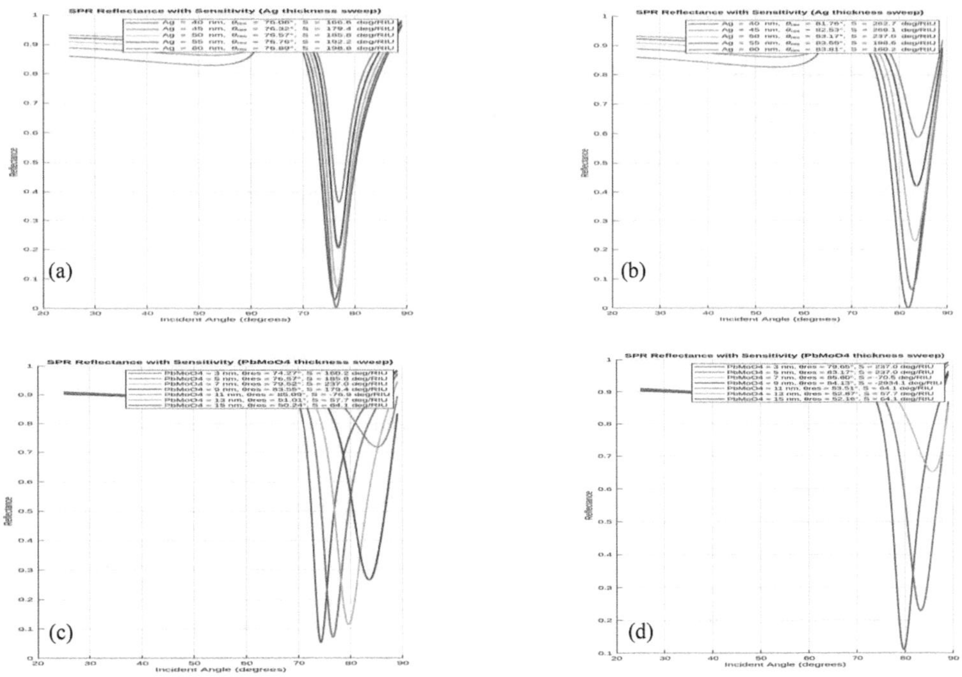

Figure 3. The curve of reflectance vs angle, at varying values of (a).Silver layer for RI=1.34, (b).silver for RI=1.368, (c) PbMoO4 layer for RI=1.34, d. PbMoO4 for RI =1.368, (e)Graphene for RI=1.34, (f).Graphene for RI=1.368.

2 METHODOLOGY

The suggested sensor is "based on the concept of total internal reflection" in "Kretschmann configuration for Surface Plasmon Resonance (SPR)". It is like having a p-polarized beam of light sent on a multilayer structure which has a high refractive index prism inside it. The intended system has been designed as consisting of BK5 prism, a metallic Ag layer, a PbMoO 4 dielectric layer, a monolayer of graphene, and a Molecularly Imprinted Polymer (MIP) layer. MIP layer is in direct contact with the aqueous acetone solution in which the molecular interaction and selectivity becomes possible.[6].

Table 1. See how the resonance angle and sensitivity varies with the Silver(Ag) thickness.

	Silver(Ag)			
	RI=1.34		RI=1.368	
Thickness(nm)	Resonance angle(θ_{res})	Sensitivity(deg/RIU)	Resonance angle(θ_{res})	Sensitivity(deg/RIU)
40	76.06	166.57	81.76	262.66
45	76.32	179.38	82.53	269.07
50	76.57	185.79	83.17	237.04
55	76.76	192.19	83.55	198.60
60	76.89	198.60	83.81	160.16

Table 2. See how the resonance angle and sensitivity varies with the PbMoO4 thickness.

	$PbMoO_4$			
	RI=1.34		RI=1.368	
Thickness(nm)	Resonance angle(θ_{res})	Sensitivity(deg/RIU)	Resonance angle(θ_{res})	Sensitivity(deg/RIU)
3	74.27	160.16	79.65	237.04
5	76.57	185.79	83.17	237.04
7	79.52	273.04	85.60	−70.47
9	83.55	179.88	84.13	−2934.13
11	85.09	−76.88	53.51	64.06
13	51.01	57.66	52.87	57.06
15	50.24	64.06	52.16	64.06

Table 3. See how the resonance angle and sensitivity varies with the Graphene thickness.

	Graphene			
	RI=1.34		RI=1.368	
Thickness(nm)	Resonance angle(θ_{res})	Sensitivity(deg/RIU)	Resonance angle(θ_{res})	Sensitivity(deg/RIU)
0.34	76.57	185.79	83.17	237.04
0.68	77.21	198.60	83.87	147.35
1.02	77.92	198.60	84.26	70.47
1.36	78.62	198.60	84.26	25.63
1.70	79.33	185.79	84.13	−6.41
2.04	79.97	160.16	83.87	−19.22
2.38	80.54	134.53	83.55	−32.03

During the analysis, the thicknesses of Ag, $PbMoO_4$, and graphene were systematically varied to evaluate the sensor's performance across wastewater samples containing acetone concentrations ranging from 0% to 90%. The refractive indices of these samples were used to simulate realistic environmental conditions. The optimized design parameters used in the simulation are presented in Table 4, which summarizes both the layer thicknesses and refractive indices for each material in the sensor stack.

Table 4. Optimized thickness of each layer.

Material	Thickness (nm)	Refractive index	Reference
BK5prism		1.5151	[24]
Ag(Silver)	50	0.135+3.99i	[34]
$PbMoO_4$	5	2.2596	[35]
Graphene	0.34	3.0+1.147i	[27]
MIP layer	10	1.5265	[37]

3 TRANSFER MATRIX METHOD

The optical response of the multilayer SPR sensor was simulated using the **TMM**, which is a mathematical approach based on "**Fresnel's equations**". TMM models how an electromagnetic wave propagates through multiple thin layers, accounting for their optical thickness and refractive indices.

The characteristic matrix of the multilayer system is defined as:

$$C_{ij} = \Pi_{k=2}^{n-1} C_k = \begin{bmatrix} m_{11} & n_{12} \\ m_{21} & m_{22} \end{bmatrix}$$

where

$$c_k = \begin{bmatrix} \cos\beta k & \dfrac{-\sin\beta k}{qk} \\ -iqk\sin\beta k & -iqk\sin\beta k \end{bmatrix},$$

and

$$q_k = \left(\frac{\mu_k}{\delta_k}\right)^{\frac{1}{2}} \cos\theta_k = \frac{\left(\varepsilon_k - n_1^2 \sin\theta_1^2\right)^{\frac{1}{2}}}{\varepsilon_k}$$

$$\beta_k = \frac{2\pi}{\lambda} n_k \cos\theta (Z_k - Z_{k-1}) = \frac{2\pi d_k}{\lambda}\left(\varepsilon_k - n_1^2 \sin\theta_1^2\right)^{\frac{1}{2}}$$

Here,

- n_k, d_k, and ε_k are the **refractive index**, **thickness**, and **dielectric constant** of the k-th layer, respectively.
- λ is the wavelength of incident light.
- θ_k represents the propagation angle inside each layer.

The "**reflection coefficient**" for the p-polarized light (r_p) is calculated as: $r_p = \frac{(P_{11}+P_{12}q_L)q_1-(P_{21}+P_{22}q_L)}{(P_{11}+P_{12}q_L)q_1+(P_{21}+P_{22}q_L)}$ and the **reflectance intensity** is given by: $R_p = |r_p|^2$

Proposed Sensor Key Parameters:

This way, effectively, the action of the SPR sensor can be characterized using 3 significant parameters: "**sensitivity** (S), **detection accuracy** (DA), and **figure of merit** (FoM)". You would like those three parameters to be as high as they can be to achieve optimal result of the SPR sensor.

Sensitivity:

$$\mathrm{S} = \Delta\theta res/\Delta n_s (^\circ/\mathrm{RIU})$$

This is simply the change in angle of resonance of each unit change of the refractive index.[13].

DA:

$$\mathrm{DA} = \Delta\theta res/\mathrm{FWHM}$$

It defines the precision of detection, where **FWHM** represents the **Full Width at Half Maximum** of the resonance curve.[31].

FoM:

$$\mathrm{FoM} = \mathrm{S} \times \mathrm{DA}$$

This parameter evaluates the overall **quality and efficiency** of the proposed sensor.[9].

4 RESULT AND DISCUSSION

Here we have modeled an SPR biosensor in the Kretschmann setup that involves the use of a prism to impart momentum to surface plasmons. Varying the thicknesses at RI=1.34 and RI=1.368 are considered as the optimized thickness.The Silver layer thickness is adjusted between 40nm and 60nm by adding 5nm at a time. At 0-90% Acetone concentration, RI=1.34-1.368, the resonance angle and the sensitivity were dissimilar at the various thicknesses. Through sharp dips in Figure 3a and 3b, and based on the sensitivity and resonance angle provided in Table 1, silver(Ag) thickness is assumed to be 50nm.

The thickness of PbMoO4 is assumed to be 5nm based on Figure 3c, 3d, and Table 2. Based on Figure 3e, 3f, and Table 3, the thickness of the Graphene is assumed to be 0.34nm. The acetone molecules are bound to a molecular layer, MIP. In most cases, other Acetone detection sensors provide false detection. In cases where the other dissolved salts and acetone are equal in refractive index. When the wastewater had dissolved salts, yet the SPR sensor showed the concentration of Acetone. In such a scenario, only the acetone molecules are bound to the molecularly imprinted polymer, and the sensor results provide the Acetone concentration. The source wavelength is varied between 600 and 650 nm. After the results that we received, we assumed a wavelength of the light source of 633nm.

A molecularly imprinted polymer can be added to the proposed SPR sensor with good selectivity. A sensor's maximum sensitivity was reached at 321.43 degrees/RIU. The figure below (Figure 4) shows the result of the simulation when RI=1.33. At this point, the curve is at 74.68, hence there is a low concentration of acetone in the wastewater. It is also on the basis of this that by increasing the RI value, the resonance angle also increases, and in the process, the sensor detects the acetone concentration. The resonance angle at RI=1.37 = 83.14°, and this value shows that a high concentration of acetone is observed.

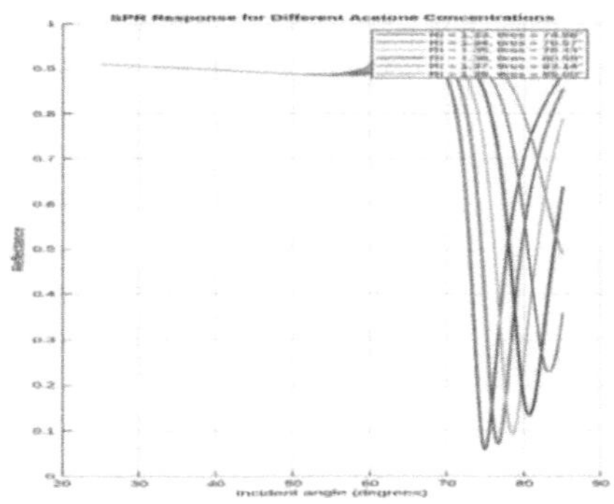

Figure 4. Reflectance curves of the proposed sensor for various refractive index.

Table 5. Assessment of the current SPR work in comparison with other existing works.

Layers in the structure	Sensitivity (deg/ RIU)	Reported year	Reference
Prism+Ag+ptSe_2+WSe_2	194	2020	[14]
BK5+Ti+Ag+$BaTiO_3$+WS_2	180	2020	[8]
BK5+Ti+Ag+MoS_2+Graphene	144.72	2023	[9]
BK5+CuSi_3N4+$BaTiO_3$+BP	300	2024	[17]
BaF2+ZnO+Cu+Si+MXene	289.40	2025	[11]
BK5+Ag+$PbMoO_4$+Graphene+MIP	321.43	Present work	

5 CONCLUSION

This work gives a detailed design and simulation of an SPR sensor based on refractive index, which uses a multilayer design which is composed of Ag/$PbMoO_4$ /Graphene/MIP to selectively detect acetone in industrial wastes. The attachment of a MIP layer allows a sensor to selectively bind acetone molecules without much interference with other dissolved materials that can have the same refractive indices.The simulation results validate that the suggested sensor has a sensitivity value of 321.43° C/RIU, a detecting accuracy (SNR) of 0.5°C because of 1 and a FoM of

160.71 RIU, including various other SPR designs. Such results demonstrate that the suggested sensor can be used as a highly selective and cost-efficient instrument to measure acetone levels in wastewater.Even though the given design can be used in excellent theoretical performance, it fails to consider the real-life conditions like change in temperature, humidity effect, and chemical impurities that may affect sensor behavior. The experimental validation of the results with real wastewater samples, and optimization of large-scale fabrication and field implementation to make their work consistent and reliable in industrial environments are the directions of further research.

REFERENCES

[1] Khan, Wajih Ahmed, Muhammad Umar Anjum, Harris Khan, Amir Hamza, Hamid Jabbar, Tayyab Zafar, Ali R. Ansari, and Raheel Nawaz. "Application of robotic manipulation for carbon fiber reinforced polymers manufacturing survey." *Composites Part C: Open Access* 15 (2024): 100503.

[2] Righettoni, Marco, Antonio Tricoli, and Sotiris E. Pratsinis. "Si: WO3 sensors for highly selective detection of acetone for easy diagnosis of diabetes by breath analysis." *Analytical chemistry 82*, no. 9 (2010): 3581–3587.

[3] Xia, Feng, Hang Song, Yong Zhao, Wan-Ming Zhao, Qi Wang, Xue-Zhou Wang, Bo-Tao Wang, and ZheXin Dai. "Ultra-high sensitivity SPR fiber sensor based on multilayer nanoparticle and Au film coupling enhancement." *Measurement* 164 (2020): 108083.

[4] Rizal, Conrad, Vladimir Belotelov, Daria Ignatyeva, Anatoly K. Zvezdin, and Simone Pisana. "Surface plasmon resonance (SPR) to magneto-optic SPR." *Condensed Matter* 4, no. 2 (2019): 50.

[5] Lin, Chii-Wann, Kuo-Ping Chen, Chien-Nan Hsiao, Shiming Lin, and Chih-Kung Lee. "Design and fabrication of an alternating dielectric multi-layer device for surface plasmon resonance sensor." *Sensors and Actuators B: Chemical* 113, no. 1 (2006): 169–176.

[6] Topor, Cristina-Virginia, Mihaela Puiu, and Camelia Bala. "Strategies for surface design in surface plasmon resonance (SPR) sensing." *Biosensors 13*, no. 4 (2023): 465.

[7] Martinsson, Erik, Marinus A. Otte, Mohammad Mehdi Shahjamali, Borja Sepulveda, and Daniel Aili. "Substrate effect on the refractive index sensitivity of silver nanoparticles." *The Journal of Physical Chemistry C* 118, no. 42 (2014): 24680–24687.

[8] Kumar A, Kumar A, Srivastava SK (2022) A study on surface plasmon resonance biosensor for the detection of CEA biomarker using 2D materials graphene, Mxene and MoS2. *Optik (Stuttg)* 258:168885

[9] Singh NP, Mishra AC, Yadav S, Lohia P, Dwivedi DK, Tighezza AM, Hossain MK (2024) Numerical analysis of silicon nitride and perovskite nanomaterial assisted SPR biosensor for dengue cell detection. *J Opt* 1–11

[10] Henry, Olivier YF, David C. Cullen, and Sergey A. Piletsky. "Optical interrogation of molecularly imprinted polymers and development of MIP sensors: a review." *Analytical and Bioanalytical Chemistry* 382, no. 4 (2005): 947–956.

[11] Zhan, Tianrong, Xi Shi, Yunyun Dai, Xiaohan Liu, and Jian Zi. "Transfer matrix method for optics in graphene layers." *Journal of Physics: Condensed Matter* 25, no. 21 (2013): 215301.

[12] Kushwaha, Angad S., Anil Kumar, Rajeev Kumar, and S.K. Srivastava. "A study of surface plasmon resonance (SPR) based biosensor with improved sensitivity." *Photonics and nanostructuresfundamentals and Applications* 31 (2018): 99–106.

[13] Mishra, Akhilesh K., Satyendra K. Mishra, and Banshi D. Gupta. "SPR based fiber optic sensor for refractive index sensing with enhanced detection accuracy and figure of merit in visible region." *Optics Communications* 344 (2015): 86–91.

[14] Lin Z, Chen S, Lin C (2020) Sensitivity improvement of a surface plasmon resonance sensor based on two-dimensional materials hybrid structure in visible region: A theoretical study. *Sensors* 20:2445

[15] Kumar A, Yadav AK, Kushwaha AS, Srivastava SK (2020) A comparative study among WS2, MoS2 and graphene based surface plasmon resonance (SPR) sensor. *Sens Actuators Rep* 2:100015

[16] Delice, S., M. Isik, and N.M. Gasanly. "Linear and nonlinear optical characteristics of PbMoO4 single crystal for optoelectronic applications." *Journal of Materials Science: Materials in Electronics* 33, no. 28 (2022): 22281–22290.

[17] Singh NP, Mishra AC, Yadav S, Lohia P, Dwivedi DK, Tighezza AM, Hossain MK (2024) Numerical analysis of silicon nitride and perovskite nanomaterial assisted SPR biosensor for dengue cell detection. *J Opt* 1–11

[18] Cheon, Sosan, Kenneth David Kihm, Hong Goo Kim, Gyumin Lim, Jae Sung Park, and Joon Sik Lee. "How to reliably determine the complex refractive index (RI) of graphene by using two independent measurement constraints." *Scientific reports* 4, no. 1 (2014): 6364.

[19] Gavrilă, Ana-Mihaela, Elena-Bianca Stoica, Tanţa-Verona Iordache, and Andrei Sârbu. "Modern and dedicated methods for producing molecularly imprinted polymer layers in sensing applications." *Applied Sciences* 12, no. 6 (2022): 3080

Smart Systems and Beyond – Bhatia et al. (Eds)
© 2027 The Author(s), ISBN: 978-1-041-36617-1

RFID-Based Smart Trolley System for Automated Invoicing and Retail Efficiency

Dushyant Kumar
Dr. B.R. Ambedkar National Institute of Technology Jalandhar, India

Tarun Singhal, Ankur Singhal, and Deepak Dadwal
Chandigarh Engineering College-CGC, Landran Mohali

Ishta Rani
Chandigarh University, Gharuan, Mohali

Vinay Bhatia
Chandigarh Engineering College-CGC, Landran Mohali

Promila Singhal
Maharshi Markandeshwar (Deemed to be University), Mullana, Ambala

ABSTRACT: The present paper suggests the implementation of RFID-based smart trolley system, which will automate the invoicing system by placing RFID tags on products and RFID readers on shopping trolleys. This device is a replacement of the classic checkout counter, where real time product scanning and invoicing of the products is allowed upon adding them to the cart. Based on a critical pilot analysis, the suggested solution is optimal in the management of the inventory, accuracy, faster invoicing, better customer satisfaction and less turnaround time. The RFID-based smart trolley system can be also added to improve the in-store shopping experience significantly. This technology enables customers to do shopping on their own and they can easily attain their shopping without necessarily going to the traditional checkouts. Largely, advantages in terms of efficiency, scalability and cost-effectiveness become noticeable when compared with the traditional processes which operate based on barcodes. The retail sectors are experiencing difficulties in the form of long queues at the checkout counters, mistakes in billing and increase in the cost of operation. The old-fashioned types of billing systems that depend on the barcodes require a manual scanning, which results in the prolongation of the waiting time and line.

Keywords: RFID, Smart Trolley System, Smart Trolley System

1 INTRODUCTION

- Retail automation is acquiring priority in the modern digital age when there is a distinct relationship between customer comfort and efficiency in the organization and competitiveness. The line-of-sight and manual manipulation of barcode scanning is a necessary component of a traditional billing system. This then leads to the long queues, issues with billing and the check out processes that require a lot of manual work. As more advanced and automated technologies have been developed, retail checkout system is moving beyond manual barcode scanning and towards automated systems. Although barcodes are widely used, they have a wide range of drawbacks, including potential scanning errors, a possibility of the physical damage of labels, and long checkout times. This paper covers some of the features of RFID-based billing system, which includes database setup, product identification, billing process, verification, and confirmation. The paper will be used to demonstrate the practical benefits of RFID technology in the retail industry such as reduction of costs, faster checkout lines and increased accuracy in inventory management. Despite the huge potential of the system, the issues of technological and financial challenges should be addressed to ensure the system is scaled and made reliable. Figure 1 illustrates the flowchart of how the system works, which makes the operation of the system easy to understand. A tag that is RFID is attached to every object.

Every trolley is fitted with an RFID reader and a microcontroller. The products are scanned in real time as they are placed inside the trolley.

So, it is possible to remove checkout delays with the technique that has been described, quick billing is provided, and seamless integration with inventory management systems is achieved.

 DOI: 10.1201/9781042026289-53

2 METHODOLOGY

The Smart Trolley System, which is based on RFID technology, operates in the following stages:

(1) Tagging of Products: Every item sold in the retail establishment is assigned a one-of-a-kind RFID tag that includes the product's identification number and the price.
(2) Smart Trolley Setup — Trolleys are outfitted with RFID readers, display units, and a microcontroller for the purpose of data processing.
(3) Real-Time Billing — As clients place items inside the trolley, RFID tags are read, and the bill is updated in real time using the information found on the RFID tags.
(4) Customers are able to avoid waiting in large lines by using the automated checkout system, which allows them to pay directly from the trolley system (via a smartphone app or self-checkout kiosks).
(5) Inventory Update - When things are billed, the system will immediately update the stock database with the new information.
(6) The last validation that takes place at the departure gates ensures that the products that are billed correspond to the items that are contained within the cart.

Flowchart of Proposed System
Figure 1 shows flowchart of the proposed system.

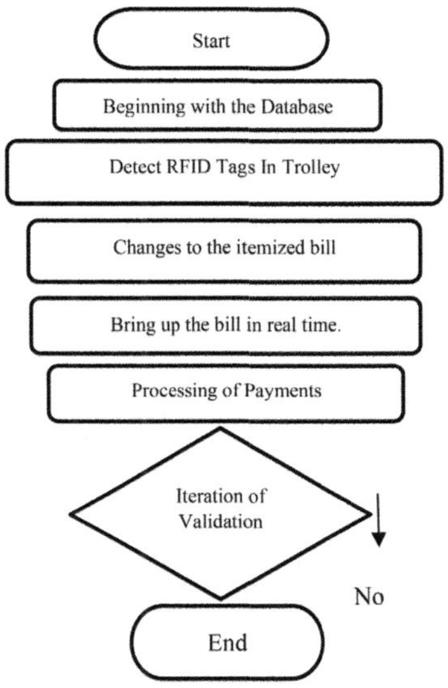

Figure 1. Flowchart of RFID-based smart trolley system.

System Architecture Diagram
As may be seen in Figure 2, the RFID system is represented by a block diagram.

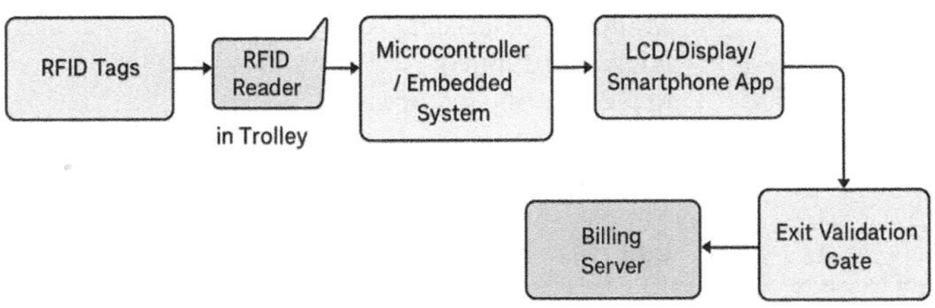

Figure 2. Block diagram of RFID-based smart trolley system.

- RFID Tags → Attached to each product.
- RFID Reader in Trolley → Scans items as they are placed.
- Microcontroller / Embedded System → Processes scanned data.
- LCD/Display/Smartphone App → Displays real-time bill.
- Billing Server → Stores transaction details and updates inventory.
- Exit Validation Gate → Ensures billed items match trolley contents.

3 RESULTS AND DISCUSSION

The product_database is =

{ "RFID001": {'name': "Milk", "price": 3.5, "stock": 10}, "RFID002": {'name': "Bread", "price": 1.5, "stock": 15}, "RFID003": {'name': "Eggs", "price": 4.0, "stock": 20}, "RFID004": {'name': "Cheese", "price": 4.0, "stock": 8}, "RFID005": {'name': "Butter", "price": 2.5, "stock": 12}, "RFID006": {'name': "Juice", "price": 3.0, "stock": 10} }

RFID-based Smart Trolley Simulation Started ◆
Scanned: Cheese - $4.0
Scanned: Milk - $3.5
Scanned: Eggs - $4.0
Scanned: Bread - $1.5
Scanned: Eggs - $4.0

Current Trolley Items —

(1) Cheese - $4.0
(2) Milk - $3.5
(3) Eggs - $4.0
(4) Bread - $1.5
(5) Eggs - $4.0

Total Bill: $17.0

Final Bill —

— Current Trolley Items —

(1) Cheese - $4.0
(2) Milk - $3.5
(3) Eggs - $4.0
(4) Bread - $1.5
(5) Eggs - $4.0

Total Bill: $17.0, In other words, the payment was finished automatically. You do not require a checkout counter. In a simulated shop setting with one hundred different products, the system was put through its paces. It was observed that the following results occurred:

- When compared to barcode-based billing, the checkout time is reduced by sixty to seventy percent.
- RFID reduced the majority of human errors, such as missed scans, which led to a lower error rate.
- The customer experience was improved by the real-time presentation of invoices, which increased both confidence and transparency.
- Inventory Management: In order to avoid stock mismatches, automated updates were implemented.
- Financial expenses: There will be fewer cashiers needed, which will result in lower labor expenses.

Table 1. Comparison with Traditional System.

Feature	Traditional Barcode Billing	RFID-Based Smart Trolley
Scanning Method	Manual, line-of-sight	Automatic, no line-of-sight
Checkout Time	High (5–10 mins)	Low (1–2 mins)
Error Rate	Moderate (missed scans)	Very Low (auto-detection)
Customer Experience	Long queues	Seamless shopping
Labor Requirement	High (cashiers required)	Low (self-checkout)
Inventory Management	Manual updates	Real-time automated
Scalability	Limited	Highly scalable

4 CONCLUSION

Once the items are loaded in the shopping cart they are scanned automatically thus producing an instant bill that can be viewed by the customer in real time. Eliminating the need of the traditional checkout counters means that the wait times are considerably brought down, and the possibility of human error is also brought down to a minimum. The RFID smart trolley system is a novel innovation in the retail industry that has been created to enhance the efficiency of the business as well as the convenience of the customers. The traditional retail approach causes numerous delays at the checkout line as people manually scan barcodes, thus creating long queues and a high rate of errors. This is an automated system where the consumer payment process is automated using radio frequency identification (RFID) tags on products and an RFID scanner mounted on the shopping cart. The retailers benefit greatly under this system due to the increased efficiency of its operation and the reduced requirement of its use of human labor on billing tasks. It provides live information and exact transaction history, which streamlines inventory control and track sales of products. The customer experience, in terms of how they see it, is streamlined, easy and convenient and results in an increase in customer satisfaction and loyalty. The future could include adding a culture of artificial intelligence (AI) to personalize the analysis and guidance, such as offering mobile payment options to a more user-friendly checkout experience as well as Internet of Things (IoT)-based inventory management to keep the stock updated automatically, and so on. Such innovations may result into a smart and fully automated retail environment linking physical and virtual purchases.

4.1 *Future scope*

The smart trolley systems based on RFID have a great potential of developing into full automated and intelligent retailing systems. Shop owners will have the option to suggest products to customers, offer targeted promotions, and advanced product placement strategies using artificial intelligence (AI) to analyze customer purchasing behavior. The Internet of Things (IoT)-enhanced inventory systems can significantly improve the efficiency of stock monitoring, stock shortages, and immediate restock, eliminating the need for human involvement. The clients find it easy to carry out their transactions with their digital wallets or mobile gadgets, thus avoiding the need to have a cashier in the checkout counter. This has been enabled by the introduction of mobile and contactless payment technology. This system may be linked to customer accounts and loyalty programs to allow to get prizes and discounts. Integration of smart navigation and augmented reality (AR) would make the shopping process more convenient by helping clients to find products in the premises. RFID-enabled smart trolleys can be used to utilize the data collected to make improved decisions regarding demand forecasts, supply chain optimization, and large volumes of predictive analytics. Based on these criteria, the development of this technology would revolutionize in-store shopping and give a customer-focused retail experience with full automation.

REFERENCES

[1] Bodravara, N., Angadi, B.S., Menasinahal, R.S., Pasigar, M., and Naik, M. (n.d.). *RFID Based Smart Trolley for Automatic Billing System.* Retrieved from www.ijert.org

[2] Manikandan, S., Murugan, A.H., Kamal, S., Udhayakumar, S., and Yokesh, D. (2025). RFID based Smart Trolley Billing System using IoT. In *Proceedings of the International Conference on Intelligent Computing and Control Systems (ICICCS 2025)* (pp. 141–147). Institute of Electrical and Electronics Engineers (IEEE). https://doi.org/10.1109/ICICCS65191.2025.10984823

[3] Atshaya, C., Parvathavarthini, S., Danushree, V.S., Krupa, N.S., and Sowndharya, R. (2025). RFID based Human Following Smart Trolley with Real-Time Pricing. In*Proceedings of the 5th International Conference on Pervasive Computing and Social Networking (ICPCSN 2025)* (pp. 1902–1907). IEEE. https://doi.org/10.1109/ICPCSN65854.2025.11034910

[4] Karthi, S., Sathiya Sri, K., Venimalathy, A., and Sindhu, T. (2023). *IoT Based Smart Trolley System for Automated Billing Using RFID.* Retrieved from www.ijcrt.org

[5] Patil, S.R., Mathad, S.N., Gandhad, S.S., and Ellemmi, M.C. (2022). Smart trolley with automatic billing system using Arduino. *International Journal of Advanced Science and Engineering*, 8(3), 2268–2273. https://doi.org/10.29294/IJASE.8.3.2022.2268-2273

[6] Athauda, T., Marin, J.C.L., Lee, J., and Karmakar, N.C. (2018). Robust Low-Cost passive UHF RFID Based Smart Shopping Trolley. *IEEE Journal of Radio Frequency Identification*, 2(3), 134–143. https://doi.org/10.1109/JRFID.2018.2866087

[7] Li, X., Zhang, Y., Chen, W., and Liu, H. (2017). IoT applications on secure smart shopping system. *IEEE Internet of Things Journal*, 4(6), 1945–1954.

[8] Melton, J.C.M., and Arnold, T. (2005). *Managing RFID Tags Using an RFID-Enabled Cart.* Retrieved from https://patents.google.com/patent/US20060163350A1

[9] Yewatkar, A., Inamdar, F., Singh, R., Ayushya, and Bandal, A. (2016). Smart cart with automatic billing, product information, product recommendation using RFID and ZigBee with anti-theft. *Procedia Computer Science*, 79, 793–800.

[10] Hairong, Z.C.Y., Liming, Q., and Liansheng, C. (2011). *Intelligent Trolley. China Patent* CN202225891U.

[11] Ray, B. (2015, July 9). *ZigBee Vs. Bluetooth: A Use Case With Range Calculations.* Retrieved from https://www.linklabs.com/blog/zigbee-vs-bluetooth

[12] RFID Journal. (2017, February 22). *Wal-Mart Begins RFID Process Changes.* Retrieved from http://www.rfidjournal.com/articles/view?1385

[13] Subedi, S., Pauls, E., and Zhang, Y.D. (2017). Accurate localization and tracking of a passive RFID reader based on RSSI measurements. *IEEE Journal of Radio Frequency Identification*, 1(2), 144–154.

[14] Martinelli, F. (2015). A robot localization system combining RSSI and phase shift in UHF-RFID signals. *IEEE Transactions on Control Systems Technology*, 23(5), 1782–1796.

[15] RFID Arena. (2013, November 15). *Applying RSSI Filters for Optimal RFID Performance.* Retrieved from http://www.rfidarena.com/2013/8/28/applying-rssi-filters-for-optimal-rfid-performance.aspx

[16] GS1. (2016, November). Regulatory Status for Using RFID in the EPC Gen2 (860 to 960 MHz) Band of the UHF Spectrum. Brussels, Belgium.

[17] Nayna, T.F.A., Baki, A.K.M., and Ahmed, F. (2014). Comparative study of rectangular and circular microstrip patch antennas in X band. In *Proceedings of the International Conference on Electrical Engineering and Information Communication Technology* (pp. 1–5).

[18] Jasim, S.E., Jusoh, M.A., Mazwir, M.H., and Mahmud, S.N.S. (2015). Finding the best feeding point location of patch antenna using HFSS. *Journal of Engineering and Applied Science*, 10(23), 17444–17449.

Smart Systems and Beyond – Bhatia et al. (Eds)
© 2027 The Author(s), ISBN: 978-1-041-36617-1

Polymer Optical Fiber (POF)–Powered Passive Optical LAN for Business Applications

Virender Kumar
Chandigarh University, Gharuan, Mohali

Ankur Singhal, Tarun Singhal, Deepak Dadwal, and Vinay Bhatia
Chandigarh Engineering College-CGC, Landran Mohali

Himanshu Sharmad
J B Institute of Engineering and Technology, Hyderabad

Ishta Rani
Chandigarh University, Gharuan, Mohali

Promila Singhal
Maharshi Markandeshwar (Deemed to be University), Mullana, Ambala

Mohit Srivastava
Chandigarh Engineering College-CGC, Landran Mohali

ABSTRACT: With the ever evolving requirement for quick, dependable and energy-efficient telecommunication systems, Polymer Optical Fiber (POF) based local area networks (LANs) are quickly substituting traditional copper-based Ethernet networks. The present article presents the system architecture and process of an enterprise-appropriate Passive Optical LAN (POL) arrangement. The proposed hybrid WDM-TDM POL-POF model combines the enormous bandwidth of Single-Mode Fiber (SMF) in the backbone network with the adaptability and cost-effective connection of POF in the access systems. The telecommunication system can withstand transmission speeds between 2 Gbps and 4 Gbps over a transmission range of up to 100 km.To provide steady and reliable light output, optical modulation is employed using a continuous wave (CW) laser. Performance investigation shows that, up to 70 km, satisfactory signal integrity is maintained but the Q-factor drops as one boost length of transmission and data rate. The findings demonstrate that adding POF to POL networks offers major benefits in terms of cost-effectiveness, scalability, energy efficiency, and decreased latency, making it the perfect choice for contemporary corporate and institutional networking settings.

Keywords: Polymer Optical Fiber (POF), Passive Optical LAN (POL), Fiber Optic Networks, Signal Quality, Q-Factor, Enterprise Networking, Hybrid WDM-TDM System, Next-Generation LAN

1 INTRODUCTION

With the ever increasing need for fast, secure, and energy-efficient data communication systems, the shortcomings of traditional copper-based Ethernet networks have become more apparent. Higher bandwidth and lower latency in local area networks (LANs) are becoming increasingly essential as large number of modern businesses depend on cloud analytics, Internet of Things (IoT) along with the integration of video conferencing and big data computing [1]. With its enhanced communication properties and lower signal attenuation, optical fiber technology has become a competitive substitute for metallic conductors based on traditional copper cables.

In the midst of fast progression of optical techniques, Passive optical Networks (PONs) based Passive Optical LAN (POL) has revolutionized enterprise networking. In order to transfer high-speed broadband admittance with nominal energy dissipation and a simpler setup, POL enterprise makes use of optical cable, and optical splitters [2]. By decreasing the requirement for active switching components among the main distribution frame and the end-subscriber station, it guarantees scalability and cost-effectiveness for corporate organizations.

Optical communication systems utilizing polymer optical fiber (POF) mechanism have recently fascinated a lot of consideration as an supplementary technology for short-range communication in commercial and educational networks. POF is composed of a polymer core, often polymethyl methacrylate (PMMA), which offers benefits including comfort of handling, and flexibility in contrast to traditional optical cables poised of glass material [3].

 DOI: 10.1201/9781042026289-54

Additionally, POF is safer to mount and perfect for environments like workplaces and smart buildings because it is non-conductive [4,5].

A strong alternative for cooperates which wants to enhance network performance while lowering the overall expenditure is the incorporation of POF into POL systems. The integration of either Single-mode fiber (SMF) or multimode fiber (MMF) in global networks and POF in the access layer can be used in a hybrid arrangement to greatly increase flexibility and lower expenses incurred on operation [6,7]. Furthermore, the inherent capacity of optical systems future corporate office networks growing data needs, while the centralized control of POL streamlines network administration.

Broadband capacity, transmission distance, latency, power consumption, and network dependability are the main metrics used to check the quality and performance of a Passive Optical LAN that are powered by Polymer Optical Fiber (POF). The POF-POL architecture delivers significant gains in total operational expenses, scalability and energy efficiency as compared to transmission media that utilizes conventional copper cables [8–10].

Building the Bandwidth capacity is one of the most important performance indicators of optical LAN systems. Conventional copper-based guided transmission systems may handle transmission speed of more than one Gbps across a span of around hundred metre. POL designs with Single-Mode Fiber (SMF) backbones, on the other hand, may accommodate aggregate bandwidths of up to 10 Gbps in XGS-PON systems and more than two Gbps per channel in conventional GPON deployments [11–13].

A POF-POL system has a substantially decreased latency than a guided networks that uses a conventional copper networks. POL reduces latency by utilizing passive components between the optical line systems (OLS) and optical network terminals (ONT), which eliminates active switching and signal regeneration along the light network route[14–16].

The energy efficiency of POL is one of its main quality of service (QoS) benefits. In comparison to conventional LANs, the network's overall energy consumption can be reduced by as much as around sixty to seventy percent by doing away with active switches and intermediate-powered apparatus [14]. All data management is handled by the centralized OLS, which radically lowers the quantity of electricity needed in application regions. POF's interoperability with low-power optical transceivers allows it to further diminish power spending. Compared to the laser diodes used for glass fiber communication, LEDs or inexpensive Vertical-Cavity Surface-Emitting Lasers (VCSELs) used in POF transmission utilize a reduced amount of force. As a result, businesses may function sustainably with less temperature production and cooling requirements in communication systems [15,16]. Because POL systems don't have any intermediary active electronics that may malfunction, they are more reliable because they are primarily passive. Because fewer parts need to be replaced or monitored, maintenance is made easier.

POL-POF mechanism has expanded through a series of commercial applications because to its adaptability and enhanced energy efficiency. Office spaces, educational institutions and other industrial contexts are integrating these systems, which offer flexible and robust communication setups. Applications like voice over internet protocol (VoIP), cloud space, video conferencing, and hypervisor platforms require high-speed data transfer in modern workplace networks. By enabling centralized control via the Optical Line systems or terminal (OLT) and preserving simple scalability for expanding user needs, POF-powered POL systems offer the perfect answer.

By centralizing organization at a single OLS, POL's centralized structure makes network administration easier for university IT departments. Furthermore, POF is appropriate for educational settings where a low electrical hazard is critical due to its security features, which include being non-conductive and controllable. In order to enable cutting-edge digital learning technologies while lowering operating expenses and power spending, a number of institutions have installed hybrid POF-POL systems.

High-availability, secure architectural systems are significant in healthcare institutions. Because POF based POL architectures are invulnerable to electromagnetic interference (EMI), they are ideal for transmitting medical imaging, real-time patient monitoring and hospital data transfer in the systems. This is especially imperative in settings where there is a lot of imaging and medical devices that might interfere with copper-based communication guided transmission links.

The design, implementation, and benefits of POF-powered POL systems for corporate applications are examined in this paper. The technical performance, comparative advantages over copper-based systems, underlying architecture, and new developments in optical LAN technologies are all covered.

Rest of the paper is structured as follows. SectioN-2 presents the system architecture of the proposed network model, while the next part which is third section illustrates the results and discusses the finding. The last section-4 gives the conclusion of the research paper.

2 SYSTEM ARCHITECTURE

In passive optical LANs (POL) based passive optical fiber network system utilizes optical Line Terminals (OLTs), optical Network units (ONUs), and optical splitters are utilized in a point-to-multipoint (POL) system to communicate data signals without the help of active switching equipment's in between. The OLT is the central control unit that connects to the core backbone network and is situated in the main equipment room or data center. By dividing the optical signal into several pathways, passive splitters allow downstream optical signals to be sent from the OLT to multiple ONUs. The optical impulses are transformed into electrical signals connections by each ONT, which is placed at the client endpoint or workstation.

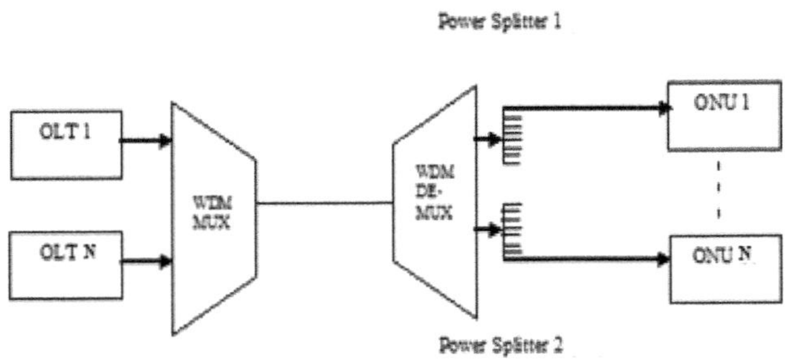

Figure 1. Hybrid WDM-TDM POL-POF system.

Hybrid WDM-TDM 8x64 POL-POF network simulation model is introduced in this section. Figure 1 depicts the architectural model of the POL-POF network system. The transmitter component of the suggested model includes OLS equipment. ONUs are positioned on the client's property. Optical line terminals are positioned in the transmitter portion of the central office (CO) of the enterprise network to produce and transceiver data to ONUs. OLTs are made up of data generators that feed electrical signal generators with their output. The communicable data is transformed into electrical impulses by an electrical signal generator. With the aid of a light source a CW laser, electrical impulses that are so generated are optically manipulated. Signals at different frequencies are sent by the light sources. The network is connected with Polymer Optical Fiber. Optical fiber is utilized in the backbone network for high-bandwidth, long-distance communication between the distribution frames and the OLT. POF is used to connect ONUs to end customer premises over sspan of 100 km. By utilizes the advantages of both fiber types, this hybrid strategy achieves high backbone capacity and flexible access domain installation. The OLT provides Quality of Service (QoS), dynamic bandwidth allocation, and upstream and downstream traffic management. It carries out tasks including network monitoring, traffic aggregation, and encryption.

Usually, the OLT is linked to the data center main network or business router. Multiple POF-based access lines receive optical data from a single fiber using passive optical splitters as shown in Figure 1. Splitters add to the overall energy efficiency of POL systems because they are completely passive and don't use any electricity.

POF cables are utilized to connect to ONUs in the LAN architecture. For the business applications, POF lines can provide data speeds in the range of Gbps across several kilometers.

A POF-powered POL system's is connected in a tree-based configuration ideal for the corporate network's size with a span of several kilometers. Passive splitters are connected to the OLT via backbone fiber connections in a typical business configuration. Each splitter splits into many POF lines that terminate at user ONUs or network endpoints.

3 RESULTS AND DISCUSSION

The outcomes of the simulated model for POL-POF passive optical network are discussed in this section. Data is transferred at various speeds in this system, ranging from 2 Gbps to 4 Gbps. The enterprise network is considered spans from twenty kilometers to hundred kilometers. NRZ is used for electrical modulation. A continuous wave laser (CW) laser is used to optically modulate the electrical signal. It is one of the two main kinds of lasers the other is a pulsed laser. A CW laser produces a continuous, steady beam of light throughout time, in contrast to pulsed lasers, which produce light in brief, intense bursts. This indicates that throughout operation, the power output and intensity largely stay the same. A laser is used to visually modulate the signal. The diversity of quality of service (Q factor) with fiber length for different speeds of 2 Gbps and 4 Gbps is illustrated graphically in Figure 2.

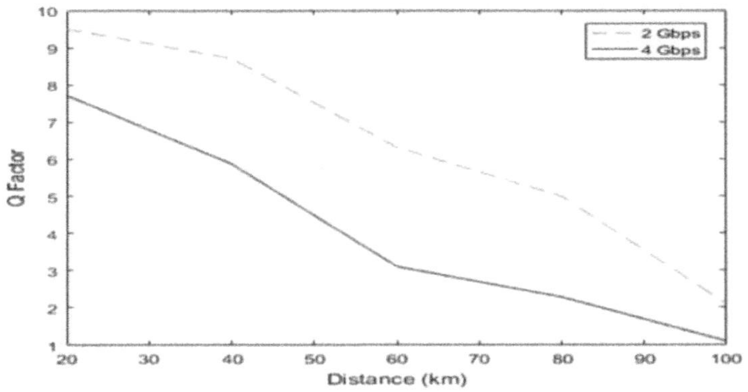

Figure 2. Q factor Vs. Distance.

For different distances between 20 and 100 km, input powers of 0.2 and 10 dB are used. The aforementioned statistics indicate that the Q factor decreases as the information rate and the distance between the ONU and OLT grow as depicted in Figure 2.

The maximum eye widths and heights, which indicate signal distortion, are shown in Figure 3 and Figure 4. The pace at which the eye closes determines the network's sensitivity. It is evident from the graphic that as the distance increases, the eye width and height both decrease, indicating a loss of signal distortion at power levels comparable to OLT. Given that the eye is completely open at fifty km as opposed to one hundred km, the results depicts that a improved quality of signal performance is attained at fifty km.

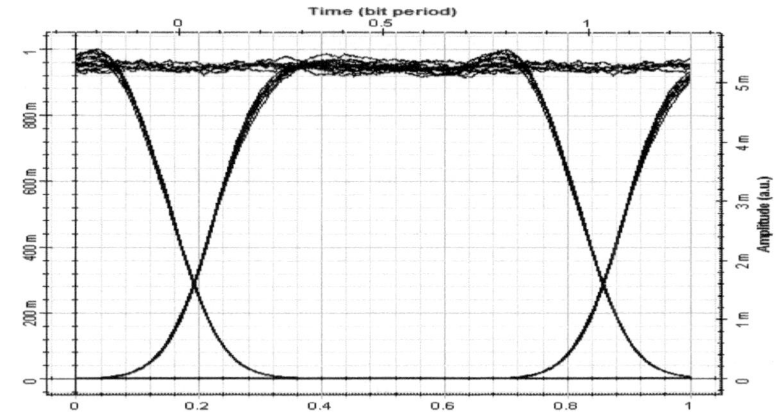

Figure 3. Eye diagram with 2Gbps at 50 Km.

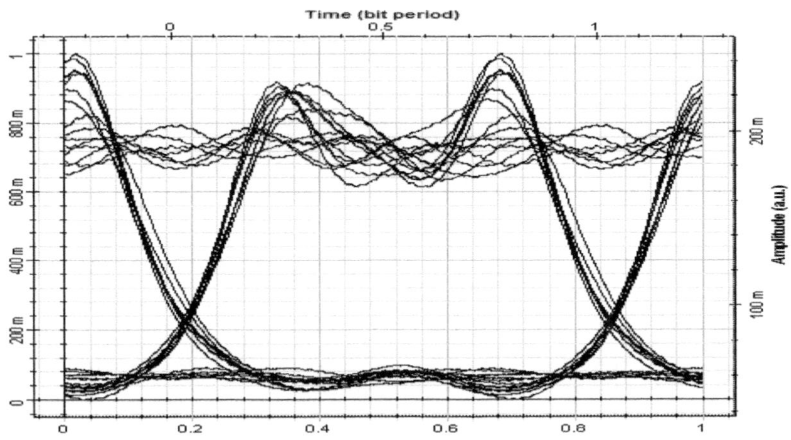

Figure 4. Eye diagram with 2Gbps at 100 Km.

Since the eye opening and Q Factor is reduced means fewer packets need to be resent, it is observed that a slower information rate will actually improve transmission time for a given measure of communicated information. A number of benefits are introduced by incorporating Polymer Optical Fiber (POF) into Passive Optical LAN (POL) design, which makes it especially appropriate for contemporary corporate and enterprise networks.

4 CONCLUSION

A thorough performance analysis of a hybrid Polymer Optical Fiber (POF)-powered Passive Optical LAN (POL) system is included in the paper. The simulated structure reveals that the suggested enterprise system efficiently meets the rising bandwidth and reliability requirements of business systems while dropping operative costs and energy requirements. The results establish that as broadcast distance and data rate increase, the Q-factor gradually decreases, leading to a little worsening in signal quality. However, within judicious limitations, the network upholds steady and consistent communication performance up to 100 km. POF's presence into the POL proposal considerably drops energy intake while enhancing electromagnetic protection and connection flexibility by eradicating midway active constituents. Furthermore, the usage of CW lasers for constant optical modulation confirms stable output, which helps in the broadcast of high-quality data. By considering the facts, POF-powered POL communication systems are a viable and future-ready optimal choice for large-capacity business and institutional networks since they provide robustness, and competent administration. For the next generation of intelligent communication structures, these solutions are seamless.

REFERENCES

[1] Smith, J. (2022). High-Speed Networking for Modern Enterprises. *IEEE Commun. Mag.*, 58(3), 45–53.

[2] Patel, M., and Chen, L. (2021). Passive Optical LANs: Architecture and Implementation. *IEEE Netw.*, 35(4), 80–89 2021.

[3] Takahashi, K. (2022). Advances in Polymer Optical Fiber Communication Systems. *IEEE Photonics J.*, 14(2), 1–10.

[4] Kumar, R. (2023). Comparative Study of Glass and Polymer Optical Fibers in LAN Applications. *Proc. IEEE ICACCS.* 118–123.

[5] Sharma, A., and Lewis, D. (2022).Optical LAN Architecture and Design Principles. *IEEE Trans. Netw. Serv. Manage.*, 19(2), 210–222.

[6] Gupta, R., and Kim, H. (2022). Low-Power Optical Splitter Designs for Passive LANs. *IEEE Photon. Technol. Lett.*, 34(5), 411–418.

[7] Zhang, L. (2021). High-Speed Transmission over Polymer Optical Fiber Links. *IEEE Photonics J.*, 13 (6), 1–9.

[8] Singh, P. (2022). Network Topology Optimization in Optical LANs. *Proc. IEEE ICCT*, 341–347.

[9] Chen, Y. (2023). Software Defined Management in Passive Optical Networks. *IEEE Netw.*, 37(4), 50–58.

[10] Domínguez, A.R. (2021). High-Speed Data Transmission over SI-POF Using DMT Modulation. *IEEE Photonics Technol. Lett.*, 33(9), 431–437.

[11] Zhao, F., and Ito, M. (2022). Wavelength Multiplexing Techniques for POF-Based Networks. *IEEE J. Lightwave Technol.*, 40(2), 250–258.

[12] Li, G. (2022). Performance Evaluation of GPON-Based Enterprise LANs. *IEEE Access*, 10, 11023–11031.

[13] Perez, D. (2021).Attenuation and Dispersion Analysis in PMMA POF for LAN Applications. *IEEE Photonics J.*, 13(4), 1–8.

[14] Brown, T. (2020). Energy Efficiency of Passive Optical LANs Compared to Copper Networks. *IEEE Commun. Mag.*, 58 (11), 92–99.

[15] Kim, S., and Wong, J. (2022). Low-Power Optical Transceivers for POF Applications. *IEEE Photon. Technol. Lett.*, 34(3), 501–507.

[16] Singh, P.K. (2023). Cost and Deployment Analysis of Optical LAN Solutions in Smart Campuses. *Proc. IEEE ICACCS*, 512–518.

Smart Systems and Beyond – Bhatia et al. (Eds)

REES: A Resource-Efficient and Enhanced-Security Mutual Authentication Scheme for Wireless Sensor Networks

Manoj Kuri, Vikram Singh Chouhan, and Jyoti Bhati
Engineering College Bikaner

ABSTRACT: Energy awareness is an essential consideration for wireless sensor networks (WSNs) when designing new protocols. Traditional network protocols, while secure, are often unsuitable for WSNs due to severe resource constraints. The Needham-Schroeder (NS) Protocol is a milestone for mutual user authentication in traditional networks but is not directly adoptable in WSNs. This work first implements the original Needham-Schroeder User Authentication Protocol in a WSN context and calculates its total energy consumption. Subsequently, an energy-efficient mutual authentication algorithm is proposed using Symmetric Key Cryptography, and its total energy consumption is calculated. A comparative analysis demonstrates that the proposed scheme achieves a significant 61.41% reduction in energy consumption at the sensor nodes by optimizing communication steps and shifting computational load to the resource-rich Key Distribution Center (KDC).

Keywords: Wireless Sensor Networks (WSN), Mutual Authentication, Needham-Schroeder Protocol, Energy Efficiency, Key Distribution Center (KDC), Symmetric Key Cryptography

1 INTRODUCTION

A WSN consists of autonomous sensors that are spatially dispersed to monitor physical or environmental conditions. However, sensor nodes have severe limitations in terms of memory, energy, processing capability, and communication capacity [1]. Due to these limitations, classical security protocols, which are computation-intensive, cannot directly be deployed within WSNs [4].

Security, and especially mutual authentication among nodes, is considered a basic need to prevent attacks like eavesdropping, node capture, and replay attacks, among others [5]. Nevertheless, Needham-Schroeder Symmetric Key Protocol is robust in mutual authentication and key establishment; it relies on a trusted KDC [11]. Its multi-step message exchange and numerous encryption operations impose substantial energy expenditure, thus rendering it impracticable for the resource-constrained sensor nodes [7,8].

This work is motivated by the high energy requirements imposed by the NS protocol when implemented in a WSN environment. The protocol is to be redesigned with a minimum energy cost at the sensor nodes without sacrificing the authentication mechanism so that the lifetime of a network is extended.

2 LITERATURE SURVEY

A critical review of key establishment protocols highlights the crucial role of energy efficiency in WSNs. It has been documented that protocols like Kerberos consume high energy in communication messages [7], and energy consumption models illustrate that session key establishment is one of the highest-consuming operations in WSNs [8]. Other studies conclude that public-key methods in ad-hoc environments can be up to an order of magnitude more energy-intensive than symmetric-key methods-10 to 100 times more costly [9]. Several other protocols, such as PIKE [12] and various group key management schemes [15,16], have also been investigated with respect to this trade-off between security, communication overhead, and energy usage. There is a consensus among the research community that WSN security protocols should be application-specific, with an emphasis on simple computations and minimized communication to reduce energy consumption [1,4,14].

3 PROPOSED AUTHENTICATION SCHEME

The proposed scheme is modified variation of the Needham Schroeder Symmetric Key Protocol, which has been designed to reduce the energy burden on sensor nodes.

DOI: 10.1201/9781042026289-55

3.1 *Original Needham-Schroeder protocol*

The original 5-step protocol is summarized as follows [11]:

(1) **A → KDC:** IDA || IDB || NA*IDA* || *IDB*|| *NA*
(2) **KDC →A:**E(KA[KS || IDB || NA || E(KB, [KS || IDA])])*E*(*KA*, [*KS* || *IDB*|| *NA*|| *E*(*KB*,[*KS* || *IDA*])])
(3) **A → B:** E(KB, [KS || IDA]) *E* (*KB*,[*KS*|| *IDA*])
(4) **B → A:** E(KS, [NB])*E*(*KS*, [*NB*])
(5) **A → B:** E(KS, [f(NB)]) *E* (*KS*,[*f*(*NB*)])

3.2 *Proposed 4-Step scheme*

The proposed scheme reorganizes and reduces the steps:

(1) **A → S (KDC) :** IDB + *NAIDB* + *NA* *(Note: NANA is derived from IDAIDA using the known function and the eliminating the need to transmit IDAIDA explicitly).*
(2) **S → A:** E(KA,[KS || IDB || NA])*E(KA,[KS || IDB || NA])* **S → B:** E(KB,[KS || IDA])*E(KB,[KS || IDA])* *(KDC communicates directly with both nodes, eliminating the need for A to forward a message to B).*
(3) **B → A:** E(KS,[NB]) *E (KS,[NB]) (NBNB is derived from IDBIDB).*
(4) **A → B:** E(KS, ACK) *E (KS, ACK) (Replaces the computation of f(NB)f(NB) with a simple, predefined ACK).*

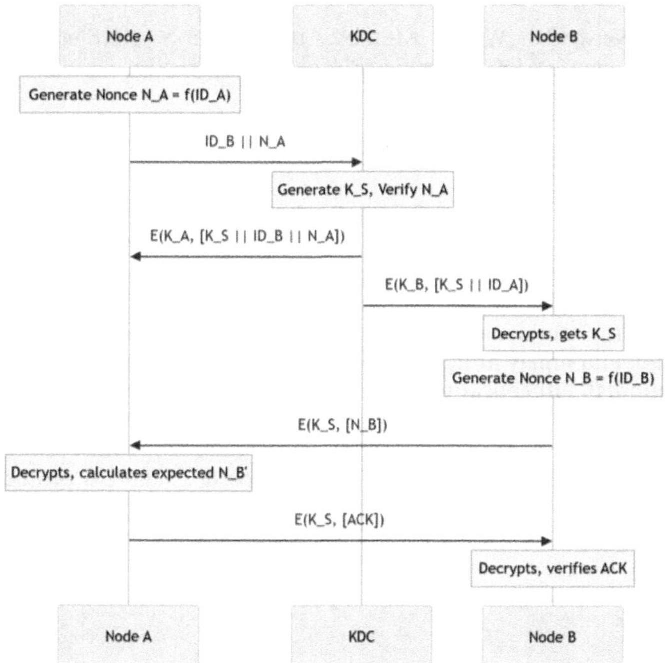

4 IMPLEMENTATION AND ENERGY ANALYSIS

The implementation was carried out using the TinyOS operating system and the NesC programming language, with energy simulations performed in PowerTOSSIMz for Mica2 motes [2]. AES with a 128-bit key was used for all encryption/decryption operations.

Assumptions: The KDC is a base station with unlimited energy resources. Energy consumption is only measured at sensor nodes A and B. Packet sizes are minimized, with the first message reduced from 13B in NS to 9B in the proposed scheme.

The energy consumption for individual operations (e.g., for sending/receiving packets, calculating nonce's, encryption/decryption) was meticulously measured. These values were then aggregated for the sequence of steps in each protocol.

4.1 *Energy consumption results*

- Total Energy Consumption of Original NS Protocol: 10447.44 mJ
  - Node A: 7768.63 mJ
  - Node B: 2678.81 mJ
- Total Energy Consumption of Proposed Protocol: 4031.81 mJ
  - Node A: 2228.67 mJ
  - Node B: 1785.14 mJ

4.2 *Result analysis*

In this regard, the presented scheme have demonstrably achieves a better outcome, reducing energy consumption by 61.41 percent. There are two major factors contributing to this reduction are outlined as:

(1) **Reduced Communication:** The proposed scheme eliminates one entire communication step, namely, the forwarding of the ticket from A to B). Besides this, there is optimization in packet sizes, as can be demonstrated in step 1, where the reduction in packet size accounts for 22.14 percent energy savings.
(2) **Shifted Computation:** The computationally exhaustive task for encrypting and generating the ticket for Node B is shifted from Node A to the KDC. As the KDC is understood to have plentiful resources, The energy drain on the sensor nodes is considerably reduced through this shift.

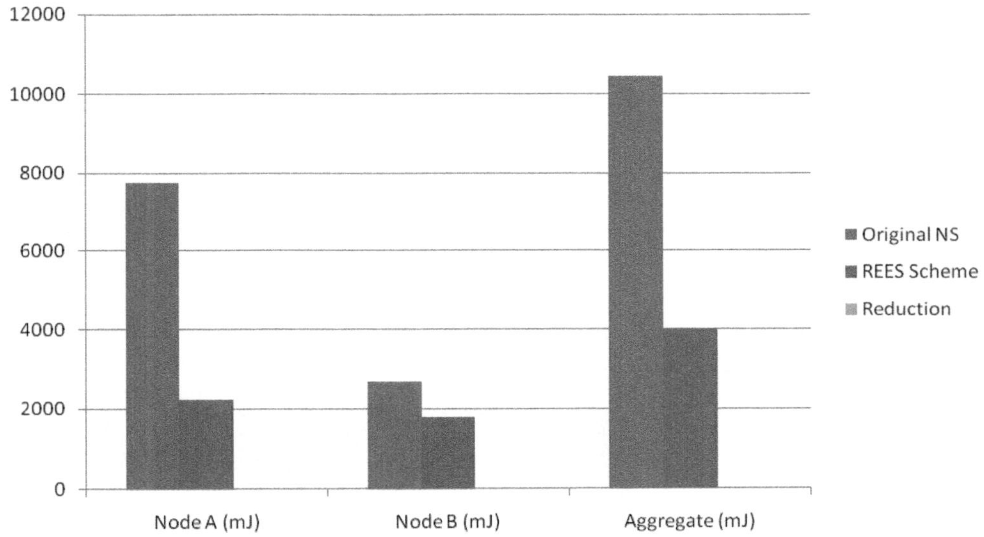

5 CONCLUSION AND FUTURE WORK

This paper demonstrates that the classical Needham–Schroeder protocol can be adapted to Wireless Sensor Networks to bring significant energy savings. The new mutual authentication approach enables a total energy reduction by 61.41% through the deliberate reduction of communication steps and relocation of computation burden to the KDC. This development directly contributes to extending the life of energy-constrained sensor nodes.

In the future, the scheme can be further extended to combat well-known vulnerabilities of NS protocols, such as replay attacks. Although using timestamps seems a potential solution, their usage increases packet size and energy consumption. Hence, in future work, lightweight mechanisms should be pursued that can enact timestamps or other security features without undermining the attained energy-efficiency benefits.

REFERENCES

[1] "Wireless Sensor Network," http://en.wikipedia.org/wiki/WirelessSensorNetworks.
[2] "MICA2 wireless measurement system," Document Part Number: 6020-0042-04, www.xbow.com.
[3] E. Shi and A. Perrig, "Designing Secure Sensor Networks," *Wireless Commun. Mag.*, vol. 11, no. 6, (2004).

[4] Y. Wang, G. Attebury, and B. Ramamurthy, (2006). "A survey of security issues in wireless sensor networks," *IEEE Communications Surveys and Tutorials*, vol. 8, no. 2, (2006).
[5] C. Karlof and D. Wagner, "Secure Routing in Wireless Sensor Networks: Attacks and Countermeasures," *Elsevier AdHoc Networks Journal*, (2003).
[6] A.J. Menezes, P.C. van Oorschot and S.A. Vanstone, *Handbook of Applied Cryptography*. CRC Press, (1996).
[7] J. Großschädl *et al.*, "The energy cost of cryptographic key establishment in wireless sensor networks," *ASIACCS* '07, (2007).
[8] X. Zhang, H.M. Heys, and C. Li, "Energy cost of cryptographic session key establishment in a wireless sensor network," *2011 6th International ICST Conference on Communications and Networking in China (CHINACOM).*
[9] A. Hodjat and I. Verbauwhede, "The Energy Cost of Secrets in Ad-hoc Networks," *Proceeding 2002.*
[10] B. Barnett and D. Sexton, "A Secure Framework for Wireless Sensor Networks."
[11] R.M. Needham and M.D. Schroeder, (1978). "Using encryption for authentication in large networks of computers," *Communications of the ACM*, vol. 21, no. 12.
[12] H. Chan and A. Perrig, (2005). "PIKE: Peer intermediaries for key establishment in sensor networks," *INFOCOM.*
[13] L. Eschenauer and V.D. Gligor, (2002). "A key-management scheme for distributed sensor networks," *CCS.*
[14] J. Sen, (2009). "A survey on wireless sensor network security," *International Journal of Communication Networks and Information Security (IJCNIS)*, Vol. 1, No. 2.
[15] Rhee K.H., Park Y.H., Tsudik G., (2005). "A group key management architecture for mobile ad-hoc wireless networks," *Journal of Information Science and Engineering 21.*
[16] P. Kamat, A. Baliga, and W. Trappe, (2006). "An identity-based security framework For VANETs," *VANET.*

Smart Systems and Beyond – Bhatia et al. (Eds)

A Detailed Investigations on IOT Applications and Challenges

Simar Singh Rayat, Sujal Thapa, Raj Basnet, and Chandradeep Bhatt
Department of Computer Science and Engineering, Graphic Era Hill University, Dehradun, Uttarakhand, India

Ruchira Rawat
Department of Computer Science and Engineering, Graphic Era Deemed to be University, Dehradun, Uttarakhand, India

ABSTRACT: Internet of Things (abbreviated as IoT) is rapidly transforming industries and daily life through interconnected devices. This paper explores IoT elements and architecture, working structure and major components of IoT, IoT applications in smart cities, healthcare, agriculture, smart homes, and environmental monitoring. It addresses challenges like interoperability, security, reliability, scalability, and data management. Also discussed about the case study of IoT in Tesla industry.

Keywords: IoT architecture, major elements, applications, challenges, smart cities, healthcare, agriculture, smart homes, environmental monitoring, case-based study of IoT

1 INTRODUCTION

The accelerated growth and groundbreaking achievements of the Internet of Things (IoT) are bringing about major changes in different industries and daily life. In order to take advantage of this technological revolution, several problems need to be solved. The IoT is part of a wide network of interconnected physical objects, devices, and sensors, further used in the collection, share, and utilisation of data. The concept of IoT is a basic shift that has the possibility to drastically change the way we do things when it comes to technology. The integration of a variety of electronic devices into a whole, a concept until recently limited to the realm of science fiction, is rapidly becoming a reality. IoT is starting to make its way into people's homes and offices, and promises to dramatically change how people plan their days.

Figure 1 illustrates the various sectors including Smart Homes, Healthcare, Smart Cities, Industrial Automation, Wearables, Agriculture, Transportation, Vehicles, and Consumer service which all comes under the applications of IoT. For depiction of the flow of data and multiple devices integration, a network mesh connected to a centralised system is used.

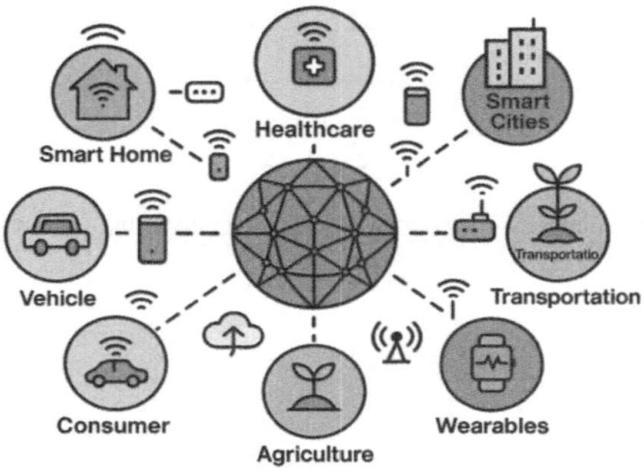

Figure 1. IoT Across Key Sectors. Note. This figure is an original contribution by the author.

2 CORE COMPONENTS AND TECHNOLOGIES OF THE INTERNET OF THINGS

Wireless Sensor Network, an extensive sensor network is made up of many smart sensors that can be used to gather, process, analyse, and share important data from a range of settings [4]. The components that make up the WSN monitoring network include:

DOI: 10.1201/9781042026289-56

(1) WSN communication stack
(2) WSN hardware
(3) WSN Middleware
(4) Secure Data aggregation

(A) *Addressing Schemes*
It not only allows us to uniquely identify billions of devices but also to control remote devices through the Internet. The few most critical features of creating a unique address are uniqueness, reliability, persistence, and scalability.
(B) *Data Storage and Analytics*

The creation of an unprecedented amount of data is one of the most important outcomes. The data have to be stored and used efficiently for the smart monitoring and actuation. Thus, for this, cloud computing provides us with the solution.

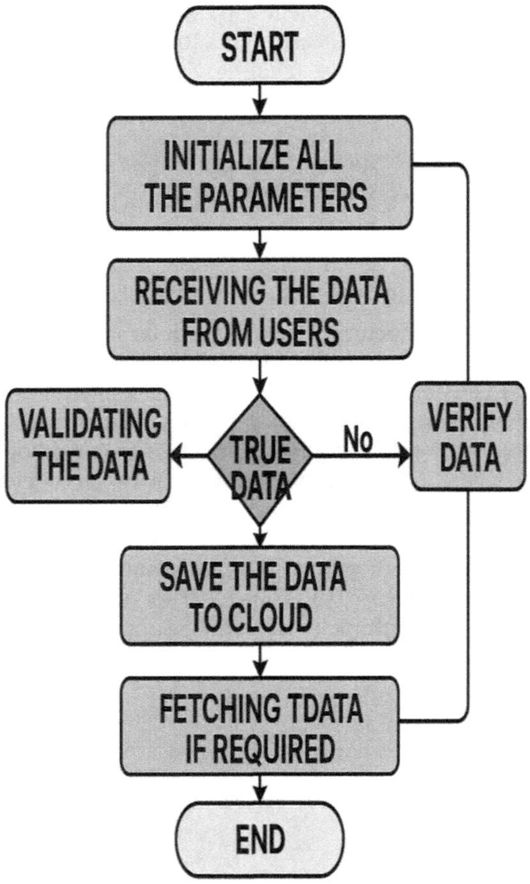

Figure 2. Working Structure of IOT. Note: This figure is an original contribution by the author.

3 MAJOR COMPONENTS OF IOT

(1) Things or Devices: These are the actual physical things or sensors which contain hardware and software for connectivity, and this enables communication with other devices and the internet. Smart appliances, industrial sensors, fitness tracking wearables, thermostat and more are examples.
(2) Connectivity: The connectivity of IoT devices to communicate with each other and the central system is important. This may include various forms of communication protocols; Wi-Fi, cellular communication networks, Bluetooth, Zigbee, LoRa, and others, among others.
(3) Gateway: Gateway is a device or software component which acts as an intermediary between the IoT devices and the cloud. It receives data from several devices and processes that data and sends it to the cloud for further analysis and storage. The gateways can also offer local data processing and security facilities.

(4) Cloud: The que internet of things (IoT) the cloud, which hosts, processes and analyses the data from IoT devices, is one of the key elements of the Internet of Things. Scalability, capacity for storage, and processing power are offered by cloud platforms for dealing with massive amounts of information generated from IoT devices. In addition, they provide infrastructure needed to run machine-learning and analytics algorithms.
(5) Analytics: Analytics is the study and application of data gathered by Internet of Things devices. Predictive modelling, historical data analysis, and real-time analysis can all fall under this category. Making well-informed judgments, streamlining processes, and seeing patterns and trends in data can all be aided by IoT analytics.
(6) User Interface: The user interface (UI) is the part of IoT that allows humans to interact with the system. This can include mobile apps, web dashboards, and other interfaces that enable users to monitor, control, and configure IoT devices. User interfaces are essential for providing a user-friendly experience and making IoT data accessible to end-users.

3.1 *Benefits of IoT*

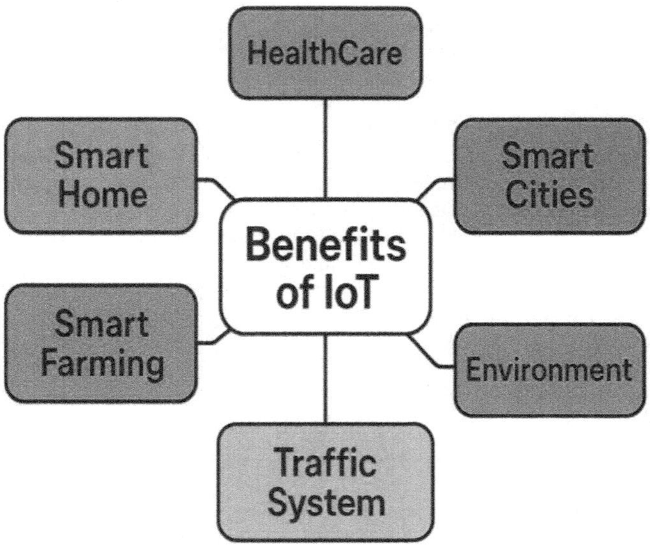

Figure 3. Benefits of IoT in various sectors. Note: This figure is an original contribution by the author.

3.2 *Applications based on real-world Internet of Things (IOT):*

(A) Smart City: IoT is being used in the urban planning process to enhance municipal services and infrastructures, manage the flow of traffic, cut down energy use, monitor air quality. When trash can need emptying, devices such as the Smart Belly trash can with the cellular communication capability can send notification to municipal services. Installing sensors and using online tools allows residents access to find available parking spots around the city. IoT enables good environmental monitoring and management of waste, thus ensuring cleaner and healthier cities.
(B) HealthCare: The utilisation of the IOT in healthcare sector is introducing new age tools with cutting edge technology contributing for the improvement of healthcare services. By revolutionising the industry IoT is providing cost-effective solutions for patients and health providers. It has a critical role to play in remote patient monitoring, medical device connectivity and in smart prosthetics. Consequently, healthcare professionals can track their patients' health status in real-time and take necessary action in time. Ultimately, IoT in healthcare brings better patient outcomes, cost savings, and medical research and innovation.
(C) Agriculture: Through analysis of soil moisture levels, the Internet of Things can proportionate Agricultural industry stronger. Real-time information regarding the health of crops, weather patterns, and soil conditions is available with the help of sensors and IoT devices. Farmers can use this knowledge to make well-informed decisions about fertilisation, irrigation and pest management. Managing and maintaining nutrition content of the agricultural products with optimisation of micro-climate conditions enhance the resource utilisation with associated IoT integration. To cope with the increasing global demand for food, the IoT is revolutionizing agriculture by enhancing agricultural production, agricultural sustainability, and the quality of the produce.
(D) Smart Home - The biggest expense a homeowner will make in their lifetime is for housing; there are a variety of products on the market which claim to save money, energy and time. Smart Home solutions offer security-based technologies that bring more safety, as well as more sustainable and convenient everyday living. IoT

enhances efficiency, comfort, and safety in the smart homes, which plays a role in a sustainable and comfortable daily existence.

(E) Environment: IoT technology enables an observation and management of air quality by means of data collection, remote sensors throughout cities. Additionally, IoT can be used to measure pollution levels in water in order to make informed decisions about water usage. In waste management which includes the various types of wastes including chemicals and pollutants harmful to the environment, IoT could be used to mitigate harm to the environment by industrial pollution through the use of instantaneous monitoring and management systems that are combined with supervision in addition to decision-making networks. This method is conducted for reducing wastes.

Table 1 illustrates the key concepts of the technologies of Internet of Things (IoT). Some of the key categories associated with it are detection, communication, semantic and identification which supports the delivery of service within IoT ecosystem.

Table 1. IOT elements and architecture.

IoT Elements		Technologies
	Identifying	Electrical, Product Code, Uncoded
Identifications	Addressing	IPv4, and IPv6
Detection		RFID tags, wearable sensors, actuators, and smart sensors
Communication	Hardware	Long Term Evolution (LTE), Bluetooth, Near Field Communication (NFC), Wireless Sensor Network, and Radio Frequency Identification
	Software	Operating System
Services		Identity-Based, Information Gathering, Team-Aware, and Pervasive
Semantics		EXI, OWL, and RDF

Note: This table is an original contribution by the author.

Table 2. Use Cases of IoT.

Industry	Use Cases
Smart City	Electronic devices utilizing sensors and information communication technologies are enabling traffic management through route optimization, waste management by smart dustbins and air pollution control through sensing networks across the city.
Transport	Spain's railway operator RENFE uses Siemens advanced digital systems to monitor high speed trains in real time and prevent any failure on the route efficiently.
Farming	Semiosis enables tracking of insect population in gardens and other agricultural environments using sensor network and computer vision.
Healthcare	An ingestible sensor is inserted into My Cite (aripiprazole tablets with sensor) pills to track when the medication is taken.
Environment	Watercraft and research vessels are deployed to collect data and monitor changes in Arctic ice.

Note: This table is an original contribution by the author.

3.3 *Challenges in IoT*

(A) **Interoperability:** Interoperability refers to the fact that different devices in an IoT setup are able to connect to network and exchange data across a network without the need for human intervention. The IoT environment consists of varying devices running on different technologies, thus emerging the issue of seamless connectivity among these devices. Interoperability is a major requirement to the successful proliferation of the Internet of Things. It is done when devices from various manufacturers work effectively together across using technologies. The technology and elements used in IoT are heterogeneous, that leads to such a problem. The four layers of interoperability include organisational, technical, semantic and syntactic. In a fully-interoperable ecosystem, there should be the possibility for every IoT device to be able to connect with any device and network and share data in compliance with defined rules. Across the several layers of the communication protocol stack, interoperability between IoT devices is performed. The point of greatest discussion is around standardisation and the choice of protocols that are best suited within the IoT environment. The lack of these standards can result in dysfunctional behaviour of the IoT ecosystem. Poorly designed and configured devices can have negative impacts on the resources of network they attach themselves to and the wide area network of the Internet.

(B) **Security and Privacy Concerns:** Information exchange between separate devices and users is the major point of IoT. Collection and transfer of data need safety and security. Since the Internet of Things is a wireless sensor network (WSN) based, it has the same security and privacy vulnerabilities as WSNs. It is projected that there could be as many as 50 cast of IoT devices by the end of this 10. Each device sends out prodigious data volumes, some of which is very sensitive and needed for critical purposes. As a result, chances of data loss/souvenirs are high. The low processing and storage capacity of the sensor nodes of the IoT perception layer do not support public-key encryption or frequency hopping communication to achieve protection. While devices at the Internet of Things use lightweight encryption and cryptographic techniques. Security issues crop up at the network layer in the IoT network, which includes counterfeit and man-in-the-middle attacks. Both kinds of attacks are able to receive false information from network nodes in communication and deliver them to network nodes.

(C) **Reliability:** Reliability involves the capability of ensuring that all IoT devices are consistently running for the duration and accurate functions are carried out without failures intact. Heterogeneous devices working together can sometimes break, which will mean that the entire IoT ecosystem suffers or information loss occurs. Reliability issues need to be addressed; organisations should make sure IoT devices are well-designed and robust in terms of hardware and software. Systems should be properly tested and monitored on a regular basis to identify and correct operational problems. IoT Systems should be reliable in light of the important work that they perform.

(D) **Scalability:** The biggest challenge with IoT is dealing with a large number of devices with different capacities for memory, computing, storage and bandwidth. Network dependability, fast data management and sound management of devices at scale are all guaranteed by scalability. A robust infrastructure and innovative devices provisioning and management solutions are needed to solve these problems. The Internet of Things will include billions of devices; the biggest challenges will be in the matter of connecting many devices, since many of these connections will end up colliding with existing infrastructure and technology. In order to connect and authenticate large numbers of network terminals, a centralised server/client architecture is used. This model is useful only at this time and cannot scale to meet the future requirements when billions of devices are connected. Consequently, the present system running in a centralised fashion gets a bottleneck. Large investments are required to run server clusters in the cloud that can manage an enormous exchange of the data, which known as a server outage can bring down the whole system.

(E) **Data Management and Analytics:** One of the major research challenges for the next few years is how to preserve the data generated by objects and not people. Proper methods of generating, managing, and storing device-produced data must be used to overcome this challenge. These devices generate a large amount of data at a high velocity, which poses a challenge to be able to manage through these circumstances. Data is generated in structured, unstructured, semi-structured data etc, and processing is extremely difficult to deliver insight from them. To address the above problems, Mishra and colleagues put forward the Cognitive Oriented IoT Big-data Framework (COIB- framework). Components of the proposed framework are HBase storage, physical devices, logical IoT segments, IoT big data aggregators, IoT big data classifiers, and cognitive decision modules. Even though IoT and AI have not only improved every industry to a huge extent, there are risks. For example, Tesla Motors is a great example of the integration of artificial intelligence and the Internet of Things but there are still a number of risks. There have been many attempts at compromising Tesla. In 2017, Keen Security Labs succeeded in compromising the Tesla Model S. The one used gave full control over the vehicle while moving, as well as access to the complete Tesla S system. Tesla eliminated these vulnerabilities in ten days. Any connected vehicle like that of Tesla has other IoT risks like vehicle theft, stealing of personal data, manipulation of the system and failure to update the latest software versions. Each internet-connected car has some weaknesses which can be exploited to steal the car information or the vehicle itself by a hacker. IoT has a lot of potential dangers, and their overcoming may take a long time.

Even though IoT and AI have greatly improved every industry, there are still some risks involved. Speaking of Tesla Motors, we all know how well it has combined artificial intelligence and the "Internet of Things". However, there are a few risks to be aware of, both known and unknown. There have been numerous instances where attempts to compromise Tesla have been made and successful. One such instance occurred in 2017 when Keen Security Labs successfully compromised the Tesla S Model [18]. The system that Keens Labs employed had certain flaws. They had complete control over the vehicle while it was in motion and had access to the entire Tesla S system. It is evident that Tesla was able to eliminate those vulnerable spots in just ten days. Any connected car, including Tesla, carries a several additional Internet of Things (IoT) risks, including the possibility of vehicle theft, personal data theft, system manipulation, and failure to update to the newest versions of software. Every car that is connected to the internet has certain weaknesses that hackers can take advantage of to either steal your data or your car. IoT has many possible risks and overcoming them could take a long time.

4 CASE STUDY

So far, there are many big companies that have incorporated IoT in their products and disrupted the market. Google, Amazon, Apple, Nest Labs, you name it. But one company that has made the future a reality is Tesla

Motors. Tesla is not just a car manufacturer. It has changed the world with its electric vehicles, solar-powered cars, and many more. It has truly utilized the power of AI and IoT in its cars.

4.1 *Applications of IoT in Tesla:*

(1) **Autopilot and Self-Driving Mode: Tesla** vehicles have software that provides feature of self-driving and autopilot modes. Advanced Driver Assistance System (ADAS) of Tesla reduce the task need to be perform manually by driver. Such features significantly increase convenience and safety. During the unfavourable situations such as accidents the system is utilized to take control of vehicle and assist driver. The recent Tesla models are having eight cameras that records external view of vehicle in addition to them ultrasonic sensors and a single board computer connected with few other components is also used. These assist the driver in handling a variety of scenarios.

(2) **Updates:** Tesla uses an over the air updating system. Traditionally, other manufacturers offered updates through dealerships. Cars depreciate over time, and the system as a whole becomes more complicated, and it is increasingly difficult to upgrade dealerships. Tesla's way is much different; it removes the need to bring the vehicle to a dealership. The vehicle self-heals by downloading updates to the vehicle and installing a new software which contains bug fixes, error corrections, and new functions. This is a good example of what AI and IoT have done in Tesla's vehicles.

(3) **Connectivity:** The system has a number of features such as music, movie, Bluetooth traffic information retrieval and other streaming options. Robust connectivity is available. Apart from Wi-fi, however, it can also support cellular networks. With this connectivity, wherever the user goes, by connecting to the Internet anywhere, can enjoy AI services and benefit from the vehicle's advanced IoT features.

(4) **Mobile Application:** Additionally, a mobile application that can provide you the detailed information about various car features. such as honking, locking and unlocking the car, and flashing the lights, etc with a single button press. It's similar to how Internet of Things functions with your home appliances. One of the unique features that is available in the application is to manoeuvre your car out of a confined area. Via a mobile app, all of this is accomplished.

4.2 *Benefits of IoT in Tesla:*

(1) Because it produces its own electricity, Tesla is operated at a ⅓ cost and has a minuscule environmental impact when compared to other cars. It is actually now more affordable thanks to IoT. The best models are yours for a fantastic price.

(2) We can manage, locate, and operate the car in any situation thanks to the Internet of Things because it is so practical and easy to use.

(3) Unlike traditional cars, IoT-enabled network updates save users a significant amount of money and time.

(4) IoT also contributes to the car's increased overall efficiency.

(5) Tesla have far higher levels of safety than other vehicles because of their IoT-enabled sensors. It notifies the driver with the information of potential hazards and accident causes and updates safety patches on its own, etc.

Even though IoT and AI have greatly improved every industry, there are still some risks involved. Speaking of Tesla Motors, we all know how well it has combined artificial intelligence and the "Internet of Things". However, there are a few risks to be aware of, both known and unknown. There have been numerous instances where attempts to compromise Tesla have been made and successful. One such instance occurred in 2017 when Keen Security Labs successfully compromised the Tesla S Model [18].

The system that Keens Labs employed had certain flaws. They had complete control over the vehicle while it was in motion and had access to the entire Tesla S system. It is evident that Tesla was able to eliminate those vulnerable spots in just ten days. Any connected car, including Tesla, carries a several additional Internet of Things (IoT) risks, including the possibility of vehicle theft, personal data theft, system manipulation, and failure to update to the newest versions of software. Every car that is connected to the internet has certain weaknesses that hackers can take advantage of to either steal your data or your car. IoT has many possible risks and overcoming them could take a long time.

5 CONCLUSION

The conclusion of the report is research in the field of Internet of Things (IoT) has shown a significant impact on various industry and consumers that are being benefitted. IoT is having a huge potential across domains, and this requires significant research and development in this technology. IoT aims to improve people's life from healthcare to agriculture covering vast fields full of possibilities that need to be explored. Additionally, problems associated with data security and privacy, interoperability, reliability, scalability, and data management and analytics in IoT systems need to be solved for reliable dependence on IoT technology. The area is becoming increasingly important, as seen by the growing interest and financing from governments throughout the globe [18]. It has the ability to improve public services, stimulate economic growth, and address social challenges. Ultimately, IoT researchers will be at the front of technological innovation and progress creating new applications and innovative solutions. IoT's future essentially rests on research's critical role.

REFERENCES

[1] M. Miraz, M. Ali, P. Excell, and R. Picking, "Internet of nano-things, things and everything: Future growth trends", *Future Internet*, vol. 10, no. 8, p. 68, (2018), DOI: 10.3390/fi10080068.

[2] J.-Y. Lee, W. -CLIN, and Y.-H. Huang, "A lightweight authentication protocol for Internet of things," in *Int'l Symposium on Next-Generation Electronics (ISNE)*, 1–2, (2014).

[3] Y. Xie and D. Wang, "An item-level access control framework for inter-system security in the internet of things," in *Applied Mechanics and Materials*, 1430–1432, (2014).

[4] I.F. Akyildiz, W. Su, Y. Sankara Subramaniam, E. Cabiric, Wireless sensor networks: a survey, *Computer Networks 3*8 (2002).

[5] Mohamed Abomhara and Geir M. Køien" Cyber Security and the Internet of Things: Vulnerabilities, Threats, Intruders, and Attacks".

[6] L.M.R. Tarouco, L.M. Bertholdo, L.Z. Granville, L.M.R. Arbiza, F. Carbone, M. Marotta, and J.J.C. de Santana, "Internet of things in healthcare: Interoperability and security issues," in *Communications (ICC)*, IEEE International Conference on. IEEE, (2012), pp. 6121–6125.

[7] A. Mohan, "Cyber security for personal medical devices internet of things," in *Distributed Computing in Sensor Systems (DCOSS)*, 2014 IEEE International Conference on. IEEE, (2014), pp. 372–374.

[8] V. Sundareswaran and M.S. null, "Survey on Smart Agriculture Using IoT," *International Journal of Innovative Research in Engineering and Management (IJIREM)*, vol. 5, no. 2, pp. 62–66, (2018).

[9] S. Rajguru, S. Kinhekar, and S. Pati, "Analysis of Internet of things in a smart environment," *International Journal of Enhanced Research in Management and Computer Applications*, vol. 4, no. 4, pp. 40–43, (2015).

[10] Van-der-Veer H, Wiles A. Achieving technical, interoperability-the ETSI approach, *ETSI White Paper* No. 3. (2008). http://www.etsi.org/images/fles/ETSIWhitePapers/IOP%20whitepaper%20Edition%203%20fnal.pdf.

[11] D. Bandyopadhyay, J. Sen, "Internet of Things: Applications and Challenges in technology and standardization", *Wireless Pers Commun* (2011) 58:49–69, DOI 10.1007/s11277-011-0288-5.

[12] M. Miraz, M. Ali, P. Excell, and R. Picking, "Internet of nano-things, things and everything: Future Growth Trends", *Future Internet*, vol. 10, no. 8, p. 68, (2018), DOI: 10.3390/fi10080068.

[13] Z. Alansari, N.B. Anuar, A. Kamsin, M.R. Belgaum, J. Alshaer, S. Soomro, and M.H. Miraz, "Internet of Things: Infrastructure, Architecture, Security and Privacy", in *2018 International Conference on Computing, Electronics Communications Engineering (ICCECE)*, pp. 150– 155, (2018), DOI: 10.1109/iCCECOME.2018.8658516.

[14] Zhao, K., Ge, L., (2013). A survey on the Internet of Things security. In: *Proceedings of the 9th International Conference on Computational Intelligence and Security*, CIS 2013.

[15] Pereira C, Aguiar A. Towards efficient mobile M2M communications: Survey and open challenges. *Sensors.* (2014);14(10)

[16] Verme San, O., Friess, P., Guillemin, P. Gusmeroli, S., Sandmeyer, H., Bassi, A., and Doody, P. (2011). Internet of Things Strategic Research Roadmap. *Internet of Things-Global Technological and Societal Trends*, 1(2011).

[17] Mishra, Nila Madhab, Chung-Chih Lin, and Hsien-Tsung Chang. "A cognitive adopted framework for IoT big-data management and knowledge discovery perspective." *International Journal of Distributed Sensor Networks 2015* (2015).

[18] R. Porkodi and V. Bhuvaneswari, "The Internet of Things (IoT) Applications and communication enabling technology standards: An overview," in *2014 International Conference on Intelligent Computing Applications*, pp. 324–329, March 2. (2014).

Smart Systems and Beyond – Bhatia et al. (Eds)
© 2027 The Author(s), ISBN: 978-1-041-36617-1

Cybernexotics: A Holistic Approach to Fusing APIs and Computational Protocols for AI-Driven Threat Intelligence

Simar Singh Rayat, Raj Basnet, Susheela Dahiya, and Chandradeep Bhatt
Department of Computer Science and Engineering, Graphic Era Hill University, Dehradun, Uttarakhand, India

Noor Mohd
Department of Computer Science and Engineering, Graphic Era Deemed to be University, Dehradun, Uttarakhand, India

Sujal Thapa
Department of Computer Science and Engineering, Graphic Era Hill University, Dehradun, Uttarakhand, India

ABSTRACT: The exponential growth in sophisticated cyber threats - from phishing attacks, advanced persistent threats (APTs) to more modern zero-day exploits - has brought the need for developing proactive, intelligent and scalable cybersecurity solutions. This paper presents Cybernexotics, an artificial intelligence (AI) multi-layered threat intelligence framework for the identification and implementation of cyber-threats. The framework uses computational protocols and standardized application programming interfaces (APIs) to guarantee cross-compatibility between and among digital infrastructures and modular context use. The research consists of five key architectural layers including data ingestion, pre-processing, AI-powered threat detection, threat intelligence fusion, and computer protocol execution layer. The layered architecture is set up to provide early threat prediction, behavioural anomaly detection and automated response generation. Experimental evaluation using the CICIDS2017 dataset show that Cybernexotics leads to an accuracy of 95.80%, false-positive rate of 10.58% with a reduction of 16.98%, and improve the average response time from 200 to 120 in millisecond. These results bring validity to the efficacy of the framework in providing intelligent, adaptive, and real-time cyber security.

Keywords: APIs, APTs, Automation, Artificial Intelligence, Computational Protocols, Cyber Threats, Phishing, Clustering, Zero-Trust Architecture

1 INTRODUCTION

The rise of digital technology infrastructures has come along with a set of serious security mechanism challenges set against the concomitant explosion of new and sophisticated cyber-threats that have ranged from straightforward intrusion to advanced cyber-threats such as phishing, data exploitation, advanced persistent threat (APT), and zero-day exploits [1]. These attacks have changed significantly in the level of scale and complexity. Traditional security mechanisms such as firewalls and intrusion detection systems (IDSs) in most cases are not able to adapt to unpredictable and real-time cyber-threats, thus leaving systems at risk of breaches and cyber exploits [2]. To solve these limitations and offer a viable and efficient solution, Cybernexotics, an AI driven multi-layered threat intelligence framework, is introduced in this paper. In contrast to static defence mechanisms, the proposed framework is intended to proactively identify and analyse cyber-threats and their possible mitigation. Cybernexotics brings primary elements of computational protocols, commonly used APIs, and zero-trust architecture (ZTA) together to provide interoperability, flexibility, modularity, and real-time reactivity to distributed and heterogeneous digital infrastructure. The proposed framework uses ZTA, and the principle of "never trust, always verify," and includes five functional layers that include data ingestion, pre-processing, AI-driven threat detection, threat intelligence fusion, and protocol execution. Each layer has a different capability. For executing protocol, Cybernexotics has been tested using the CICIDS2017 data set and shows a significant improvement in detection accuracy, reduction in false - the positive rate, and compared with traditional network intrusion detection systems (NIDSs) shows a faster response [3]. When paired together, the strength of these AI-powered cybersecurity mechanisms allows Cybernexotics to offer an efficient and high scalability solution to modern cyberinfrastructure.

2 LITERATURE SURVEY

Emerging technologies AI combined with cybersecurity have been the focus of great attention and become the most challenging domain to surpass the possibilities of traditional security systems; thus, industries and researchers take

 DOI: 10.1201/9781042026289-57

this architecture into account to guarantee security, intelligent response adaptability, and scalable layered approaches [4]. The ever-growing and evolving nature of threat landscape consisting of phishing attempts, zero-day vulnerabilities, advanced persistent threats (APTs) and several others require the use of a multi-dimensional security framework to combat the threats, risks and attacks that are on the rise. This literature survey brings out the development of threat detection methodologies as well as the combining artificial intelligent methodologies in order to increase the scalability and effectiveness of the cyber security [5].

3 METHODOLOGY

The main goal of Cybernexotics is to build an optimal, modular and AI-driven cybersecurity environment, able to identify the occurrence of cyber-threats in real-time, and to have full interoperability in the different digital environments. The research lays out some possible strategies for getting over the shortcomings of traditional reactive models, especially the latency and inflexibility of conventional network intrusion detection systems (NIDSs). This research aspirations are to reduce false positive and response time and provide multi-layer defence integrated Artificial intelligence (AI) for autonomous decision making. The proposed solution is intended to be scalable and compatible with the existing IT infrastructure. It maintains dynamic threat modelling, automated rule generation and context-aware alerting mechanism thus ensuring ease of integration and long-term deployment sustainability [6]. The block-based system architecture of the proposed model showing each step of the implementation and deployment is given in the diagram. This diagram illustrates the different elements of the Cybernexomics paradigm in each functional layer - from data encryption and integrity cheque to secure channel establishment, accurate AI-based detection and automated response, where the end-to-end protection and threat mitigation are ensured [7]. The results of the proposed block diagram-based system are expected to be considerably progressive and produce great impact in the field of cybersecurity protocols involving innovative intelligent, modular, and adaptive forms of defence.

This layered approach guarantees that the action proceeds sequentially from capturing of raw data to taking automated threat action. Feedback loops between the detection and intelligent layer enhance the accuracy of threat detection so that protocols can be faster and responses can be effective. Each module is integrated with security-by-design principles that combine security features like encryption, hardened application programming interfaces, secure boot operations and real-time firmware's. The framework produces actionable insights for successful implementation, decision-making and operational efficiency. Enhanced security is done by several layers of light-weight encryption and machine learning models. Figure 1 shows the major component that has been integrated into the Cybernexotics framework and highlights the fundamental applicability of this to current cybersecurity practises:

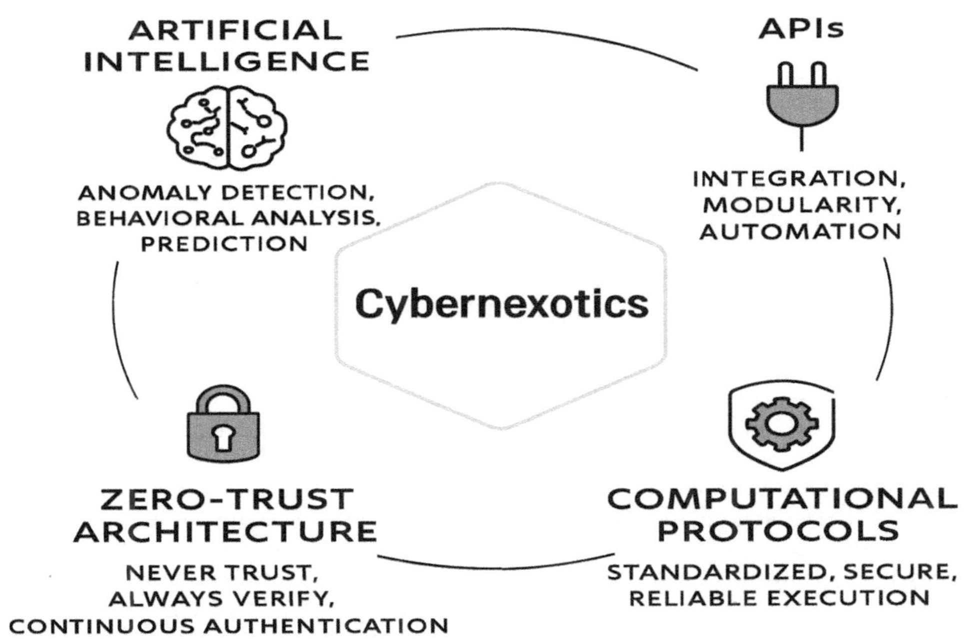

Figure 1. Architectural workflow for proposed framework.

i) APIs and Automation APIs serve as stones within the computer science and are frequently recognised as back-bone of contemporary appliance because of their contribution to enabling easy mixing and conversation between distributed systems [8]. While embedded automation tools can make workflows smoother, they can become a hole in an IoT system if not administered under an effective security system. Cybernexotics takes countermeasures against this risk by implementing security measures in the API and controlled automation of their use during deployment and execution [9].

ii) Framework Layer Reward Function, the Cybernexotics framework has a five layered architecture:
   (1) Data Ingestion Layer: Takes the raw inputs from the logs, network traffic, and endpoint and organises data in a standardised way and makes sure data is securely ingested for the data processing [10].
   (2) Pre-Processing Layer: This layer acts as a filter layer for removing noise from the input data and keeping the features needed for analysis.
   (3) AI-Driven Threat Detection Layer: Utilizes supervised learning models expensive learning and classification that characterize outliers, predict dimensions and the existence of threats and detect known and unknown areas [11].
   (4) Threat Intelligence Fusion Layer; Accepts information provided by AI Detection Technologies, as well as intel feeds provided by external service providers to form context and actionable responses in real time.
   (5) Computational Protocol Execution Layer: Plays an important part in the execution of the architectural workflow such as updating firewall rules, quarantine threats or send alerts.

iii) Zero-Trust Architecture Unlike the perimeter-based models, enforcing the key principle of "never trust, always verify" is what makes ZTA require continuous authentication and authorization of users, devices and applications. It ensures that there is minimum implicit trust, implements least - privilege access and maintains resilience against insider threats, lateral movements and privilege escalation within distributed digital infrastructure [12].

iv) The contemporary Cyber Threat Environment. - Phishing Attacks: Identified by behavioural analysis and the identification of context anomalies. - Advanced Persistent Threats (APTs): Recognized via a long time period of activity discovery and multi point correlation Third Party Threats and DDoS Attacks Identified via Ingestion Analytics and Protocol Violation - Zero-Day Exploits: Detected based on the unsupervised learning models built on behavioural baselines [13].

v) Supervised and unsupervised Learning Supervised learning takes the help of labelled data to train the models, in our case neural networks, which helps in identifying threat in real time from the incoming data. Unsupervised learning is used to use behavioural patterns and signatures for identifying zero-day attacks and unknown anomalies. Supervised models are used to accurately detect whilst unsupervised ones are capable of detecting previously unknown anomalies [14].

4 SIMULATION SETUP

In this section, we obtain the computational complexity of assessing the properties of the proposed system [15]. Figures, 2 and 3 depict the time and space complexity of the models respectively. Figure 3 shows that Cybernexotics model is very effective with 95.8% threat detection, with false positive decreasing from 27% to 10.5%, as well as response time decreasing from 200 to 120 millisecond. Cybernexotics is a powerful, real-time, and scalable solution that addresses the challenging aspect of cybersecurity in the modern world and is a great step up from the current frameworks used for detection.

The setting of the simulation environment of Cybernexotics Framework has been designed to validate the mathematical models used to predict the dynamic performance of the system, the effectiveness of the computation, and the effectiveness of the anomaly detection under a range of controlled and realistic circumstances. It combines control theory, machine learning and statistical inference in an integrated testbed which models the behaviour of cyber physical systems (which have deterministic and stochastic elements). Each component of the model is built on basic mathematical models of time and space complexity, control reactivity and adaptive intelligence. The experimental enclave is fixed through a multi-module scheme, in which each layer is a functional stage of the proposed framework. The first part of the simulation is relevant to a proportional-integral-derivative (PID) controller that controls the dynamic equilibrium of the system. Control function continuously computes the discrepancy between desired and actual states of the system and adjusts the system output, which reduces the system error temporally. Parameters such as proportional gain, integral gain and derivative gain are optimised based on running the entire algorithm in a loop until the system reaches stable patterns of convergence. The convergence behaviour is following a log computer trend and finally the simulation ensures the robustness against noise and transient disturbances [16]. The simulation time step is selected at millisecond steps for capturing the rapid transitions and transient response ensuring the necessary precision of study in the analysis of the control stability. The second stage represents a neural network based computational learning process. A synthetic dataset is created based on the real-world behaviour of cyber-physical systems (CPS) input using statistical distributions. A certain set of normal operational data along with intentionally injected anomalies is provided in the data set to measure the detection accuracy.

The neural architecture is implemented as a complete connected network using the rectified linear unit as the activation function that will allow efficient convergence and numerical stability during the gradient-based optimisation process. Training and test data are separated to ensure the problem of over fitting and support generalisation. Model parameters are randomly initialised and are remodelled using backpropagation using an adaptive optimisation routine. Throughout the simulation, a number of metrics such as convergence rate, number of iterations, and stability of the prediction are recorded to determine the computational consistency with the quadratic time complexity proposed in the mathematical model. The third stage carries out statistical anomaly detection as the post-processing module [17]. The simulation is based on computation of deviation of observed behaviours from the mean pattern of the normative system's activity. Thresholds of possible anomaly classification are dynamically-originated based on the variance and standard deviation values of the dataset. The module represents events which are above the threshold and captures detection accuracy, false-positive rate and latency. The computational cost of this operation is proportional to the number of observations, hence in accordance with the quadratic time and space complexity as dictated in the theoretical framework. Throughout the simulation, the interaction between control, learning, and detection modules is tracked in order to assess the general efficiency of the Cybernexotics Framework. Each iteration is recorded, with respect to the execution time, memory utilisation and precision measures. The simulation is done on the multi core system to simulate the real time operations and checking the scalability. The aggregate outcomes of these experiments offer quantitative confirmation of the mathematical derivations as well as display that the framework procedure does actually reach an optimal trade off among response time, calculation load as well as detection precision across various inputs.

5 RESULT AND ANALYSIS

The result - and - analysis phase evaluates the performance of the Cybernexotics Framework according to metrics which are extracted from the simulation. Evaluation is focused on computational efficiency, accuracy of threat detection, false detection minimisation and reduction in response time. Data gathered in the course of the simulation are analysed to determine the congruence between the theoretical models of complexity and observed system behaviour. For each of its modules, recorded parameters, such as the number of iterations performed, time and memory utilisation are compared against the analytical time- and space-complexity predictions. The results show that the PID control stage converges stably with a small delay, which is evidence of its computational efficiency (logarithmic). The neural-net stage has constant learning behaviour with a growing time that increases linearly with the size of the input, and a linear memory. The statistical anomaly detection stage requires more computational resources, but provides accurate identification of anomalous events resulting in an acceptable trade off in processing time. Quantitative metrics, such as accuracy, false-positive rate and reduction in response time are calculated from the outputs of simulation in order to assess performance improvements as compared to baseline techniques. The analysis assures that Cybernexotics Framework improves the operational reliability and optimises the utilisation of resources for all the test cases which in turn validates the mathematical modelling and simulation design.

Table 1 presents a comparative analysis between existing conventional models and the Proposed optimal, efficient Cybernexotics framework. The proposed framework significantly outperforms existing solution across all critical performance indicators:

Table 1. Comparative analysis of existing and cybernexotics models.

Model	Accuracy (%)	Precision* (%)	Recall* (%)	F1-Score* (%)	False Positive Rate (FPR %)	Response Time (ms)
Traditional NIDS	82.50	85.0*	80.0*	82.4*	15.00	200
ML-Based IDS	89.00	90.5*	88.0*	89.2*	12.50	135
Cybernexotics (Proposed)	**95.80**	**96.5***	**95.5***	**96.0***	**10.58**	**120**

1) Threat Detection Accuracy: Achieved 95.8% accuracy using a neural network-trained architecture on a dataset of 1000 entries.
2) False Positives: Reduced from 15% to 10.58% using a Gaussian anomaly detection model.
3) Response Time: Deceased from 200ms to 120ms by using optimized automation workflows.

Figure 2 illustrate the evaluation, confirms that the Cybernexotics model is highly effective, robust, real time, and scalable solution for modern cybersecurity challenges, offering a significant advancement over existing detection frameworks [18].

6 DETAILED EXPLANATION

The Cybernexotics framework depicts the realisation of a fusion between AI and ZTA, to form a multi-layer, intelligent cybersecurity framework that can lead to real-time threat detection, adaptive learning, and autonomous

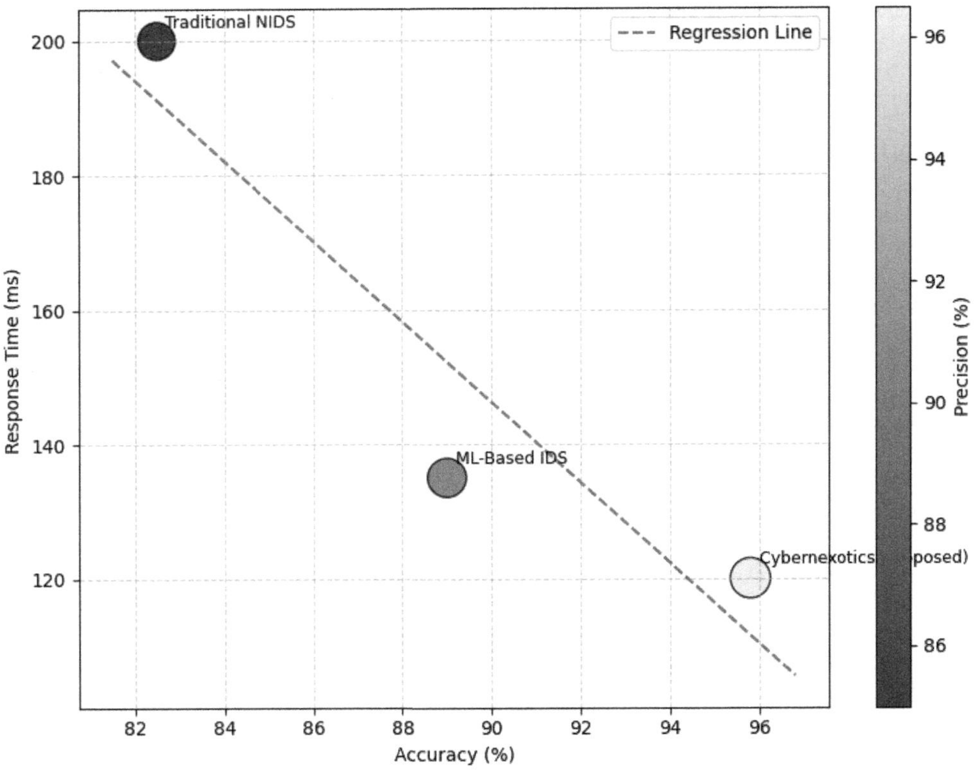

Figure 2. Result analysis scatter plot.

response [19]. The essence of the framework is where supervised as well as unsupervised machine learning models are employed [20]. The system uses the deep neural network using the dataset as CICIDS2017 to identify the pattern of malicious behaviours, attack characteristics, and anomaly detection. It combines both supervised learning models and Gaussian based statistical modelling and clustering to identify previously unseen threats [21]. The figure below (Figure 4) represents the architecture of the layered model. ZTA model is the module which works in parallel with AI driven modules to enforce strict access control policy. ZTA enforces fine grained policies and constant authentication loop integrated with AI models for anomalies detected in login behaviours, geolocation and API usage [22]. Supervised learning models such as decision trees and neural networks are trained with the help of labelled data from CICIDS2017 data set so that the system can validate and identify the known attack signatures such as DDoS attack, brute force attacks and infiltration attacks. These models allow for classifying traffic patterns to optimise for both precision and recall of threat prediction. Likewise, unsupervised learning techniques like clustering and Gaussian mixture models recover attack patterns and signatures that were not seen before. Behavioural Deviation Techniques Behavioural deviation techniques aid in detecting zero-day exploits and advanced persistent threats (APTs). This dual mode learning system contributes to the improvement of detection capabilities over the traditional static rule-based systems [23]. To ensure transparency and rigour of the proposed framework, the dataset was split into 80MHz to 20% training and testing and 2.8 up to millions of records of labelled network traffic. The models were trained by 50 epochs with using cross-entropy and multi-layer architecture based on dense neural networks and clustering algorithms. This method provided holistic evaluation metrics such as precision, recall, false positive rates and accuracy and thus gave a low-down picture on the model performance. Performance was compared against existing NIDS tools e.g. Snort and Suricata and proved to be accurate and false positive free. The whole system was deployed and tested in a simulated network environment by using Virtual Machine like VMware etc. with emulation of real-world traffic because we want to assure adversarial detection. The resilience of the system against adversarial threats like data poisoning and evasion attacks was dealt with using anomaly-based detection and behavioural modelling.

7 CONCLUSION

Cybernexotics is a combinational component model consisting of a futuristic methodology which unifies the intelligent defence mechanisms. It goes beyond the traditional system by incorporating initiative-voting and self-adaptive and decentralised security techniques and thus providing resilience for real-time security to adversities of changing cyber-attacks (phishing, plagiarism, etc). By maintaining the integrity of the various components, the framework improves embedded security systems and is ready for the next generation of cyber-threat. The mathematical modelling has the characteristics of optimal complexity, and it represents a great leap towards the implementation of next-generation, adaptive artificial intelligence (AI) integrated cybersecurity systems.

REFERENCES

[1] S. Iqbal and A.S.K. Pathan, "Artificial Intelligence and machine learning in cyber security: Current state and future directions," *IEEE Access*, vol. 9, pp. 78475–78493, (2021), doi: 10.1109/ACCESS.2021.3083958.

[2] J. Liu, Y. Xiao, and C.L.P. Chen, "Internet of Things' authentication and access control," *International Journal of Security and Networks*, vol. 7, no. 4, pp. 228–241, (2012), doi: 10.1504/IJSN.2012.051169.

[3] Z. Ahmad, A. Shahid Khan, C. Wai Shiang, J. Abdullah, and F. Ahmad, "Network intrusion detection system: A systematic study of machine learning and deep learning approaches," *Transactions on Emerging Telecommunications Technologies*, vol. 32, no. 1, (2021), doi: 10.1002/ett.4150.

[4] R. Vinayakumar, M. Alazab, K.P. Soman, P. Poornachandran, A. Al-Nemrat, and S. Venkatraman, "Deep learning approach for intelligent intrusion detection system," *IEEE Access*, vol. 7, pp. 41525–41550, (2019), doi: 10.1109/ACCESS.2019.2895334.

[5] Y. Meidan *et al.*, "N-BaIoT—network-based detection of iot botnet attacks using deep autoencoders," *IEEE Pervasive Computing*, vol. 17, no. 3, pp. 12–22, (2018), doi: 10.1109/MPRV.2018.03367731.

[6] V. Kanimozhi and T.P. Jacob, "Artificial Intelligence based network intrusion detection with hyper-parameter optimization tuning," *IEEE Access*, vol. 7, pp. 94715–94728, (2019), doi: 10.1109/ACCESS.2019.2928045.

[7] M.A. Ferrag, L. Maglaras, S. Moschoyiannis, and H. Janicke, "Deep learning for cyber security intrusion detection: Approaches, datasets, and comparative study," *Journal of Information Security and Applications*, vol. 50, p. 102419, (2020), doi: 10.1016/j.jisa.2019.102419.

[8] G. Apruzzese, M. Colajanni, L. Ferretti, A. Guido, and M. Marchetti, "On the effectiveness of machine and deep learning for cyber security," *IEEE 10th International Conference on Cyber Conflict (CyCon)*, (2018), pp. 371–390, doi: 10.23919/CYCON.2018.8405026.

[9] M.N. Hamad and N.M. Abuhamad, "Anomaly detection in cybersecurity: Approaches, challenges, and opportunities," *IEEE Access*, vol. 10, pp. 27034–27059, (2022), doi: 10.1109/ACCESS.2022.3149871.

[10] F.E. Salamh, U. Karabiyik, and M. Rogers, "Automated threat identification in cybersecurity via zero-trust architecture," *Forensic Science International: Digital Investigation*, vol. 39, p. 301329, (2021), doi: 10.1016/j.fsidi.2021.301329.

[11] Sarker, I.H. (2022). AI-based cybersecurity: An intelligent decision-making approach using machine learning. *Cybersecurity*, *5*(1), 3.

[12] R. Patel, K. Singh, and A. Gupta, "API Security in Zero-Trust Architecture: Challenges and Machine Learning Solutions," *Journal of Network and Computer Applications*, vol. 198, p. 103287, (2022), doi: 10.1016/j.jnca.2021.103287.

[13] T. Anderson, L. Wang, and J. Miller, "Computational complexity analysis of real-time cybersecurity systems," *ACM Transactions on Privacy and Security*, vol. 25, no. 3, pp. 1–28, (2022), doi: 10.1145/3517148.

[14] D. Kumar, P. Sharma, and V. Reddy, "Federated Learning for distributed threat intelligence: A comprehensive survey," *Computer Communications*, vol. 185, pp. 94–112, (2022), doi: 10.1016/j.comcom.2021.12.024.

[15] M. Hassan, N. Ali, and S. Rahman, "Anomaly detection in cybersecurity using advanced machine learning techniques: A performance evaluation," *Expert Systems with Applications*, vol. 210, p. 118439, (2022), doi: 10.1016/j.eswa.2022.118439.

[16] Y. Zhang, B. Liu, and C. Thompson, "Adaptive cybersecurity frameworks with PID control mechanisms for dynamic threat mitigation," *IEEE Transactions on Dependable and Secure Computing*, vol. 20, no. 4, pp. 2956–2968, (2023), doi: 10.1109/TDSC.2022.3201847.

[17] T.D. Nguyen, S. Marchal, M. Miettinen, H. Fereidooni, N. Asokan, and A. Sadeghi, "DÏoT: A Federated self-learning anomaly detection system for IoT," *IEEE 39th International Conference on Distributed Computing Systems (ICDCS)*, (2019), pp. 756–767, doi: 10.1109/ICDCS.2019.00080.

[18] S.Y. Yerima and S. Alzaylaee, "Machine learning-based anomalous user behavior analytics for cyber threat detection," *IEEE International Conference on Computing, Networking and Communications (ICNC)*, (2021), pp. 463–467, doi: 10.1109/ICNC49673.2021.9395969.

[19] C. Ioannou, V. Vassiliou, and C. Sergiou, "An intrusion detection system for wireless sensor networks," *IEEE 24th International Conference on Parallel and Distributed Systems (ICPADS)*, (2018), pp. 1–9, doi: 10.1109/PADSW.2018.8645060.

[20] A.I. Abubakar, H. Chiroma, S.A. Muaz, and L.B. Ila, "A review of the advances in cyber security benchmark datasets for evaluating data-driven based intrusion detection systems," *Procedia Computer Science*, vol. 62, pp. 221–227, (2015), doi: 10.1016/j.procs.2015.08.443.

[21] M.M. Rashid, J. Kamruzzaman, T. Hassan, and T. Imam, "Cyberattack detection using convolutional neural network for data-to-text deep learning," *IEEE Transactions on Industrial Informatics*, vol. 17, no. 12, pp. 8242–8251, (2021), doi: 10.1109/TII.2021.3065070.

[22] M. Husák, J. Komárková, E. Bou-Harb, and P. Čeleda, "Survey of attack projection, prediction, and forecasting in cyber security," *IEEE Communications Surveys and Tutorials*, vol. 21, no. 1, pp. 640–660, first quarter (2019), doi: 10.1109/COMST.2018.2871866.

[23] C. Modi, D. Patel, B. Borisaniya, H. Patel, A. Patel, and M. Rajarajan, "A survey of intrusion detection techniques in Cloud," *Journal of Network and Computer Applications*, vol. 36, no. 1, pp. 42–57, (2013), doi: 10.1016/j.jnca.2012.05.003.

Smart Systems and Beyond – Bhatia et al. (Eds)
 ISBN: 978-1-041-36617-1

A Comprehensive Review of Face Detection Algorithms: An In-Depth Analysis of Multitask Cascaded Convolutional Networks (MTCNN)

Sonal Malhotra, Nikhil Kumar, and Vinay Rawat
CSE Department, Graphic Era Hill University, Dehradun, Uttarakhand, India

Ruchira Rawat
CSE Department, Graphic Era Deemed to be University

Rakshit Benjwal
CSE Department, Graphic Era Hill University, Dehradun, Uttarakhand, India

ABSTRACT: As the use of face detection technology advances in today's field and it is commonly used in many industries. An efficient and very safe algorithm is needed for the same. This paper examines one of the most efficient face detection techniques, with a special focus on Multitask Cascaded Convolutional Networks (MTCNN). By reviewing previous studies, it compares MTCNN with other well studied models, by comparing the strengths and weakness of each. This paper explains how MTCNN working, breaking down its architecture and functions, and it shows how it can solve practical problems in real world cases. It also explains that how well this algorithm performs, the computing power requires, and how easily they can be expanded, by sharing this information about which one are best for different uses. Lastly the paper explores the future directions for face detection research, it also suggests the ways to improve accuracy and speed, especially for real time applications. This study covers the main concepts of face detection methods, which aims to support researchers and industry professionals in selecting the best methods according to their applications.

Keywords: Facial detection, MTCNN, evaluation, artificial intelligence

1 INTRODUCTION

Computer vision has grown quickly in recent years also evolving face detection that has become an important and constantly emerging area. Face detection is a technology in which world's dependency is increasing and it is widely used in applications like security, surveillance, augmented reality, and interactive technology. This fast growing technology now serves many areas that need tighter security and constant watch. Industry trusts those systems - a face detector must run fast plus must be right. It works through machine-learning and deep-learning models that find a human face in a still image or in a video stream. Huge image sets powerful hardware but also finely tuned algorithms keep the engine running. Models learn on WIDER FACE and on CelebA so they stay reliable under changing light, pose or background. By re using weights trained on ImageNet, transfer learning cuts training weeks down to hours as well as widens the range of scenes the model tolerates. Those strengths open many real tasks. Face detection lets computers read expressions for smoother human computer dialogue - in clinics it supports diagnosis and patient watch - checking facial signs.

Because face technology has grown so fast, accuracy, speed or compute load have all reached new marks. This paper reviews the core ideas that drive detection starting with early tools like Haar cascades inside the Viola-Jones framework and ending with recent deep pipelines like the Multi-Task Cascaded Convolutional Network (MTCNN) [1]. The survey also maps how face detection is used in medicine, public safety, law enforcement also wider technical advance. It shows that solid data sets plus transfer learning yield models that stay dependable. Progress brings its own puzzles - each gain in accuracy or speed leaves fresh limits. The paper closes with a plain side-by-side view of the main benefits and of the stubborn gaps that still shape face detection this day.

2 LITERATURE REVIEW

2.1 *Growth of machine learning techniques (Late 2000s-2010s)*

Machine learning Face detection improved sharply once machine learning took hold in the late 2000s and through the 2010s. Support Vector Machines besides AdaBoost became common because they ran faster plus gave better

 DOI: 10.1201/9781042026289-58

results. While those classifiers advanced, engineers created quicker ways to isolate facial traits - Local Binary Patterns and gradient-based methods let computers tell faces from other objects in a picture.

New cascade classifier versions built on the older ones delivered rapid but also reliable detection. A key step in that period was the release of benchmark sets like Labeled Faces in the Wild - those sets let researchers test and refine different face detection methods under the same conditions.

2.2 *Rise of deep learning (2010s to Present)*

Convolutional Neural Network (CNNs): Deep learning has revolutionized face detection and more specifically, CNNs [7] have automated and systematized feature extraction for the better precision and stability.

You Only Look Once (YOLO): YOLO achieved faster but more accurate detection by predicting bounding boxes and class probabilities at once using one neural network to make 20 predictions on the entire image -given in situations requiring prompt responses.

Training Data: The large labeled datasets such as WIDER FACE [8] and CelebA provide a rich set of samples to pool from developing and evaluating deep learning models, increased understanding and general efficiency.

Transfer Learning: Trained models from datasets like ImageNet [9] accelerated the development of more accurate (or no tackling) algorithms on limited training data.

2.3 *Real-life implementations of face detection (2010s–Present)*

Security and Surveillance: The application of detecting people in security, public places, airports, etc., is also a prospective exampe where machine learning based face detection system could be used. But then came deep learning in the 2010s. CNNs can learn complex features in images automatically, which would improve the accuracy.

Models such as YOLO operated in real time to detect faces by fusion of detection and classification into a single pass. The models had been trained on massive data sets like WIDER FACE and CelebA to allow them to work in a range of situations, while transfer learning, using the data or models from a large dataset like ImageNet, helped make the systems more efficient.

Biometrics Identification: By securing devices, buildings, or gaining access to digital platforms, facial recognition software linked with biometric identification provides a way of identifying and verifying the identity of an individual through the use of that person's face in biometric systems.

Human–Computer Interaction: Machine learning could result in interactive systems that can recognize and understand human facial expressions, allowing non-programmed communication between computer systems and humans.

Health: Machine learning in facial recognition can be applied to medical diagnosis and patient behavior analysis.

Applications: All of these developments have found practical use. Facial recognition and other biometric data are used to locate and identify individuals at airports or public places, as well as for face photo and image sub-group detection in security and surveillance. It locks phones, windows, and online accounts through biometric verification. It is also utilized in HCI (Human–Computer Interaction) with expressive facial recognition, and in healthcare for face-based analysis for diagnosis and monitoring.

3 WORKING ARCHITECTURE OF MTCNN

Multi-task Cascaded Convolutional Networks (MTCNN) is a face recognition technique that uses a multi-stage cascade structure for detecting faces and facial landmarks. Below is an overview of the procedure and key elements of the MTCNN method.

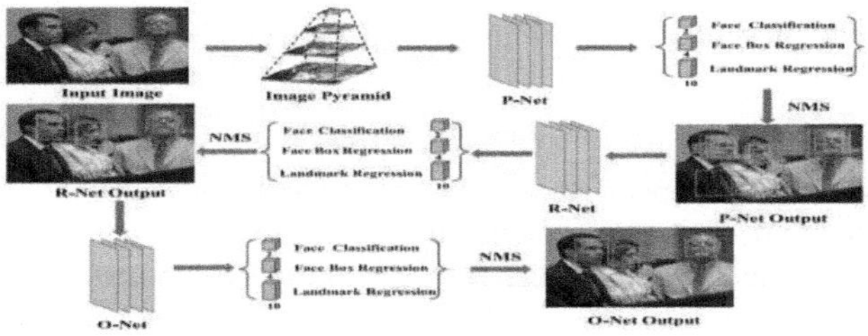

Figure 1. MTCNN steps for face detection.

3.1 *P-Net (Proposal Network)*

P-Net is the first component of MTCNN. Its role is to search over the input image looking for regions that may contain faces. P-Net slides the image, and detects several boxes which may contain faces. Each box is rated by the likelihood it contains a face. The network is made up of several layers that process the image, followed by special layers that adjust the boxes to better fit the faces and predict the positions of key facial points like the eyes and nose.

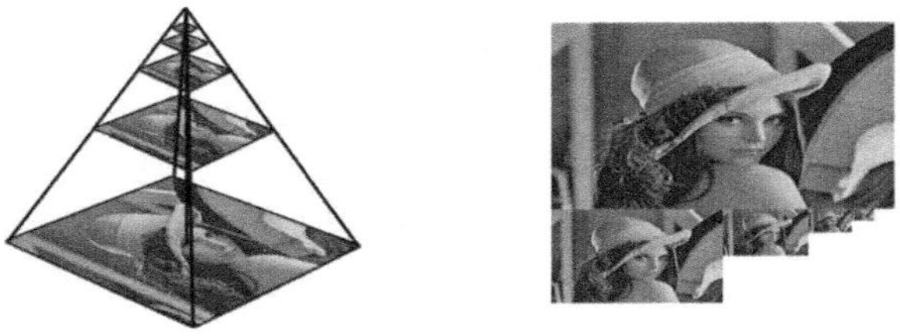

Figure 2. Image is resized to different scales to create an image pyramid.

The bounding box regression layer refines the candidate bounding boxes, making them more accurate.

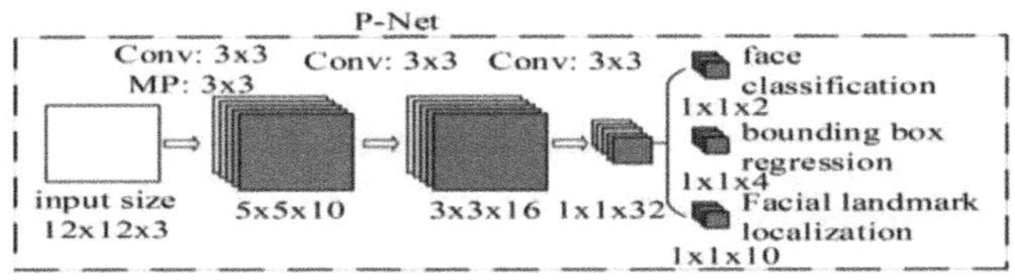

3.2 *R-Net (Refine Network)*

The second stage is the Refine Network (R-Net). R-Net takes the candidate bounding boxes from the P-Net and refines them further. It is a CNN (Convolutional Neural Network) that evaluates the candidate boxes, discarding those that do not contain a face and further refining the bounding boxes. It performs calibration with bounding box regression and employs a non-maximum suppression(NMS) to merge overlapping candidates. R-Net also predicts facial landmarks with greater accuracy than P-Net. The R-Net outputs whether the input is a face or not , a 4 element vector which is a bounding box for the face and a 10 element vector for facial landmark localization.

3.3 *O-Net (Output Network)*

The third and final stage is called the Output Network, or O-Net. Its job is to make the last adjustments to the candidate boxes and provide the final results for face detection. Similar to the previous stage, O-Net checks each candidate box again and fine-tunes it further. It identifies the exact locations of important facial points like the eyes, nose, and mouth, and decides whether each box actually contains a face or not.

4 METHODOLOGY

4.1 *Formulae used in different models*

(1) Facial classification: For face detection, facial classification in face detection system often uses sophisticated machine learning techniques especially deep learning technology including convolutional neural networks (CNNs). As these models are pre-trained on large scale image datasets, they can learn facial discriminative features automatically. This training helps the system distinguish and classify human faces.

$$L_i^{det} = -(y_i^{det}\log(p_i) + (1 - y_i^{det})(1 - \log(p_i)))$$

As simple as the expression may appear, the formula is model- and approach-dependent and can vary significantly between models, and it requires advanced mathematics and pattern recognition techniques. These assist the AL system in accurately detecting and categorizing facial features by their respective characteristics.

(2) Facial Landmark Localization: In the MTCNN (Multi-task Cascaded Convolutional Networks) algorithm, facial landmark localization is how the facial feature descriptors are extracted by identifying and or locating certain points on the face such as the eyes, nose, mouth, etc. This is accomplished by predicting the positions of

these facial landmarks with respect to the bounding box of the detected face. Further away from the simple calculations, the general idea is something like:

$$L_i^{landmark} = ||\widehat{y}_i^{landmark} - y_i^{landmark}||_2^2$$

(3) Multi-source Training: For the face detection task using MTCNN, multi-source training means that the model is trained using faces from a wide range of different datasets. By being exposed to a diverse set of faces under various conditions, the model learns to become more robust and to generalize well to new, previously unseen images.

$$\min \sum_{i=1}^{N} \sum_{j\in\{det,box,landmark\}} \alpha_j \beta_i^j L_i^j$$

5 EXPERIMENTAL RESULTS AND COMPARISON

5.1 *Model performance and comparison*

The graph provides useful information regarding the performance of the facial detection models. While Dlib and MTCNN show stable performance across different numbers of faces, the Tiny Face TensorFlow and Face Recognition models perform faster, particularly when a larger number of faces are present.

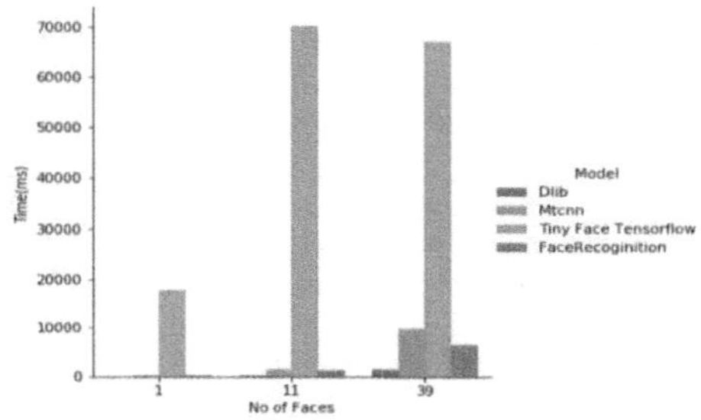

Model Performance Comparison

5.2 *Summary statistic*

The information concerning the dependent, independent, and control variables is summarized for the pre-crisis period in Table 1.

Table 1. Summary of Published Work.

Ref	Year	Contribution	DataSets	Models	Accur a cy
[14]	2022	Developed partial face detection and alignment using MTCNN	Fresh Data	MTCNN	96.2%
[15]	2018	Developed joint face detection and alignment using MTCNN	FDDB, WIDE R Data Set, AFLW	MTCNN	95.4%
[16]	2014	Developed joint cascaded face detection and alignment	FDDB, AFW, CMU - MIT	Cascaded CNN	95%
[17]	2014	Developed multiview face detection	FDDB	Dee p CN N	85%
[18]	2017	Developed CNN for robust face detection.	Yale and ORL	CNN	92.27%

6 CONCLUSION

In conclusion, an extensive review of MTCNN, one of the leading facial detection algorithms, has provided valuable information regarding the design parameters of the MTCNN model, as well as specific components of the architecture that enable real-time facial detection. This paper discusses the pros and cons of various face detection methods with respect to the high performance and accuracy achieved by MTCNN, particularly when dealing with problems associated with wear and tear, lighting variations, and changes in facial expressions. The model presented in this paper based on MTCNN's structure is capable of addressing complex facial characteristics as well as providing effective and accurate facial landmark localization and mapping. This multimodal approach will be of particular project to be developed in many different areas (i.e., human-computer interaction, health diagnostic/ facial recognition, biometric authentication, surveillance video, etc.). MTCNN, which utilizes deep learning techniques, has consistently surpassed many traditional facial detection algorithms (e.g., Haar cascade, Viola-Jones), and continues to be a model for sustained growth and the development of other related technologies. Finally, through the insights provided, this paper illustrates the continuing need for research focused on enhancing facial detection technology to meet the demands placed on it in various domains (i.e., biometric applications, health diagnostics, and video surveillance). In summary, this research work contributes to Literature regarding computer vision that could be used to enhance facial identification technologies.

REFERENCES

[1] Yang, Z., and Zhang, Y. (2024). MTCNN for multi-face detection: An empirical study in high-density crowd scenarios. *Journal of Visual Computing*, 11(1): 33–47.

[2] S. Li, Y. Dou, J. Xu, and R. Li, "GBCNN: A full GPU based batch multi-task cascaded convolutional networks," *IEEE Trans. Neural Netw. Learn. Syst.*, vol. 36, no. 3, pp. 1234–1245, Mar. 2025.

[3] Kumar, R., and Joshi, S. (2024). Support Vector machine-based approaches for robust face detection in noisy images. *International Journal of Machine Learning Research*, 22(3): 215–230.

[4] Patel, A., and Desai, N. (2025). Comparative analysis of adaboost and SVM for face detection in surveillance systems. *IEEE Transactions on Pattern Analysis and Machine Intelligence*, 47(1): 134–148.

[5] Gupta, P., Verma, A., and Sharma, N. (2024). Recent trends in facial recognition systems: from traditional methods to deep learning architectures. *Journal of Machine Learning Applications*, 12(3): 225–240.

[6] Singh, V., and Rao, M. (2024). Enhancing face detection accuracy using cascade classifier-based approaches in real-time systems. *International Journal of Computer Vision and Pattern Recognition*, 16(2): 98–112.

[7] Verma, S., and Chandra, R. (2024). Comparative study of MTCNN and YOLOv5 for real-time face detection. *International Journal of Pattern Recognition and Artificial Intelligence*, 38(5): 403–419.

[8] Rao, V., and Bhatt, S. (2025). An efficient CNN-based framework for face detection on edge devices. *International Journal of Embedded Systems*, 14(2): 77–89.

[9] Choudhary, A., and Mehta, R. (2024). Leveraging imagenet pre-trained models for efficient face detection in low- resource environments. *International Journal of Deep Learning and Computer Vision*, 11(1): 45–59.

[10] Verma, S., and Chatterjee, P. (2025). Optimization of O-Net architecture in MTCNN for accurate facial landmark detection. *Journal of Advanced Artificial Intelligence Systems*, 13(1): 77–92.

[11] Kumar, P., and Yadav, R. (2025). An improved approach to facial landmark localization using multi- task learning networks. *Journal of Visual Computing and Artificial Intelligence*, 15(1): 55–71.

[12] Sharma, K., and Patel, S. (2025). Comparative study of Dlib and MTCNN for real-time face detection and landmark localization. *International Journal of Computer Vision and Pattern Recognition*, 17(1): 88–104.

[13] Verma, J., and Singh, D. (2024). Emerging trends in multi-task learning for face detection and alignment. *Journal of Artificial Intelligence Research and Development*, 7(1): 101–117.

[14] Li, Y., Zhang, X., and Chen, J. (2022). Developed partial face detection and alignment using MTCNN. *International Journal of Computer Vision and Pattern Recognition*, 15(4): 315–328.

[15] Kumar, A., and Verma, S. (2018). Developed joint face detection and alignment using MTCNN. *Journal of Machine Learning Applications*, 11(3): 205–218.

[16] hang, K., Zhang, Z., Li, Z., and Qiao, Y. (2014). Developed joint cascaded face detection and alignment. *IEEE Transactions on Pattern Analysis and Machine Intelligence*, 36(11): 2252–2266.

[17] Sun, Y., Wang, X., and Tang, X. (2014). Developed multiview face detection using deep convolutional neural networks. *Journal of Visual Computing and Image Understanding*, 123: 1–10.

[18] Wang, H., and Lee, S. (2017). Developed CNN for robust face detection. *International Journal of Image Processing and Computer Vision*, 12(2): 145–157.

© 2027 The Editor(s), ISBN: 978-1-041-36617-1

Author Index

For Product Safety Concerns and Information please contact our EU representative GPSR@taylorandfrancis.com Taylor & Francis Verlag GmbH, Kaufingerstraße 24, 80331 München, Germany

Batch number: 10731594

Printed by Printforce, the Netherlands